CONCORDIA CUM VERITATE

Hydraulic fracturing and geothermal energy

Monographs and textbooks on mechanics of solids and fluids
Editor-in-chief: G.Æ. Oravas

Mechanics of elastic and inelastic solids
Editor: S. Nemat-Nasser

Books published under this series are:

1. G.M.L. Gladwell
   Contact problems in the classical theory of elasticity
2. G. Wempner
   Mechanics of solids with applications to thin bodies
3. T. Mura
   Micromechanics of defects in solids
4. R.G. Payton
   Elastic wave propagation in transversely isotropic media
5. S. Nemat-Nasser, H. Abé, S. Hirakawa (Eds.)
   Hydraulic fracturing and geothermal energy

Hydraulic fracturing and geothermal energy

Proceedings of the First Japan-United States Joint Seminar on Hydraulic Fracturing and Geothermal Energy, Tokyo, Japan, November 2-5, 1982, and
Symposium on Fracture Mechanics Approach to Hydraulic Fracturing and Geothermal Energy, Sendai, Japan, November 8-9, 1982

Edited by

Siavouche Nemat-Nasser
Professor of Civil Engineering and Applied Mathematics
Northwestern University, Evanston, Illinois, U.S.A.

Hiroyuki Abé
Professor of Mechanical Engineering
Tohoku University, Sendai, Japan

Seiichi Hirakawa
Professor of Mineral Development Engineering
University of Tokyo, Tokyo, Japan

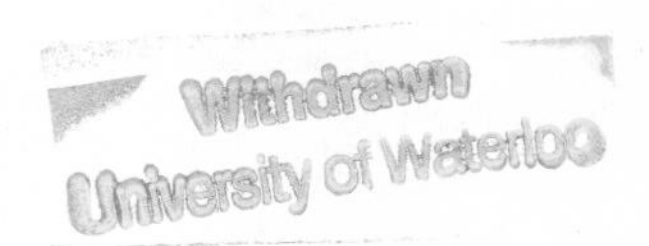

1983 **MARTINUS NIJHOFF PUBLISHERS**
a member of the KLUWER ACADEMIC PUBLISHERS GROUP
THE HAGUE / BOSTON / LANCASTER

Distributors

for the United States and Canada: Kluwer Boston, Inc., 190 Old Derby Street, Hingham, MA 02043, USA
for all other countries: Kluwer Academic Publishers Group, Distribution Center, P.O.Box 322, 3300 AH Dordrecht, The Netherlands

Library of Congress Cataloging in Publication Data

Main entry under title:

Hydraulic fracturing and geothermal energy.

(Mechanics of elastic and inelastic solids ; 5)
"Japan-United States seminar in Tokyo sponsored by Japan Society for the Promotion of Science, United States National Science Foundation; symposium in Sendai sponsored by Mining and Metallurgical Institute of Japan (MMIJ), the Geothermal Research Society of Japan, Japan Geothermal Energy Association and New Energy Foundation."
1. Geothermal engineering--Congresses. 2. Hydraulic fracturing--Congresses. I. Nemat-Nasser, S. II. Abe, Hiroyuki, 1925- . III. Hirakawa, Seiichi, 1923- IV. Japan-United States Joint Seminar on Hydraulic Fracturing and Geothermal Energy (1st : 1982 : Tokyo, Japan) V. Symposium on Fracture Mechanics Approach to Hydraulic Fracture and Geothermal Energy (1982 : Sendai-han, Japan) VI. Nihon Gakujutsu Shinkōkai. VII. Nihon Kōgyōkai. VIII. Series.
TJ280.7.H93 1983 621.44 83-8325
ISBN 90-247-2855-X

ISBN 90-247-2855-X (this volume)

Book information

Japan-United States Seminar in Tokyo
Sponsored by Japan Society for the promotion of Science
United States National Science Foundation

Symposium in Sendai
Sponsored by Mining and Metallurgical Institute of Japan (MMIJ)
The Geothermal Research Society of Japan
Japan Geothermal Energy Association and New Energy Foundation

PRINTED IN THE NETHERLANDS

PREFACE

Hydraulic fracturing has been and continues to be a major technological tool in oil and gas recovery, nuclear and other waste disposal, mining and particularly in-situ coal gasification, and, more recently, in geothermal heat recovery, particularly extracting heat from hot dry rock masses. The understanding of the fracture process under the action of pressurized fluid at various temperatures is of fundamental scientific importance, which requires an adequate description of thermomechanical properties of subsurface rock, fluid-solid interaction effects, as well as degradation of the host rock due to temperature gradients introduced by heat extraction.

Considerable progress has been made over the past several years in laboratory experiments, analytical and numerical modeling, and in-situ field studies in various aspects of hydraulic fracturing and geothermal energy extraction, by researchers in the United States and Japan and also elsewhere. However, the results have been scattered throughout the literature. Therefore, the time seemed ripe for bringing together selected researchers from the two countries, as well as observers from other countries, in order to survey the state of the art, exchange scientific information, and establish closer collaboration for further, better coordinated scientific effort in this important area of research and exploration.

This book is the proceedings of the First Japan-United States Seminar on Hydraulic Fracturing and Geothermal Energy, held in Tokyo, November 2-5, 1982, under the auspices of the Japan Society for the Promotion of Science, with assistance by the Mining and Metallurgical Institute of Japan (MMIJ), and the National Science Foundation of the United States, It also includes the proceedings of a post-seminar Symposium on Fracture Mechanics Approach to Hydraulic Fracturing and Geothermal Energy, held at Tohoku University in Sendai, November 8-9, 1982.

The Seminar focused attention on the following four related and complementary areas:

1. Subsurface structure and hydraulic fracturing.
2. Rock mass properties in the presence of pressurized fluid at elevated temperatures.
3. Mapping of subsurface fractures.
4. Simulation of geothermal reservoirs.

These research areas bear strongly on all technological activities that require hydraulic fracturing, e.g. nuclear waste disposal, gas, oil, and coal recovery and mining, and in-situ coal gasification. While these applications were not ignored, greater emphasis was placed on geothermal heat extraction processes, including both "wet" and "dry" heat reservoirs.

Both the United States and the Japanese sponsoring agencies restrict the number and the distribution of the participants in this kind of seminar. Therefore, the scientific committee made a serious effort to include a representative cross-section of scientists from universities, industry, and government laboratories. As a result, the seminar, as well as the post-seminar symposium, provided an excellent forum for scientific exchange between the two countries and observers from France, Germany, and Australia. While the participants obtained better appreciation of the scientific achievements in hydraulic fracturing, geothermal energy extraction processes, and related topics, they also developed international friendship and closer relations with their peers, which we expect will bear significant fruit for many years to come. The present book, which includes the articles presented at the meetings, is published in an effort to disseminate some of the latest achievements in hydraulic fracturing, rock mechanics, and mining of the earth's heat energy.

We wish to thank the United States National Science Foundation and the Japan Society for the Promotion of Science who provided support for the Seminar. We also thank the Mining and Metallurgical Institute of Japan (MMIJ), the Geothermal Research Society of Japan, Japan Geothermal Energy Association, the New Energy Foundation, and the New Energy Development Organization for sponsoring the Sendai Symposium and for their assistance with the two meetings. The support of many Japanese industries is also gratefully acknowledged.

In addition, we are grateful to Mr. Satoru Suda, Secretary General of MMIJ, and his two charming secretaries who, with the help of the MMIJ staff, selflessly assisted and artfully coordinated many necessary details. Also, thanks are due to Professors Kazuo Hayashi, Toshio Kawashima, Hideaki Niitsuma, and Hideaki Takahashi of Tohoku University, who helped with the local organization of the meetings. Finally, we wish to thank Mrs. Erika Ivansons and Mrs. Éva Nemat-Nasser, who helped with the coordination of this book and with various editorial tasks.

Siavouche Nemat-Nasser
Northwestern University

Hiroyuki Abé
Tohoku University

Seiichi Hirakawa
University of Tokyo

TABLE OF CONTENTS

Topic 4: Simulation of Geothermal Reservoirs

SYMPOSIUM ON FRACTURE MECHANICS APPROACH TO HYDRAULIC FRACTURING AND GEOTHERMAL ENERGY, Sendai, Japan, November 8-9, 1982

SUMMARY

It is useful to provide a summary of the proceedings of the First Japan-United States Seminar on Hydraulic Fracturing and Geothermal Energy, and the associated post-seminar Symposium on Fracture Mechanics Approach to Hydraulic Fracturing and Geothermal Energy. However, the range of interdisciplinary topics covered by the participants renders the task essentially impossible. The pre-printed lectures included contributions ranging from the most fundamental study of rock fracture and properties to the geological subsurface structure of geothermal reservoirs and the associated design procedures for effective heat extraction. Balanced theoretical and laboratory experimental results were presented. Therefore, so many significant technical points were examined that identification of individual contributions would not be possible in the form of a short summary. It was therefore decided to restrict comments to a brief survey of the diverse nature and status of the main subject of the seminar, namely "hydraulic fracturing and geothermal energy." To provide an overview, Fig. 1 was prepared. Though possibly deficient in draftsmanship, we trust it will accentuate the major areas that were discussed at the two meetings. The following comments may help to bring out some of the features implied by this illustration.

Subsurface Structure and Hydraulic Fracturing: The feasibility of a fracture mechanics approach to the design of hydraulic fracturing in geothermal applications was extensively discussed, though no consensus on a design philosophy was achieved.

Rock Mass Properties in the Presence of Water at High Temperatures: The significance of experimental investigations of rock fracture behavior in the presence of water at high temperatures was demonstrated. It was emphasized that the experimental procedure for determining the fracture toughness of rocks urgently requires standardization.

Fracture Mapping: Hot dry rock geothermal energy extraction requires accurate fracture mapping. Refined three-dimensional mapping techniques must be developed, in order to provide detailed subsurface data.

Simulation of Geothermal Reservoirs: Theoretical models of reservoirs were extensively discussed, with the main focus on three-dimensional cracks in layered subsurface media and in geological faults. In addition, theoretical and experimental results on growth regimes, stability, and general configurations of thermally induced cracks in hot dry rock masses were discussed, and their relation to the efficiency and life expectancy of the reservoir was considered.

In reference to Fig. 1, one observes a crowd of rock mechanicians, geoscientists, and geothermal engineers making unintelligible noises when they do not fully comprehend what Mother Nature seems to be suggesting to them. Nonetheless, some in the crowd have come away better informed and certainly inspired, as a result of the Seminar and the follow-up Symposium.

Hiroyuki Abé
Hideaki Niitsuma
Hideaki Takahashi

Tohoku University

Summary - Fig. 1

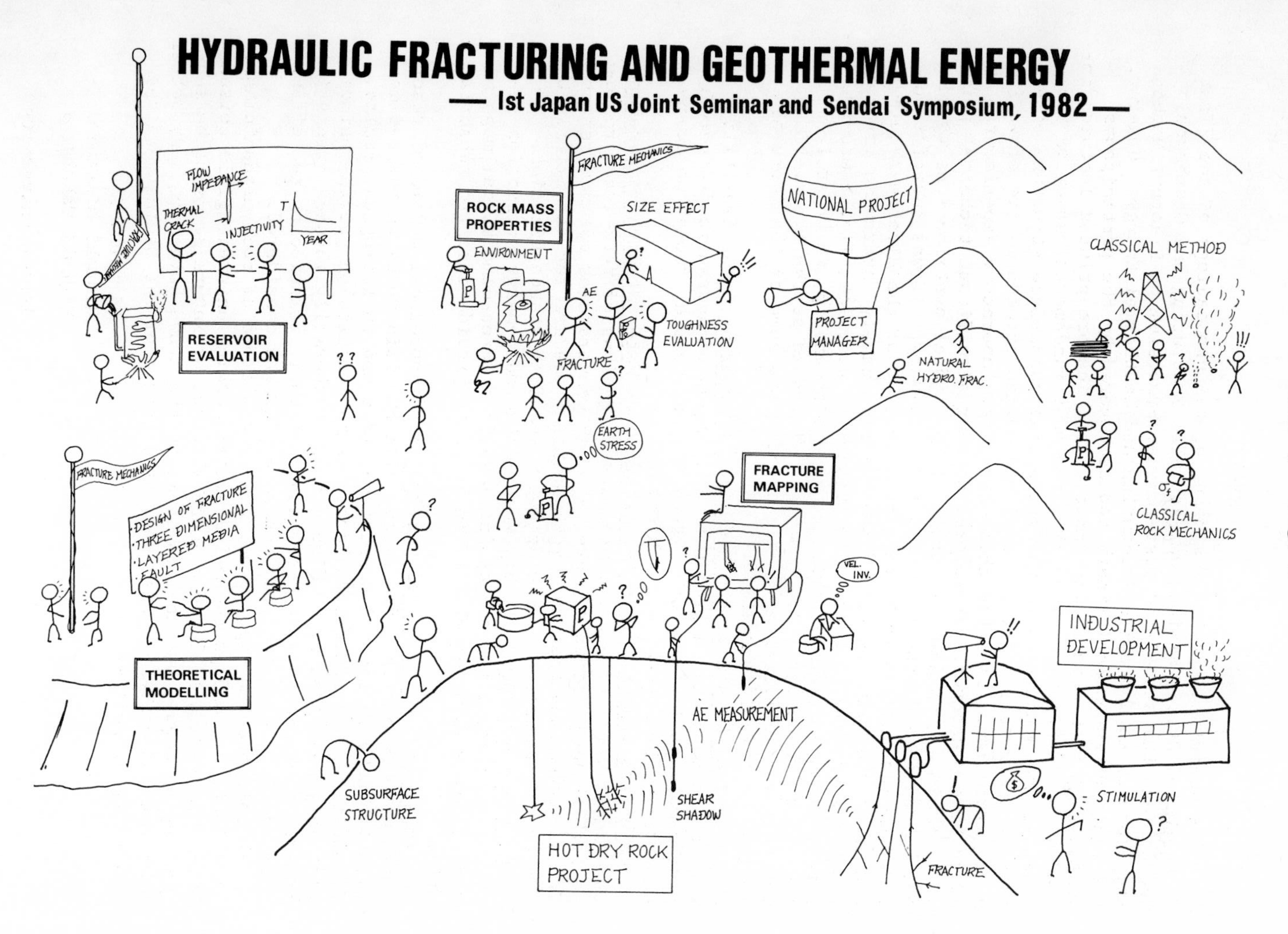

GEOTHERMAL ENERGY DEVELOPMENT IN JAPAN

by Suyama, J. and K. Ogawa

Geological Survey of Japan
Tsukuba, Ibaraki

INTRODUCTION

Japanese energy consumption has been constantly growing with the development of industry and the advance of living level of the people. As a result, Japan imports more than 99% of oil and gas consumed in the country (domestic oil and gas production is only 2 million tons per year in 1980).

Based on the energy situation above, Japanese government started the program to accelerate the development of alternative energy in the last decade. Geothermal energy is considered to be the most realistic alternative energy available with present technology and under current economic situation. According to the Japanese long term energy policy issued in 1982, it is expected that, in 1990, the total installed capacity of the geothermal power plant will increase to 3,000 MW and account for 1.3 percent of total electricity production.

However, present total installed capacity of the geothermal power plant is only 215.0 MW. The target of the energy supply policy seems not so easily attained. So the Japanese government is promoting various national geothermal projects as well as assisting geothermal developers by granting of subsidies and low interest loans.

GEOTHERMAL RESOURCES AND ITS COMMERCIALIZED UTILIZATION

Japan is located within the Circum Pacific Volcanic Belt and has a lot of geothermal fields closely related to the Quaternary volcanic activity. From ancient times, Japanese have utilized their geothermal energy for bathing. Japanese have an undissoluble tie to geothermal energy.

After World War II, Geological Survey of Japan started the fundamental research works of the nation's geothermal energy stimulated by the success of utilization of geothermal energy for power plants in New Zealand and Italy. The research works revealed the existence of many promising geothermal fields in Japanese Islands and led the active effort by the government and private sector for the construction of geothermal power plants.

The first geothermal plant was constructed at Matsukawa, in Iwate Prefecture, northern Japan in 1966. The success of the Matsukawa Plant, although it is small scale (22 MW), gave the strong impact to the private sector, especially electric power companies, and hastened the exploration activity.

As shown on Table 1, eight geothermal power plants are now in operation. All plants make use of high enthalpy water or steam from the hydrothermal convection system related to the Quaternary volcanic activity. It is commonly agreed that only small portions of the high enthalpy water or steam are now in use.

Table 1. Geothermal power plants in operation

name of station	location	capacity (MW)	starting operation date
Matsukawa	Iwate	22.0	1966, Oct.
Otake	Oita	12.5	1967, Oct.
Onuma	Akita	10.0	1974, June
Onikobe	Miyagi	12.5	1975, Mar.
Hatchobaru	Oita	55.0	1977, June
Kakkonda	Iwate	50.0	1978, May
Suginoi	Oita	3.0	1981, Aug.
Mori	Hokkaido	50.0	1982, Nov.
Total		215.0	

The intermediate and low enthalpy water of temperature ranging 60 ~ 80°C related to hydrothermal convection system is also expected to be inexhaustible. The utilization of low temperature hot water (50 ~ 90°C) for non-electric purposes such as agricultural and residential uses is also being attempted but still is in small scale or non-profitable.

The utilization of intermediate temperature water for electricity generation may greatly expand the geothermal potential. For this purpose, the experimental study of the development of a binary cycle power generation system using a low temperature boiling working fluid was implemented and 1 MW of electricity was successfully generated. A large scale (10 MW) binary test plant will be constructed in a few years under governmental sponsorship. However, the commercial binary cycle power plant is not yet planned because of its unreasonable cost under the present level of technology.

As to the energy extraction from the hot dry rock, various basic research works including a field experiment in Yakedake, central Japan, are in progress under the governmental sponsorship.

The non-electricy utilization of low temperature hot water existing in relatively deep strata in sedimentary basin area is also under experimental stage. Recently a pilot plant was constructed in Wada basin, Akita Prefecture, north-western Japan and hot water of 85°C was successfully obtained at 1500 meters depth from the Miocene sediments. The commercial use of this geothermal water is expected in several years.

GEOTHERMAL AREAS IN JAPAN

Japan has about one hundred seventy Quaternary volcanoes, among which sixty-seven are active. These volcanoes are located within two major volcanic zones, i.e., eastern and western Japan zone by Sugimura (1959) as shown in Fig. 1. The eastern zone shows a typical volcanic activity of an arc-trench system and has following characteristics of 1) distribution of numerous active volcanoes, frequent deep earthquakes and high energy release of shallow earthquakes per unit arc length, and 2) having a close relation between the depth of earthquakes and the characteristics of colcanic activity. Volcanoes are located densely along the west-side of the volcanic front which is running parallel to a trench keeping nearly constant distance (200 ± 50 km), while no volcano exists in the east of the front. It may be said that, in the area along the west-side of the front, the condition for the generation of magma in the upper part of the mantle is suitable and, as a result, the heat transportation with the upward migration of magmas is superior, which leads to the occurrence of active geothermal fields.

Contrary to the eastern zone, the western zone, except the southern Kyushu which has characteristics similar to the eastern zone, is featured by 1) poor relation between earthquakes and volcanoes, 2) relatively weak volcanic activity and 3) an indistinct volcanic front.

The extensive geothermal activity is indicated by the existence of numerous hot springs and fumaroles up to 2237, as shown in Fig. 2. The hot springs with temperature above 25°C are mostly located in the Quaternary volcanic zones. However, several high temperature (> 50°C) hot springs are distributed independently of the volcanic zones. The heat source of this type of hot springs is still not made clear.

Fig. 3 shows the distribution of high temperature (>90°C) hot springs and steam fumaroles, most of which are located in the quaternary volcanic zones. Fig. 4 shows 85 areas of hydrothermal systems selected by Geological Survey of Japan based on the definition, 1) discharge of hot water with temperature more than 75°C or 2) relatively low temperature hot springs ranging from 50 to 75°C with extensive ($>1 km^2$) distribution of hydrothermal alteration. These hydrothermal systems are estimated to have temperatures greater than 180°C.

The areas existing within deep strata in sedimentary basins expected to produce low to intermediate temperature water which is heated up by conductive heat flux are also shown in Fig. 5 (Geological Survey of Japan, 1980).

GEOTHERMAL EXPLORATION ACTIVITIES

The government has made various efforts for the acceleration of development of the nation's geothermal resources in the last decade. The Office of Sun-Shine Project was established inside the government (MITI) to promote alternative energy including geothermal energy in 1975, and several large-scale research projects were conducted. In 1980, New Energy Development Organization (NEDO) was established under the sponsorship of the government, and a lot of big-scale geothermal projects were initiated. The total governmental budget for geothermal projects reached 18 billion yen (70 million dollars) in 1982. The Projects by Sunshine and NEDO are listed on Table 2.

The following is an outline of some projects.

a) Nation-Wide Survey of Geothermal Resources

The remote sensing data including Landsat and micro-wave radar imagery to analyse fracture systems and aeromagnetic data to estimate the Curie point (575°C) depth surface are acquired all over the Japanese Islands. In addition to this, gravity data to determine the distribution of basement rock are collected in the area of possible hydrothermal convection systems. These data will be interpreted independently and combined with presently existing data, and, finally, integrated to select a number of appropriate regions for further regional exploration, which will in turn be used to determine the specific sites for deep exploratory drilling testing, resource evaluation, and field development planning.

The data thus accumulated are filed in a computerized data base system to assess the nation-wide geothermal resources. This work is being implemented by the Geological Survey of Japan.

Table 2. Projects by Sun-Shine and NEDO

a) Sun-Shine Project

1. Technology for exploration and extraction of geothermal energy
   1) Confirmation study on the effectiveness of prospecting techniques for deep geothermal resources
   2) Research on exploration technology of deep geothermal resources
   3) Basic study on nation-wide and regional geothermal assessment
   4) Development of drilling mud to be used in geothermal circumstances
   5) The development of geothermal well cement to be used in geothermal circumstances up to 350°C
   6) Development of drilling techniques for high temperature formations
   7) Development of logging technology for geothermal wells
   8) Studies on physicochemical mechanism of hot water injection

2. Development of material for geothermal circumstances
   1) Study on geothermal material development

3. Technology for hot dry rock power generation system
   1) Studies on the drilling and fracturing techniques of hot dry rocks for extracting of geothermal energy
   2) Feasibility study on the power generation system from hot dry rocks

4. Technology for multi-purpose utilization of geothermal energy and environmental preservation
   1) Research and development on the technology for preventing adhesion of scale precipitation from geothermal hot water
   2) Development of hydrogen sulphide removal technology

5. Technology for power generation utilizing hot water
   1) Development of geothermal hot water power generating plant

b) NEDO's Project

1. R & D for new energy technology
   1) Geothermal hot water supply system
   2) Binary cycle power plant

2. Geothermal surveys
   1) Survey of large-scale deep geothermal development with regard to environmental conservation (Hohi Project)
   2) Confirmation survey on the effectiveness of prospecting techniques for deep geothermal resources
   3) Survey to identify and promote geothermal development
   4) Nation-wide survey of geothermal resources

b) Hohi Project for Development of Deep Resources

This project was initiated in 1978 by the government (the Fire Power Section, MITI) to make clear the deep geothermal reservoir existing from 2500 to 3000 meters in depth. The experimental site was selected in Hohi area where the Otake and Hachobaru power stations are located. In the first two years various geological, geophysical and geochemical exploration works including shallow heat holes were carried out and the target area for the drilling was selected. In the next two years ten shallow (500 meters) wells were drilled to examine a result obtained by the surface surveys, and the sites for intermediate wells were selected. Six intermediate wells (1500 meters) were drilled during this period, and the hydrothermal system down to this depth was roughly estimated. Temperature pattern obtained from six wells can be classified into two types, that is, 1) high temperature at shallow depth and constant temperature at the deeper depth, which infers upflow of hot water and 2) low temperature at shallow depth but rapidly increased temperature at the deeper depth, which infers downflow on or near heat source.

Based on the data obtained by the intermediate wells as well as the previously obtained surface data and shallow well data, the sites of three deep (3000 meters) wells were determined. In 1981 these three wells were drilled. The first well hit two reservoirs at the depth of 2000 and 2600 meters and produced 10 tons of steam and 100 tons of hot water. However, the reservoir temperature (about 200°C) is thought to be a bit low for commercial utilization. Both the second and third wells reached the crystalline basement rock (mixture of granite and gneiss) before the scheduled depth. The temperature in the basement rock is high but shows a typical conduction type heat transfer, that means no hydrothermal circulation (dry). Now another deep well is scheduled near the first well in order to hit the higher temperature part of the reservoir whose existence has been confirmed.

This ambitious project would bring a fruitful new idea about the deep hydrothermal system in Japan.

c) Survey to identify and promote geothermal development

This survey is conducted by NEDO to promote geothermal development in the areas that are geothermally hopeful but are still risky for commercial development due to lack of data. Several regional surface syrveys are carried out and five shallow structural wells (1000m) are drilled in the first year of the survey to assess a regional geothermal potential. In the second year, two intermediate wells (1500 m) are drilled in the high potential area estimated from a result of the first year survey to evaluate the productivity of geothermal fluid. Based on the results of the two-year survey, the possibility and recommendable location of the future commercial development is finally evaluated.

Eight areas as shown in Fig. 6 are being surveyed in 1982.

CONCLUSIONS

It is generally agreed that the geothermal resource base is huge. The full potential of electric and nonelectric applications of geothermal resources available with current technology has not yet been realized. The development of geothermal resources for the purpose of contributing significantly to the future national energy supply requires a major effort including the improvement of existing exploration and assessment technology; the application of such technology to accelerate the identification of major types of geothermal resources; the development and demonstration of appropriate technologies for utilizing these resources in a cost-effective and environmentally acceptable manner; and the alleviation of many institutional constraints.

REFERENCE

Sugimura, A., Bull. Volcan. Soc. of Japan, Ser. No. 2, Vol. 4, No. 2, pp. 77–103, 1959.

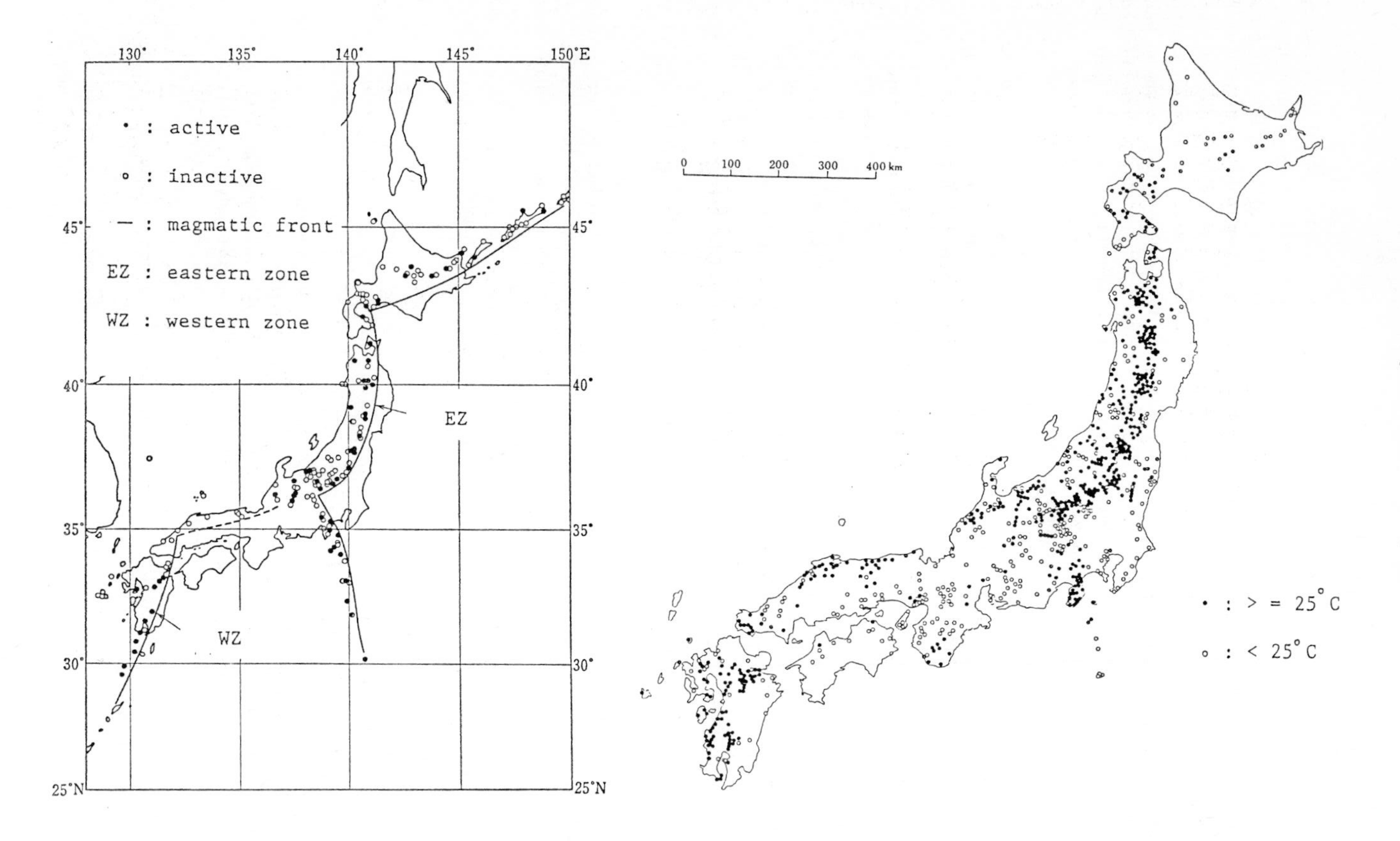

Fig. 1 Volcanoes in Japan

Fig. 2 Hot Springs in Japan

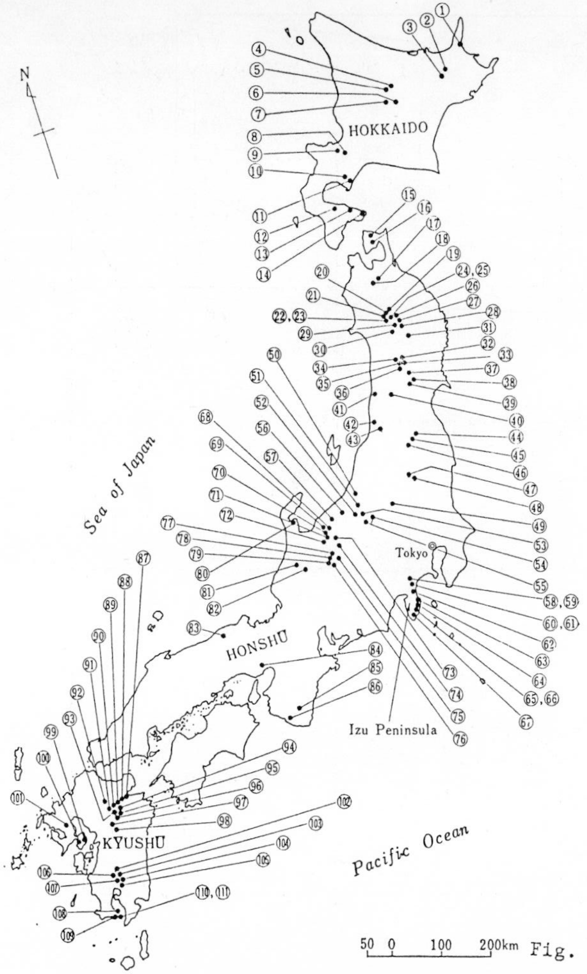

No.	Name	No.	Name	No.	Name
1	Yunosawa	38	Narugo	75	Nakanoyu
2	Atosanupuri	39	Nakayamadaira	76	Hirayu
3	Wakoto	40	Yunohama	77	Shinhotaka
4	Sōunkyō	41	Hijiori	78	Gamada
5	Daisetsukogen	42	Senami	79	Hitoegane
6	Tomuraushi	43	Kira	80	Wakura
7	Tokachidake	44	Tsuchiyu	81	Iwama
8	Jōzankei	45	Noji	82	Oshirakawa
9	Toyoha	46	Bandai-funkayu	83	Yumura
10	Kitayuzawa	47	Santogoya	84	Arima
11	Noboribetsu	48	Myōbanzawa	85	Yunomine
12	Nigorikawa	49	Oigami	86	Shirahama
13	Shikabe	50	Matsunoyama	87	Beppu
14	Esan	51	Nozawa	88	Tsukahara
15	Shimoburo	52	Yudanaka	89	Yufuin
16	Osorezan	53	Hoppo	90	Noya
17	Sukayu	54	Kusatsu-sessōgawara	91	Tsuetate
18	Okiura	55	Manza	92	Amagase
19	Zenikawa	56	Myōko-jigokudani	93	Takenoyu
20	Kami-toroko	57	Renge	94	Otake
21	Sakebizawa	58	Owakudani	95	Hatchōbaru
22	Tamagawa	56	Sōunzan	96	Kurokawa
23	Onuma	60	Kowakudani	97	Yunotani
24	Goshogake	61	Yunohanazawa	98	Tarutama
25	Fukenoyu	62	Atami	99	Unzen
26	Tōshichi	63	Atagawa	100	Obama
27	Matsukawa	64	Shirata	101	Ureshino
28	Amihari	65	Mine	102	Shiratori
29	Takinoue	66	Yatsu	103	Ebino
30	Nyūtō	67	Shimogamo	104	Shinyu
31	Dai	68	Babadani	105	Yunono
32	Oyasu	69	Kuronagi	106	Kurinodake
33	Oyu	70	Azowara	107	Tearai
34	Kawarage	71	Sen'ninyu	108	Ibusuki
35	Doroyu	72	Tateyama-jigokudani	109	Unagi
36	Arayu	73	Kuzu	110	Yamakawa
37	Onikōbe	74	Nakafusa	111	Fushime

50 0 100 200km Fig. 3 Distribution map of geothermal fields having temperatures higher than 90°C in Japan.

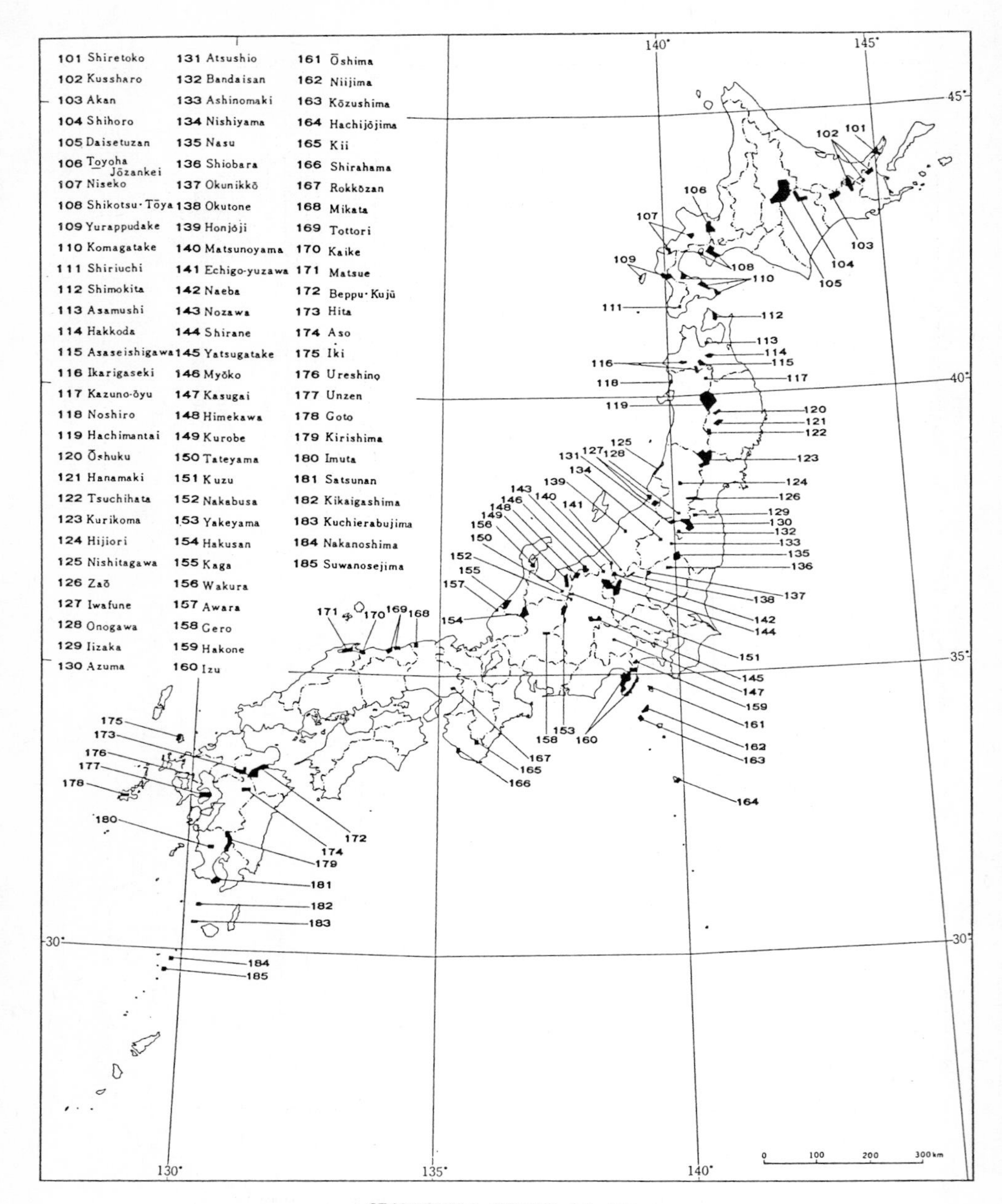

GEOLOGICAL SURVEY OF JAPAN

1977

Fig. 4 Hydrothermal Systems in Japan

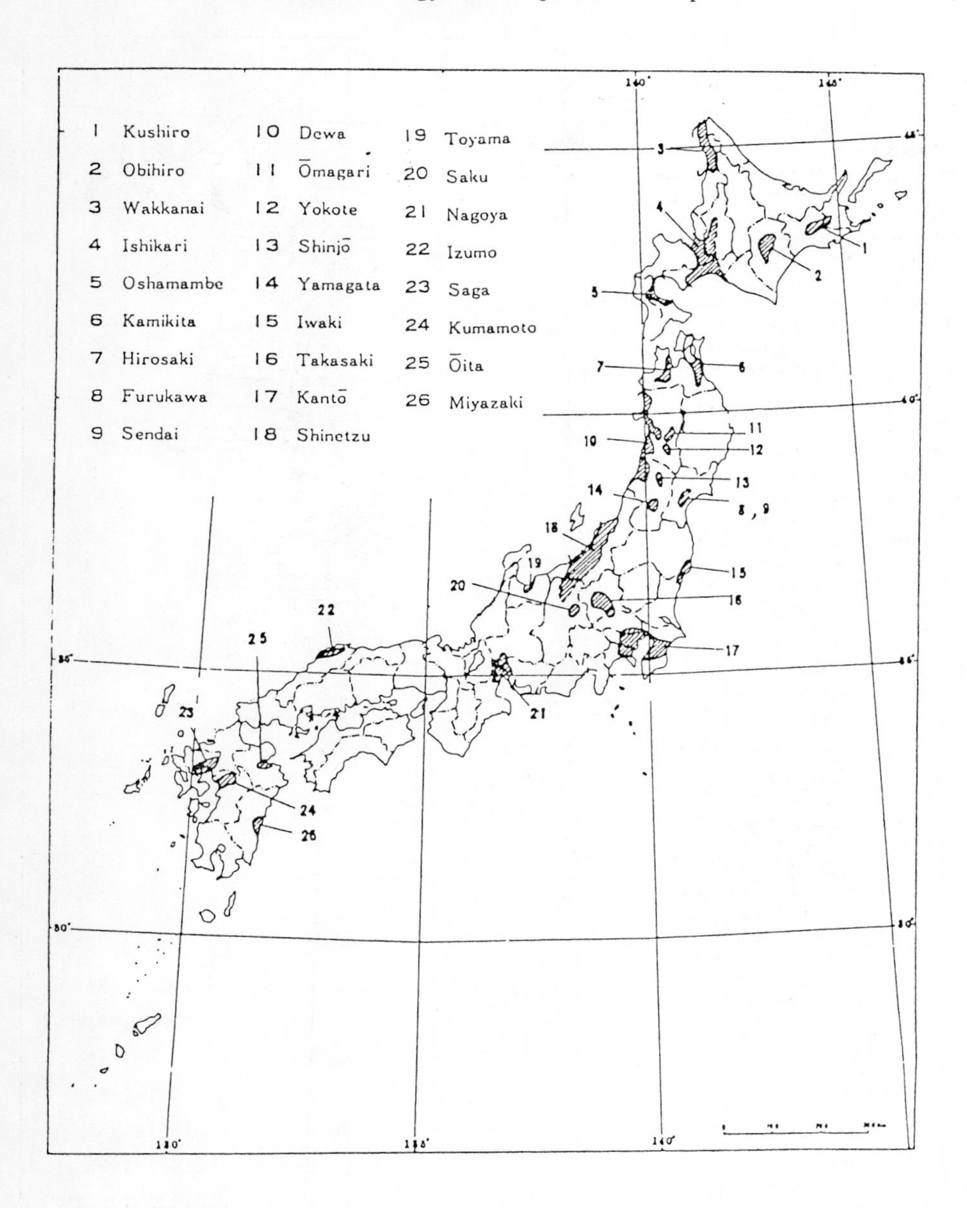

Fig. 5 Areas of low to intermediate temperature water (mainly sedimentary basin area)

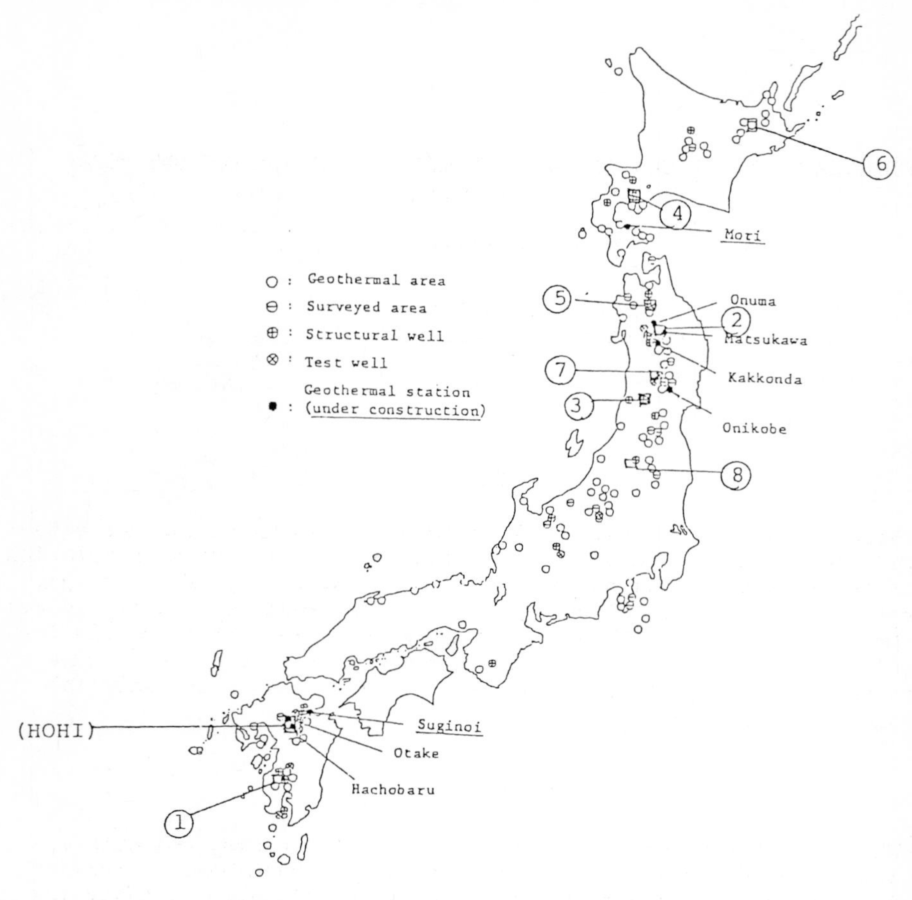

AREA NAME	PREFECTURE	AREA	YEAR
1 KURINO-TEARAI	KAGOSHIMA	40 km^2	1980-1981
2 EASTERN HACHIMANTAI	IWATE	70 km^2	1980-1981
3 DOZANGAWA	YAMAGATA	60 km^2	1980-1981
4 IBURI	HOKKAIDO	70 km^2	1981-1982
5 OKIURA	AOMORI	70 km^2	1981-1982
6 WESTERN TESHIKAGA	HOKKAIDO	65 km^2	1982-1983
7 YUZAWA-OKACHI	AKITA	65 km^2	1982-1983
8 OKU-AIZU	FUKUSHIMA	60 km^2	1982-1983

Fig. 6 Locations of survey to identify and promote geothermal development

THERMALLY INDUCED CRACKS AND HEAT EXTRACTION FROM HOT DRY ROCKS

by S. Nemat-Nasser

Northwestern University
Evanston, Illinois 60201 USA

ABSTRACT

Some recent theoretical and experimental results on thermally induced cracks in brittle solids are reviewed and discussed in relation to heat extraction from hot dry rocks. The stability of growth patterns of these cracks, as well as those of hydraulically produced connected penny-shaped cracks, are investigated. For the thermally induced cracks, universal stability charts are presented, and the important destabilizing effects of geometric imperfections and material inhomogeneities are examined.

1. INTRODUCTION

While geothermal energy in the form of hot springs, volcanic activities, and other natural manifestations may have been sources of comfort, mystery, and often fear for early civilizations, today its possible systematic utilization deserves careful exploration with modern technology. In this context, the geothermal fields that lend themselves to commercial utilization may be divided into two categories: (1) hydrothermal convection systems, and (2) hot dry rock (HDR) masses.

The concept of extracting heat from hot dry rocks for commercial purposes was originated by a group of scientists at the Los Alamos Scientific Laboratory under the leadership of Morton Smith in 1970; see Smith, Potter, Brown, and Aamodt [1], Tester, Morris, Cummings, and Bivins [2], and Blair, Tester, and Mortensen [3]. The presence of abnormally high heat flow in the Rio Grande rift which includes a superheated source associated with the formation of the Valles Caldera, provided an excellent setting for field investigation. The basic idea, patented by the Atomic Energy Commission, anticipates a hole drilled down into hot rocks, a vertical fracture, a second hole drilled to the fractured zone, and the circulation of water down from

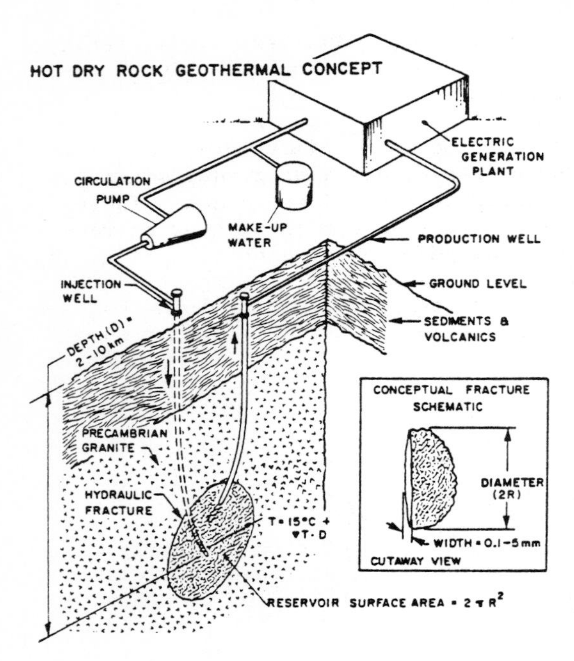

Fig. 1. Taken from Tester et al. [2].

one hole through the fractured region and up to ground level in the form of pressurized hot water. Figure 1 is a schematic representation of this idea. The actual site used is on Fenton Hill, 32 km west of Los Alamos, with adequate clearing and access. Early tests to depths of close to 10,000 feet revealed temperatures of about 200°C. Subsequent efforts over several years led finally to a successful preliminary test in September, 1977, where the entire system was operated for 96 hours of water circulation, attaining temperatures of 130°C at the surface. This then marked the first successful mining of the heat of HDR, and paved the way for obtaining commercially competitive clean energy from another earth's heat reservoir. Considerable progress has been made since this early effort (see Dash, Murphy, and Cremer [4]), and the state of the art is reviewed by Dr. Hugh Murphy of the Los Alamos Scientific Laboratory; see these proceedings.

The major technical problems in this energy extraction scheme include: (1) Generation of a large fracture or a system of fractures; (2) Circulation of water under pressure through the fractured zone and back to ground surface; (3) An assessment of the secondary cooling cracks, their growth regime, and their influence on the heat extraction process; and (4) Possible thermally induced rock fragmentation, the size of fragments, and the influence of their presence on the overall performance.

In this lecture I shall review some of the theoretical and experimental work on thermally induced cracks in brittle solids, and also briefly comment on the stability of hydraulically produced connected penny-shaped cracks. These topics are integral parts of a HDR geothermal system. Those interested in further details and other related topics may consult the references and other articles in these proceedings.

2. THERMALLY INDUCED SECONDARY CRACKS

Estimates based on elementary considerations reveal for an initial (primary) hydraulically induced vertical crack of several hundred meters in diameter, crack opening of a few millimeters. This opening increases due to thermal contraction as the faces of the primary vertical crack are cooled by heat extraction.

Figure 2 is a schematic representation of a primary crack and the state of stress. Creation of the primary crack also produces numerous microcracks perpendicular to the faces of the primary crack. These may grow into secondary cracks as the rock is cooled by heat extraction. Soon after cooling, the thermal penetration depth, δ, is of the order of millimeters. The estimate of the minimum spacing of secondary cracks requires an estimate of the temperature profile in the rock. For illustration, two possible limiting profiles are shown in

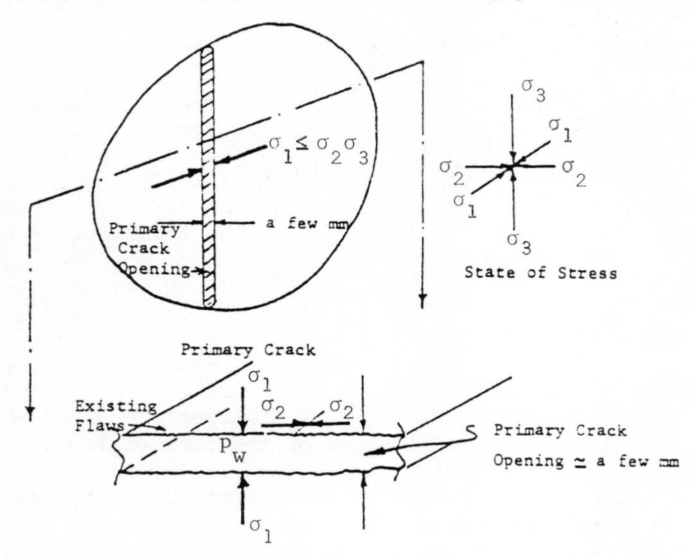

Fig. 2. Primary crack and the stress state.

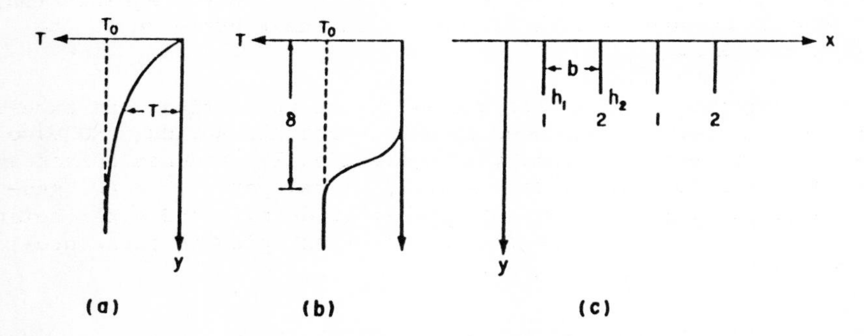

Fig. 3. (a) Temperature profile (2.2);
(b) Temperature profile (2.1) with n = 0.5;
(c) A system of equally spaced edge cracks.

Fig. 3. These are expressed by

$$T = 0 \quad \text{for } 0 \le y \le \delta/(n+1),$$

$$T = \frac{T_0}{2}\left[1 - \cos \pi \frac{y(n+1) - \delta}{n\delta}\right] \quad \text{for } \delta/(n+1) \le y \le \delta,$$

$$T = T_0 \quad \text{for } \delta \le y, \quad T_0 = T_r - T_w, \tag{2.1}$$

where $n = 0.5$ is used (see Fig. 3(b)); and

$$T = T_0 \operatorname{erf}\left[\frac{y\sqrt{3}}{\delta}\right], \quad y \ge 0, \tag{2.2}$$

where $\operatorname{erf}(x) = \frac{2}{\sqrt{\pi}} \int_0^x e^{-u^2} du$ (see Fig. 3(a)). Profile (2.1) assumes convective heat exchange which completely cools the rock to the depth $\delta/(n+1)$. Profile (2.2), the error function, involves conductive heat transfer only.

Let $f(y/\delta)$ stand for the temperature profile. The total elastic energy in a strip of unit thickness and width b is (Nemat-Nasser, Keer, and Parihar [5] and Nemat-Nasser, Sumi, and Keer [6])

$$W = K_0 \frac{\alpha^2 T_0^2 \, E \, b \, \delta}{1 - \nu}, \tag{2.3}$$

where the numerical factor K_0 depends on the form of $f(y/\delta)$: with $n = 0.5$, $K_0 = 19/48$ for profile (2.1), and for profile (2.2) $K_0 = 1/10$. In (2.3), $(1-\nu)$ must be replaced by $(1-2\nu)$ if there is no free surface (i.e., for an interior crack).

A fraction, θ, of the elastic energy is used to generate new crack surfaces, $\theta W = 2h\gamma$, and this leads to

$$b = \left[\frac{2(1-\nu)\gamma}{K_0 \alpha^2 E \, T_0^2}\right]\left(\frac{h}{\theta\delta}\right), \tag{2.4}$$

where γ is the specific surface energy, and h is the crack depth. Since $h/\theta\delta \simeq 1$, with

$$\alpha = 8 \times 10^{-6}/°\text{C}, \quad E = 4 \times 10^{10} \ \text{N/m}^2,$$

$$\nu = 0.22, \quad \gamma = 100 - 1{,}000 \ \text{J/m}^2, \tag{2.5}$$

and $T_0 = 100°\text{C}$, profile (2.1) yields $b = 16$ cm, and profile (2.2) yields $b = 65$ cm when $\gamma = 1{,}000 \ \text{J/m}^2$.

The minimum crack spacing may also be estimated by a stability consideration, see Nemat-Nasser and Oranratnachai [7]. This will be discussed later on.

3. GROWTH OF THERMAL CRACKS

As the depth, δ, of thermal boundary layer increases, secondary cracks grow. The character of this growth pattern is assessed by the examination of the stability of the configuration of cracks.

The theory of the stability of the growth pattern of tension cracks in brittle solids is developed by the present writer and coworkers; see Keer, Nemat-Nasser, and Parihar [8], Keer, Nemat-Nasser, and Oranratnachai [9], Nemat-Nasser [10,11], Nemat-Nasser _et al_. [5], and Sumi, Nemat-Nasser, and Keer [12]. The theory includes the effects of crack interaction, and material and geometric imperfections. The actual analysis is rather complicated, requiring special numerical techniques. Some of the main points are outlined below; in subsection 3.5 experimental results are also summarized and used to verify the theory.

3.1 Critical States

Consider plane strain problems, and assume edge cracks of common spacing b and common length h, which grow in a stable manner, as δ is increased; δ is used as the "load parameter." When the spacing b is large compared with the length h, there is weak interaction between adjacent cracks, and the crack growth pattern is stable. Each crack may be regarded as an isolated one, and since for a fixed δ, there is a fixed amount of elastic energy available, an extension of a crack at constant δ would release a certain amount of elastic energy, which results in a reduction of the corresponding stress intensity factor at the crack tip, and therefore crack growth will be arrested. If K is the stress intensity factor, then, in this case, $\partial K/\partial h < 0$.

Figure 4 shows two interacting cracks in a "unit cell," with h_i and $K_i = K_i(h_1,h_2;\delta)$, i = 1,2, as the corresponding length and stress intensity factor, respectively.

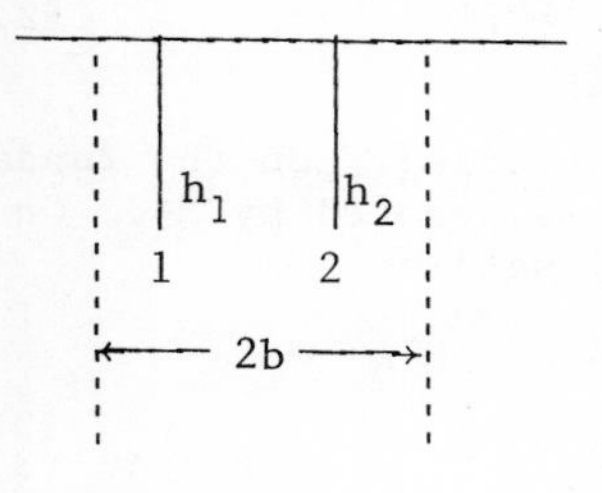

Fig. 4. Two interacting edge cracks.

With equal crack spacing and length, and in the absence of any imperfections and inhomogeneities (an ideal system), the cracks remain equal and continue to grow at a common rate. This growth pattern is defined as the _fundamental equilibrium path_, Nemat-Nasser _et al_. [6],

$$h_1 = h_2 \quad \text{and} \quad K_1 = K_2 = K_c; \qquad (3.1)$$

curves AB in Fig. 5. For an increase, $d\delta > 0$, in δ,

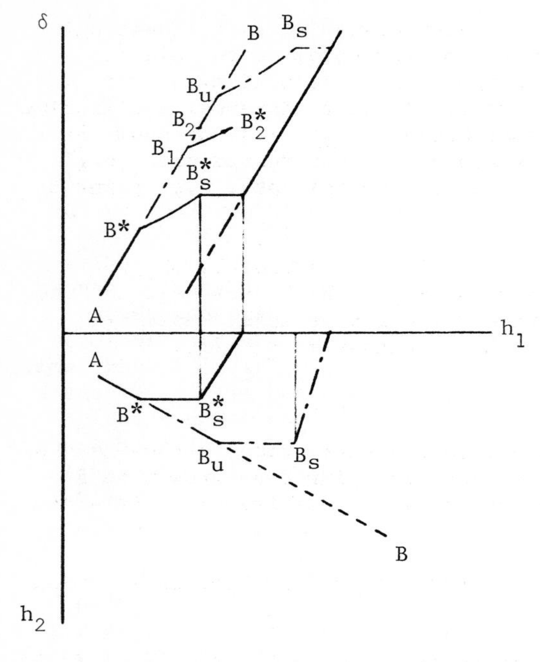

Fig. 5. Various equilibrium states for two interacting cracks: B* is a stable, and B_u an unstable bifurcation point; B_s^* and B_s are snap-through critical points.

$$dK_1 = \frac{\partial K_1}{\partial h_1} dh_1 + \frac{\partial K_1}{\partial h_2} dh_2 + \frac{\partial K_1}{\partial \delta} d\delta,$$

$$dK_2 = \frac{\partial K_2}{\partial h_1} dh_1 + \frac{\partial K_2}{\partial h_2} dh_2 + \frac{\partial K_2}{\partial \delta} d\delta, \tag{3.2}$$

where it is assumed that $dK_c = 0$, i.e. K_c = constant.

As the two cracks in Fig. 4 grow with increasing δ, their interaction becomes more important, and a point may be reached at which, with h_2 and δ kept constant, an infinitesimal extension of crack 1 decreases the stress intensity factors at crack 2 and crack 1 by an equal amount. This defines a critical point on the fundamental equilibrium path. It is a stable bifurcation point characterized by

$$\frac{\partial K_1}{\partial h_1} = \frac{\partial K_2}{\partial h_2} = \frac{\partial K_1}{\partial h_2} = \frac{\partial K_2}{\partial h_1} < 0, \quad h_1 = h_2. \tag{3.3}$$

Indeed, from (3.2),

$$dh_1 = \left(\frac{\partial K_2}{\partial h_1} \frac{\partial K_2}{\partial \delta} - \frac{\partial K_1}{\partial h_1} \frac{\partial K_1}{\partial \delta}\right) d\delta / \left\{ \left(\frac{\partial K_1}{\partial h_1}\right)^2 - \left(\frac{\partial K_1}{\partial h_2}\right)^2 \right\},$$

$$dh_2 = \left(\frac{\partial K_2}{\partial h_1} \frac{\partial K_1}{\partial \delta} - \frac{\partial K_1}{\partial h_1} \frac{\partial K_2}{\partial \delta}\right) d\delta / \left\{ \left(\frac{\partial K_1}{\partial h_1}\right)^2 - \left(\frac{\partial K_1}{\partial h_2}\right)^2 \right\}. \tag{3.4}$$

Therefore, uniqueness is lost when (3.3) is satisfied. On the fundamental equilibrium path in Fig. 5, this point is denoted by B*. On this same path, points above B* (but below B_u) satisfy

$$0 > \frac{\partial K_1}{\partial h_1} = \frac{\partial K_2}{\partial h_2} > \frac{\partial K_1}{\partial h_2} = \frac{\partial K_2}{\partial h_1}, \quad h_1 = h_2, \tag{3.5}$$

and hence correspond to stable states. Each one of these points defines a stable bifurcation point from which another equilibrium path emanates.

In the immediate neighborhood of the fundamental path, these bifurcated paths characterize states with smaller total stored elastic energy per unit cell, and hence energetically are preferred over the corresponding ones (for the same δ) on the fundamental equilibrium path. This has been proved by Nemat-Nasser et al. [6] who show the stored elastic energy, $\mathcal{E}_s$, for a state on the bifurcated path, e.g. point B_2^*, to be less than that on the fundamental path, e.g. point B_2, for the same value of δ.

In Fig. 5, states at and above B_u on AB are defined by $\partial K_1/\partial h_1 = \partial K_2/\partial h_2 \geq 0$ with the equality sign holding at B_u. These are all unstable bifurcation points at which one crack is arrested as the other one grows spontaneously, see Nemat-Nasser et al. [5].

3.2 Post Critical Response

At the state corresponding to B*, every other crack in the array of cracks shown in Fig. 3(c) stops, as the remaining ones grow with increasing δ at an initial rate twice that which existed prior to state B*; see Nemat-Nasser et al. [6].

To examine the growth pattern after B*, consider three interacting cracks (Fig. 6) for which

$$h_1 = h_3, \quad h_2 = h_2(B^*) = \text{constant}, \; K_1 = K_3 = K_c, \; K_2 < K_c. \tag{3.6a}$$

Since the spacing between cracks 1 and 3 in Fig. 6 is equal to 2b, the corresponding initial interaction is weak.

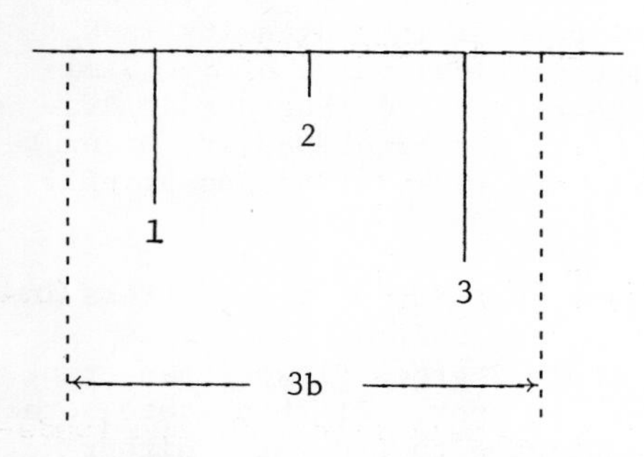

Fig. 6. Three interacting cracks.

On the bifurcated path $B^*B_s^*$, the following results hold for temperature profile (2.1); for illustration this profile is used:

$$dK_2 = \frac{\partial K_2}{\partial h_1} dh_1 + \frac{\partial K_2}{\partial \delta} d\delta < 0. \tag{3.6b}$$

Hence K_2 continues to decrease with increasing δ and h_1. At point B_s^*, K_2 is zero, and a further increase in the load parameter results in the closure of crack 2. At this point and for temperature profile (2.1), calculations show that crack 2 snaps closed while crack 1 snaps into a finitely longer length, for infinitesimally larger values of the load parameter δ. Thus point B_s^* defines a snap-through critical state. In the infinitesimally small neighborhood of this state, there are no equilibrium states corresponding to (infinitesimal) $d\delta > 0$.

From the relation between the energy release rate and the stress intensity factor in Mode I, it follows that, Nemat-Nasser et al. [5],

$$K_1 \frac{\partial K_1}{\partial h_2} = K_2 \frac{\partial K_2}{\partial h_1} . \tag{3.7}$$

Since at B_s^*, $K_2 = 0$, this point is characterized by

$$\frac{\partial K_1}{\partial h_2} = 0. \tag{3.8}$$

Actually, since $\partial K_1/\partial h_2$ is in general non-positive, if this quantity is plotted as a function of h_1, the curve would be tangent to the h_1-axis at the critical point defined by (3.8); see Nemat-Nasser et al. [5].

For some other temperature profiles, e.g. (2.2), it happens that K_2 never reaches zero. In this case, cracks 1 and 3 of Fig. 6 begin to have significant interaction, and one of them stops at a certain stable point before the stress intensity factor at crack 2 can reach zero value. For cases of this kind there is no crack closure. This has been illustrated for temperature profile (2.2) by Nemat-Nasser et al. [5] and Keer et al. [9].

3.3 Imperfections and Inhomogeneities

In actual situations, the material properties are not uniform, nor do cracks initiate at equal spacings. Such imperfections are expected to reduce the critical value of the load parameter at which the growth pattern changes. It has been shown by Nemat-Nasser et al. [6] that this indeed is the case, and that a few percent inhomogeneity in K_c may result in 20 to 40% reduction in the critical value of δ. Similarly, unequal crack spacings reduce the critical δ, but not as dramatically, as is the case for unequal K_c's; see Nemat-Nasser, Oranratnachai, and Keer [13]. Here these facts are exemplified and briefly discussed.

Effect of Material Inhomogeneity: Assume that the critical stress intensity factor at crack 1 in Fig. 4 is K_c and that at crack 2 is $(1+\varepsilon)K_c$, $\varepsilon \ll 1$. With increasing δ, crack 2 grows slower than crack 1 until point b_2^* in Fig. 7, when it ceases to grow. At this state, crack 1 is at state b_1^*, and thereafter, it continues to grow at a higher rate; curve $b_1^* b_{s1}$. For a certain temperature profile, e.g. profile (2.1), K_2 reaches zero at point b_{s2}, and then tends to become negative as δ increases. At this state crack 2 snaps closed as crack 1 snaps to a finitely longer length.

The crack lengths and K's are functions of ε, $h_i = h_i(\varepsilon)$, $K_i = K_i(h_1, h_2; \delta, \varepsilon)$, $i = 1,2$. For fixed δ write

$$h_1(\varepsilon) = h_1(0) + \Delta h_1, \quad h_2(\varepsilon) = h_2(0) + \Delta h_2, \tag{3.9}$$

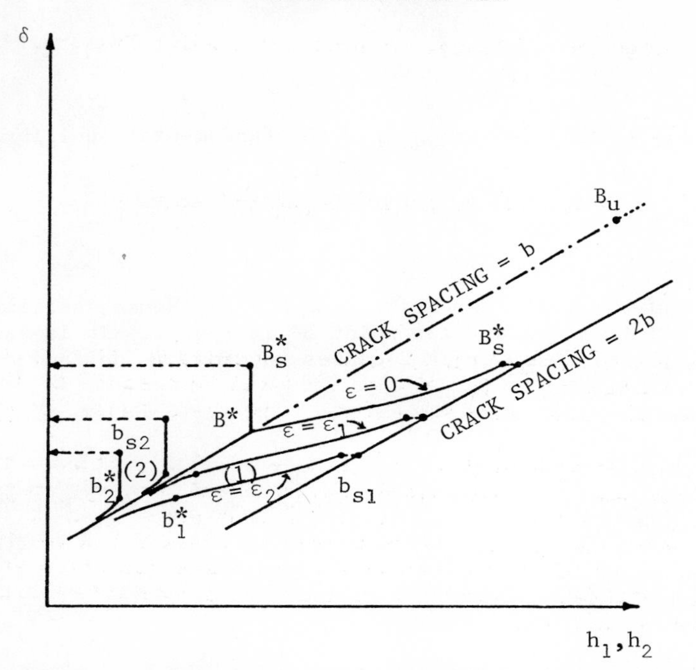

Fig. 7. Post critical response and the effect of imperfections.

and

$$K_1(\varepsilon) = K_c + \frac{\partial K_1}{\partial h_1}\Delta h_1 + \frac{\partial K_1}{\partial h_2}\Delta h_2 = K_c,$$

$$K_2(\varepsilon) = K_c + \frac{\partial K_2}{\partial h_1}\Delta h_1 + \frac{\partial K_2}{\partial h_2}\Delta h_2 = K_c + \varepsilon K_c, \qquad (3.10)$$

where on the fundamental path, $h_1(0) = h_2(0)$ and $K_1 = K_2 = K_c$; all partial derivatives are evaluated on the fundamental equilibrium path, i.e. at $\varepsilon = 0$, and it is assumed that the deviations from this path, denoted by Δh_1 and Δh_2, are small. Since on the fundamental equilibrium path $\partial K_1/\partial h_1 = \partial K_2/\partial h_2 < 0$ and $\partial K_1/\partial h_2 = \partial K_2/\partial h_1 < 0$, it follows from (3.10) that

$$\Delta h_1 \simeq -\left[K_c \frac{\partial K_1}{\partial h_2} / \left\{\left(\frac{\partial K_1}{\partial h_1}\right)^2 - \left(\frac{\partial K_1}{\partial h_2}\right)^2\right\}\right]\varepsilon > 0,$$

$$\Delta h_2 \simeq \left[K_c \frac{\partial K_1}{\partial h_1} / \left\{\left(\frac{\partial K_1}{\partial h_1}\right)^2 - \left(\frac{\partial K_1}{\partial h_2}\right)^2\right\}\right]\varepsilon < 0, \qquad (3.11)$$

where

$$D \equiv \left(\frac{\partial K_1}{\partial h_1}\right)^2 - \left(\frac{\partial K_1}{\partial h_2}\right)^2 > 0 \tag{3.12}$$

below point B*. The deviation from the fundamental equilibrium path, measured by

$$h_1(\varepsilon) - h_2(\varepsilon) = \varepsilon K_c / \left(\frac{\partial K_1}{\partial h_1} - \frac{\partial K_1}{\partial h_2}\right), \tag{3.13}$$

tends to increase with increasing δ as $D \to 0$. Hence, estimates (3.11) to (3.13) are only valid below point B* in Fig. 7. It is easy to show that point b_2^*, at which crack 2 ceases to grow, is defined by

$$\frac{\partial K_2}{\partial h_1}\frac{\partial K_1}{\partial \delta} - \frac{\partial K_1}{\partial h_1}\frac{\partial K_2}{\partial \delta} = 0 \tag{3.14}$$

which replaces condition (3.3) of the perfect system. Moreover, crack 2 snaps closed at

$$\frac{\partial K_1}{\partial h_2} = 0. \tag{3.15}$$

Unequal Crack Spacings: Both Mode I and Mode II are involved when cracks are unequally spaced. This makes the actual numerical calculations complicated. A rather complete analysis is given by Nemat-Nasser *et al*. [13]. Here an example is presented for the sake of illustration.

Figure 8 shows a unit cell with three unequally spaced cracks, where $\varepsilon \ll 1$. If Π is the total potential energy, (Nemat-Nasser *et al*. [5]), the equilibrium lengths are given by the solutions of

$$\frac{\partial \Pi}{\partial h_i} = 0, \quad i = 1,2,3. \tag{3.16}$$

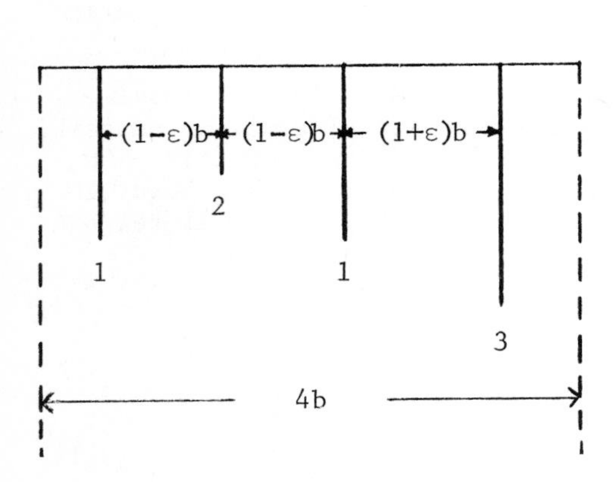

Fig. 8. Three unequal cracks with unequal spacings.

Denote by G_i the energy release rate,

$$G_i = \frac{1-\nu^2}{E}[K_{Ii}^2 + K_{IIi}^2], \tag{3.17}$$

and from the system of equations [$G_i(h_1,h_2,h_3;\delta,\varepsilon) = G_c$ = const. is assumed],

$$dG_i = \frac{\partial G_i}{\partial h_j} dh_j + \frac{\partial G_i}{\partial \delta} d\delta = 0, \quad i,j = 1,2,3 \tag{3.18}$$

(j summed), obtain

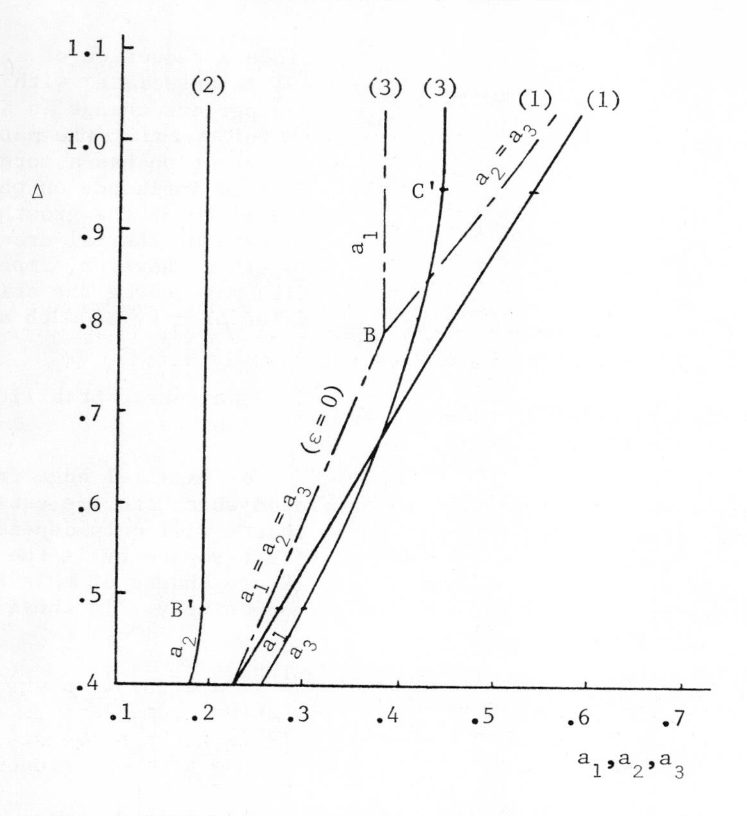

Fig. 9. Effect of geometric imperfections on unstable crack growth; temperature profile (2.2).

$$dh_i = \frac{Q_i}{D} d\delta, \tag{3.19}$$

where Q_i's and D are the relevant determinants in (3.18). Crack i ceases to grow when $Q_i = 0$.

Figure 9 shows a typical result for profile (2.2). The dimensionless parameters are

$$\Delta = \delta/b, \quad a_i = h_i/b. \tag{3.20}$$

Curves denoted by (1), (2), and (3) in Fig. 9 correspond to cracks 1, 2, and 3 of Fig. 8. Crack 2 stops at point B'. At point C', crack 3 stops, and after that only crack 1 remains active.

Figure 10 shows the change in the critical value of Δ with respect to ε; see Fig. 8 for definition of ε. Here, Δ_0 corresponds to $\varepsilon = 0$, and Δ_ε corresponds to the value of Δ for the given ε. As is seen, a reduction of 40% in critical Δ is accompanied by an ε of about 0.2.

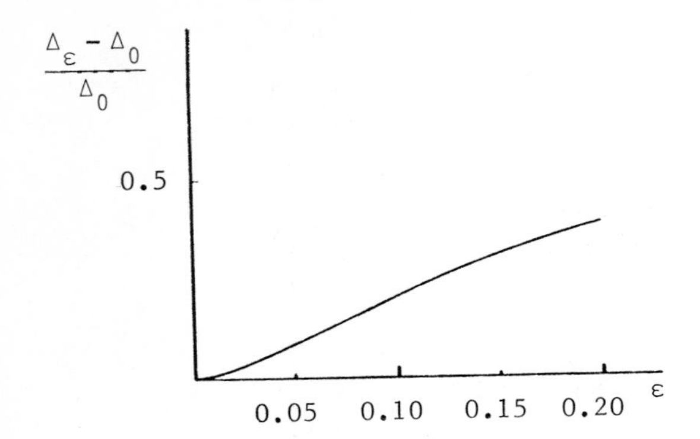

Fig. 10. Change in critical value of Δ with ε for profile (2.2).

Since a reduction of about 40% is associated with a few percent change in K_c, it follows that the material imperfection has a more pronounced influence on the stability of the growth pattern of thermal cracks. Together, however, imperfection may reduce the critical value of δ by as much as 50%.

3.4 Universal Stability Charts

For parallel edge cracks dimensionless stability charts can be developed which provide estimates for the entire crack growth regime. These charts will only depend on the temperature profile and on the degree of heterogeneity in the material, i.e. imperfection. Figures 11 and 12 are charts of this kind for temperature profiles (2.1) and (2.2), respectively. In these figures the dimensionless parameters are

$$\Delta = \delta/b, \quad a = h/b, \quad N = K_c/[\sqrt{2\pi}\ \hat{\beta}\ T_0 b], \tag{3.21}$$

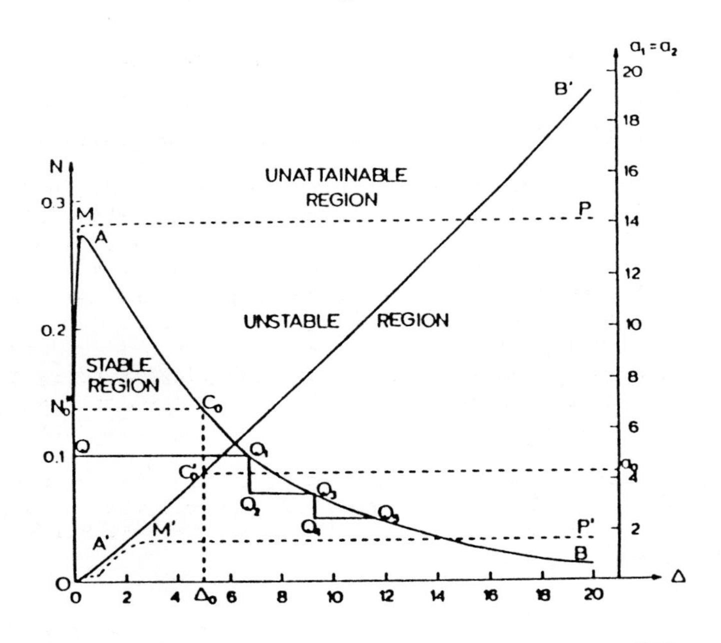

Fig. 11. Stability chart for temperature profile (2.1); $n = 0.5$, $\Delta = \delta/b$, and $a_i = h_i/b$.

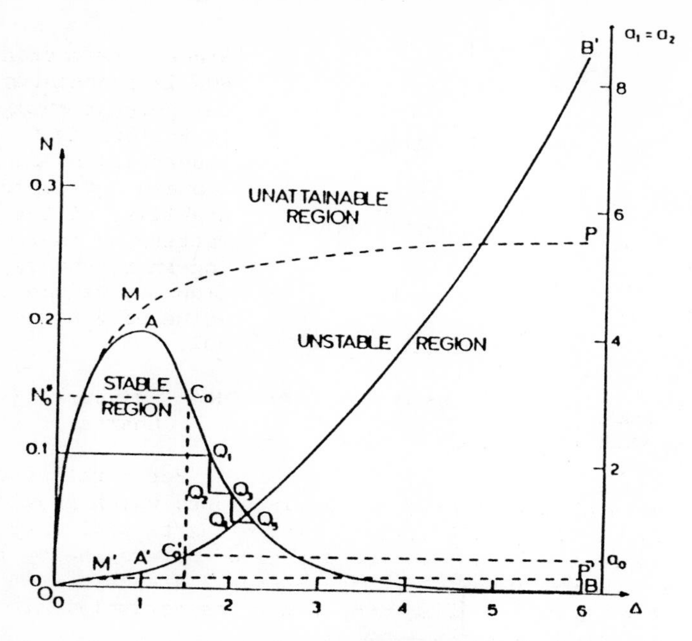

Fig. 12. Stability chart for temperature profile (2.2); $\Delta = \delta/b$, $a_i = h_i/b$.

where $\hat{\beta} = \alpha E/(1 - \nu)$ for plane strain, edge cracks;

$\hat{\beta} = \alpha E$ for plane stress, edge cracks;

and $\hat{\beta} = \alpha E/(1 - 2\nu)$ for plane strain, interior cracks.

To use these charts, first estimate the initial minimum crack spacing. Then calculate the initial value of N. With this value of N, enter the chart and find the first point on the boundary between stable and unstable regions. The abscissa of this point gives Δ, and from the chart also obtain the corresponding a. In Figs. 11 and 12, the procedure is illustrated by taking the initial N = 0.1. At Q_1 the crack spacing doubles, and this gives point Q_2 whose ordinate is $0.1/\sqrt{2}$. Extending Q_2 horizontally, the stability boundary is intersected at Q_3. At this stage the crack spacing doubles again, and this yields Q_4 with ordinate 0.1/2. The procedure is then continued.

In Figs. 13 and 14 we have added to Figs. 11 and 12 new curves which correspond to the stable bifurcation points with 0, 5, 10, and 20% material imperfections. These curves are used in exactly the same manner as Figs. 11 and 12.

3.5 Experimental Results

Geyer and Nemat-Nasser [14] report on a series of experiments which verify qualitatively and quantitatively the theoretical results dis-

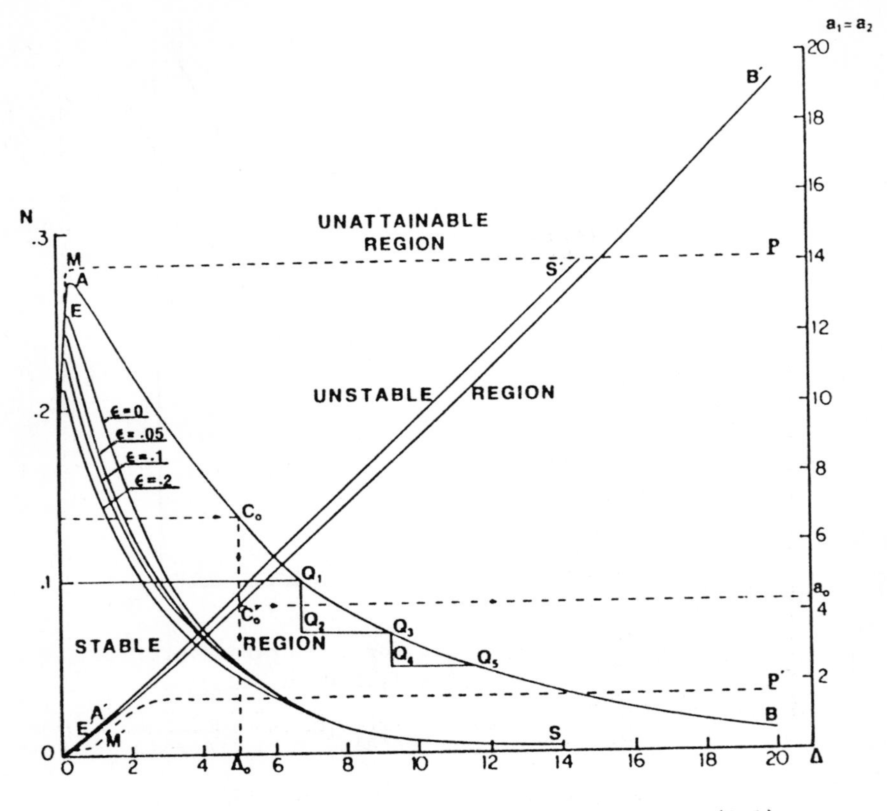

Fig. 13. Stability chart for temperature profile (2.1); ε denotes the percentage by which K_c of one crack exceeds that of the other crack.

cussed in this section. In these experiments, glass plates were heated to a uniform temperature and then brought in contact with a liquid bath cooled by dry ice. Because of the cooling of the contacting edge, a thermal boundary layer develops in the solid. The thermal contraction of the boundary layer produces tension cracks which propagate perpendicular to the cooled edge. As the depth of the thermal layer increases the cracks continue to grow until a critical state is reached. Then, alternate cracks stop, while the others continue to grow. This process is then repeated. Prior to cooling, initial cracks were created at constant spacing along the free edge by scratching the plate with a glass cutter and then tapping the crack open. In these plates stable crack growth was observed, in agreement with the theoretical predictions.

The temperature of the glass plate was monitored by thermocouples, attached at 1.3 cm intervals to the glass plate. To achieve a uniform temperature, two 0.6 cm thick aluminum plates were hung on either side of the glass in the heat bath. These plates were heated by radiation

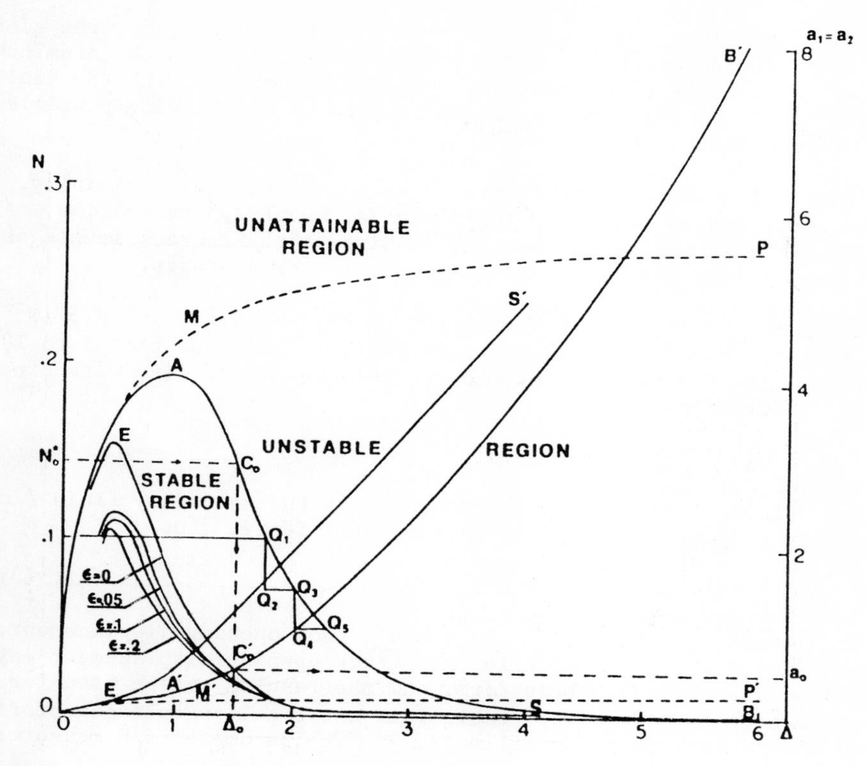

Fig. 14. Stability chart for temperature profile (2.2).

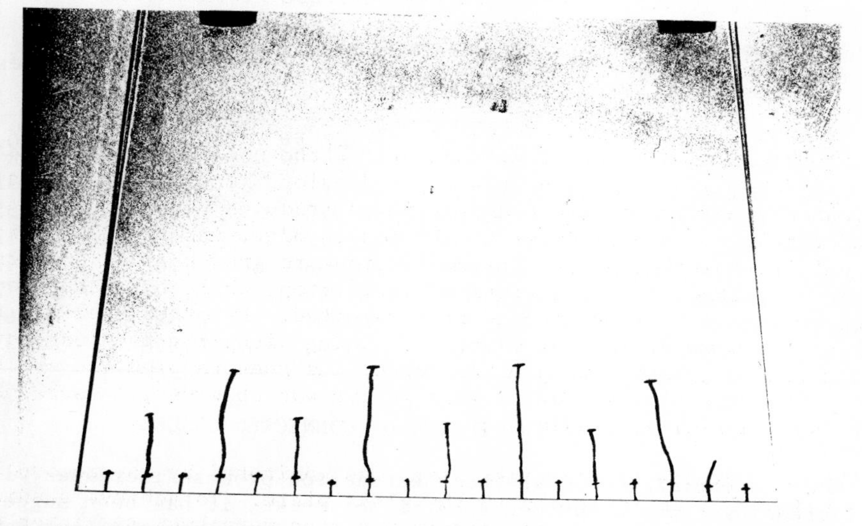

Fig. 15. Crack regime for plate with initial cracks; crack ends are marked at the end of the experiment.

from heating wires, and in view of their high conductivity, the glass plate was subjected to a uniform radiation heating from the aluminum plates. After the glass plate was brought in contact with the cooling bath of -78°C temperature, measures were taken to minimize possible air circulation around the glass plate.

Figure 15 shows a plate tested at room temperature. Initially, 0.7 cm deep cracks have been introduced at about 2.5 cm spacing. All cracks became active upon cooling, growing to three crack levels as alternate cracks were arrested in two consecutive stages.

Using Eq. (3.21) with $E = 6.9 \times 10^{10}$ N/m^2, $\nu = 0.23$, $\alpha = 8.5 \times 10^{-6}$/°C and $K_c = 0.5$ MN/m$^{3/2}$, Geyer and Nemat-Nasser [14] show that 5 to 10% geometric imperfections result in estimates quite in line with experimental observations.

3.6 Opening of Secondary Cracks

The opening of secondary cracks controls the secondary fluid flow which may increase the heat extraction efficiency. Let

$$\Delta u^* = \Delta u \; E/\hat{\beta} \; T_0 b(1 - \nu^2) \tag{3.22}$$

be the dimensionless, and Δu the actual crack opening displacement. For a given temperature profile, the crack opening displacement may be represented in dimensionless form, and the results can be used for different material properties. Table 1 is a typical example for profile (2.1) and Table 2 is for profile (2.2). The parameter y/h measures length along the crack.

TABLE 1

Dimensionless Crack Opening Displacement Δu^* at Stable Bifurcation States for Temperature Profile (2.1); All Quantities Are Dimensionless

			Δu^*			
Δ	a	N	y/h = 0	y/h = 0.25	y/h = 0.5	y/h = 0.75
0.33	0.256	0.264	0.988	0.862	0.724	0.531
0.50	0.398	0.251	1.106	0.978	0.841	0.635
1.00	0.835	0.212	1.056	1.012	0.967	0.812
1.973	1.718	0.150	1.000	0.996	1.005	0.942
2.970	2.652	0.106	1.000	0.994	0.998	0.972
5.00	4.626	0.052	1.000	0.996	0.996	0.969
10.00	9.625	0.013	1.000	0.998	0.998	0.931

4. HYDRAULIC FRACTURING OF CONNECTED CRACKS

Since the efficiency of the heat extraction process increases with the increasing exposed surface area of hot rocks, it has been suggested that one may introduce several primary cracks along the same bore hole. Since these cracks contain pressurized fluid, they may become

TABLE 2

Dimensionless Crack Opening Displacement Δu^* at Stable Bifurcation States for Temperature Profile (2.2); All Quantities Are Dimensionless

Δ	a	N	Δu^* y/h = 0.0	y/h = 0.25	y/h = 0.50	y/h = 0.75
0.488	0.186	0.150	0.672	0.554	0.425	0.279
0.787	0.383	0.106	0.882	0.699	0.504	0.307
0.955	0.536	0.075	0.920	0.704	0.482	0.275
1.350	0.970	0.031	0.947	0.658	0.386	0.184
1.500	1.220	0.022	0.953	0.620	0.326	0.135
1.840	1.680	0.013	0.963	0.579	0.267	0.093
2.000	1.978	0.009	0.965	0.548	0.228	0.069

unstable in the sense that some may extend spontaneously as others close. Nemat-Nasser and Keer* [15] report results of stability analysis for an array of such cracks, which seem to support this conclusion. They consider an infinite array of co-axial penny-shaped cracks, equally spaced at distance b apart, with alternate radii of $R_j = a_j/b$, $j = 1,2$. Then with N_1 and N_2 representing suitable dimensionless ($N = K/\mu\sqrt{b}$) stress intensity factors for the corresponding alternate cracks, they obtain the results given in Fig. 16, where V is the fluid volume, μ is the rock shear modulus, and p is the fluid pressure. As is seen, the unstable critical point,

$$\frac{\partial N_1}{\partial R_1} = \frac{\partial N_2}{\partial R_2} = 0, \tag{4.1}$$

is reached when $R_1 = R_2 = 0.827$.

Table 3 shows the post critical behavior, and gives the values of p/μ for critical $N_c = 1$.

TABLE 3. Post Critical Behavior ($R_1 = 0.827$)

p/μ	R_2	$V/(1-\nu)$	$V_1/(1-\nu)$	$V_2/(1-\nu)$	N_1	N_2
1.493	0.827	2.545	1.272	1.272	1	1
1.241	1.000	2.799	0.716	2.083	0.615	1
1.114	1.200	3.650	0.302	3.348	0.311	1
1.070	1.400	4.985	0.055	4.930	0.113	1
1.056	1.600	6.686	-0.077	6.763	-0.003	1

The results show that as the critical state defined by (4.1) is exceeded, every other one of the cracks will stop growing, while the active cracks will grow at a faster rate; furthermore, the fluid of the

*This work has been completed in collaboration with Dr. A. Oranratnachai.

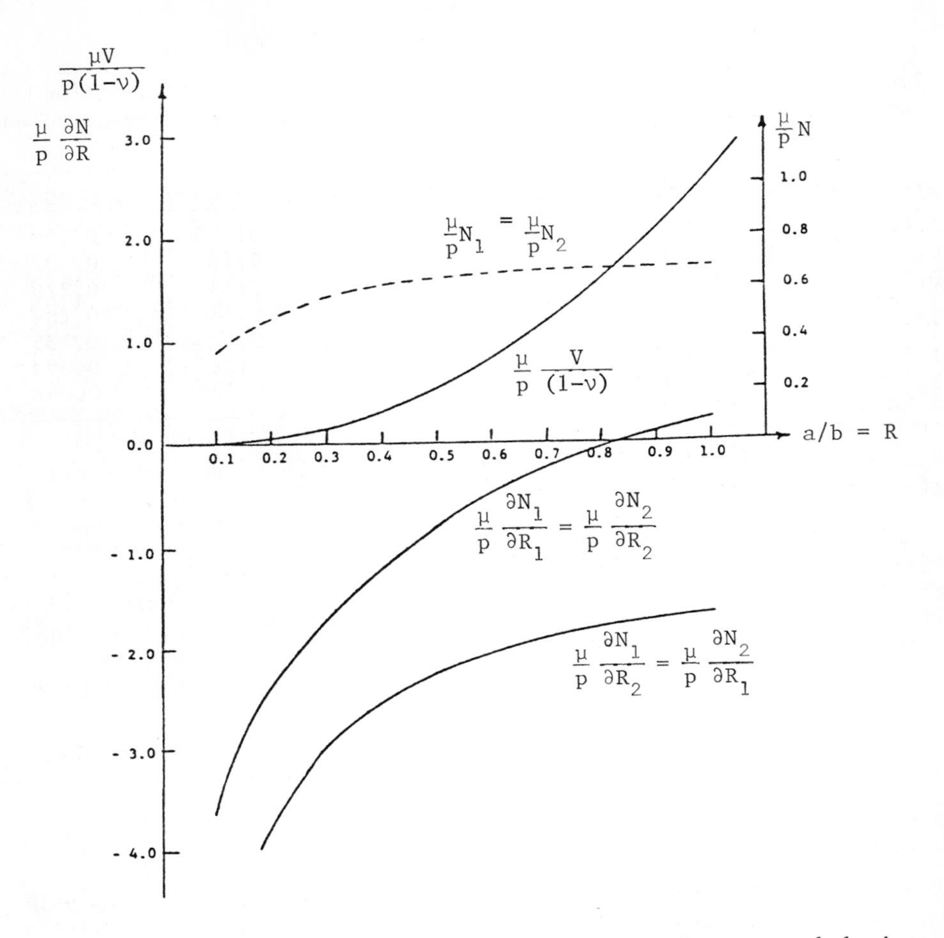

Fig. 16. Curves showing volume, stress intensity factor, and derivatives of stress intensity factor versus $R = R_1 = R_2$.

arrested crack will tend to be transferred into the growing crack; i.e. as the total volume of the system is increased, the volume of the inactive crack will decrease, eventually becoming zero. The significance of these results is that when the geometry of the crack array is such that the unstable critical point has been achieved, then for connected cracks any further growth will occur in only one of the cracks, and the fluid volume of the inactive crack will tend to be transferred to the growing crack as the total fluid volume increases. Thus, care must be taken to ensure that the cracks are not grown beyond the critical radius R = a/b = 0.827; otherwise, continued crack growth, which can occur naturally through thermal expansion and cracking, may cause a loss of fluid volume from the arrested crack and its transfer to the growing crack. The introduction of an additional crack will then have little effect in improving the reservoir efficiency.

It should be noted from Fig. 16 that conditions

$$\frac{\partial N_1}{\partial R_1} = \frac{\partial N_2}{\partial R_2} < \frac{\partial N_1}{\partial R_2} = \frac{\partial N_2}{\partial R_1}, \quad R_1 = R_2 = R \tag{4.2}$$

are always satisfied, and hence a stable bifurcation will be achieved for all values of R. Therefore, for connected cracks of equal radius, if the pressure in the cracks is large enough to cause crack growth, one crack will always grow at the expense of the other, and it would be impossible to grow two cracks simultaneously. The proper method of growing more than one crack on a single bore hole would be to grow the cracks sequentially so that only one crack at a time is active. All cracks that are intended to be inactive would have to be sealed to prevent their further growth. When the cracks have been grown, then the seals can be removed and the fluid circulated between the cracks. At this stage the stress intensity factor must be well below its critical value in order to prevent further crack growth.

Nemat-Nasser and Keer [15] also report the fluid pressures required to achieve the critical value of stress intensity factor for a single crack, as well as multiple cracks. Due to interaction, higher pressure is required for multiple cracks, and this pressure increases as the crack spacing decreases.

ACKNOWLEDGMENT

This work has been supported by the National Science Foundation under Grant No. CME-8006265 to Northwestern University.

REFERENCES

1. Smith, M. C., Potter, R., Brown, D., and Aamodt, R. L., "Induction Growth of Fractures in Hot Rock," Geothermal Energy, P. Kruger and C. Otto, eds., pp. 251-268, Stanford University Press, 1973.

2. Tester, J. W., Morris, G. E., Cummings, R. G., and Bivins, R. L., "Electricity from Hot Dry Rock Geothermal Energy: Technical and Economic Issues," Report LA-7603-MS, Los Alamos Scientific Laboratory, Los Alamos, NM, 1979.

3. Blair, A. G., Tester, J. W., and Mortensen, J. J., "LASL Hot Dry Rock Geothermal Project," Report LA-6525-PR, Los Alamos Scientific Laboratory, Los Alamos, NM, 1976.

4. Dash, Z. V., Murphy, H. D., and Cremer, G. M., eds., "Hot Dry Rock Geothermal Reservoir Testing: 1978 to 1980," Report LA-9080-SR, Los Alamos Scientific Laboratory, Los Alamos, NM, 1981.

5. Nemat-Nasser, S., Keer, L. M., and Parihar, K. S., "Unstable Growth of Thermally Induced Interacting Cracks in Brittle Solids,"

Earthquake Research and Engineering Laboratory Technical Report No. 77-9-2, Department of Civil Engineering, Northwestern University, Evanston, IL, September 1977; Int. J. Solids and Structures 14, pp. 409-430, 1978.

6. Nemat-Nasser, S., Sumi, Y., and Keer, L. M., "Unstable Growth of Tension Cracks in Brittle Solids: Stable and Unstable Bifurcations, Snap-Through, and Imperfection Sensitivity," Earthquake Research and Engineering Laboratory Technical Report No. 79-7-20, Department of Civil Engineering, Northwestern University, Evanston, IL, July 1979; Int. J. Solids and Structures 16, pp. 1017-1035, 1980.

7. Nemat-Nasser, S., and Oranratnachai, A., "Minimum Spacing of Thermally Induced Cracks in Brittle Solids: Addendum," J. of Energy Resources Technology, ASME 104, pp. 96-98, 1982.

8. Keer, L. M., Nemat-Nasser, S., and Parihar, K. S., "Growth and Stability of Thermally Induced Cracks in Brittle Solids," Geothermal Energy Research Project Report, Department of Civil Engineering, Northwestern University, Evanston, IL, September 1976.

9. Keer, L. M., Nemat-Nasser, S., and Oranratnachai, A., "Unstable Growth of Thermally Induced Interacting Cracks in Brittle Solids: Further Results," Earthquake Research and Engineering Laboratory Technical Report No. 78-2-8, Department of Civil Engineering, Northwestern University, Evanston, IL, February 1978; Int. J. Solids and Structures 15, pp. 111-126, 1979.

10. Nemat-Nasser, S., "Stability of a System of Interacting Cracks," Letters in Applied and Engineering Sciences 16, pp. 277-285, 1978.

11. Nemat-Nasser, S., "On Stability of the Growth of Interacting Cracks, and Crack Kinking and Curving in Brittle Solids," Earthquake Research and Engineering Laboratory Technical Report No. 80-5-31, Department of Civil Engineering, Northwestern University, Evanston, IL, May 1980; Numerical Methods in Fracture Mechanics, D. R. J. Owen and A. R. Luxmoore, eds., pp. 687-706, Pineridge Press, Swansea, U. K., 1980.

12. Sumi, Y., Nemat-Nasser, S., and Keer, L. M., "A New Combined Analytical and Finite-Element Solution Method for Stability Analysis of the Growth of Interacting Tension Cracks in Brittle Solids," Earthquake Research and Engineering Laboratory Technical Report No. 79-5-14, Department of Civil Engineering, Northwestern University, Evanston, IL, May 1979; Int. J. of Engineering Science 18 pp. 211-224, 1980.

13. Nemat-Nasser, S., Oranratnachai, A., and Keer, L. M., "Effect of Geometric Imperfection on Stability of the Growth Regime of Thermally Induced Interacting Edge Cracks," 1983, in preparation.

14. Geyer, J. F., and Nemat-Nasser, S., "Experimental Investigation of Thermally Induced Interacting Cracks in Brittle Solids," Earthquake Research and Engineering Laboratory Technical Report No. 80-8-35, Department of Civil Engineering, Northwestern University, Evanston, IL, August 1980; Int. J. Solids and Structures 18, pp. 349-356, 1982.

15. Nemat-Nasser, S., and Keer, L. M., "Thermally and Hydraulically Induced Tension Cracks," in "Hot Dry Rock Reservoir Characterization and Modeling; October 1, 1978-September 30, 1979; Final Report," Report LA-8343-MS, Chapter 4, pp. 52-113, Los Alamos Scientific Laboratory, Los Alamos, NM, May 1980.

HOT DRY ROCK RESERVOIR DEVELOPMENT AND TESTING IN THE USA

Hugh Murphy

Los Alamos National Laboratory, New Mexico

ABSTRACT

Two hot dry rock geothermal reservoirs have been created, or are being created, in hard impermeable rock at a site in the Jemez Mountains of New Mexico. In the first reservoir, created for research purposes, the technical feasibility of this new energy source was established during three years of intensive testing. Now the Phase II reservoir, intended as the first step in demonstrating commercial viability, is being created.

INTRODUCTION

Geothermal energy is available from several types of resources (1): (i) hydrothermal resources, where steam or hot water occur naturally; (ii) geopressured resources, where hot water containing quantities of methane lies confined under intense pressure; (iii) magma resources, typified by Kilauea volcano in Hawaii, consisting of masses of accessible molten rock; and (iv) hot dry rock resources, which exist everywhere and contain large amounts of useful heat energy but do not spontaneously produce fluids. The first three types are geographically rare, and of these, only hydrothermal resources are being exploited commercially today. In contrast to the other geothermal resources, hot dry rock resources are so widespread and broadly distributed that underlying the United States alone over 13×10^6 quads (1 quad = 10^{18} J) of thermal energy are potentially available above 150°C at depths less than 10 km (2). Basically, hot dry rock reservoirs are formed by drilling into low-permeability basement rock to a depth where the temperature is high enough to be useful, creating the necessary permeability, i.e., flow paths, by hydraulic fracturing, and then completing the circulation loop with a second well which intercepts the fractured region. Thermal power is extracted by injecting cold water down the first well, forcing the water to sweep by the freshly exposed hot rock surface in the reservoir-fracture system, and then returning the hot water in the second well to the surface, where the thermal energy is converted to electrical energy or used for other purposes. Circulating fluid pressures are maintained high enough to avoid vaporisation.

Development of hot dry rock geothermal extraction technology has been underway at the Los Alamos National Laboratory in New Mexico since 1973 (3). Currently this work is funded by the US Department of Energy, Japan's New Energy Development Organisation, and the Federal Republic of Germany, represented by Kernforshungsanlage Jülich. Total funding for 1982 was approximately $15 million, 66% from the US DOE, and 17% each from W Germany and Japan. Nearly sixty scientists, engineers and technicians, including five Japanese and three German scientists, are engaged at Los Alamos.

The ultimate goal is to determine not just the technical feasibility, but, paramountly, the economic feasibility of exploiting this form of geothermal energy. To accomplish these goals, the Los Alamos scientists have expended most of their efforts on drilling wells and creating and operating two reservoirs: the Phase I, or research, reservoir, and the Phase II, or engineering, reservoir, both of which are illustrated schematically in Figure 1.

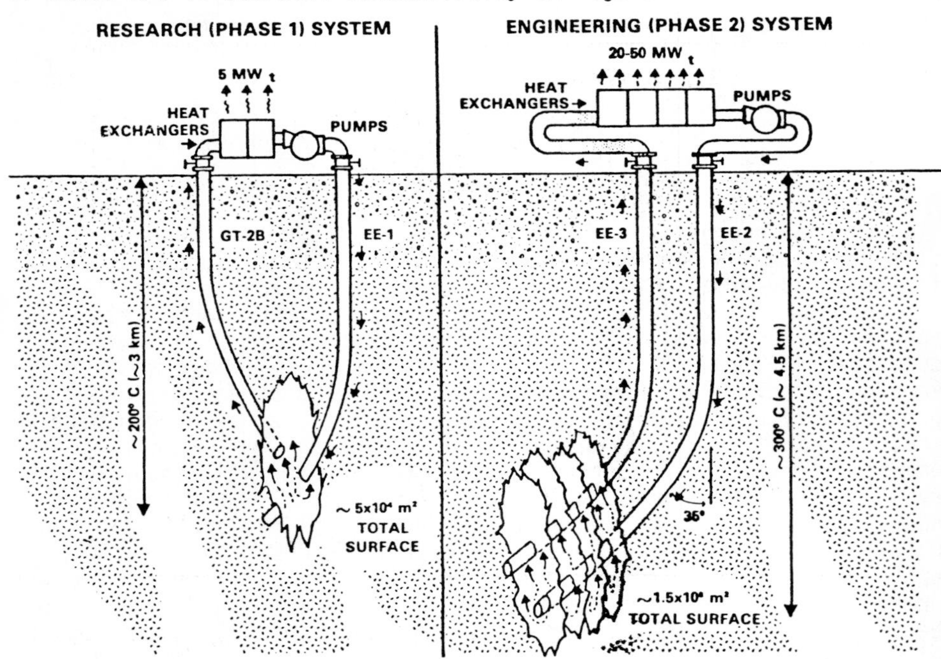

Fig 1 Comparison of the Phase I and Phase II systems.

Both reservoirs are located at Fenton Hill, on the west flank of a dormant volcanic complex, the Valles Caldera, in the Jemez Mountains of northern New Mexico (see Figure 2). The Valles Caldera lies at the intersection of the tectonically active Rio Grande Rift and the volcanically active Jemez lineament. The Fenton Hill site is 27 km west of Los Alamos and 8 km from the hydrothermal geothermal system recently being developed by Union Oil company. The relatively recent (< 0.1 to 1.4 my) volcanic activity (4, 5), coupled to a higher than normal heat flow associated with the Rio Grande Rift (6), results in an effective geothermal gradient of 100°C/km in the sediments from the surface to a depth of 720 m, while in the Precambrian, granitic rocks below, gradients ranging from 50 to 70°C/km prevail. Based upon thermal conductivity measurements (7, 8) in the reservoir host rock, essentially biotite granodiorite, these gradients correspond to a heat flow of 0.16 W/m^2 (3.8 HFU), about 2.5 times the worldwide average (9). The high thermal gradients at the Fenton Hill site, coupled to its convenient location, made the site ideal for early field investigations of the HDR concept because reasonably shallow depths were required to reach temperatures at which production of electricity from the geoheat is economically feasible (10). In the Phase I reservoir a rock temperature of 200°C was attained at 3 km and in the Phase II reservoir 325°C was attained at 4.4 km. Additional details of the site geology are provided by Laughlin (11) and results of petrological and geochemical characterisation of rock samples obtained by coring are presented by Laughlin and Eddy (12).

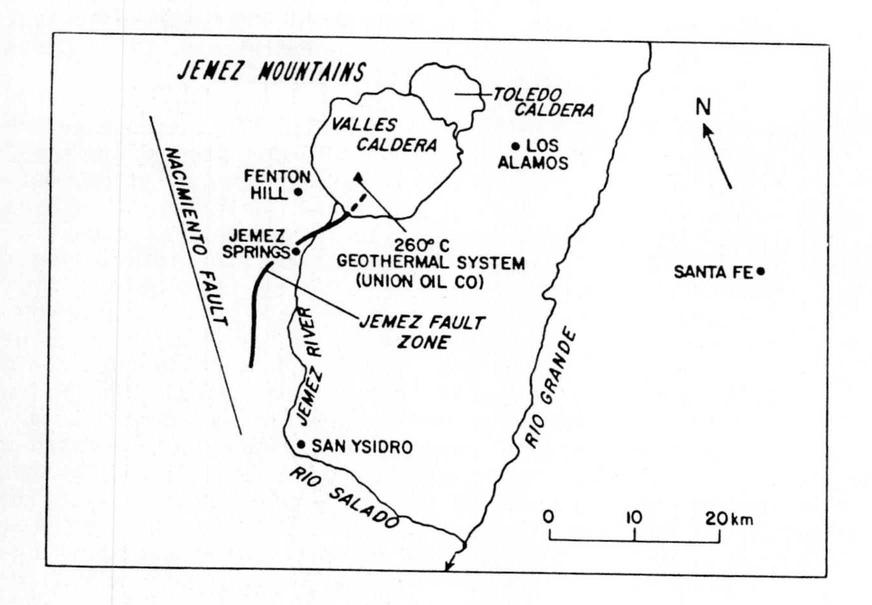

Fig 2 Location of Fenton Hill site showing relationship to Valles Caldera.

In the next section the Phase I reservoir results accumulated over the span 1977 to 1980 are summarised. In the section that follows the Phase II reservoir design, the well drilling, and the hydraulic fracturing tests are discussed.

PHASE I RESERVOIR RESULTS

The Phase I, research, reservoir was formed by connecting two wells, GT-2 and EE-1 with hydraulic fractures which are believed to be nearly vertical. Vertical fractures would be expected on theoretical grounds because the planes of hydraulic fractures are orthogonal to the minimum (least compressive) component of the tectonic earth stress. In tectonically relaxed geological settings this stress is expected to be a horizontal one at depths greater than about 1 km. The minimum compressive horizontal tectonic stress at the reservoir depth of 2.8 km is 37 MPa, about one-half the vertical overburden stress, as estimated by pressurisation tests (13). However, downhole monitoring of microseismic events induced during hydraulic fracturing with geophones indicated that the amplitudes of the shear waves were two to three times those of the compressional waves (14). Thus fracturing exhibits a strong shearing component, an observation which will become more significant when the Phase II fracturing results are reviewed. In all fracturing operations conducted, for both reservoirs, essentially pure water is used. Viscosity-increasing or loss-of-fluid additives are not normally used, nor are proppants used. Pumping tests have indicated that the fractures are self-propped - upon depressurisation the fractures remain partly open due to particle-to-particle contacts between the misaligned rough surfaces produced.

Following a series of preliminary hydraulic fracturing experiments in GT-2, the EE-1 well was drilled towards the largest of the GT-2 fractures in an attempt to complete a heat extraction system, but due to wellbore survey errors during directional drilling, it missed by 6 m, as determined by subsequent gyrosurveys as well as magnetic and acoustic ranging. Despite such close proximity, the intervening rock was so impermeable that the impedance to water flow was too high to permit a viable heat extraction experiment. Various attempts to improve flow communication between the boreholes, such as fracturing the EE-1 well, proved unsuccessful, apparently because the fractures in both wells were parallel and did not intersect (13). The EE-1 well was then cased to 2.93 km. Subsequent temperature and cement bond logs showed that the casing cement from 2.74 to 2.93 km deteriorated with time because of the high downhole temperatures as well as the thermal cycling caused by numerous flow and pressurisation experiments conducted after cementing the casing. This deterioration resulted in the formation of a water flow path in the annulus between the rock and casing, and eventually a large vertical hydraulic fracture was formed behind the casing at a depth interval centred about 2.75 km.

The first reservoir was eventually completed by deviating GT-2 at

2.5 km and redrilling it towards the top of this large EE-1 fracture. The intent was to produce a large vertical spacing between the inlet and outlet locations to maximise the effective heat transfer area, while still achieving reasonably low flow impedance. Low flow impedance is required for high rates of heat extraction as the allowable pressure loss is limited by tectonic stress considerations. Excessive pressures during heat extraction would result in uncontrolled fracture growth and very large downhole water losses. Two redrilling attempts were required: the first redrilling path apparently passed above the top of the fracture, and while the second path also did not intersect the fracture directly it did penetrate several natural fractures or major joints which communicated with the hydraulic fracture. The impedance to flow circulation was low enough to proceed with a heat extraction test. The combination of the original GT-2 wellbore and the second redrilling path is referred to as GT-2B. In subsequent testing of the reservoir EE-1 was used as the injection well and GT-2B as the extraction or production well.

Reservoir performance was first evaluated by a 75-day period of closed-loop operation beginning in January 1978. The assessment of this first reservoir in EE-1 and GT-2B is referred to later as Run Segment 2 (Run Segment 1 consisted of a 4-day precursor experiment conducted in September 1977). Hot water from GT-2B was cooled to 25°C in a water-to-air heat exchanger before reinjection. Make-up water, required to replace downhole losses to the rock surrounding the fracture, was added to the cooled water and pumped down EE-1, and then through the fracture system. Heat was transferred to the circulating water by thermal conduction through the nearly impervious rock adjacent to the fracture surfaces. The average thermal power extracted during Run Segment 2 was 3.1 MWt, with a maximum of 5 MWt. The flow impedance, a measure of the pressure loss through the reservoir per unit flow rate, was initially 1.7 GPa s/m^3 but decreased by a factor of five with time. This was due to thermal contraction and continued pressurisation, resulting in the opening of natural joints that provided additional communication with the producing well. Water losses to the rock surrounding the fracture steadily diminished, and eventually this loss rate was about 1% of the injected rate. The geochemistry of the produced fluid was benign, and the seismic effects associated with heat extraction were undetected. However, the relatively rapid thermal drawdown of the produced water, from 175 to 85°C, indicated that the effective heat-transfer area was small, about 8000 m^2.

Run Segment 3, the High Back-Pressure Flow Experiment, was run for 28 days during September and October of 1978. Purpose of this experiment was to evaluate reservoir flow characteristics at high mean-pressure levels. The high back pressure was induced by throttling the production well. The mean reservoir pressure was thus increased and the flow impedance was reduced several fold. It was discovered during Run Segment 3 that, as a result of deteriorated casing cement, the water injected into EE-1 was flowing in the annulus to depths as shallow as 760 m. This posed a potential danger to ground-water aquifers and caused increased water losses. To alleviate these problems, and also to investigate the feasibility of creating a larger fracture from the same wellbores, the EE-1 casing was recemented near its bottom at 2.93 km. An enlarged reservoir was then formed by extending a hydraulic fracture just below the casing with two "massive" fracturing experiments referred to as MHF Expts 203 and 195. In each experiment, approximately 760 m^3 of plain water was injected at a rate of approximately 0.04 m^3/s, raising the downhole pressure by about 20 MPa. The resulting large fracture propagated upward to at least 2.6 km and appeared to have a minimum inlet-to-outlet spacing of 300 m, three times that of the reservoir prior to refracturing.

Preliminary evaluation of the enlarged reservoir was accomplished during a 23-day experiment (Run Segment 4) in late 1979 (15). Longer term reservoir characteristics were investigated in Run Segment 5, which began in March 1980. Because of the large heat transfer surface area size and resulting slow thermal drawdown, a lengthy flow time, some 280 days, was necessary to evaluate the reservoir. This experiment, along with the two-day Stress Unlocking Experiment (SUE) that immediately followed Run Segment 5 are described by Zyvoloski (16) and Murphy (17).

With these research size reservoirs, no attempt was made during Run Segments 2 through 4 to use the geoheat for generating electrical energy because of the low power levels produced. The heat was simply dissipated into the atmosphere by the heat exchanger. However, during Run Segment 5 a binary cycle electrical generating unit designed by Barber-Nichols Engineering was incorporated in the circulation loop. This generator extracts energy from the water produced from the reservoir and heats the working fluid, Refrigerant 114, which is then expanded through a single-stage turbine. Problems with leakage of the working fluid prevented sustained operation of the generator, but it did produce a peak power of 60 kWe.

In the three years during which these reservoir tests were conducted, our understanding of reservoir behaviour has steadily improved. In particular, numerical modelling evolved continuously, so that simplified models developed for Run Segment 2 were significantly modified by Run Segment 5. Hot dry rock reservoirs are not static but continuously change. Following pressurisation and thermal contractions induced by heat extraction, they grow larger with time, with profound effects upon hydraulic and heat production characteristics. These reservoir changes and some of their effects are presented below.

Heat Production

The cumulative thermal energy extracted from the Phase I reservoir during Run Segments 2-5 is illustrated in Figure 3. The upper curve depicts total energy produced, reservoir plus wells, whereas the lower curve represents energy extracted from the reservoir alone. Average thermal energy extracted at the surface was 3.1 MWt (5 MWt peak value) for Run Segment 2, 2.1 MWt for Run Segment 3, 2.8 MWt for Run Segment

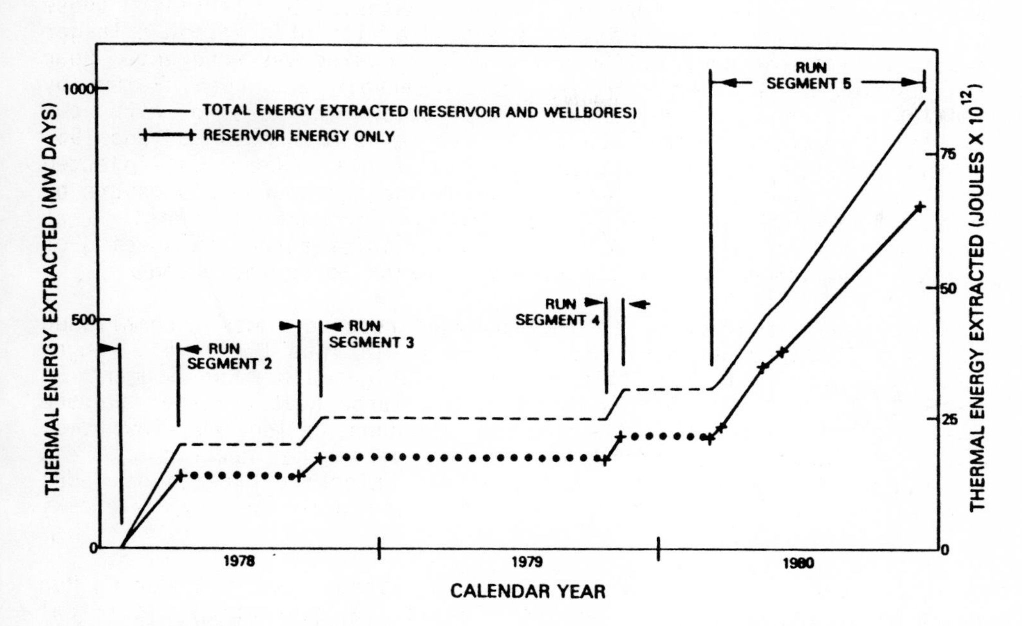

Fig 3 Cumulative thermal energy extracted from the Phase I reservoir.

4, and 2.3 MWt during Run Segment 5. During Run Segment 2 the temperature of the water exiting the reservoir dropped from 175 to 85°C. Temperatures during Run Segment 3 decreased from 135 to 98.5°C, while temperatures during Run Segment 4 remained almost constant at 153°C. Finally during Run Segment 5, the initial reservoir temperature of 156°C climbed to about 158°C after 60 days then dropped to about 149°C by the end (286 days). The rate of decline, or drawdown, of these temperatures permits estimates of the effective heat-transfer areas of the reservoirs when analysed with the heat-transfer models described below. Some parts of the total area are either inaccessible to, or inefficiently bathed by, the water flow because of fluid dynamic and geometrical considerations, leaving an effective heat transfer area which, it will be shown, is considerably smaller than the actual fracture surface area.

Heat-transfer modelling of the reservoirs has been performed with two numerical models. Both models use two-dimensional simulators in which heat is transported by conduction within the rock to the fractures but each invokes a differing geometrical description of the reservoir. Reservoir geometry is inferred from tracer, flow rate (spinner), and temperature logs, and the heat-production experiments. These data as well as the results of active and passive microseismic experiments (18, 19) indicate that the reservoir geometry is extremely complex. However, simplified abstractions of the reservoir, useful for

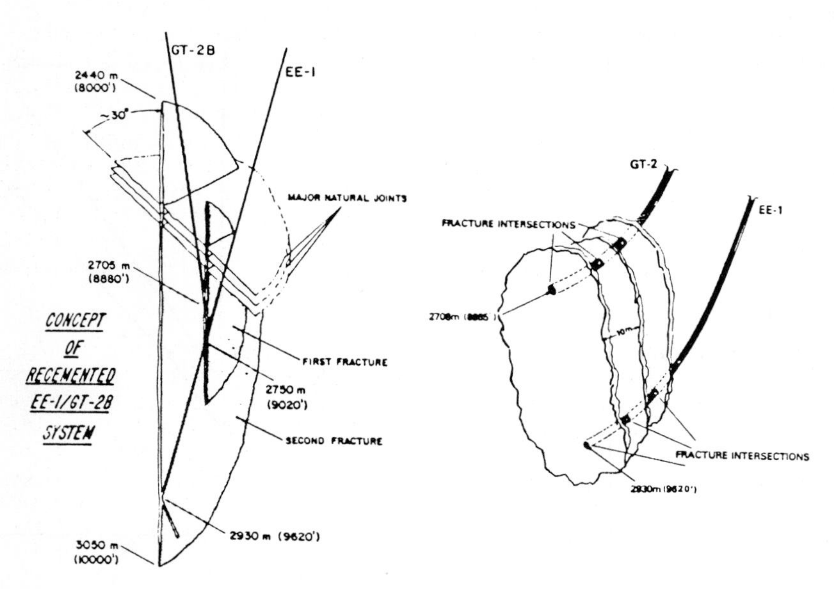

Fig 4 Inferred reservoir geometry (early concept).

Fig 5 Inferred reservoir geometry (later concept).

modelling purposes, are presented in Figures 4 and 5. Figure 4 is an early conceptual model showing both the small fracture exploited for Run Segments 2 and 3 and the enlarged fracture system of Run Segments 4 and 5. Also shown are the natural joints with a dip of about 60° from the horizontal which afford hydraulic communication between the main hydraulic fractures and GT-2B. Figure 5 is a more recent view of the system. This model is characterised by at least three vertically oriented fractures and evolved after a detailed analysis of temperature drawdown and recovery in both wells as well as the fracture residence-time distribution curves provided by tracer tests.

The independent-fractures model, based upon Figure 4, assumes that the fractures are circular and allows proper local positioning of the inlet and outlets, ie, the point-like intersection of the injection well with the fracture can be modelled, as can the intersection of the main hydraulic fractures and the slanting joints that provide the connections to GT-2B. In view of this more faithful representation of inlet and outlets, and the fact that a complete two-dimensional solution to the Navier-Stokes fluid dynamic equations is incorporated, the independent-fractures model results in a more realistic assessment of the effect of fluid dynamics and sweep efficiencies upon heat extraction. However, in the present two-dimensional version of the code, thermal interaction as the temperature waves in the rock between fractures overlap cannot be represented realistically, as in the multiple-fracture model.

The most recently developed model, the multiple-fracture model, based upon the geometry in Figure 5, assumes that the fractures are parallel rectangles and that, for simplicity, flow is distributed uniformly along the bottom of each fracture and uniformly withdrawn from the top of each fracture. The flow is thus one-dimensional, and the streamlines are straight vertical lines. Consequently fluid dynamic considerations do not directly enter into the heat-extraction process - the sweep efficiency is implicitly assumed to be 100%. However, a rigorous two-dimensional heat-conduction solution is incorporated for the rock between the fractures, and this permits valid consideration of thermal interaction effects between the fractures.

Independent-Fractures Modelling The first application was to the early research reservoir during Run Segment 2, when only a small single hydraulic fracture existed. Estimates of the thermal drawdown were calculated with the model for trial values of fracture radii and a value of 60 m resulted in a good fit to the measurements. A radius of 60 m implies a total fracture area (on one side) of 11 000 m^2; however, because of hydrodynamic flow sweep inefficiencies the net area effective in heat exchange was only 8000 m^2. Further details are given by Murphy et al (15).

During Run Segment 3, the high back-pressure experiment, thermal drawdown suggested that the effective heat area was nearly the same as that determined in Run Segment 2. However, flow rate surveys in GT-2 indicated that because of the higher pressure level most of the flow was entering GT-2B at positions that averaged 25 m deeper than during Run Segment 2. In effect the reservoir flow paths were shortened about 25%. It was concluded that while pressurisation did indeed result in some shortening of the streamlines, it also resulted in a notable decrease in impedance, which afforded better fluid sweep and bathing of the remaining area.

As mentioned earlier, the reservoir was enlarged during MHF Expts 195 and 203 and evaluated during Run Segments 4 and 5. For Run Segment 4 it was found that the old fracture had an effective heat-transfer area of 15 000 m^2 and the new fracture had an effective area of at least 30 000 m^2. Thus the old fracture area was at least twice that determined in Run Segment 2, a growth attributed to thermal stress cracking effects (20). An improved estimate of the total effective heat-transfer area of both fractures was obtained in Run Segment 5. As shown in Figure 6, the thermal drawdown data are fit very well by a model with a combined area of 50 000 m^2, some 5000 m^2 greater than the area tentatively estimated during Run Segment 4.

A summary of the heat-exchange areas determined with the independent-fractures model is presented in Figure 7, and a steady increase, from 8000 to 50 000 m^2, is shown. The question marks in Figure 7 indicate that the area increase due to the MHF experiments (Expts 195 and 203), is uncertain. The heat-transfer area was not measured until the later stages of Run Segment 4. Consequently, the area increase

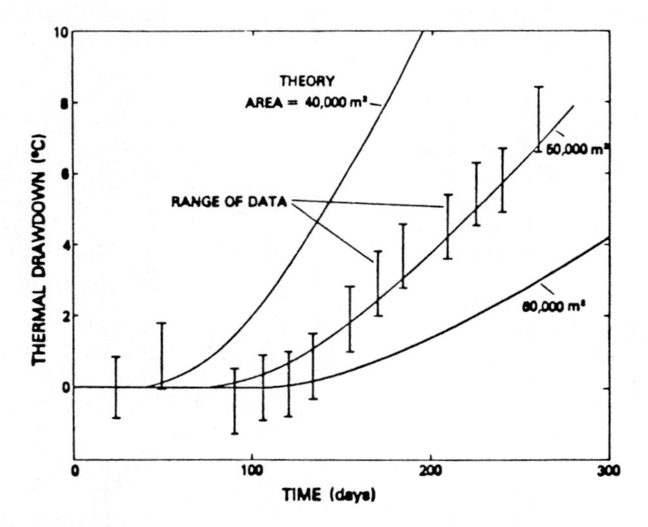

Fig 6 Comparison of field data with independent fracture heat transfer model for Run Segment 5.

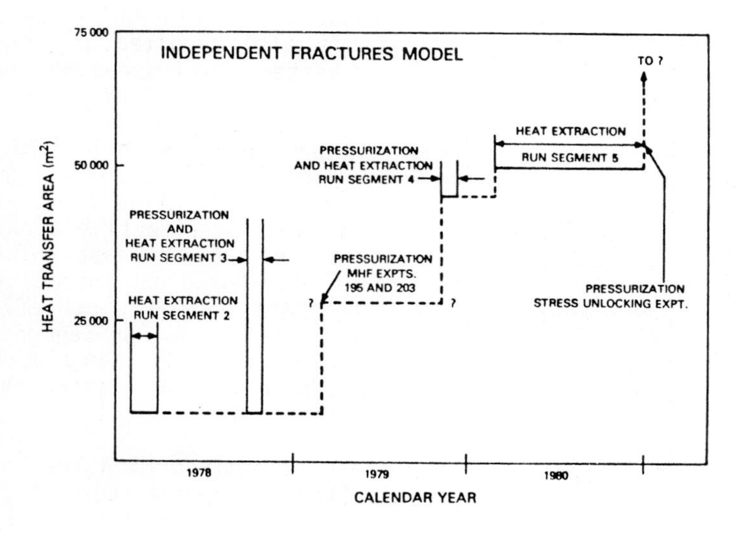

Fig 7 Effective heat-transfer area growth determined by the independent fractures model.

measured is due to the combined effects of all the fracturing and Run Segment 4 operations, and cannot be individually ascribed to the separate operations.

Multiple-Fracture Modelling Figure 8 summarises the growth of the heat-exchange area according to the multiple-fractures model. The initial area was 7500 m^2 and grew to 15 000 m^2 during Run Segment 2. As indicated earlier, the high back pressure of Run Segment 3 caused a redistribution of flow and shortening of streamlines. The newer model indicates that the initial heat-exchange area was actually less than that of Run Segment 2, starting at 6000 m^2; but then grew to 12 000 m^2 during Run Segment 3. After Run Segment 4, the EE-1 temperature logs indicated that between 6000 and 9000 m^2 had been added to the lower part of the reservoir by the recementing and pressurisation prior to and during Run Segment 4. Modelling of Run Segment 5 indicates that the area increased continuously, and at the end of the test the area was 45 000 m^2, 10% less than the area derived with the older model.

In summary, there exists some difference in detail but both heat transfer models yield the same overall trends: an effective heat transfer that continuously grows in response to pressure- and thermal-induced changes of the reservoir. The tracer results, discussed next, also support this conclusion.

Tracer Studies and Fracture Volume Growth

Unlike the determination of effective heat transfer, which can require months and perhaps years of operation in order to attain a thermal drawdown sufficient for modelling, the determination of

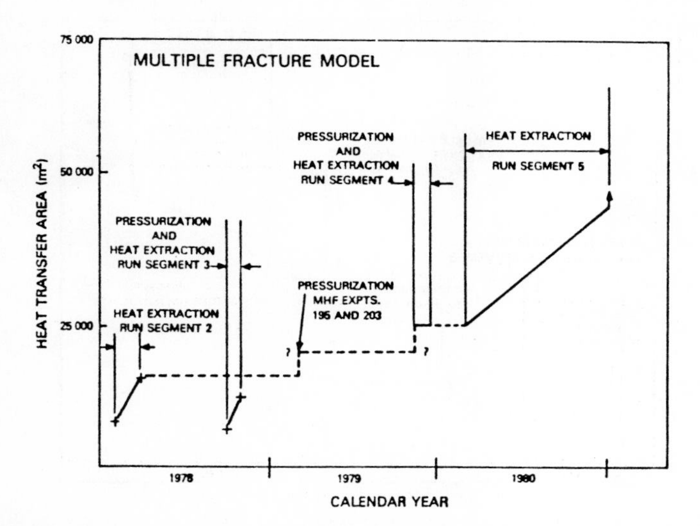

Fig 8 Effective Heat-transfer area growth determined with multiple fracture model.

fracture volume in a reservoir takes only a few hours to perhaps several days when using a tracer. Unfortunately, the effective heat transfer area is required for estimates of future heat production and reservoir lifetime. However, if a relationship between volume and area can be established, tracer studies would be an extremely convenient means of characterising a reservoir. Two tracers are used in the Los Alamos work. The first is a visible dye, sodium fluorescein, which is monitored in the produced fluid with a UV spectrophotometer. The other tracer, radioactive Br^{82} (half life equals 36 h) injected as ammonium bromide, is monitored in the produced fluid with a flow-through gamma counter. The radioactive tracer is not temperature sensitive and therefore does not undergo thermal decomposition as does the sodium fluorescein in the higher temperature portions of the reservoir.

The fracture modal volume, considered the most reliable indicator of reservoir volume change (21), is simply the volume of fluid produced at GT-2B from the time the tracer pulse is injected until the peak tracer concentration appears in the produced fluid. The wellbore volumes are subtracted from the total volume produced to give the true fracture modal volumes. When the curve of modal volume vs time, in Figure 9, is compared to the corresponding heat-transfer areas plotted in Figures 7 and 8, the similarities of the growth of area and volume are quite striking. This can be further quantified by considering the relationships between area, volume, and aperture (or effective fracture opening). The volume, V, is simply the product of the area, A, and the mean aperture, w, ($V = Aw$). During heat extraction and pressurisation, the area and aperture can both vary; therefore the

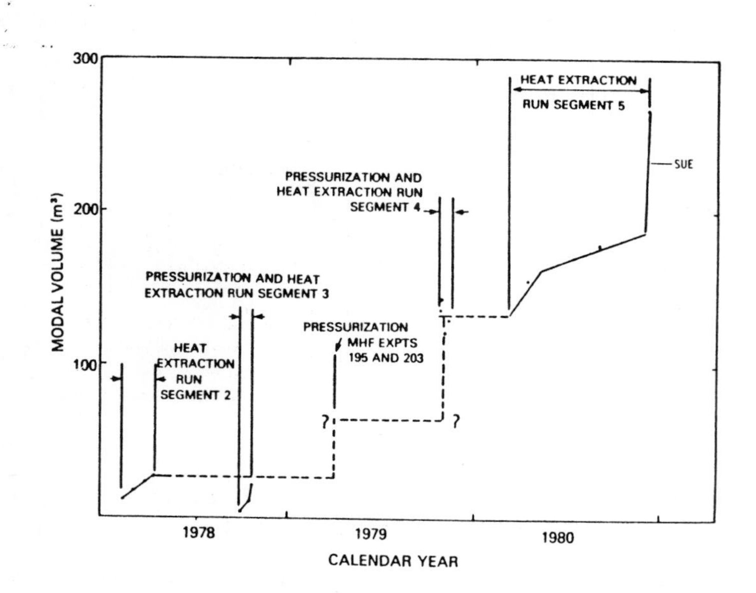

Fig 9 Growth of tracer modal volume in the Phase I reservoir.

volume is a function of two variables rather than one. For constant aperture, the tracer volumes should scale directly with heat-transfer area, and a suggestion of this behaviour is seen in the data from Run Segment 2, shown in Figure 10, where the area vs modal-volume curve has an intercept that corresponds to a constant 1.7 mm aperture. Subsequent pressurisation, particularly MHF Expts 195 and 203, has apparently increased the aperture to about 4 mm.

The reservoir growth due to heat extraction alone is a thermal-contraction effect - as the rock surrounding the fractures shrinks, the fractures, and consequently, the measured modal volume expands. However, the observed thermal contraction is always less than simple estimates of the free thermal contraction. For example, the modal volume during Run Segment 5 increased linearly with the thermal energy extracted, but the increase was only about one-tenth that expected from thermal contraction of unconfined granitic rock. This disparity results from the tectonic stresses which tend to constrain the fractures from opening fully.

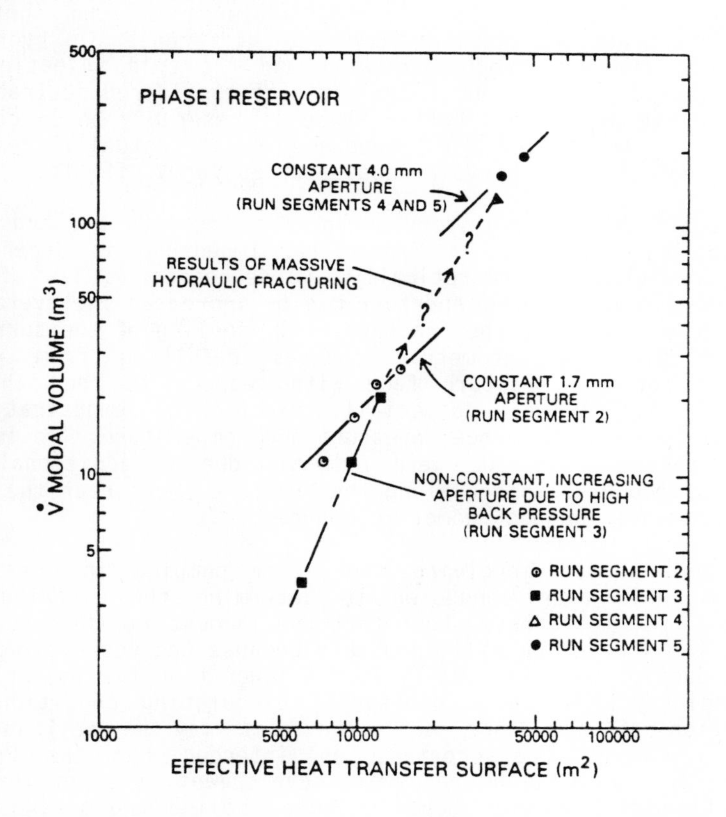

Fig 10 Correlation of increases in effective heat-transfer surface area with modal volume increase.

The Stress Unlocking Experiment (SUE) conducted at the end of Run Segment 5 relieved some of the residual constraining stresses by rapidly pressurising the reservoir to pressures in excess of the minimum compressive earth stress, and then releasing the pressure and restoring equilibrium conditions (17). The reservoir volume was immediately increased from 10% to 30% of the free thermal contraction value. The flow resistance decreased also. In addition, the sudden relaxation of stresses resulted in a burst of micro-earthquake activity from the thermally-depleted regions, resulting in a micro-seismic map of the effective heat transfer area. This area was 60 000 m^2, within 20 to 35% of the areas determined at the end of Run Segment 5 by heat transfer modelling.

Impedance Characteristics

The overall impedance, or flow resistance, of a circulating geothermal reservoir is usually defined as the pressure drop between the inlet and outlet of the fracture divided by the exit volumetric flow rate. As pressures are usually measured at the surface, a "buoyancy" correction is made for the difference in hydrostatic pressures in the hot production well and the cold injection well. Corrected impedances of about 1 GPa s/m^3 are considered desirable. For example, in the deeper and hotter Phase II reservoir being completed now, such a low value of impedance could actually result in "self-pumping" of the reservoir because of buoyancy effects.

The overall impedance history during Run Segments 2-5 and the SUE experiment is summarised in Figure 11. Impedance is dependent on fracture aperture, w. Theoretically, it decreases as $1/w^3$ for both laminar and turbulent flow. Aperture may be increased in several ways: (i) by pressurisation of the fracture, (ii) cooling of the surrounding rock, and (iii) by geometric changes resulting from relative displacement of one fracture face with respect to the other. Run Segments 2 and 3 were especially useful in demonstrating the correlation between impedance, pressure and temperature. The impedance changes observed after SUE were probably due to additional "self-propping" caused by slippage along the fracture faces near the exit or by other pressure-induced geometric changes.

Additional tests, involving short-term pumping into each well separately, have been conducted to determine the distribution of impedance. These tests have shown that the largest impedance component occurs at the production well, probably because the reservoir pressure at the production well is usually 10 MPa lower than the injection well during ordinary, "low back pressure", circulating conditions. This concentration of impedance near the exit may be desirable when multiple fractures are created, as intended for the Phase II reservoir. In this mode of reservoir development, the possibility of unstable flow distribution (one fracture cooling and taking much of the flow) exists, and the exit impedance concentration will prevent this until reservoir cooling has been extensive.

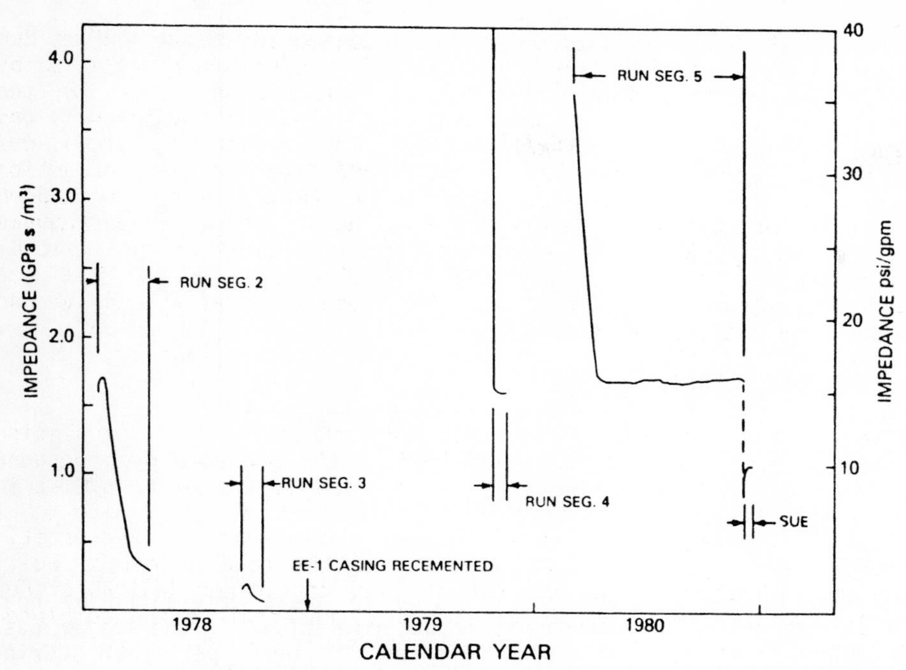

Fig 11 Flow impedance behaviour in the Phase I reservoir.

Water Losses

Water for HDR reservoirs must be provided by external sources, so water losses can have important economic implications. The total water loss is the difference between that injected and that recovered at the production well, and is presented in Figure 12. Run Segment 4, only 23 days long, was excluded from this comparison because of the disparate conditions under which it was conducted. Direct comparison indicates that the water loss for Run Segment 5 is approximately 40% higher than that of Run Segment 2 at comparable times after the beginning of heat extraction. However, because the operating pressure was 10% higher during Run Segment 5, the water loss for Run Segment 2 should be scaled up by 10% as in curve 2. Then it is seen that the Run Segment 5 water loss is only 30% higher than Run Segment 2, despite a several-fold increase in heat-transfer area and volume. An obvious conclusion is that the heat-exchange system utilises only a small portion of a much larger fracture system that controls water loss.

The possibility that the ultimate heat production capacity of an HDR reservoir could be increased several-fold, if better access to the remoter areas of the fractures could be gained by redrilling, or by cyclic and high buoyancy modes of heat production (22), is so intriguing that a series of hydraulic tests were performed to provide

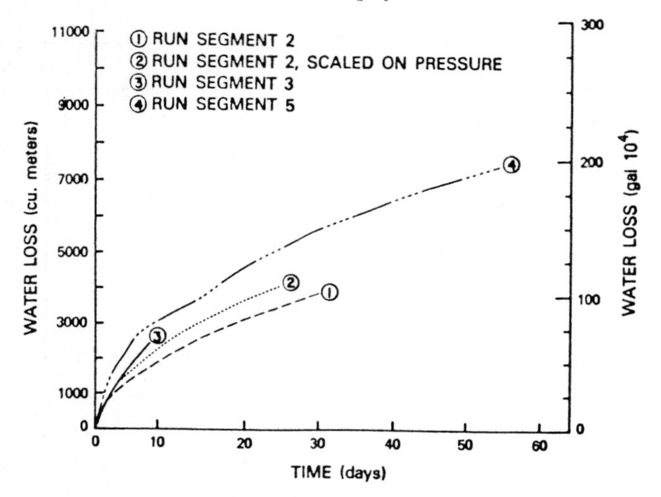

Fig 12 Cumulative water losses.

estimates of <u>total</u> fracture area. The results of these tests are presented in Table I and are compared to areas determined by heat transfer modelling and microseismic determination. In the latter case the microseismic measurements referred to were performed during hydraulic fracturing experiments, not SUE, which excited seismicity primarily in the thermally depleted zone. The seismic results are presented in terms of N, the number of fractures, which must be at least two and more likely three for the enlarged reservoir. The extremely large areas in Table I resulting from fracture inflation tests result from an assumed aperture of 2 mm. In the section on tracer studies values from 1.7 to 4 mm were derived, but these pertained to that part of the reservoir effective in heat production, so they would be expected to be much larger than the average aperture of the total fracture area. However, even if the aperture were taken as as large as 4 mm the inflation-derived areas would still be

TABLE I Comparison of reservoir fracture areas
Area (m^2)

Run segment	Heat transfer (Independent fractures model)	Seismic	Inflation	Venting	Diffusion
2	8 000		3×10^6	40 000	~250 000
3	8 000		3×10^6		
4	45 000	75 000 x N			
5	50 000	90 000 x N	10×10^6	>250 000	~350 000

enormous, enough, in fact, to exceed the Phase II reservoir goals. Another hydraulically-inferred area comes from measurement of the volume of water released during sudden, rapid depressurisations (venting). In this case an aperture of 4 mm was assumed. Finally, the diffusion areas were determined from analysis of water losses using the methods of Fisher (23). In all cases the evidence is clear: although the heat transfer area has continued to grow due to pressurisation and thermal contraction, it still represents only a small fraction of that potentially available.

Fluid Geochemistry

Analyses of the concentrations of dissolved rock minerals and gases indicate the existence of parallel flow paths: (i) a fracture-dominated flow path, perhaps consisting of multiple fractures, that includes the effective heat-transfer surfaces, and (ii) a high-impedance flow path consisting of connected microfractures and pores in the rock surrounding the heat-extraction portion of the reservoir. Displacement of the indigenous pore fluid contained in this high-impedance flow path is discussed by Grigsby et al (24). Steady-state concentration of total dissolved solids was typically 2500 mg/ℓ - similar to water used for human consumption in the U.S.A. The pH of the water was 6.5, nearly neutral, and problems with corrosion or deposition upon surface equipment such as piping, heat exchangers, and pumps were minimal.

Seismicity

In addition to the diagnostic monitoring already discussed, seismic monitoring was conducted to evaluate environmental hazards. A surface seismic array and downhole geophone packages positioned in the reservoir vicinity were utilised. The largest event detected in Run Segment 4 with the downhole package had a magnitude of -1.5. The energy release of a -1.5 magnitude microseismic event is roughly equivalent to that of a 10 kg mass dropped 3 m. Furthermore, this event occurred during the high back-pressure stage. During the low back-pressure stage, more typical of ordinary heat-extraction conditions, the largest event was -3.

THE PHASE II RESERVOIR

The reservoir presently being created at the Fenton Hill site is intended to serve as a first demonstration of commercial viability. Essentially a pilot-plant reservoir, performance goals call for the production of thermal power with no more than 20% drawdown of production fluid temperature after 10 years.

Reservoir Design

Heat transfer modelling indicates that one million square metres of effective heat transfer area (again counting only one side of each fracture) is required. A single fracture would require, if circular, a radius of 580 m, which is beyond established fracturing technology in HDR reservoirs. We have instead adopted the conservative philosophy that the Phase II fractures will not be much larger than the Phase I fractures. The largest of these fractures, when judged by the inlet-to-outlet vertical separation distance of 300 m, has an effective heat-transfer area of about 50 000 m^2, as established by its thermal-drawdown characteristics. However, this fracture was the result of the modest attempt in which only 1500 m^3 of water, without additives of any sort, were injected as the fracturing fluid. Fracturing capabilities will be expanded for the Phase II reservoir, but even so, it was planned conservatively to create fractures with an inlet-to-outlet separation of 370 m. As the effective heat transfer area scales with the square of the separation, values for these fractures should be 75 000 m^2 each, so that approximately 14 such fractures eventually will be required. Fractures will be created in two stages: an interim step, scheduled for 1982, in which one to three fractures will be produced, seeking the best method; and a final stage in which the remaining fractures will be created.

Drilling

Because the horizontal earth stresses at depth are usually considerably smaller than the vertical, or overburden, stress, the fracture planes were expected to be vertical. In order, then, to accommodate 14 fractures with reasonable horizontal separation distance between fractures, it was necessary to deviate the wells from the vertical direction in the hot downhole region, as shown in Figure 13. The wells were drilled in the east-north-east direction as this was perpendicular to the expected orientation of the fracture planes. An elevation view projecting the traces of both wells on a plane oriented east to west is shown. Deviation too far from the vertical is impractical because it becomes difficult to centre and set casing, and even more difficult to run logging tools; therefore, 35° was chosen as a compromise.

To avoid significant heat production deterioration because of thermal interaction between the fractures, they must be horizontally separated by approximately two times the thermal diffusion distance, $\sqrt{\kappa t}$, where κ is the rock thermal diffusivity and t is time. For 10 yr the total horizontal distance is about 500 m, and a reservoir vertical length of 700 m is required. For high-quality energy production purposes, reservoir temperatures in excess of 200°C are preferred, which, considering the geothermal gradients at Fenton Hill, correspond to depths greater than 3 km. Thus both wells were drilled vertically to about 3 km and then directionally drilled until a deviation of 35° from vertical was attained. Finally each well was drilled at this angle for an additional 1000 m, corresponding to a true vertical penetration of about 800 m, providing some margin of conservativeness over the 700 m requirement. During the drilling the wells were maintained at a vertical separation of 360 m, nearly as planned. The deeper of the two wells, EE-2, which will serve as the injection well,

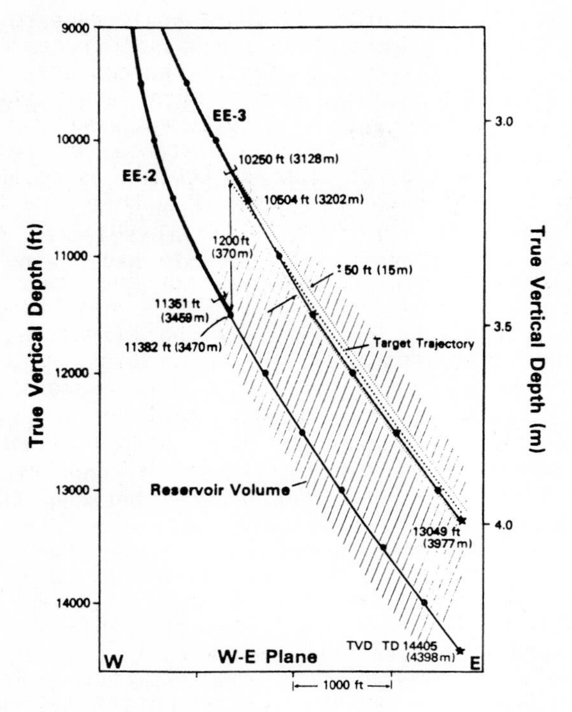

Fig 13 Elevation of Wells EE-2 and EE-3 projected into WE plane.

was completed in 1980 to a true vertical depth of 4.4 km where the temperature was 325°C. The designated production well, EE-3, was completed in 1981 to a depth of 4.0 km. Drilling to such great depths in very hot, hard rock is difficult under any circumstances and the necessity to control precisely the trajectory of each well along the 35° slanted interval required advanced directional drilling and surveying technology (25). As a consequence of these factors and drilling problems, each well required over one year to drill and cost about $10 million. This cost is about twice that of the conventional oil or gas well drilled non-directionally to the same depths, and is comparable in cost to other geothermal wells to these depths (26, 27). Despite expensive drilling, electrical power generated from Phase II prototype reservoirs is still anticipated to be competitive with electricity from coal-fired or nuclear plants (28).

Fracturing

At times the torsional drag in the slanted interval approached 10 revolutions of drillstring twist, which made orientation of downhole tools and directional drilling motors from surface indications almost impossible. To rectify this problem lubricants were added to the drilling fluid. While these lubricants reduced the drag by 50% (25)

they also contaminated the fluid in the wells, and it was feared that they might also adversely affect fracturing operations. Consequently, fracturing did not begin until mid-1982, after the wells were cleaned. Fracturing is being conducted in EE-2, with a view toward propagating fractures upwards towards EE-3, in the direction of lower earth stress. Both wells were cased so as to leave bottom sections of open hole some 1.1 km in length, in which fractures could be made. The most difficult operational problem so far has been obtaining adequate zone isolation, so that only the depth interval selected is exposed to fracturing pressure. To date, three methods have been attempted and these are discussed below in chronological order.

Open Hole Inflatable Packers Despite the low probability of operating a packer at high temperature and obtaining an adequate seal against the rough walls of the well, two attempts were made because packers are the most inexpensive and fastest means of achieving zone isolation. Single packers were utilised because of the anticipated difficulties. Unfortunately in this mode of operation the single packer must resist, via frictional shear between the inflatable bladder and the rock wall, the entire upward thrust, about 10^6 N, generated by the fracturing pressure. In the two-packer straddle configuration, this thrust is instead reacted by structural members connecting the upper and lower packers. In two attempts the packers were successfully set and were initially able to sustain the fracturing pressure, but eventually began to slip along the well and failed. This may be a consequence of the anomalously high fracturing pressures required, about 80 MPa, which is 75% of the overburden pressure. Despite their ultimate failures, both packers performed remarkably well considering the harsh environment, and it now apppears that if they are rebuilt for the straddle mode of operation an excellent means of fracturing may be achieved.

Cemented Liner The second major attempt consisted of placing a short length of steel casing, or "liner", about 130 m long, near the bottom of the well and cementing it in place with specially developed, high silica cement for high temperature applications. The liner was fitted with a polished bore receptacle with seals at the upper end, so that drillpipe or a separate fracturing string could be mated to the liner, exposing only the open hole below the liner to fracturing pressure. Accomplishing this seal, at 4 km depth, is a difficult task. The injection of the cool fracturing fluid from the surface results in a thermal contraction of the fracturing string, so that the seal must accommodate as much as 3 m of relative motion while resisting a differential pressure of 35 MPa without leaking. After several preliminary attempts which were beset by seal problems, a fracturing operation was launched which resulted in the injection of 4900 m^3 of water at a flow rate of 0.1 m^3/s. The course of fracture propagation was monitored by detecting microseismic events with 3-axis geophones and accelerometers located 2.8 km downhole. Distances to the micro-earthquake hypocentres (almost all with magnitudes less than 1 on the extrapolated Richter scale) were estimated from compressional and shear wave arrival time differences, and the directions inferred from the first arrivals at the 3-axis geophone package. A crudely drawn envelope around the events gives the situation shown near the bottom of Figure 14. Instead of being vertical the fractures appear to be inclined about 45° from the vertical and thus passed below EE-3. Only the early occurring hypocentres were used in Figure 14 because these are believed to be a more reliable indication of the propagation of the main fractures. Later events are more diffuse and scattered, reflecting the propagation of joints and secondary fractures. Non-vertical fracturing suggests a shearing rather than a tensile failure. One possibility is that the fractures are dominated by the reservoir's proximity to the collapsed magma chamber of the Valles Caldera, and are constrained to be parallel to an inclined ring fault (29, 30). Another is that shear failure results because of the high confining pressure and differential tectonic stresses, coupled with a slow hydraulic injection rate (31).

It is hoped that future funding will allow deepening EE-3 to intercept the inclined fractures below the liner. In hindsight, it would have been advisable to drill EE-3 deeper in the first place, and in fact this was proposed during the drilling campaign, but at that time the frictional torque and drag mentioned earlier were beginning to exceed the capabilities of the drilling rig, so deepening could not be carried out.

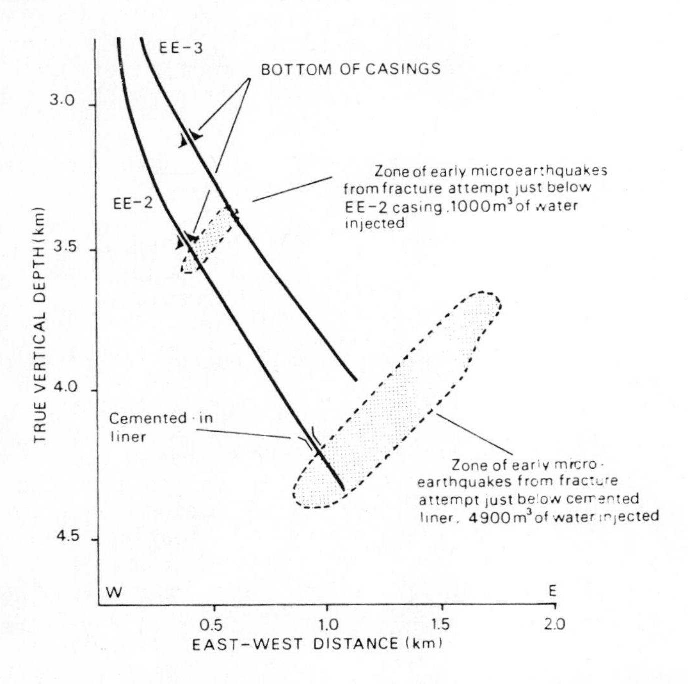

Fig 14 Inclination of fractures determined from microseismic hypocentres.

Casing Packer With the observation that fractures were inclined, the liner at the bottom of EE-2 was temporarily abandoned and the next fracturing attempt took place at the top of the open hole section, where inclined fractures had an excellent probability of intercepting EE-3. A short interval just below the bottom of the EE-2 casing was isolated by temporarily filling the bottom of EE-2 with sand to within 130 m of the casing bottom. Unfortunately the inner casing in EE-2 had been weakened at 2 km by wear and grooving of the inner wall as a consequence of drilling operations after the casing was set, and to avoid casing rupture a casing packer was set below the weakened section. Thus the pressurised string consisted of drillpipe to the packer, the full-strength casing below that, and the open hole between the casing and the top of the sand. The latest fracture attempt, conducted in August 1982, resulted in the injection of only 1000 m^3 of water before the casing packer failed. Fracture propagation was again monitored with downhole seismometers and an inclined fracture propagation was observed, as shown near the top of Figure 14. While the injection volume is considered too small to result in the formation of a low impedance hydraulical connection to EE-3, the microseismic events did propagate to distances as far away as EE-3, and the events seemed to be on the verge of enveloping EE-3.

Considering the results above, it is believed that once a reasonably large volume of water is injected, a successful connection with EE-3 will be obtained. With a successful formula thus obtained, fracturing will be repeated one or two more times to create the interim Phase II reservoir, and reservoir testing will then ensue.

CONCLUSIONS

Two hot dry rock geothermal reservoirs have been, or are being, created at Fenton Hill NM. The first reservoir, intended for research purposes, has been tested intensively over a three-year period and has exhibited particularly encouraging characteristics. These include:

- Production of 9×10^{13} joules of thermal energy, at rates of 2-5 MWt

- Low flow resistance, with impedances approaching or lower than 1 GPa.s/m^3, which could result in self-pumping in the Phase II reservoir

- Low water losses, with loss rates declining with time to values as low as 1% of the injection rate

- Benign geochemistry, with circulated water almost of drinking water quality and no significant corrosion or deposition problems

- Negligible seismicity, with largest magnitudes being less than -1, and this during hydraulic fracturing. During normal heat production testing $M_L < -3$.

Perhaps the most encouraging aspect of Phase I reservoir testing has been the continuous growth of effective heat production capacity, and the potential for even further increase. This growth was due not just to pressurisation and intentional hydraulic fracturing, but also to thermal contraction effects during normal operation. Two separate heat transfer models resulted in effective areas of about 8000 m^2 at the start, increasing to about 50 000 m^2 at the end, a six-fold increase. This was confirmed by tracer-measured volumes, which grew 25-fold. Despite the growth noted already, the potential exists for even greater increases, as evidence for microseismic and hydraulic experiments suggest a total fracture area many times greater than the present effective heat transfer area.

Having established technical feasibility with the Phase I reservoir, the next step toward commercial viability is the Phase II reservoir. Two deeper and much hotter (325°C) wells have been drilled at the desired inclination from vertical and the distance between these wells has been precisely maintained. While early fracturing attempts were not successful due to either operational difficulties or non-vertical inclination of the fractures, later attempts have been initiated from positions where such fractures will still result in fracture connection with the second well. In fact, microseismic observations suggested that the latest attempt was on the verge of such a connection when the packer failed, so that future attempts with greater injection volumes should result in a completed reservoir.

REFERENCES

1 Muffler, L. J. P. (ed.), "Assessment of Geothermal Resources of the United States - 1978", U.S. Geol. Surv. Circ. 790, 1979.

2 Tester, J. W., and C. O. Grigsby, "Geothermal Energy", in Encyclopedia of Chemical Technology, 3rd ed., John Wiley, New York, pp. 746-790, 1980.

3 Smith, M.C., R. M. Potter, D. W. Brown, and R. L. Aamodt, "Induction and Growth of Fractures in Hot Rock", Geothermal Energy, P. Kruger and C. Otte, eds., Stanford University Press, Stanford, Calif., 1973.

4 Bailey, R. A., and R. L. Smith, "Volcanic Geology of the Jemez Mountains, Guidebook to Rio Grande Rift in New Mexico and Colorado", Circ. N. M. Bur. Mines Miner. Resour. 163, 190, 1978.

5 Doell, R. R., G. B. Dalrymple, R. L. Smith, and R. A. Bailey, "Paleomagnetism, Potassium-Argon Ages and Geology of Rhyolites and Associated Rocks of the Valles Caldera, New Mexico", Mem. Geol. Soc. Am. 116, 211-247, 1968.

6 Reiter, J., C. Weidman, C. L. Edwards, and H. Hartman, "Sub-Surface Temperature Data in Jemez Mountains, New Mexico", Circ. N. M. Bur. Mines Miner. Resour. 151, 1976.

7 Murphy, H. D., and R. G. Lawton, "Downhole Measurements of Thermal Conductivity in Geothermal Reservoirs", J. Pressure Vessel Technol., 4, 607-611, 1977.

8 Sibbitt, W. L., J. G. Dodson, and J. W. Tester, "Thermal Conductivity of Crystalline Rocks Associated with Energy Extraction from Hot Dry Rock Geothermal Systems", J. Geophys. Res., 84, 1117-1124, 1979.

9 Lee, W. H. K., "Review of Heat Flow Data", in Terrestrial Heat Flow, Geophys. Monogr. Ser., 8, W. H. K. Lee, (ed.), Amer. Geophys. Union, Washington D. C., 1965.

10 Milora, S. L., and J. W. Tester, Geothermal Energy as a Source of Electric Power, MIT Press, Cambridge, Mass., 1976.

11 Laughlin, A. W., "The Geothermal System of the Jemez Mountains, New Mexico and Its Exploration", Geothermal Systems: Principles and Case Histories, L. Rybach and L. J. P. Muffler (eds.), John Wiley, New York, 295-320, 1981.

12 Laughlin, A. W., and A. C. Eddy, "Petrography and Geochemistry of Precambrian Rocks from GT-2 and EE-1", Los Alamos Scientific Laboratory report LA-6930-MS, (August 1977).

13 Murphy, H. D., R. G. Lawton, J. W. Tester, R. M. Potter, D. W. Brown, and R. L. Aamodt, "Preliminary Assessment of a Geothermal Energy Reservoir Formed by Hydraulic Fracturing", Soc. Petrol. Eng. J., 17, 317-326, 1977.

14 Albright, J. N., and C. M. Pearson, "Location of Hydraulic Fractures using Microseismic Techniques", paper presented at 55th Annual Tech. Conf. and Exhibition, Soc. Petr. Engr., Dallas, Sept. 1980.

15 Murphy, H. D., J. W. Tester, C. O. Grigsby, and R. M. Potter, "Energy Extraction from Fractured Geothermal Reservoirs in Low-Permeability Crystalline Rock", J. Geophys. Res., 86, 7145-7158, 1981.

16 Zyvoloski, G. (ed.), "Evaluation of the Second Hot Dry Rock Geothermal Energy Reservoir: Result of Phase I, Run Segment 5", Los Alamos National Laboratory report LA-8940-HDR (August 1981).

17 Murphy, H. D. (ed.), "Relaxation of Geothermal Stresses Induced by Heat Production", Los Alamos National Laboratory report LA-8954-MS (August 1981).

18 Fehler, M., "Changes in P Wave Velocity during Operation of a Hot Dry Rock Geothermal System", J. Geophys. Res., 86, 2925-2928, 1981.

19 Aki, K., M. Fehler, R. L. Aamodt, J. N. Albright, R. M. Potter, C. M. Pearson, and J. W. Tester, "Interpretation of Seismic Data from Hydraulic Fracturing Experiments at the Fenton Hill, New Mexico, Hot Dry Rock Geothermal Site", J. Geophys. Res., 87, 936-944, 1982.

20 Murphy, H. D., "Thermal Stress Cracking and Enhancement of Heat Extraction from Fractured Geothermal Reservoirs", Geotherm. Energy, 7, 22-29, 1979.

21 Tester, J. W., R. M. Potter, and R. L. Bivins, "Interwell Tracer Analysis of a Hydraulically Fractured Granitic Geothermal Reservoir", paper SPE 8270 presented at 54th Annual Mtg. of Soc. Petr. Engrs. of AIME, Las Vegas, NV, September 23-26, 1979.

22 Dash, Z. V., H. D. Murphy, G. M. Cremer, "Hot Dry Rock Geothermal Reservoir Testing: 1978-1980", Los Alamos National Laboratory report LA-9080-SR, pp 54-58 (November 1981).

23 Fisher, H. N., and J. W. Tester, "The Pressure Transient Testing of a Man-Made Fractured Geothermal Reservoir: an Examination of Fracture Versus Matrix Dominated Flow Effect", Los Alamos National Laboratory report LA-8535-MS (September 1980).

24 Grigsby, C. O., J. W. Tester, P. E. Trujillo, D. A. Counce, J. Abbott, C. E. Holley, and L. A. Blatz, "Rock-Water Interactions in Hot Dry Rock Geothermal Systems: Field Investigations of In Situ Geochemical Behaviour", submitted to J. of Volcanology and Geothermal Research, 1982.

25 Brittenham, T. L., J. W. Neudecker, J. C. Rowley, and R. E. Williams, "Directional Drilling Equipment and Techniques for Deep, Hot Granite Wells", J. Petr. Tech., 34, 1421-1430 (July 1982).

26 Anon. "1979 Joint Association Survey on Drilling Costs", Amer. Petr. Inst., Washington D.C. (February 1981).

27 Carson, C. C., and Y. T. Lin, "Geothermal Well Costs and Their Sensitivities to Changes in Drilling and Completion Operations", Proc. Int. Conf. on Geothermal Drilling and Completion Technology, Albuquerque N.M., January 21-23, 1981.

28 Murphy, H., R. Drake, J. Tester, and G. Zyvoloski, "Economics of a 75 MW(e) Hot Dry Rock Geothermal Power Station Based upon the Design of the Phase II Reservoir at Fenton Hill", Los Alamos National Laboratory report LA-9241-MS, (February 1982).

29 Lauglin, A. W., Los Alamos National Laboratory, personal communication, 1982.

30 Anderson, E. M., "The Dynamics of Faulting", Trans. Geol. Soc. Edin., 8, 387, 1905.

31 Lockner, D., and J. D. Byerlee, "Hydrofracture in Weber Sandstone at High Confining Pressure and Differential Stress", J. Geophys. Res., 82, 2018-2026, 1977.

ACKNOWLEDGEMENTS

The information summarised here is the product of the entire staff of the Los Alamos Hot Dry Rock geothermal energy development program, including participants from the Federal Rupublic of Germany and Japan. The financial support of the US Department of Energy, the US National Science Foundation, the Government of Japan, and the Federal Republic of Germany is gratefully acknowledged. Finally, the hospitality and support offered by personnel at the Camborne School of Mines HDR geothermal project in Cornwall, UK, is also sincerely appreciated, particularly A. Booth who assisted in preparation of the manuscript.

THE FALKENBERG GEOTHERMAL FRAC-PROJECT: CONCEPTS AND EXPERIMENTAL RESULTS

by F. Rummel and O. Kappelmeyer

Professor, Institute of Geophysics
Ruhr-University, Bochum, FRG

Division Head, Federal Institute of Geoscience
and Natural Resources (BGR), Hannover, FRG

ABSTRACT

Similar to the Cornwall and the French Geothermal Frac-Projects the purpose of the Falkenberg HDR project was to set up a field laboratory to study hydraulic fracture propagation in crystalline rock at shallow depth and to investigate fluid circulation and heat exchange within such an artificial fracture. The test site selected is situated within the Falkenberg granite massif, NE Bavaria, FRG. Since 1978 6 boreholes were drilled each to a depth of 300 m on a surface area of about 10^4 m². While 3 of the holes served for site survey and seismic observations, an about 20 m diameter hydraulic fracture was introduced from a central borehole. Fracture propagation was carefully monitored by passive seismic investigations. This allowed to intersect the fracture by 2 other boreholes and to circulate water through the system.

The vertical fracture is exactly aligned with the regional stress field as determined from hydraulic fracturing stress measurement at the eastern boundary of the Southwest German block. During the experiments the fracture width, flow resistance, fluid losses and the effective heat exchange area were determined.

OBJECTIVES

Hydraulic fracturing operations are routine practice for oil and gas well stimulation in deep-seated sedimentary formation at moderate temperatures. In most cases they lead to a considerable increase of oil and gas production in rock formations of low matrix permeability. This indicates that considerable mass flow takes place along such artificial fractures. However, even today our understanding of the hydraulic behaviour of such fractures is limited. In particular this concerns the pressure distribution within the fracture, frac geometry, fracture width, flow impedance, the flow path, fracture stability on formation boundaries, the effect of tectonic stresses on fracture propagation or chemical corrosion effects. Fracturing in crystalline rock formation at great depth and high temperatures is even more complicated. This is due to the fact that borehole logging equipment is not available for fracture observation at high temperatures, say above 250° C at great depth, and because of the interaction of the fracture with naturally existing joint systems. The LASL HDR experiments have clearly demonstrated such difficulties and have shown that further experimental effort is needed to solve open questions.

With this in mind, the objective of the Falkenberg Frac-Project was to set up a field laboratory where fracturing experiments and fluid circulation can be carried out in a known geological environment and under controlled experimental conditions, which even may allow to manipulate environmental parameters such as to influence the regional stress field locally and deviate a propagating fracture, or to alter the hydraulic properties of a fracture by mechanical or chemical means. With respect to contribute to the understanding of deep HDR operations originally the following experiments were planned:

- generation of a planar fracture in a nearly intact granitic rock mass with respect to the tectonic stress field;
- determination of the fracture plane by seismic, geoelectric and magnetic methods;
- intersecting the fracture by several boreholes to conduct fluid circulation experiments;
- fluid circulation experiments to measure hydraulic fracture parameters and to determine the effective heat exchange area;
- experiments on natural joints to study their hydraulic behaviour and to understand fluid losses;
- manipulation of the hydraulic effectivity of artificial fractures by detonating fluid explosives within the frac, and reducing the flow impedance at the intersection frac/borehole by ultra-high temperature jet burners;
- manipulation of the hydraulic effectivity of fractures by propping or by chemical treatment;
- manipulation of the direction of frac growth by influencing the acting tectonic stress field at the test site by pressurization of parallel fractures;
- estimation of the fracture geometry by surface tilt measurements.

This experimental program required the following field site characteristics and test site investigations:

- The in-situ laboratory should be installed in a granitic body, which was not subjected to major tectonic events. Because of technical (instrumental) and economic reasons (drilling costs) as well as because of measurements from the surface (geoelectric magnetic or tilt fracture detection) the fracture should be induced at rather shallow depth, however deep enough to avoid near surface effects (weathering of the near surface layers, minimum principal

stress is given by the overburden pressure).
- The rock should be homogeneous and isotropic from surface to depth to have so-called "laboratory conditions" and to avoid anisotropy effects (fracture follows planes of weaknesses, drillhole deviations).
- The stress field at test depth should be near to lithostatic with $S_h < S_V < S_H$, to allow stress field manipulations and the growth of vertical fracs.
- Determination of the present regional tectonic stress field and the local gradient of horizontal principal stresses with depth;
- detailed determination of the existing natural joint systems in the vicinity of the test site from surface mapping, core inspection and borehole logging; distribution and spatial orientation of joints in all boreholes;
- evolution, geological history, geochemistry and mineral composition, and petrographic analysis of the rock and the rock mass;
- determination of the physical properties of the rock mass (seismic velocity/depth relation from seismic refraction measurements, sonic logs and acoustic borehole to borehole transmission tests; conventional geophysical borehole logs and surface measurements; permeability logs; rock mass tensile strength) and the rock (ultrasonic velocity as a function of pressure ; strength and friction data, in particular hydraulic fracturing tensile strength; fracture mechanics data such as fracture toughness or specific fracture surface energy; resistivity; magnetic properties; density, matrix permeability; micro-crack distribution and internal crack surface area);
- determination of chemical parameters with respect to corrosion by natural or the circulation fluid.
- Consideration of environmental impacts (rock burst triggering, ground water pollution).

The project is a cooperation between the following research organizations and scientists
- Federal Institute for Geosciences and Natural Resources (BGR), Hannover: O. Kappelmeyer, R. Jung, H. Keppler, G. Leydecker. Infrastructure, frac-operation, circulation tests, frac-geometry, seismic experiments.
- Niedersächsisches Landesamt für Bodenforschung (NLfB), Hannover: R. Hänel, H. Rodemann, S. Jobst. Heat exchange experiments, theoretical studies on thermal problems, borehole logs.
- Institute of Geophysics, Ruhr-University, Bochum: F. Rummel, H.J. Alheid, J. Baumgärtner, U. Heuser, Th. Wöhrl, St. Teufel, D. Kassel. In-situ permeability and hydraulic fracturing stress measurements, physical properties, fracture mechanics, petrology and geology.
- Institute of Geophysics, University, Braunschweig: G. Musmann, F. Kuhnke. Frac location by geoelectric methods.
- Institute of Geophysics, University, Clausthal: J.R. Schopper, H. Pape, L. Riepe. Physical rock properties.
- Institute of Geophysics, University, Munich: J. Pohl. Magnetic borehole measurements, rock magnetism.

- Prakla-Seismos, Hannover: Seismic field measurements for frac location, seismic in-situ velocity measurements.

The project started in 1978 and is supported by the German Federal Ministry of Research and Technology and by the European Community. The project will continue until 1984.

GEOLOGICAL SETTING

The crystalline basement outcrops in NE-Bavaria. Primarily, it consists of pre-permian metamorphites of various composition, into which a number of large granitic bodies intruded. Their average age is about 300 million years (Köhler et al., 1974; Kassel 1980). They generally form the topography of this area today. The Falkenberg test site is located at the northern edge of one of these granitic intrusions, the Falkenberg granite massif (Fig. 1).

Although the cores from the drillholes on the test site yield 4 different types of granite 96 percent of the granite is coarse-grained porphyric with large (8 cm) idiomorphic potassium feldspar crystals (microcline, 30 percent, $Or_{90.2}$ $Ab_{9.5}$ $An_{0.3}$) within a medium-grained ground mass of quartz (32 percent), plagioclase (26 percent), biotite (9 percent), muscovite (2 percent) and xenomorphic potassium feldspars (Kassel, 1980).

According to detailed mapping of joints in all boreholes (by TV-logs, ultrasonic-televiewer logs, core-mapping) as well as on granite outcrops in the vicinity 3 major subvertical joint sets could be identified: NS and 2 EW oriented inclined joint systems. Above 150 m horizontal joints dominate due to crustal uplift, subsequent erosion and corresponding unloading (Teufel, 1982).

STRESS FIELD SITUATION

The test site is located on the eastern boundary of the SW-German block which forms the major tectonic unit in Central Europe. While the western block boundary (Rhine-Graben) is seismotectonically active, the eastern boundary is presently inactive. Stress measurements and earthquake fault plane solutions for the western boundary indicate an approximately NNW-SSE direction of horizontal compression which is related to the Alpine system on the southern block boundary. In comparison, hydraulic fracturing stress measurements on the eastern block boundary yield approximately EW horizontal compression which could originate from the Karpathian range further to the east (Fig. 2; Rummel and Alheid, 1979; Rummel et al., 1982; Rummel and Baumgärtner, 1982; Baumgärtner, 1982). The hydro frac stress measurements have been conducted in various boreholes within the Falkenberg granite as well as in its vicinity prior to the large scale frac operation at the test site. The direction of the large-scale

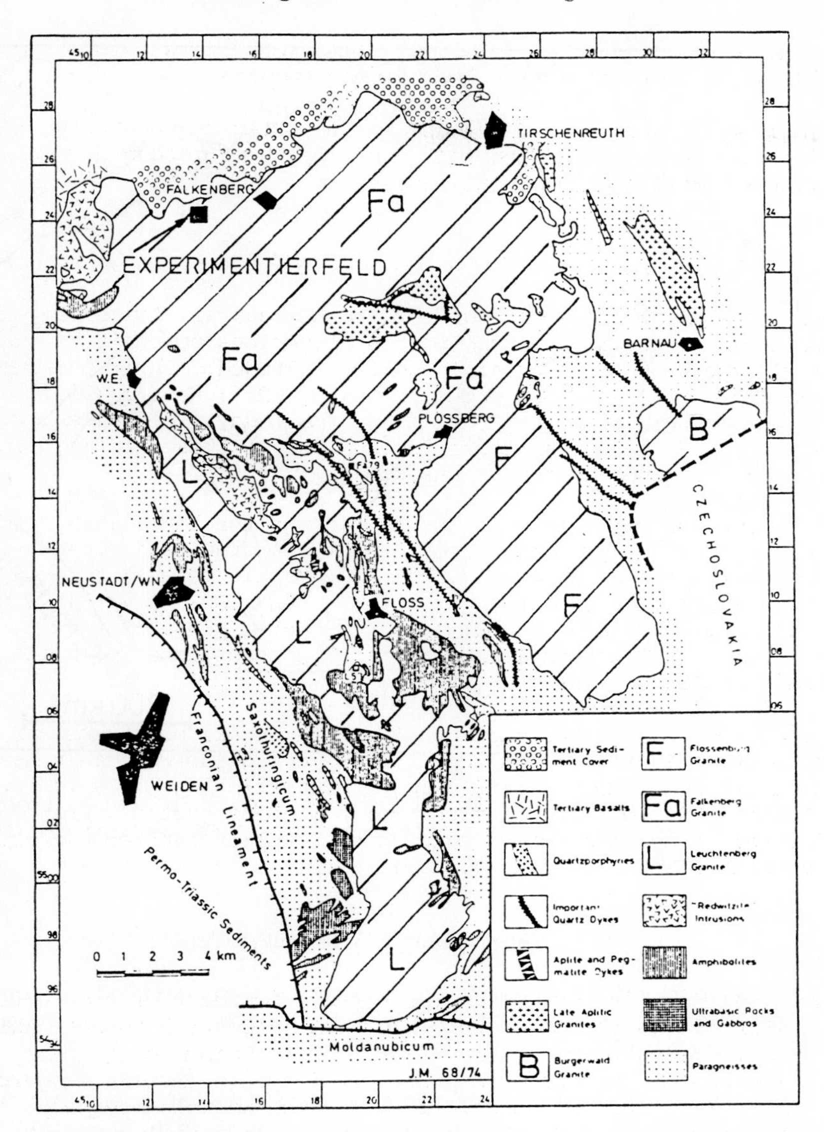

Fig. 1: Geological setting of the Falkenberg granite massif within the crystalline basement of NE Bavaria.

frac induced in the central borehole at the test site in 1979 (Jung, 1980) at a depth of about 250 m is parallel to the direction of maximum horizontal compression as determined by previous stress measurements. The hydrofrac stress measurements also revealed information on the increase of horizontal stress with depth, indicating that the minimum horizontal stress becomes the least principal stress at a depth of about 150 m (Rummel, 1980).

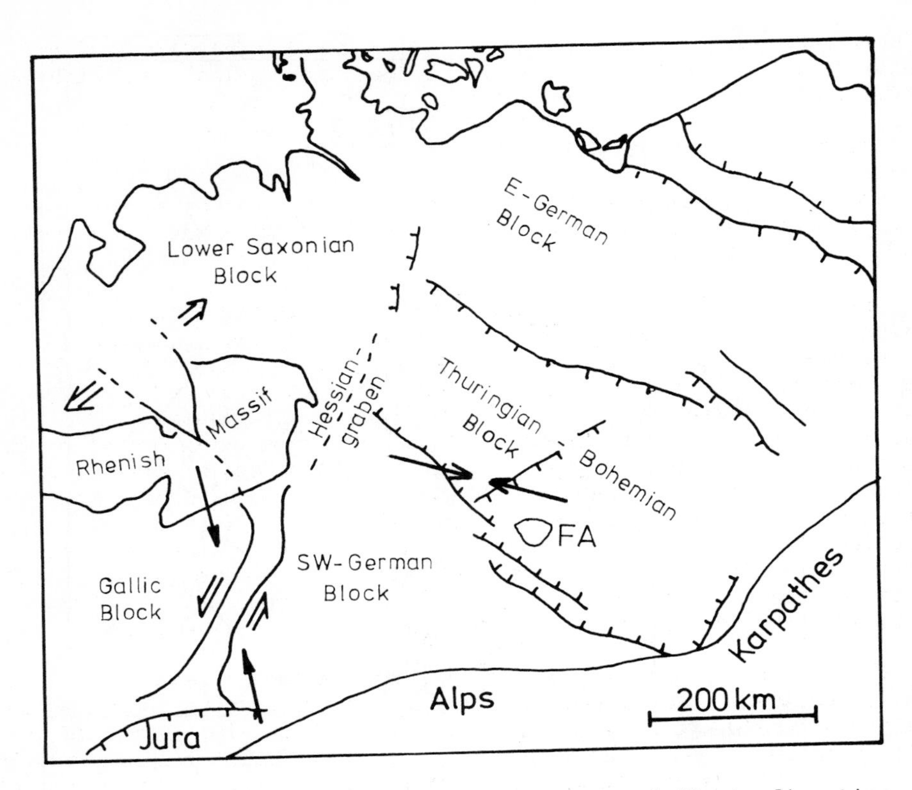

Fig. 2: Stress regime in central Europe. Arrows indicate direction of maximum horizontal compression. FA Falkenberg granite massif.

PHYSICAL PROPERTIES OF THE GRANITE

Extensive physical property measurements were carried out on the core material of all boreholes at the test site as well as by all kinds of geophysical logs in the boreholes.

Ultrasonic velocity measurements on minicores at pressures up to 2 Kbars yielded typical granite velocities of $v_p = 6.22 \pm 0.14$ km s^{-1} and $v_s \approx 3.5$ km s^{-1} (P = 2 kbar). At low pressures (P < 1.4 kbar) the velocity is strongly pressure dependent indicating the effect of microcracks (Rummel, 1980; Wöhrl, 1981). The same effect is noticed from seismic attenuation measurements. In-situ velocity measurements from acoustic logs (1 m distance between the transducers) and seismic transmission experiments between different boreholes yielded values of v_p between 4 and 5.9 km s^{-1}, a v_p/v_s-ratio of 3 and a Poisson ratio of 0.25 (Leydecker, 1981). As seen from seismic vertical profiling on the cores and by the acoustic logs the velocity values increase almost linearly with depth from surface to a depth of about

150 m indicating vertical stress relief and the existence of open microcracks within the rock matrix. Below this depth the velocity is constant if we neglect alteration and fracture zones.

The rock density is 2.58 gcm^{-3}.

The elastic rock parameters derived from static compression tests and from seismic measurement are $E = 3$ to $7 \cdot 10^5$ bars and $\nu = 0.21$ to 0.32 (Rummel, 1980).

Shear fracture of intact rock samples in triaxial experiments is given by the criterion $\tau_C(kb) = 0.76 + 0.6\ \sigma(kb)$ for high normal stresses ($\sigma > 1.5$ kb) and by $\tau_C(kb) = 0.2 + 1.08\ \sigma(kb)$ for low normal stresses. The residual friction strength is given by $\tau_R = 0.31 + 0.6\ \sigma$ for high and by $\tau_R = 0.98\ \sigma$ for low normal stresses, respectively. Hydraulic fracturing laboratory experiments on minicores subjected to confining pressures up to 800 bars yield a breakdown pressure/confining pressure relationship p_C (bar) $= 166 + 1.04\ p_m$ (bar). The hydraulic fracturing tensile strength of the granite is 166 $\pm$ 30 bars. The in-situ hydraulic fracturing tensile strength of the rock mass in intact borehole sections is about 70 bars. Fracture toughness values derived from notched beam bending tests range between $K_{IC} = 0.78$ to 0.86 $MNm^{-3/2}$. The specific surface energy is $\gamma \approx 7\ Jm^{-2}$ (Rummel, 1980).

The matrix permeability of intact rock samples is less than 100 μD, while specimens from altered granite near fracture zones show permeability values up to 3 mD (Schopper et al., 1981). Similarly, the in-situ granite permeability derived from pressurization test in intact borehole sections (Rummel et al., 1980; Heuser, 1982) shows values of $K < 25\ \mu D$, while borehole intervals intersected by joints yield effective rock mass permeabilities up to 10 mD (Jung, 1978).

FRACTURE GENERATION

The central 132 mm diameter borehole HB4a at the test site (Fig. 3) is intersected by only few joints. The borehole intervals 113 to 193 m and 222 to 272 m are free of any natural joints. Accordingly frac operations were performed at intervals from 154 to 161 m, 258 to 300 m and 252 to 255 m (Jung, 1980; Kappelmeyer, 1982). The pumping system used had a capacity of 280 bars at a flow rate of 200 $l\ min^{-1}$. Injection was achieved through 550 bar water-proof drill steel pipes into a Lynes double packer system.

During the frac tests at the intervals 154 to 161 m and 258 to 300 m pre-existing fractures were opened at rather low breakdown pressures (65 bars at 154 - 161 m). Water losses were considerable (60 percent after 4 m^3 injection at 154 - 161 m; 100 percent after injection of 20 m^3 at 258 to 300 m). The experiments demonstrated the hydraulic communication of various open joint systems.

(a)

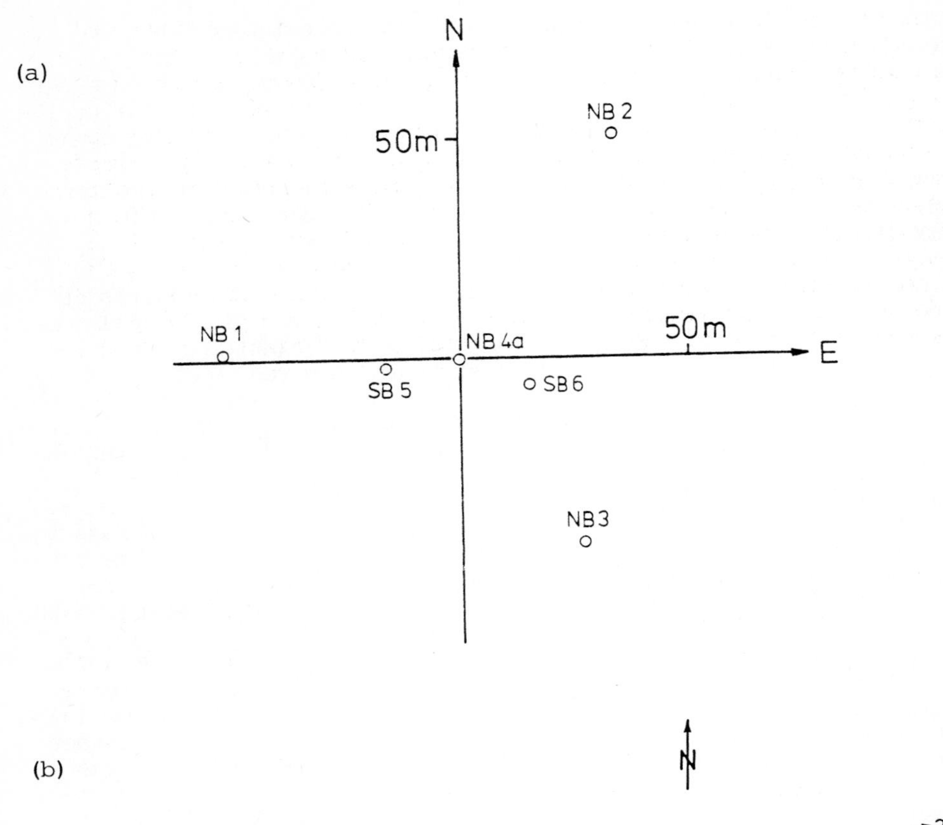

(b)

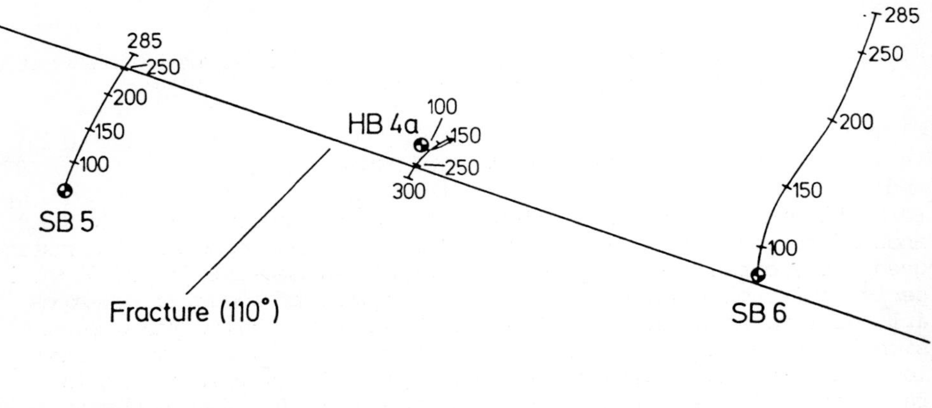

Fig. 3: Drillhole locations at the Falkenberg test site (a) and intersection drillhole deviations as a function of depth (b). HB4a central drillhole for frac generation, NB observation drillholes, SB frac intersection drillholes.

Fracturing in the interval 252 - 255 m initiated at a downhole breakdown pressure of 203 bars at a pumping rate of 200 $l\ min^{-1}$. During subsequent repressurizations the refrac pressure decreased continuously (Fig. 4). The frac extension pressure is 46 bars. During the first 3 tests (Oct. 17, 1979) 0.2, 1.0 and 5.7 m^3 water were injected. The water loss during the third test was 2.4 m^3, indicating a frac volume V of about 3 m^3. A similar value may be estimated from elastance measurements ($\Delta V/\Delta P \approx 100\ l \cdot bar^{-1}$ at P = 21 bars at the end of the third test). Assuming a penny-shaped crack this would result in a fracture with a radius of about 60 m and a fracture width of about 0.2 mm. During subsequent tests during 1980/1981 the frac was considerably enlarged (injected volumes up to about 30 m^3). Today the observed elastance of the system is about 1500 $l \cdot bar^{-1}$.

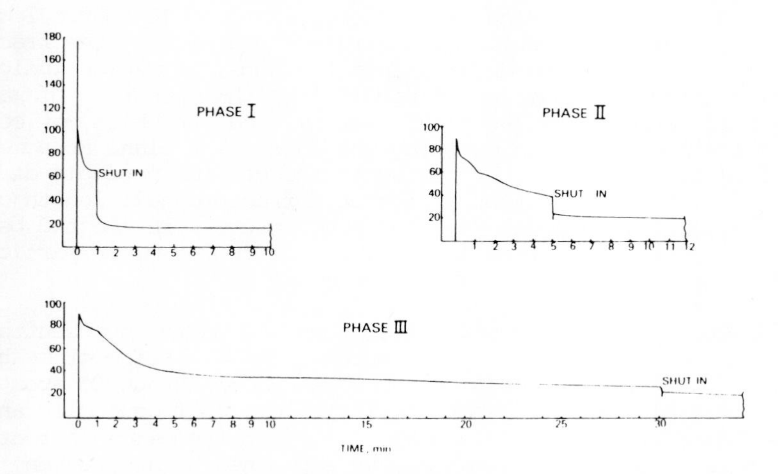

Fig. 4: Pressure/time plots during the first 3 frac initiation tests in borehole HB4a at 252 - 255 m on Oct. 17, 1979.

SEISMIC FRAC LOCATION

During the first 3 fracturing tests at the interval 252 - 255 m induced seismic events were recorded by 4 three-component high frequency piezo-electric geophone-systems with a sensitivity of 5 Volts per g, a 4 $\mu\ V_{eff}$ noise level and a resonance frequency of about 4.5 kHz (Prakla Seismos development; Leydecker, 1981). The geophone signals are preamplified (56 db) and recorded on a 16 channel analogue tape recorder with a tape velocity of 1 7/8 ips, a dynamic range of 30 db and a frequency range of 0 - 7 kHz. The geophone systems were placed in the 3 observational boreholes NB1 to NB3 around the central borehole HB4a (Fig. 3) at depths between 190 and 290 m. (Additional seismic stations were set-up on the surface and in shallow boreholes. However, no event was recorded due to attenuation within the near surface weathered granite layers and operation noise).

The system was calibrated downhole by artificial shocks using a sparker source, which demonstrated a precision in source location of about 5 m.

Data analysis was conducted using the Frac Hypocenter program with the Least Square Routine program as sub-program. The Frac-Hypocenter program consists of parts of the Hypolayer (Eaton, 1969) and the Icelog-program (Deichmann, ETH Zürich, 1978). The program is described in detail by Leydecker (1981).

During a period of 165 minutes recording time (36 minutes pumping time for the 3 first experiments) a total of 60 seismic events were recorded on at least 2 seismic stations. 30 events could be used in the data analysis for source location. No events were observed during breakdown and the first 5 minutes of pressurization. 16 source locations could be identified which demonstrate the growth of the fracture with time. They may be devided into 2 groups, one which indicates a vertical fracture plane oriented parallel to the direction of maximum horizontal compression ($\theta = 115°$ over E), and one which may correspond to the opening of an existing inclined joint plane at some distance from hole HB4a, which intersects the central borehole at a depth of 290 m. The first group indicates a fracture half length of about 20 m (Fig. 5). Unexpectedly, all hypocenters were located below the depth of the injection interval at 252 to 255 m. In particular, this is true for the events of the second group.

The average signal duration was less than 100 ms for hypocentral distances between 45 and 85 m. Power spectras of the seismograms between 130 and 3000 Hz demonstrated a frequency content up to about 2000 Hz, with a plateau-type characteristic for low frequencies and a typical corner frequency of 1000 Hz. Different frequency content of signals with similar travel times may be explained by absorption effects due to the distribution of water-filled joint systems. However, a shear-shadow experiment (LASL-HDR Rep., 1978) carried out to investigate this effect was not succesful and did not yield further information on the induced fracture (Leydecker, 1981).

GEOELECTRIC FRAC LOCATION

The geoelectric frac location was based on the classical 'mise à la masse' method, using the fluid within the frac as one electrode and placing the second electrode on the surface at a distance of about 2.5 km away from the injection borehole. Salt was added to the injection fluid to reduce the water resistivity in the frac to a few ohm-meters, compared to the granite resistivity of about 1000 ohm-meters. An alternating current of 1 ampere at various frequencies up to 12 kHz was applied to both electrodes (the primary electrode was located within the injection interval). The induced potential field and the potential gradient were measured on circles of 75, 100, 150 and 200 m radius around the injection borehole, both in the original state

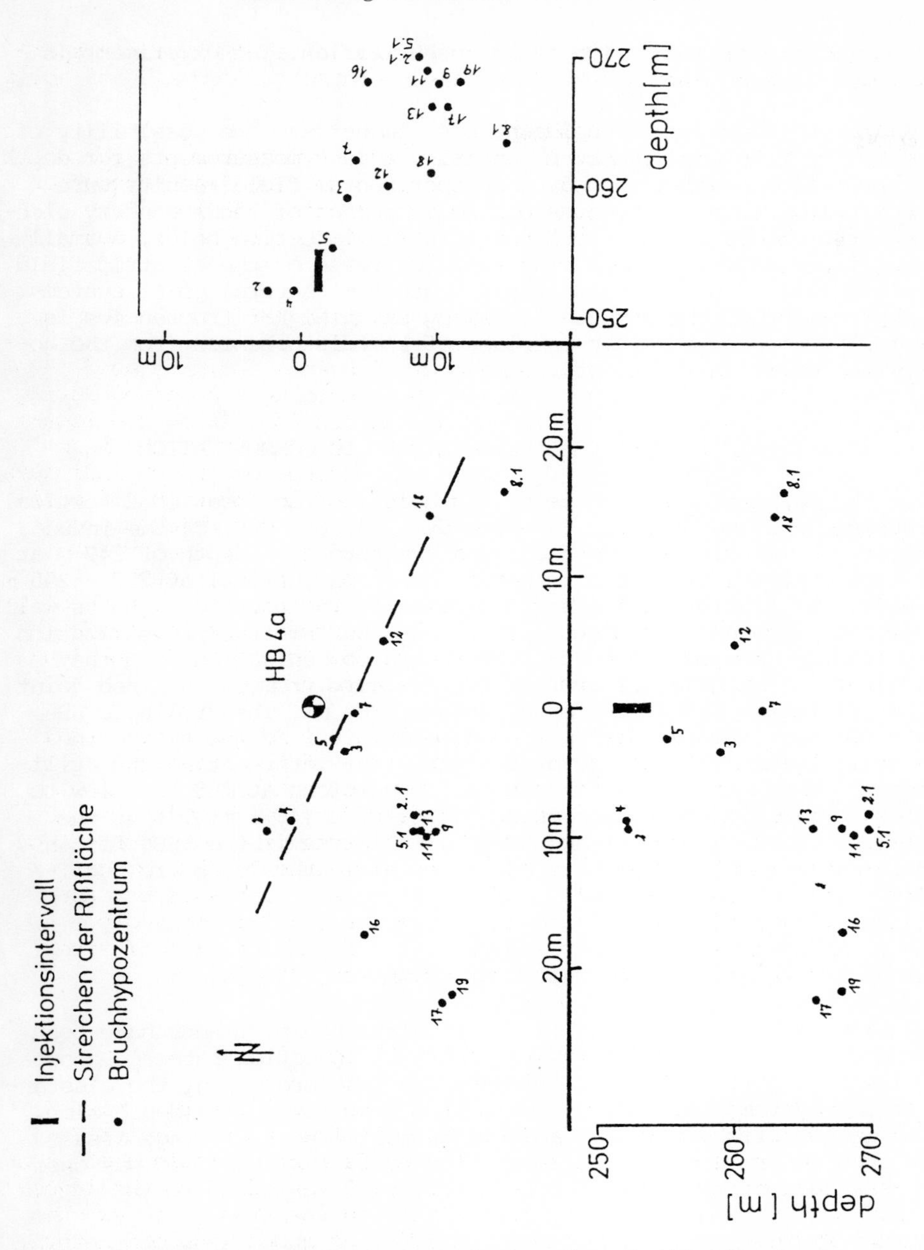

Fig. 5: Seismic source locations during frac initiation tests on Oct. 17, 1979.

(closed fracture) and during frac pressurization. The experiment is described in detail by Kuhnke (1982).

Although theoretical considerations demonstrate the possibility of geoelectric frac location by potential gradient measurements for depth to frac ratios smaller than 5, the experimental field results were discouraging. This was because of the influence of the secondary electrode, the effect of the steel pipes in the injection holes, a small signal to noise ratio due to the rapid decrease of the electric field intensity with depth, and the fluid content of natural joint systems. An improvement of the induced signal by using higher frequencies is generally compensated by an increase of the disturbances from the horizontal cables (skin-effect).

INTERSECTION DRILLHOLES AND HYDRAULIC COMMUNICATION

On the basis of the results of the seismic frac location 2 vertical boreholes (SB5 and SB6, Fig. 3) were drilled to intersect the induced fracture. Drillhole SB5 intersected a fracture at a depth of 249 m at a direct distance of 13.8 m from the injection interval at 252 - 255 m in HB4a. The fracture was also identified in the core from SB5 as well as from TV-logs and temperature/flow logs. During frac pressurization this fracture showed significant hydraulic communication with the frac interval in HB4a (minor flow also occurred from an inclined joint which intersects SB5 at 260 m and HB4a at 289 m). The drillhole SB6 deviated away from the frac plane, however, it is connected to the hydraulic system of the frac by a joint plane intersecting the drillhole at a depth of 247 m. At open hole conditions at SB5 the flow to SB6 is negligible. At back pressure (flow rate 145 l min^{-1}) at the well-head of SB5 a flow of about 40 l min^{-1} occurred from SB6 for an injection rate of 188 l min^{-1}. Fluid losses in this case are negligible.

CIRCULATION EXPERIMENTS

During 1980 to 1982 a number of circulation experiments have been conducted over a period of several hours at injection rates of about 200 l min^{-1}. During these experiments the fracture width, the elastance of the system (fracture volume), flow resistance, residence time, fluid losses, flow impedance and the effective heat exchange area have been determined. During 2 experiments in 1981 the frac was once more enlarged considerably.

Fracture Width

The fracture width was measured after the frac system was enlarged in July 1981. The measurements were carried out using a high-resolution (10^{-2} mm) caliber probe at the intersections of the frac in HB4a and SB5. The caliber probe consists of 3 pairs of spring-loaded pistons which are extended to the borehole wall. Piston displacements are measured by inductive displacement transducers. The instrument can be used within the pressurized interval.

The fracture already opens at very low pressures ($p < 7$ bars). Then the frac width increases quadratically with increasing pressure. At 22 bars (frac extension pressure) the frac is fully opened; the maximum frac width in HB4a (injection) amounts to 0.4 mm, and 0.9 mm in SB4. No distinct frac opening pressure can be identified.

Elastance and Frac Volume

The elastance was determined during the shut-in phase of most pressurization experiments by venting a certain fluid volume and observing the corresponding stress drop within the frac ($\Delta V/\Delta P$). According to theory the elastance is a measure of the actual frac volume.

Although during all circulation experiments the fluid pressure was kept well below the frac extension pressure the elastance increased linearly with pressure for pressures above 6.5 bars. This results in a quadratic increase of the frac volume for such intermediate pressures. At pressures below 6.5 bars the elastance was constant. For typical frac experiments the elastance increases considerably. For the original fracture (after the 3 experiments on Oct. 17, 1979) the frac volume was about 5 m^3, today the frac volume is about 20 m^3. Assuming an average fracture width, as determined experimentally, the frac area is estimated to $2 \cdot 10^4$ m^2 after the last frac extension experiment.

Flow Resistance

Flow resistance of the wellbore-frac system is defined as the ratio between the pressure drop within the fracture and the flow rate. Circulation experiments indicated that the flow impedance substantically decreases with increasing back pressures: For zero back pressure in SB5 and an injection rate of 138 l min^{-1} the impedance in 1980 was 8 bars per liters per second and linearly dropped to 0.5 bar · s · l for a back pressure of 17 bars. As indicated from shut-in tests during the circulation experiments as well as from pumping tests at closed outlets on SB5 and SB6, the flow impedance is mainly determined by the borehole/frac intersections, while the frac merely contributes to the impedance value. Detailed studies verify that for injection rates less than 30 liters per minute and small fluid pressures ($p < 6$ bars) the pressure drop at the inlet is linear to the flow rate and independent of the fluid pressure. This results in a laminar flow and a constant fracture width. At higher flow rates (> 42 l min^{-1}) the impedance increases quadratically and the flow becomes turbulent.

Fluid Losses

Fluid losses were estimated from the difference between the flow rates at the injection and extraction intervals during stationary flow conditions as well as from injection test at constant injection pressure, again at stationary conditions.

The measurements demonstrate that loss rates at pressures below 7 bars increase linearly with pressure. Even at zero well head pressure fluid losses occur since the natural water table is at 20 m below surface. Above 7 bars water losses seem to increase although this could be the effect of the frac opening. The evaluation of shut-in pressure curves does not indicate increasing fluid losses. At pressures near frac extension fluid losses were about 60 $l\ min^{-1}$ independent of time for injection periods of several hours.

Heat Exchange Experiments

One major heat exchange experiment was conducted in early 1980 when the water injected from a nearby river was well below the rock mass temperature at depth. The experiment is described in detail by Hänel and Rodemann (1981).

The temperature gradient at the test site was determined as 0.023° $C\ m^{-1}$. The thermal conductivity of the granite was measured from cores and yielded a value of 3.4 ± 0.2 W/mK. Both data yield a heat flow density of 0.078 $W\ m^{-2}$ ± 6 %. In comparison, the average heat flow density of the general area is about 0.065 $W\ m^{-2}$. If one corrects the heat flow for heat generation one obtains a heat flow value of 0.079 Wm^{-2} ± 6 %. The in-situ experiments showed a rock mass thermal conductivity of 0.59 W/mK which corresponds to water, if one applies an analysis presented by Ramey (1962).

For the heat exchange experiment water from a nearby river was pumped over a distance of 935 m and a vertical elevation of 48 m into the tank at the test site and then injected into borehole HB4a at an average rate of 200 $l \cdot min^{-1}$. Temperatureprobes were located in HB4a and in SB5, both at the well heads as well as at the intersections with the fracture at depth. The temperature probes consist of thermistors with an absolute error of $3 \cdot 10^{-2}$° C. The undisturbed rock temperature at a depth of 250 m was about 13° C. The injected water temperature was about 8° C. The experiment lasted from 23:00 to about 8:00 in the morning, thus over a period of about 9 hours.

Data analysis on the basis of numerical models results in an effective frac radius of 7 to 8 m or an effective heat exchange area of about 175 m^2, which is considerably smaller than the frac size determined from pumping tests or from seismic locations. This indicates that only a small part of the induced frac-plane is used during fluid circulation at stationary conditions. Channelling effects may dominate fluid flow within the induced fracture. During the early stage of the experiment the geothermal power was about 40 kW, after 8.5 hours of circulation it dropped to about 19 kW.

ACKNOWLEDGEMENT

We gratefully acknowledge the contributions of all participants of the project, the advise of the Bavarian Geological Survey and the management assistance of the Federal Institute for Geoscience and Natural Resources, Hannover. The project is supported by grants from the Federal Ministry of Research and Technology (ET 4150) as well as from the European Community (EG-498-78-1). Project managing was provided by the PLE/KFA Jülich.

REFERENCES

1. Baumgärtner, J., "Gebirgsspannungsmessungen mit der Methode des Hydraulic Fracturing in Sedimenten im östlichen Bereich der Südwestdeutschen Scholle und im Kristallin NE Bayerns," Diploma thesis, RU Bochum, 1982.

2. Hänel, R., and H. Rodemann, "Theoretische und experimentelle thermische Untersuchungen an einem künstlich erzeugten Wärmeaustauschsystem im Falkenberger Granit," NLfB-Rp., No. 89725 and 91599, Hannover, 1981.

3. Heuser, U., "Das Druckstoß-Test-Verfahren zur Bestimmung der Gebirgspermeabilität in Bohrungen," Diploma thesis, RU Bochum, 1982.

4. Jung, R., "Permeabilitätsmessungen im Falkenberger Granit-Massiv," BGR-Rep. No. 80443, Hannover, 1978.

5. Jung, R., "Hydraulische Experimente 1979 im Experimentierfeld bei Falkenberg," BGR-Report No. 86151, Hannover, 1980.

6. Kappelmeyer, O., "Experiments in an artificially created frac at shallow depth," BGR-Rep. No. 92336, Hannover, 1982.

7. Kassel, D., "Petrologische Untersuchung am Granit von Falkenberg/Opf. anhand von Bohrkernen," Diploma thesis 1980 and Internal Rep. No. 7084404 Ru 80/1, 1980.

8. Köhler, H., D. Müller-Sohnius and Camman, "Rb-Sr Altersbestimmungen an Mineral- und Gesteinsproben des Leuchtenberger und Flossenbürger Granits, NE Bayern," N.Jb. Miner. Abh., 123, 63-85, 1974.

9. Kuhnke, F., "Die Ortung künstlicher Rißflächen im Untergrund nach der Mice à la Masse Methode mit Gleich- und Wechselstrom," Bericht Techn. Hochschule Braunschweig, 1982.

10. LASL-Hot Dry Rock Annual Report, Fiscal Year 1977, LA-7109 PR, p. 97-100, Los Alamos, 1978.

11. Leydecker, G., "Seismische Ortung hydraulisch erzeugter Brüche im Geothermik Frac-Projekt Falkenberg," BGR Rep. No. 86549, Hannover, 1981.

12. Ramey, H.J., "Well bore heat transmission," J. Petrol. Techn., 427-435, 1962.

13. Rummel, F., and H.J. Alheid, "Hydraulic fracturing stress measurements in SE-Germany and tectonic stress pattern in Central Europe," Proc. Int. Conf. Intra-Continental Earthquakes, 33-65, Lake Ohrid, Yugoslavia, 1979.

14. Rummel, F., H.J. Alheid, and U. Heuser, "Modellkurvenatlas für die Auswertung von Druckabfallkurven zur Permeabilitätsbestimmung in Bohrungen," RU Bochum, Rep. No. 7084404 Ru (0/2, 1980.

15. Rummel, F., "Gesteinsphysikalische Daten des Falkenberger Granits," Annual Report, RU Bochum, No. 7084404 Ru 79/4, 1980.

16. Rummel, F., and J. Baumgärtner, "Spannungsmessungen im östlichen Bereich der Südwestdeutschen Scholle," Int. Report, RU Bochum, No. 7084408-82-3, 1982.

17. Rummel, F., J. Baumgärtner, and H.J. Alheid, "Hydraulic Fracturing Stress Measurements along the Eastern Boundary of the SW German Block," USGS Open-File Report on the 1981 Monterey Workshop on Hydr. Frac. Stress Measurements, 1982.

18. Schopper, J.R., H. Pape, and L. Riepe, "Petrophysikalische Untersuchungen an Kernen aus Flachbohrungen," Annual Rep., TU Clausthal, 1981.

19. Teufel, St., "Klüftung und Quarzmikrogefüge im Granit von Falkenberg/Opf.," Diploma thesis 1982 and Internal Rep. No. 7084404 Ru 81/1, RU Bochum, 1981.

20. Wöhrl, Th., "Geschwindigkeitsuntersuchungen am Falkenberger Granit," Diploma thesis 1981 and Project Rep. RU Bochum, No. 7084404 Ru 81, 1981.

SHALLOW DEPTH EXPERIMENTATION ON THE CONCEPT OF ENERGY EXTRACTION FROM DEEP HOT DRY ROCKS

by F.H. Cornet, J.M. Hosanski, F. Bernaudat and E. Ledoux

Institut de Physique du Globe, Université P & M Curie, 4 place Jussieu, 75230 PARIS CEDEX 05

Compagnie Française des Pétroles, 204 Rond-Point du Pont de Sèvres, 92516 Boulogne Billancourt

Ecole Nationale Supérieure des Mines de Paris, 35 rue Saint Honoré, 77305 Fontainebleau

ABSTRACT

Hydraulic fractures may not develop perpendicularly to the least principal stress direction when the rock is anisotropic with respect to its strength. This has been taken to advantage to develop two heat exchangers in a shallow granite. First a hydraulic fracture was developed at 186 m in a first borehole ; its orientation was defined by inspecting the borehole with a T.V.camera and using shut-in pressure measurement as well as a preliminary stress determination to show that the fracture had not turned away from the wellbore. A second well, bored 30 m away from the first one, intersected the fracture at the 156 m depth. The second heat exchanger was prepared by developing, from a third borehole, a second hydraulic fracture which intersected the first one somewhere between the two first boreholes.

The thermal and hydraulic charateristics of these heat exchangers have been investigated thru a 56 days circulation test for the first geometry and a 48 days test for the second one.

The mathematical model used for interpreting results from these two tests has been applied to the computation of cooling abacuses useful for the design of a deep hot dry rock heat exchanger.

1. INTRODUCTION

The concept of heat extraction from deep hot dry rocks, as pioneered by researchers from the Los Alamos Scientific Laboratories (see e.g. Murphy et al. 1981) has been investigated at shallow depth in the granite rock mass of Le Mayet de Montagne, 25 km south east from Vichy in central France.

The purpose of this work was to investigate the feasibility of establishing an artificial circulation loop between two, 200 m deep, boreholes and then to characterize, hydraulically and thermally, this heat exchanger. The program encompassed three stages :

- First a 100 m deep borehole was drilled to determine the local stress field as well as the thermal characteristics of the rock mass.
- In a second stage, the borehole was deepened to 200 m and hydraulically fractured at 186 m depth. This fracture was then intersected by a second borehole, drilled 30 m away from the first one, and water was circulated between both wells.
- In a third stage efforts were made to characterize the strength anisotropy of the granite in another borehole so as to make use of this anisotropy for developing a hydraulic fracture in such direction that it would intersect the fracture developed during the second stage of the tests. Water was then circulated between the three boreholes.

This paper outlines the main results obtained during these shallow depth heat extraction tests. First the development of the two heat exchangers are described ; then results from the two long term circulation tests are presented.

2. DEVELOPMENT OF A HEAT EXCHANGER IN AN IMPERVIOUS ROCK

A test site in a homogeneous granite has been chosen at Le Mayet de Montagne to develop a shallow depth doublet made of two boreholes linked by a hydraulic fracture. For doing so the following scheme was adopted :

- drill a first borehole for stress measurement purposes ;
- develop a large hydraulic fracture from this borehole and determine its geometry ;
- drill a second borehole to intersect the fracture.

Before describing the results obtained during these various phases it is necessary to recall some elements of fracture mechanics.

2.1 The mechanics of hydraulic fracture propagation

It is usually assumed that a hydraulic fracture develops perpendicularly to the minimum principal stress direction.

Indeed, according to Griffith's fracture criterion (Griffith 1921) when a borehole is pressurized, a hydraulic fracture occurs when :

$$\Delta W(\underline{ds}) - \Delta U(\underline{ds}) \geq \Delta D(\underline{ds}) \quad (1)$$

where $\Delta W(\underline{ds})$ is the work done by the fracturing fluid during crack growth $\underline{ds}$ ($\underline{ds} = \underline{n}\, da$) , $\Delta U(\underline{ds})$ is the strain energy variation associated with crack growth $\underline{ds}$ and $\Delta D(\underline{ds})$ is the surface energy absorbed by the formation of $\underline{ds}$. When there is strict equality, the propagation process is quasistatic whilst when there is inequality some kinetic energy is generated which dissipates partly thru seismo-acoustic activity.

Griffith supposed that the formation of a new surface corresponds to an increase in potential energy so that, for quasistatic conditions, of all possible crack extension configurations, that which actually occurs maximises the potential energy variation of the system.

$$(\Delta U(\underline{ds}) - \Delta W(\underline{ds}) + \Delta D(\underline{ds})) - (\Delta U(\underline{ds}_o) - \Delta W(\underline{ds}_o) + \Delta D(\underline{ds}_o)) \geqslant 0 \quad (2)$$

where $\Delta W(\underline{ds})$ is the work of external forces associated with any virtual crack growth $(\underline{ds})$; $\Delta U(\underline{ds})$ is the strain energy variation associated with crack configuration $(\underline{ds})$. The configuration $(\underline{ds}_o)$ corresponds to the actual extended crack ; $\Delta D(\underline{ds})$ represents the quantity of surface energy dissipated by fracture extension configuration $\underline{ds}$.

Now, if the rock is isotropic with respect to its strength $(\Delta D(ds) = \Delta D(ds_o) = \gamma da$, where γ is the free surface energy per unit area, as defined by Griffith (1921)), equation (4) becomes :

$$(\Delta U(\underline{ds}) - \Delta W(\underline{ds})) - (\Delta U(\underline{ds}_o) - \Delta W(\underline{ds}_o)) \geqslant 0 \quad (3)$$

from which it can be shown that the hydraulic fracture extends perpendicularly to the minimum principal stress direction. But if the rock is anisotropic with respect to its strength $(\Delta D(ds) = \gamma(\underline{n})da)$, orientation of the fracture may not coincide with that of the major principal stress. Here the concept of anisotropy must be taken in its broadest sense ; it may refer either to a rock matrix property or to a more or less recemented preexisting joint. In the later case, even though the rock matrix may be isotropic with respect to its stregth, the rock mass is not ; this further implies that the rock mass strength anisotropy is not a homogeneous property : it varies from point to point in the rock volume.

Results decribed here after suggest that the anisotropy can be taken to advantage for developing from one borehole, fractures in different azimuths.

2.2. In-situ stress measurements at Le Mayet de Montagne

On site, the granite is covered by a less than 5 m thick soil layer. Natural fractures were mapped on 12 large outcropping areas nearby the test site (within 10 km from the site ; Drogue et al, 1979). Four major fissure orientations have been identified, namely N 30° E + 10° , N 60° E + 10° , N 100° E + 10° , N 155° E + 5°.

All these fractures are subvertical (dip lies between 70 and 85°). Thermal, electrical and video loggings were used to identify fractured zones as well as quartz veins in the borehole. Their orientations, determined with a borehole T.V camera, were found to be similar to those identified by the surface mapping.

For stress determination purposes, seven hydraulic fractures were generated, with an inflatable straddle packer, in area thought to be homogeneous or where a light quartz vein had been identified. Results are presented in table 1. It can be noticed that orientation of these fractures is quite variable and that for very shallow depth (27 m and 42 m) more than one fracture occurred.

TABLE 1. Data from hydraulic fracturing tests at Le Mayet de Montagne test site (25 km SE of Vichy, in the center of France) for borehole INAG 3-2

Depth (m)	27	42	54	65	84	90	174	186
breakdown pressure MPa	22.3 33.3	frac by packer	frac by packer	frac by packer	10	34.7	15.1	29.5
reopening* pressure MPa	-	5.4	5.9	-	8.5	5.1	-	-
shut-in pressure MPa	-	2.1	4.2	3.2	4.6	4.4	5.6	5.4
injected volume in m^3	-	4.88	1.19	4.58	2.54	3.59	.02	13.0
recovered volume in m^3	-	.04	.24	.12	.26	.14	-	2.
pumping rate $10^{-3} m^3/min$	60	60	60	60	60	60	1	320
fracturing fluid	water	water	gel	water	water	water	water	gel
orientation at wellbore: dip	multiple		82°	80°	-	80°	82°	80°
orientation at wellbore: strike	fractures		N60°E	N46°E	-	N160°E	N50°E	N57°E

*Pressure required to reopen the fracture generated during the first injection (measured after the pore pressure has dropped back to its original value).

Since, locally, topography is very mild and since the rock is very homogeneous, in-situ principal stresses cannot be assumed to change direction rapidly so that the variety of fracture directions must be attributed to local strength anisotropy. Thus this precludes application of the classical theory of stress determination from hydraulic fracturing tests and a new technique has been developed.

If $\underset{\sim}{\sigma}$ is the local stress tensor and $\underline{n}$ the unit normal to the fracture plane the shut-in pressure P_s is simply a direct measurement of the normal stress exerted on the fracture plane :

$$P_s = \underset{\sim}{\sigma}\, \underline{n} \cdot \underline{n} \tag{4}$$

Since topography is very mild on site, it can be assumed that stress variation with depth is linear and that one of the principal stresses is vertical and equal to the weight of overburden :

$$(\underset{\sim}{\sigma}) = \begin{pmatrix} \sigma_1 + \alpha_1 z & 0 & 0 \\ 0 & \sigma_2 + \alpha_2 z & 0 \\ 0 & 0 & \rho g z \end{pmatrix} \tag{5}$$

where σ_1 and σ_2 are supposed to be independent of depth ; z is depth ; ρg is the vertical stress gradient ; α_1 and α_2 are two unknown coefficients. For the small depth interval for which measurements were made it was assumed that this linear variation of stress with depth was caused only by gravity so that $\alpha_1 = \alpha_2 = \alpha$.

Only four unknowns are left in equation (3), namely σ_1 ,σ_2 , α and η the orientation of one of the horizontal principal stresses :

$$\left[\frac{\sigma_1+\sigma_2}{2} + \frac{\sigma_1-\sigma_2}{2} \cos 2(\beta-\eta)\right] (1-n_3^2) + \alpha\, z(1-n_3^2) = P_s - n_3^2 \rho g z \tag{6}$$

where β is the strike of the normal to the fracture plane and n_3 its direction cosine with respect to the vertical axis (for subvertical fractures $n_3 \simeq 0$).

When four or more fracture planes are available equation (6) can be used to determine the local stress field : after setting $X = \sigma_1 + \sigma_2$, $Y = (\sigma_1-\sigma_2)\text{Cos } 2\eta$, $Z = (\sigma_1-\sigma_2)\sin 2\eta$,the linear system is solved by a least squares technique.

This method has been applied to the four tests available before the large hydraulic fracture was developed at 186 m . Results are $\sigma_1 = 6.3$ MPa , $\sigma_2 = 1.5$ MPa , $\eta = 17°$E, $\alpha = 0.0108$ MPa/m. The striking feature is that none of the observed fractures fits with the major principal stress direction ; rather they all coincide with one of the directions observed for the natural fissures.

2.3 Determination of the geometry of large hydraulic fractures

In the initial scheme envisaged for developing a heat exchanger at Le Mayet de Montagne, hydraulic fractures were to be intersected with the second borehole after their exact geometry had been precised.

Analysis of shut-in pressure variations with various fracture length can help determine whether the fracture has changed orientation away from the injection well (Cornet 1982). However if the fracture does indeed turn, only a class of possible geometries can be defined by this technique so that additional geophysical means are required to locate precisely the fracture.

Two such means have been investigated :

- Survey of ground surface electrical potential variations ; (Mosnier 1981)
- mapping of seismo-acoustic events generated by fracture propagation.

For the survey of surface electrical potential variation, one electrode was placed at infinity (500m away from injection well) whilst the other one was installed on the injection line. Salty water was injected so that a large resistivity contrast existed between the rock matrix and the fracture. These tests failed completely. It was observed, later on, that 95 % of the injected current was circulating within the upper 15 m of soil and altered granite because of their very low resistivity (soil was saturated).

Two types of seismo-acoustic events were expected : high frequency events (between 1 Khz and 10 Khz) associated to local instabilities during fracture propagation and lower frequency events due to induced seismicity generated by water percolation in preexisting fractures inclined with respect to the principal stresses directions.

For the 1 Hz - 125 Hz frequency range, 12 three components seismometers were installed on a circle centered on the injection well and with a 200 m radius. The voltage output of these sensors was proportional to ground velocity (117 V/cm/sec.) ; it was further amplified with a 2^8 gain. For the 100 Hz - 5 Khz domain, 10 hydrophones were installed either in shallow boreholes reaching water level, 6 m below ground surface, and located from 10 to 60 m from injection well or in another well, some 50 meters away from injection point. In this well one hydrophone was placed at the very same depth as that of the straddle packer used for the injection. The sensibility of these hydrophones was - 205 db for a 1 V/μPa reference ; after amplification, the apparent sensibility was equal to 4.2 V/100 Pascal.

Despite very low background noise for the deep hydrophones (about 4 millivolts) no signals were recorded. The seismometers have recorded a few low frequency signals which are yet unexplained because of the low velocity observed for their propagation.

Fortunately, the shut-in pressure measurements observed after injecting 13 m^3 of gel mixed with sand was very stable (it did not vary during more than two hours after injection had stopped). This value (see table 1) coincided with that computed from the fracture direction as observed at the wellbore and from the stress field determined with the above mentioned technique. Thus it was concluded that the fracture had not changed of orientation during its propagation and this information was used to implant the production well. Luckily enough, this well intersected the fracture at the 156 m depth as is discussed later.

During subsequent injection tests strings of geophone have been set up on ground surface so as to filter the noise of the pumping unit and to sum on the signal. No success was achieved.

In the final stage of the experiment a third borehole was drilled down to 250 m, 30 m away from the two first ones. The purpose of this stage was to demonstrate that by careful preliminary loggings, it was possible to select a depth were the fracture would grow in a chosen direction.

This aspect is discussed later but during these tests some new efforts were made to map the fracture by accurate location of the seismo-acoustic activity (Leydecker 1982). This time four three components geophones (accelerometers) with frequency range between 500 Hz and 2000 Hz were installed at depth (180 m and 130 m) in the two boreholes drilled for the initial stage of the program.

Signals were analog recorded on a 16 channel FM tape recorder. Events were played back on an ultra-violet paper recorder with a 4 m/sec. paper velocity. Accuracy on first arrival times was of the order of 0.1 millisecond. Corrections were made to take into account the misalignment of recording heads on the tape recorder.

Injection lasted 10 minutes at a 0.520 m^3/min. flow rate. The water level rose by two meters in one of the boreholes and some increasingly high continuous noise was recorded on the lower geophone in this borehole. In addition more than hundred signals were picked by the four geophones. Some 30 of them have been located with accuracy (see figure 1).

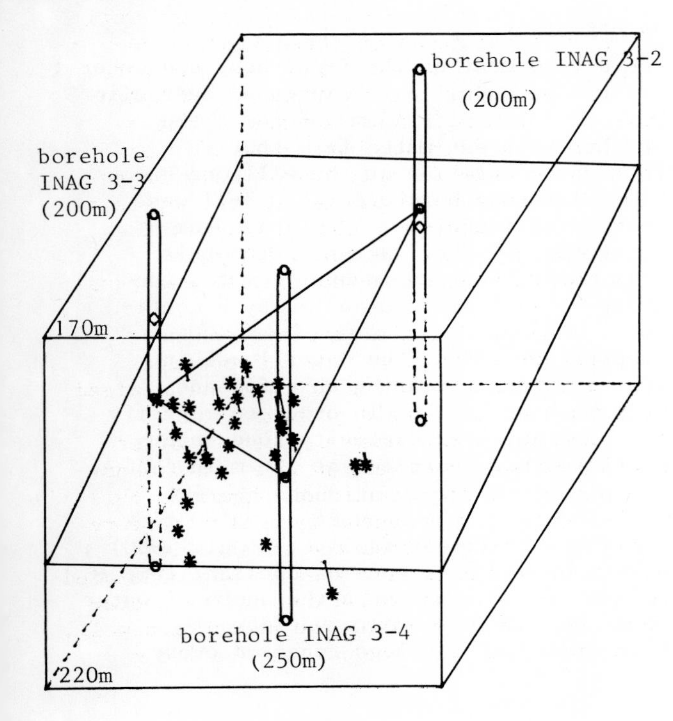

Fig. 1. Location of seismo-acoustic events associated with a fracture initiated from INAG 3-4 at 197m ; fracture plane strike N 73.6 E dip 16.5°

The first one occurred some 15 meters away from injection point two minutes after injection started ; the last one was observed 40 minutes after injection had stopped.

A least squares technique has been used to define the best plane passing thru these events. A plane with a N 74 E strike and a 16° dip was identified ; the standard deviation is 2.9 m which shows that the plane is well defined (see figure 1). It does not fit with any of the computed principal stress directions.

Simultaneously a three components geophone was anchored in a fourth well, 50 m deep, 30 m away from injection well. Its frequency range was 1 Hz - 250 Hz ; filters prevented any higher frequency from being recorded. Altough this geophone was way below the altered zone (this one has been evaluated to be 15 m thick from attenuation and velocity measurements) no signals were observed whilst injection occurred. Rather, low frequency events were observed as soon as injection stopped. However the source of these signals has not been identified yet for the signals propagation velocity appears to be very low (evaluated from P-S arrival time). They seem to have travelled partly thru the upper altered zone ; they are probably of similar origin as the events observed with the seismographs during the earlier stages of the experiments. They are truly associated with the injection process and not with human activity at the surface : they occur only after water-injection has stopped and no surface noise was recorded before and after the tests.

2.4 Preparation of a heat exchanger.

Let us now return to the development of the first heat exchanger : A first borehole (INAG 3-2) has been drilled ; a propped hydraulic fracture has been developed at 186 m with 13 m^3 of gel ; the orientation of the fracture has been estimated from shut-in pressure measurements, fracture orientation at the wellbore and in-situ stress determination. A second borehole (INAG 3-3) was drilled by the downhole percussion technique, 30 m away from INAG 3-2 in the expected direction of the fracture. With this boring technique the hole is maintained continuously empty, the incoming water and the boring debris being expelled by a continuous flow of compressed air.

When the hole reached a depth of 156 m, the water level in INAG 3-2 started to be lowered, thus indicating an efficient hydraulic connection between both boreholes. In order to identify the depth at which this high hydraulic conductivity zone intersected INAG 3-2, a straddle packer was installed at 156 m in INAG 3-3 and water was injected. Simultaneously continuous thermal logging was run in INAG 3-2. Indeed in the absence of water flow the thermal gradient was known to be linear (Jolivet et al. 1980) ; any inflow of water would then appear as a zone of constant temperature. The lower point of such a zone indicates the depth of water production. Results are shown on figure 2 where variation of temperature with depth is indicated for logs repeated every 20 minutes.

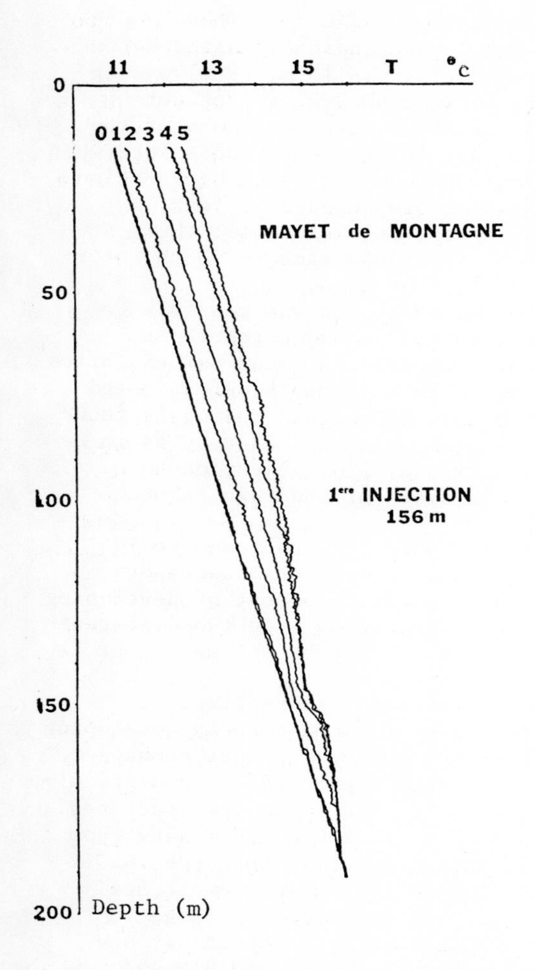

Fig. 2 Temperature logs conducted in INAG 3-2 whist injecting water in INAG 3-3 at 156 m.

Clearly the main water production occurs at 186 m where the propped hydraulic fracture was created. A secondary production zone appears at 150m. It was thought possible that the hydraulic fracture, developed at 186m intersected a horizontal fracture at the 150 m depth once it had grown slightly so that INAG 3-3 had in fact intersected this horizontal fracture. However, first of all no water flew out from INAG 3-2 during the development of the deep hydraulic fracture ; secondly the shut-in pressure remained stable during more than two hours, after pumping stopped. In addition the measured shut-in pressure was 5.4 MPa whilst the weight of overburden at 150 m is 4.0 MPa.

Accordingly it seems quite certain that, although the main hydraulic fracture did intersect preexisting fissures, it was not deviated from its orientation by these fissures. Further, as long as the hydraulic fracture was pressurized so as to balance the normal stress exerted by the in-situ stress field, those natural fissures did not act as significant drains. However, for lower fluid pressures, the tangential compressive stress, parallel to the fracture surface, was less important so that preexisting fissures could act as drains.

Other similar injection tests have been repeated with the

straddle packer set-up at various depth in INAG 3-3. These depth were chosen from observation of logs which indicated altered zones. In none of the subsequent injections (169 m, 113 m, 69 m) was any significant water production noticed in INAG 3-2, except for an injection at 143 m which yielded results similar to those of the 156 m deep test. In fact this test was fairly unfortunate for, when injection at 156 m were undertaken later on for the long duration heat exchange investigation, this fracture acted as a leak, a feature which had not occured during the initial test : the injection at 143 m reactivated a preexisting fracture which intersected the main hydraulic fracture further down.

It must be emphasized that the reason why it was possible to intersect the hydraulic fracture with the second borehole is because the fracture plane was not perfectly vertical but exhibited an 85° dip. If we assume the fracture to be 3 mm thick(estimated from T.V. camera observations) and penny shaped, its radius is found to be about 40 m for the 13 m^3 injected volume. This results in a height of about 80 m and leaves, 30 m away from injection hole, an error margin of about 3 m on both sides of the main fracture direction.

Had the linking procedure failed, hydraulic fractures would have been developed in INAG 3-3 in such directions that they intersect the original propped fracture. This concept of developing hydraulic fractures in two different directions from two different boreholes was the basis for the second heat exchanger set up at Le Mayet de Montagne.

Indeed, such a geometry not only is easier to realize, it provides in addition more area to the heat exchanger. By developing two such connections, at two differents depth, the heat exchanger area can be greatly enhanced without encountering all the difficulties associated with circulating water in many parallel fractures.

A third well INAG 3-4, was bored down to 250 m, 30 m away from both INAG 3-2 and INAG 3-3 and in such a location that the INAG 3-2 - INAG 3-4 direction be parallel to the maximum horizontal principal stress direction. The goal was to develop a large hydraulic fracture from INAG 3-4 which intersects the first one (linking INAG 3-3 to INAG 3-2) somewhere between INAG 3-2 and INAG 3-3.

For doing so, thermal, chemical, sonic, electric and video loggings were run to identify preexisting fractures and determine their orientation. The thermal log (Jolivet 1981) revealed to be not very reliable for identification or natural fissure except when these are subjected to large water circulation. It outlined a strong anomaly (0.2° off the mean gradient) at 154 m which was not confirmed by other logs as being the site of a hydraulically conductive zone. Chemical logs (Michard 1981) are run by determining ions content in the borehole water at various depth (every 10 meters). It revealed very efficient to locate the connections between INAG 3-3 (at 156 m) and INAG 3-2 (at 186 m and 154 m) by plotting the variation of the ratio of Ca^{++} content versus SR^{++} with depth. This log did show light anomalies at 100, 150,

170,220 and 230 m in INAG 3-4. Sonic logs (Le Breton 1981) measured the variation with depth of velocity between a source and a receiver placed 1.50 m apart. Analysis of both direct arrival time and variation in the amplitude of the signal (Darcilog) are used to locate open fractures. It revealed very efficient, as checked later on by direct videologging with a borehole T.V camera. The electrical log (Mosnier 1981) uses an electrode at infinity (500 m from injection well) and a specially designed electrode in the (uncased) borehole to insure that the current flows radially from the electrode into the rock formation. The quantity of current which flows into the rock is measured for 12 different azimuths. Thus not only does the tool provide an indication on the existence of local electrical conductors, it also gives the dip and strike of the conductor when it is plane.

Combination of these logs with direct video loggings was used to identify depth where natural fractures with the desired dip and strike intersected the borehole.

First a test was conducted at 197 m for developing a large hydraulic fracture. Although the fracture initiated in the desired direction at the wellbore, it soon intersected a sub-horizontal preexisting fracture which stopped its further extension, as observed from the location of seismo-acoustic events. Further this sub-horizontal fracture plane intersected another preexisting fissure which acted as a short circuit, so that after 10 minutes of pumping all the injected water was produced back into injection well, above the straddle packer.

A new test was run at 163 m. This time, shortly after injection had started (less than 5 minutes) the water level rose in INAG 3-3 thus indicating a link between INAG 3-4 and INAG 3-3. Repeated thermal logs during a new water injection indicated that water was flowing at 156 m, the depth were the original propped hydraulic fracture intersected INAG 3-3.

Clearly the newly created hydraulic fracture had intersected a preexisting fissure which in turn was in direct connection with the first propped hydraulic fracture ; when INAG 3-3 was plugged with an inflatable packer above the production zone, INAG 3-2 started to produce water at the 186 m and 154 m depth, and only at those depth.

2.5 Discussion

These results illustrate the fact that rock strength anisotropy may influence very significantly the direction of fracture propagation. This implies that the a priori hypothesis on rock strength isotropy made with the classical stress determination technique must always be verified a posteriori by using preexisting fractures or, alternatively, fractures developed in other directions than that of the computed maximum principal stress, in order to ascertain the validity of the determination.

In the stress determination technique proposed here above a very restrictive hypothesis is made namely that both horizontal principal stresses vary by the same amount with depth. This assumption was made in order to keep linear the inversion problem.

A generalized least squares technique (Tarantola and Valette 1982) has been developed for inverting non linear problems. This new mathematical technique has been adapted to the present problem (Valette and Cornet 1982) so that no a priori assumption is made with respect to stress variation with depth : this variation is simply assumed to be a stochastic process ; one of the principal stresses is still supposed to be parallel to the borehole axis. It has been applied to all hydraulic fracturing results obtained at Le Mayet de Montagne. Only those results for which the fracture geometry was well defined were retained (fracture orientation was determined with a radial view borehole T.V. camera ; only those fractures which were identified by at least six points were retained). Corrections were made to take into account the fact that the camera was not centered in the borehole but laid against one generatrix. These results are given in table 2.

TABLE 2. Results from injection tests at Le Mayet de Montagne

Depth (z) in meters		fracture strike (ϕ) in degrees		fracture dip (θ) in degrees		Instantaneous shut-in pressure (p_s) in Mpa	
z	ε_z	ϕ	ε_ϕ	θ	ε_θ	p_s	ε_{p_s}
56	0.5	58	5	85	3	4.05	0.2
90	0.5	155	5	83	3	4.4	0.3
113	0.5	25	10	75	5	2.7	0.1
143	0.5	29	5	84	3	3.5	0.1
163	0.5	155	5	84	3	5.3	0.2
174	0.5	69	5	84	3	5.65	0.1
186	0.5	55	5	85	3	5.45	0.1

The horizontal stress field variation with depth is indicated on figure 3.Clearly results are fairly different from those computed with equation (6), although the maximum principal stress direction is not significantly altered (N 19°E versus N 17°E).

Then one may wonder that INAG 3-3 intersected the deep hydraulic fracture. In fact, first of all this shows that our fracture orientation determination at the wellbore wall was correct. Secondly since only four fractures had been used for the initial stress determination and since one of these fractures was precisely one with the same orientation as that of the 186 m deep fracture and was in addition only 10 meters above it, the correct value for the 186 m deep fracture shut-in pressure was in fact built in the solution. Interestingly enough,corrections for the fact that the camera was not well centered have yielded a different orientation for this 174 m deep fracture .

Determination of the location of the seismo-acoustic events observed during the 197 m deep fracture raises another intriguing

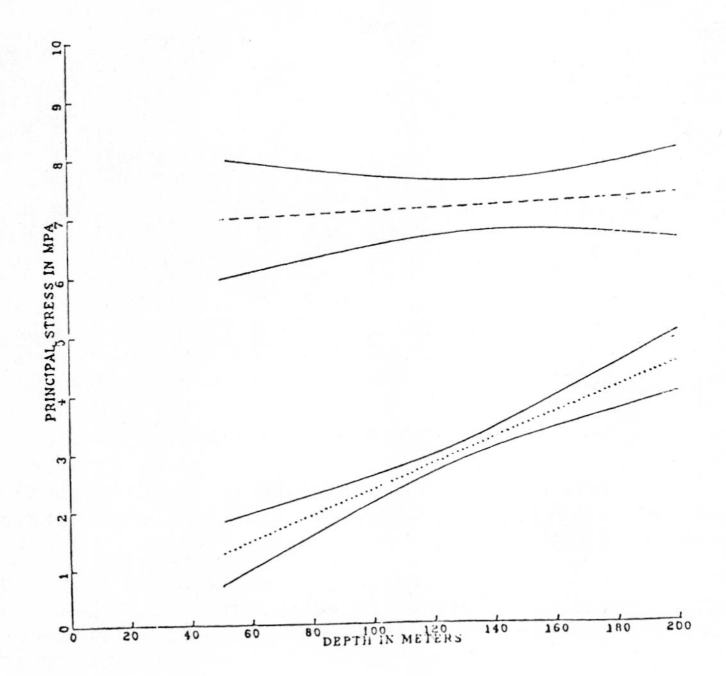

Fig. 3. Stress determination from shut-in pressure measurements with the generalized least square technique. Dotted lines are mean values, plain lines are mean value ± one standard derivation.

question : despite a sub vertical initiation at the wellbore wall are these fractures not linked together by a subhorizontal fracture plane at the 155 m depth somewhere away from the well ?

Elements for answering negatively to this question are as follows :

1) during both hydraulic fracturing tests at 186 m in INAG 3-2 and at 163 m in INAG 3-4, no significant variations of water level were observed in the injection well. This shows that if the fracture got linked to a horizontal fracture, this fracture did not intersect the borehole.

2) Shut-in pressure observed for the 186 m test was stable during more than two hours and equal to a value larger than the weight of overburden. Had the fracture been linked to a sub-horizontal fracture this one would have been placed under unstable mechanical conditions.Thus it should have extended and give rise to a decrease in fluid pressure. Since this did not occur, it can be concluded with certainty that the 186 m deep fracture remained

subvertical.

3) For the 163 m deep fracture the instantaneous shut-in pressure was larger thanthe weight of overburden so that it is certain that no horizontal fracture existed at the wellbore at the straddle packer level. Since the fracture was linked by a preexisting fissure to INAG 3-3, the shut-in pressure did not remain stable at the end of injection. Nevertheless it is believed that, if the 163 m vertical frac had intersected a horizontal fracture at the 155 m depth, water should have flowed out the injection well. Since this did not occur, it seems reasonable to assume that the fracture remained subvertical.

3. THERMAL CHARACTERIZATION OF HEAT EXCHANGERS

3.1. The single well test

The single well test is designed to determine in-situ the rock thermal characteristics. In this test, cold water is injected in one fracture thru a straddle packer. The fracture is chosen from preliminary geophysical logs investigation. After a certain amount of cold water (2 to 4 m^3) has been injected under constant flow rate (3 to 4 10^{-4} m^3/sec.) the well is shut-in for about 20 minutes. When it is opened, some of the injected water flows back into the well. Downhole temperature measurements as well as surface measurements of flow rate are used to determine the thermal characteristics of the formation. Flow rates during production are highly variable. ; generally they are exponentially decreasing with time. The recovery percentage varied from 10 % for a man made hydraulic fracture to 45 % for a natural fissure.

Results are interpreted with a finite difference model. Assumptions implicit in the model are :

- At the scale of the test, the fracture is supposed planar and infinite.
- Because of the injection rate, the flow is assumed to be radial, at least in the neighbourhood of the well where most of the heat exchange occurs. Flow into the rock matrix is neglected (we are dealing with granite) as well as the temporal variation of water storage into the fracture.

The heat transfer is assumed to be purely convective into the fracture and purely conductive into the rock. Since the fracture is very thin, the heat stored into the water in the fracture as well as that lost thru conduction in the water is neglected.

Accordingly, heat transfer is governed by equations (7) :

$$\gamma_f Q \frac{\partial \theta}{\partial r} - 4 \pi r \lambda_g \left. \frac{\partial \theta_r}{\partial y} \right|_{y=0} = 0$$

$$\frac{\partial^2 \theta_r}{\partial y^2} = \frac{\gamma_g}{\lambda_g} \frac{\partial \theta_r}{\partial t} \qquad (7)$$

where Q is the injected, or produced, flow rate ; θ is the temperature in the fracture ; θ_r is the temperature, at distance r, in the granite ; λ_g is the thermal conductivity of granite ; γ_g is the heat capacity of granite ; γ_f is the heat capacity of water ; r is the radial distance from the borehole axis and y is the coordinate perpendicular to the fracture plane.

It is noteworthy that the thickness of the fracture, which would be very difficult to apraise, does not come into figure in equations 7.

The parameters λ_g and γ_g are chosen so as to produce the best fit between observed and computed results. A typical curve is shown on figure 4. The best fit is obtained for $\lambda_g = 3.3$ w/m/°c and $\gamma_g = 2.5\ 10^6$ j/m^3/c° . As expected, the model does not represent accurately the heat exchange observed during the waiting time before discharge : heat conduction thru the tubing is not taken into account. Results obtained with this technique compare well with those obtained from laboratory measurements on core samples (3 W:m/°c).

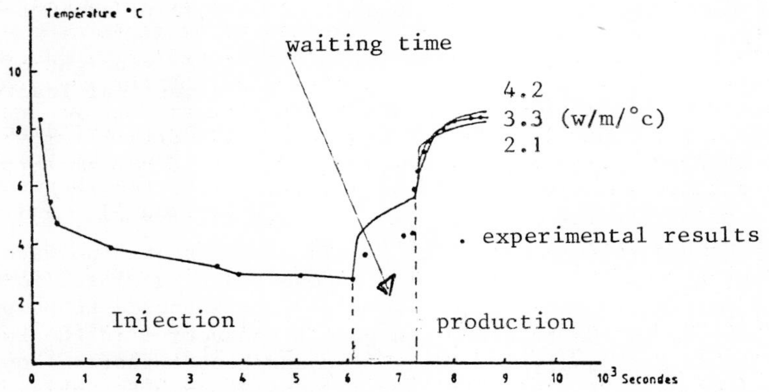

Fig. 4. Results from a single well test on a horizontal fracture ; injected volume was 2.4 m^3 , recovered volume was 0.765 m^3

3.2. Long duration circulation tests.

Two long duration circulation tests have been conducted (56 days for the first heat exchanger, 48 days for the second one) to characterize the heat exchange area in both cases. A straddle packer was used for injection (in borehole 3-3 for test 1 and borehole 3-4 in test 2) whilst a single inflatable packer was set above the production zone in the production well (3-2 for test 1 ;

3-3 for test 2), as identified by the thermal logs. A flexible, large diameter, hose was connected to this packer so as to bring the water flowing into the well below the packer up to the surface. 4.600 m^3 of water were injected at an average rate of 3.4 m^3/h during the first test with a production recovery of 65 % whilst 4.800 m^3 were injected during the second test with a 97.5 % recovery at production. During this last test flow rate was set at 8.3 10^{-4} m^3/sec for the first 12 days and then to 14. 10^{-4} m^3/sec for the remaining part.

Borehole 3-2 had been plugged with an inflatable packer set on 140 m so as to prevent any production in this borehole ; the local water pressure was recorded continuously. During the first 12 days injection pressure stabilized at 4 MPa ; when the flow rate was increased the injection pressure first rose to 6 MPa but then progressively decreased down to 3.2 MPa (see figure 5). This pressure decrease associated with an increase in flow rate is interpreted as the result of a lateral fracture extension together with a decrease of impedence at the crossing of the two fractures which connect INAG 3-4 to INAG 3-3.

Interpretation of the results has been conducted with a discrete numerical model built on hypothesis similar to those of the single well test. The fracture is supposed to be planar, non deformable but limited transversely.

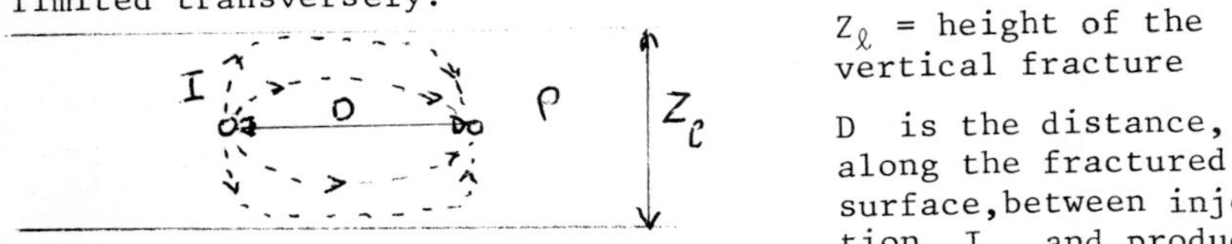

Z_ℓ = height of the vertical fracture

D is the distance, along the fractured surface, between injection I and production P

The flow is supposed to be pseudo steady state (no water accumulation in the fracture) and is represented by a closed form solution (Hosanski 1980). Heat exchanges are supposed to be purely convective in the fracture and purely conductive in the rock.

Dimensional analysis of the flow and heat transfer shows that temperature at the production well depends upon two parameters :

$$t_{ad} = \frac{Q^2 \gamma_f^2}{\lambda_g \gamma_g D^4} t = \text{dimensionless time}$$

$$A = \frac{Z_\ell}{D} = \text{shape factor for the effective fracture}$$

D is the distance along the fractured surface between injection and production point. The model is fitted to experimental results by chosing properly t_{ad} and A . Since t_{ad} varies as a function of $1/D^4$, measurement of the arrival time of the thermal perturbation at the production well provides a good determination of D . Results for the second heat exchanger are shown on figure 5.

The temperature curve observed during the first 12 days has been used to determine t_{ad} and A ; this leads to D = 40 m and Z_{ℓ} = 24 m. Interpretation of the remaining part of the curve involved a variable shape factor A as function of flow rate so as to fit the computed temperature curve with that observed at the production well.

These results suggest that doubling injection rate led to a 25 % increase of the heat exchange area. However this area increment does not explain the pressure drop observed during this phase. This decrease may be attributed, as a first supposition, to the change of impedance at the intersection of the two fractures that link INAG 3-4 to INAG 3-3.

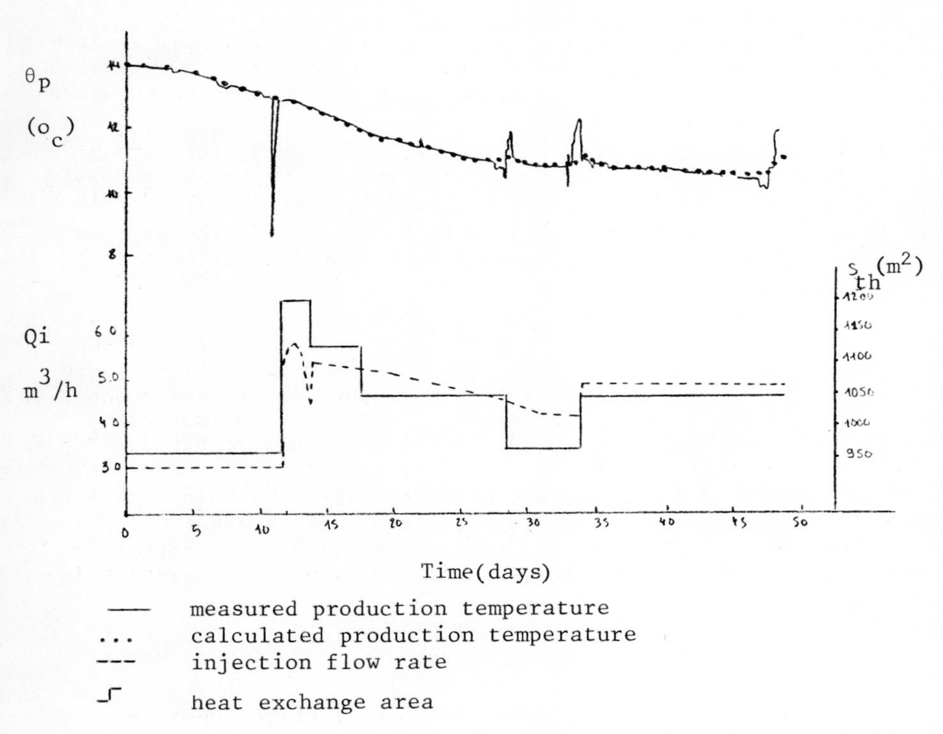

Fig. 5 Results of the second long duration circulation test. The upper curve corresponds to temperature variation whilst the lower curve indicates the proposed variations of heat exchange area.

CONCLUSION

These results show that hydraulic fractures may develop in directions different from those perpendicular to the least principal stress direction because of the strength anisotropy of the rock mass. This has been taken to advantage to develop hydraulic fractures from two different wells in two different directions so as to generate a heat exchanger.

Efficiency of this heat exchanger has been investigated by first determining in-situ the thermal characteristics of the granite and then by long duration circulation tests. Temperature variations observed at the production well suggest that the effective heat exchange area depends on flow rate.

These results have been used to construct cooling abacuses for large heat exchangers at depth ; they provide means to determine, for various flow rates and fracture shape factors, the decrease of production power as a function of the dimensionless time.

ACKNOWLEDGMENTS

Fracture plane determination from seismo acoustic events was conducted by L. Martel from Institut de Physique du Globe de Paris. This work was supported by E.E.C. Research grants from the Geothermal Research fund as well as by A.T.P. " Transfer de flux de chaleur dans l'écorce terrestre " from Institut National d'Astronomie et de Géophysique.

REFERENCES

Bernaudat, F. ; 1982 ; " Le Mayet de Montagne, Interpretation des essais hydrauliques et thermiques " ; Rapport LHM/RD/82/36, Centre d'informatique géologique ; Ecole des Mines de Paris.

Cornet, F.H. ; 1982 ; " Interpretation of hydraulic injection tests for in-situ stress determination " ; workshop on hydraulic fracturing stress measurements, Monterey 1981 ; co-sponsored by U.S.G.S. and U.S. National Committee for Rock Mechanics ; Haïmson and Zoback editors.

Drogue, C. ; J.C. Grillot and M. Razack 1980 ; " Site du Mayet de Montagne, Etude de la fracturation " ; Internal Report " Projet Energeroc 1980 " . Institut National d'Astronomie et de Géophysique.

Griffith, A.A. ; 1921 ; " The phenomenon of rupture and flow in solids " , Phil. Trans. Roy. Soc. London ; A 221.

Hosanski, J.M. ; 1980 ; " Contribution à l'étude des transferts thermiques en milieu fissuré " ; Thèse de Docteur-Ingénieur, Ecole Normale Supérieure des Mines de Paris.

Jolivet, J. ; M.G. Bertaux and G. Bienfait ; 1982 ; " Thermometrie au Mayet de Montagne " ; Internal report, Institut National d'Astronomie et de Géophysique.

Lebreton, M. ; 1981 ; " Diagraphies soniques au Mayet de Montagne " Internal Report, Institut National d'Astronomie et de Géophysique.

Leydecker, G. ; 1982 ; " Seismiche ortung hydraulisch erzeugter Brüche im franzosischen Geothermik Project Le Mayet " ; Bundesanstalt für Geowissenschaften und Rohstoffe - Hanover ; Archive nb 92 297.

Michard, G. and C. Beaucaire ; 1982 ; " Site du Mayet de Montagne chimie des eaux " ; Internal Report ; Laboratoire de Géochimie des eaux ; Université Paris VII.

Mosnier, J. ; 1982 ; " Detection electromagnetique des directions de fissuration naturelle ou provoquée dans un sondage ; Application à la géothermie des roches chaudes sèches " . Laboratoire de Géophysique Appliquée. Orléans la Source.

Murphy, H.D., J.W. Tester, C.O. Grisby and R.M. Potter ; 1981 ; " Energy Extraction from fractured geothermal reservoirs in low permeability crystalline rock " ; Jou. Geophys. Res. vol. 86 , nb B 8 , p. 7145.

Tarantola, A. and B. Valette ; 1982 ; " Generalized non linear inverse problem solved using the least squares criterion " Rev. Geophys. Space Phys. vol. 19 ; pp 219-232.

Valette, B. and F.H. Cornet ; 1982 ; " The inverse problem of stress determination from hydraulic injection tests data " submitted for publication to the Jou. Geophys. Res.

HYDRAULIC FRACTURING EXPERIMENT AT NIGORIKAWA AND FRACTURE MECHANICS EVALUATION

by Katsuto Nakatsuka*, Hideaki Takahashi** and Morihiko Takanohashi***

* Department of Mining and Mineral Engineering, Faculty of Engineering, Tohoku University
** Research Institute of Strength and Fracture of Materials, Faculty of Engineering, Tohoku University
*** Japan Metals and Chemicals Co., Ltd.

ABSTRACT

Massive hydraulic fracturing carried out at Nigorikawa hot-water dominated geothermal field in the Fall of 1978 and Spring of 1980 has been very successful in the economic sense. This paper outlines this successful experience, where details of geological subsurface structure are discussed from the viewpoint of a connection between a natural fracture zone as a reservoir and a well, in terms of a hydraulically formed fracture.

A fracture evaluation procedure based on linear elastic fracture mechanics makes it possible to estimate the crack extension process during hydraulic fracturing numerically in a quantitative manner, provided that the fracture toughness of a rock mass under geothermal reservoir environment, the location and size of preexisting fractures and the stress distribution underground are known. In this paper some simple examples demonstrate that the fracture mechanics procedure can give useful information on the evaluation of crack extension during hydraulic fracturing.

INTRODUCTION

Massive hydraulic fracturing has attracted world-wide notice as a promising well stimulation method in geothermal energy development, not only for hot dry rock but also for traditional natural geothermal reservoirs[1-5]. The goal of stimulation in geothermal energy development by means of hydraulic fracturing for natural geothermal reservoirs will, ideally be, not only to connect geothermal wells to reservoirs, but also to achieve the reinjection of separated hot water, of which almost all Japanese geothermal plants produce a large amount[6][7], into the ground in order to elongate the fields by extracting heat energy through a natural or a manmade circulation system with reasonable efficiency, thereby maintaining a hydrological balance. For this purpose, new fields of technology will be essential for designing, observing and controlling subsurface cracks in addition to the exploration of subsurface structures with high temperature. The fracture mechanics approach is expected to play an important role in these new fields[8-14].

Recently some noteworthy reports have appeared to examine the feasibility of applying hydraulic fracturing to hot-water type geothermal fields[2][4]. But these developments have just started, and cases resulting in commercial success are relatively unknown. Hydraulic fracturing experiments at the Bacca field have been unsuccessful because they were unable to provide a sufficiently conductive pathway between the wellbore and the highly productive reservoir[4]. The Nigorikawa field experience is the first case in the world where the application of hydraulic fracturing to a hot water dominated geothermal source has resulted in commercial success.

HYDRAULIC FRACTURING AT NIGORIKAWA GEOTHERMAL FIELD, HOKKAIDO[1][5]

Field Development

Dohnan Geothermal Energy Co.,Ltd., a subsidary of Japan Metals and Chemicals Co.,Ltd., has been developing the Nigorikawa field since 1971 to ultimately produce 50 MW of electricity for the Mori area in Hokkaido. To make wells commercial, hydraulic fracturing treatments were performed in lieu of redrilling. To date, several wells have been treated in two separate projects. Seven were treated during the fall of 1978. In the spring of 1980, four were treated and several additional wells were treated for the first time. Although these fracturing jobs are the first known in geothermal wells, the enhanced production or injectivity that resulted in some cases has encouraged similar applications in other geothermal areas.

The Nigorikawa field is within the caldera-fill of an extinct volcano. The caldera and surrounding areas are composed of Neogene Tertiary altered andesitic lava, black shale, tuff and sandstone: and Quaternary terrace deposits, pyroclastic flow deposits and alluvium deposits. Slate, chert, limestone and taff of pre-Tertiary formations and caldera-fill sediments were found in drilling. The caldera is funnel-shaped with steep walls at angles between 60° and 70°. The caldera wall is distinct and when wells are drilled through caldera fill deposits, circulation usually is lost suddenly. This phenomenon, along with observation of core samples, suggests there are many natural fractures at the caldera and the relative position of wells that have been drilled, Fig. 1.

Temperature profiles from wells drilled into the caldera show a characteristic curve. Temperatures will not exceed 100°C until 689 meters below the surface, where the thermal gradient is only 4.1°C per 100 meters. This low temperature zone corresponds to the post-caldera lake and alluvium deposits and the upper half of the unconsolidated soft fall-back deposits. Then the temperature rises abruptly to reach about 185°C at 1082 meters. Therefore, the thermal gradient of this zone is 25.4°C per 100 meters. However, below this thermal gradient zone, the temperature will rise gradually at 4.1°C per 100 meters. If holes being drilled penetrate into pre-Tertiary formations, the thermal gradient becomes high again, reaching 10.2°C per 100 meters.

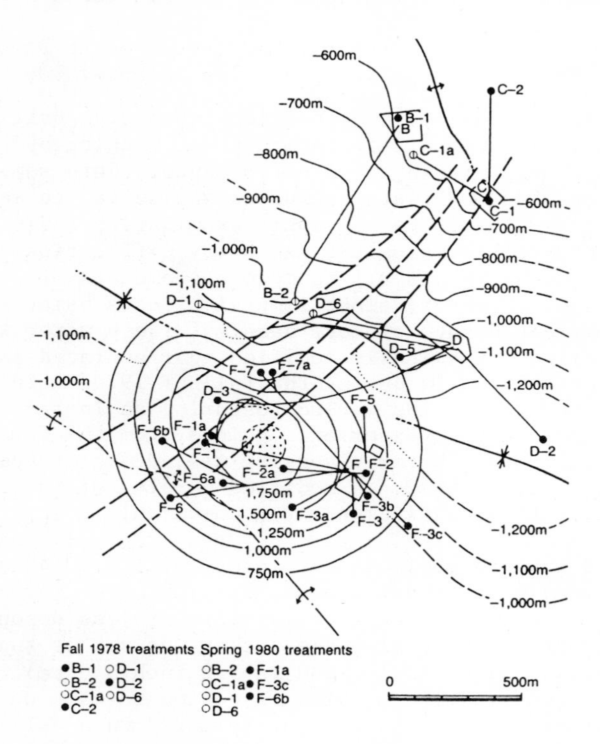

Fig. 1 An isopach and well deviation map of the Nigorikawa field (∘:production well, •:reinjection well)

Since the Nigorikawa field produces wet steam (about 5.5 : 1, hot water to steam ratio), conventional mud and drilling techniques are used. Since steam zones are underpressured, lost circulation intervals are carefully monitored during drilling. If sufficient natural fractures are not intersected to establish what is believed to be commercial steam production, the well is plugged back, sidetracked and drilled in another direction. Two sidetrack wells drilled at a single well site are quite common. Lost circulation occurring in a shallow zone normally is considered for injection use only. Wells with lost circulation in deeper zones, particularly in the fracture zones at the caldera wall with the bottom hole temperature exceeding 200°C, are normally the best candidates for steam and hot water production. Communication between upper and lower zones is carefully avoided. The basis for an economic well in the Nigorikawa field is a minimum steam production rate of 25 tons per hour at 0.7 MPa or a minimum brine injectivity of 200 tons per hour at 0.55 MPa. Considerations that hinder commercial field development include the following: (i) In several wells, steam volume was insufficient and wellhead pressure was low. (ii) Injection pressure was too high in some injection wells. (iii) Since brine produced is toxic, the excess could not be dumped into local streams and rivers.

Fracturing Program

After studying various stimulation techniques with Halliburton Limited, Dohnan decided on a stimulation program using hydraulic fracturing to improve existing well performance. Core samples from wells D-2, C-1, B-1 and D-3 were evaluated in the lab to select a suitable fracturing fluid. Petrographic examination indicated none of the cores had any visible porosity, and X-ray diffraction analysis showed them to contain very small to moderate amounts of chlorite and illite(mica) clays. However, since immersion tests using aqueous fluid exhibited no apparent damage of core samples, fresh water was chosen as the propping carrier. Chemically modified guar hydrated in fresh water was used to treat the wells during the fall of 1978. In the spring of 1980, a patented fracturing method featuring an inorganic fluid was used for these treatments. Experience gained from the first series had shown that a fluid with greater viscosity was needed at the extremely high bottom-hole temperatures. A modified version of the inorganic material satisfied this requirement at a comparable cost.

Wells treated in both projects were propped with 20/40 mesh sand. Jobs were performed in at least two stages with diverting material used between stages. Design pumping rate was more or less dependent on available equipment. The retrievable packer was set as shallow as possible, usually 60 meters below the 13-3/8 inch(340 mm) casing shoe, to minimize friction pressure at high injection rates. During both fracturing programs, treatments were performed down a 4-1/2 inch, 16.6 ppf drill pipe.

First Program (1978) For the first fracturing project, seven wells were selected for treatment. They had been completed earlier and production and injectivity tests revealed they were not economic. The design for each well called for 320 kL of fluid to carry 28 tons of 20/40 mesh sand in three stages. Fresh water from a nearby river was transferred to two 100 kL pits via deep well pumps. Water pH was adjusted with acetic acid, and chemically modified guar at a concentration of 5 kg per 1 kL was added continuously during the treatments. Diverting agents were fed into a proportioner where they were blended with the base fracturing fluid.

Second Program (1980) The second fracturing program had two objectives. The first was to stimulate three recently completed wells in the F area (Fig. 1) and the second was to retreat wells that responded best to the initial fracturing program. To accomodate considerations set forth after evaluating results of the first program, an inorganic gelling agent was chosen to make the fracturing fluid. It was prepared in 2 %

KCl water. The same chemically modified guar used in the first program was continuously added to the inorganic base solution as a co-gelling agent at a concentration of 2.5 kg per 1 kL. This time, treatment volume was set at 840 kL to carry 81 tons of 20/40 mesh sand in two stages. The base fluid was prepared in working pits using local river water. Deep well pumps were used for mixing and transfer of the new gell to two transit tanks before being pumped away by two independently operated blenders. The complexer was added simultaneously into both blender tubs and the system pH was closely monitored. Nine tons of flake benzoic acid were used to divert between stages and all materials were proportioned through the two blenders according to the designed pumping schedule(Table 1). A total of nine fracturing pumps and two blenders were connected to a combination suction/discharge, trailer-mounted manifold. All treatments were pumped to the wellhead through 3-inch discharge lines connected to a twin port fracturing head. Treating rates varied since all of the fracturing pumps were not available for each job. A typical location for this second series of seven treatments is illustrated in Fig. 2 and treatment data for these jobs are contained in Table 2.

Table 1 Pumping schedule - Spring 1980

Volume(kL)	Fluid type	Proppant /Diverter	Concentration(kg/L)	Proppant total(kg)
38	Prepad	-	-	-
151	Pad	-	-	-
114	Proppant	20/40 mesh sand	0.12	13,600
114	Proppant	20/40 mesh sand	0.24	27,200
38	Diverter	benzoic acid	0.24	-
151	Pad	-	-	-
114	Proppant	20/40 mesh sand	0.12	13,600
114	Proppant	20/40 mesh sand	0.24	27200
-	Flush	-	-	-
834				81,600

Results

Examination of Table 3 reveals that hydraulic fracturing in 1978 was successful. Extensive redrilling quite possibly would have been required to obtain comparable improvement. These initial treatments were simple and required minimum equipment, materials and personnel to obtain an economical increase in four of the seven wells, namely: B-2, C-1a, D-1 and D-6. Moreover, since every treated well improved, a more extensive fracture system could possibly have provided even greater improvement.

TABLE 2 Treatment data - Spring 1980

Date	April 14, 1980	April 17, 1980	April 19, 1980	April 21, 1980	April 27, 1980	May 1, 1980	May 5, 1980
Well No.	D-6	F-3c	F-1a	F-6b	B-2	D-1	C-1a
Well depth (m)	2,200	1,130	2,000	2,380	1,790	2,400	1,770
Open hole interval (m)	1,275-2,200	690-1,130	1,500-2,000	1,500-2,380	788-1,790	1,502-2,400	904-1,770
Bottom hole static temperature (°C)	264	228	222	238	241	242	229
Packer depth (m)	747	549	664	671	457	692	549
Average rate (kL/min)	5.4	6.0	6.9	6.8	7.9	7.7	8.4
Average pressure (MPa)	15.2	15.5	11.7	11.0	13.3	14.8	14.5
Total treatment volume (kL)	616	759	867	855	893	810	802
Quantity of sand (kg)	72,600	81,600	81,600	81,600	81,600	81,600	81,600

TABLE 3 Well test data

Well No.	Approximate date*	Injectivity test		Production test		
		Rate (t/hr)	Pressure (MPa)	Steam rate (t/hr)	Hot water rate (t/hr)	Pressure (MPa)
B - 1	Before Fall 1978	0	3.4	-	-	-
	After Fall 1978	48	3.4	-	-	-
B - 2	Before Fall 1978	69	0.93	20	43	0.20
	After Fall 1978	100	0.59	18.8	147	0.53
	Before Spring 1980	85	0.54	20.0	118	0.77
	After Spring 1980	112	0.54	-	-	-
C - 1a	Before Fall 1978	45	0.49	10.0	59.0	0.03
	After Fall 1978	130	0.69	14.9	139.0	0.91
	Before Spring 1980	100	0.54	17.5	150.0	0.77
	After Spring 1980	92	0.54	-	-	-
C - 2	Before Fall 1978	15	1.47	-	-	-
	After Fall 1978	82	1.27	-	-	-
D - 1	Before Fall 1978	160	0.62	20.0	76	0.59
	After Fall 1978	172	0.33	46.5	107	0.49
	Before Spring 1980	255	0.54	56.0	180	0.77
	After Spring 1980	-	-	67.2	289	0.77
D - 2	Before Fall 1978	23	0.72	-	-	-
	After Fall 1978	48	0.86	-	-	-
D - 6	Before Fall 1978	103	2.4	20.0	40.0	0.20
	After Fall 1978	220	0.2	34.7	50.8	0.52
	Before Spring 1980	220	0.2	43.0	125	0.77
	After Spring 1980	-	-	61.0	168	0.77
F - 1a	Before Spring 1980	49	0.67		No continuous flow	
	After Spring 1980	220	0.67	Not measured	108	0.22
F - 3c	Before Spring 1980	-	-		No continuous flow	
	After Spring 1980	110	0.48		No continuous flow	
F - 6B	Before Spring 1980	0	0.54		No continuous flow	
	After Spring 1980	66	0.44	Not measured	94	0.11

* (indicates tests before and after treatments)

Although results from the second fracturing program (1980) have not yet been fully evaluated, tests that have been made indicate greater success, Table 3. Experience also has shown that test results appear to get better with time. Considerably more equipment, materials and personnel were required for the second series of treatments. This significantly added to the cost and will justifiably mean production and injectivity improvements will have to be proportionately greater to achieve economic success. Before plans are developed for a third fracturing program, results from the second program, although very encouraging, will have to be thoroughly analyzed. Although skepticism about fracture stimulation in geothermal environments has made these initial applications unique, the assessment of hydraulic fracturing in the Nigorikawa field has shown the technique can be economically attractive.

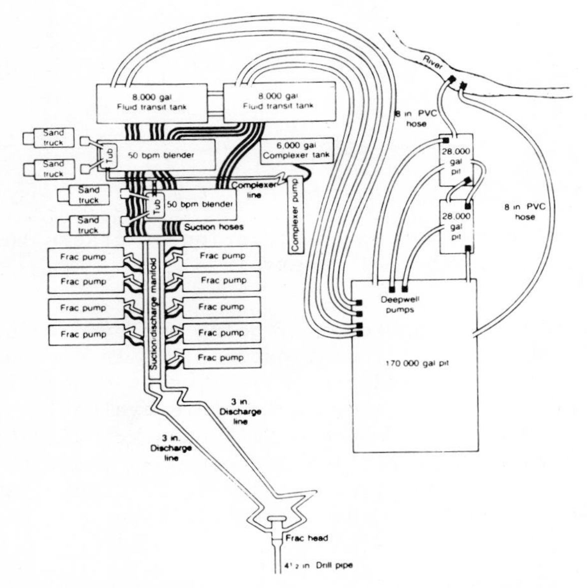

Fig. 2 A typical equipment layout for the spring 1980 fracturing program

FRACTURE MECHANICS EVALUATION OF FRACTURES IN GEOTHERMAL RESERVOIR

Figure 3 shows a general concept on the function of cracks in a natural hydrothermal reservoir. For production wells, man-made cracks or preexisting natural fractures collect geothermal fluid mass, extracting heat from surrounding rocks to the fluid. For an injection well through which hot water separated from steam is re-injected into the ground in order to maintain a hydrological balance of the field, cracks act to enable reinjection with low pressure and to distribute the water so that the water may obtain enough enthalpy until recirculated to production wells. Natural fractures and permeable leaks are to be positively used for heat exchange.

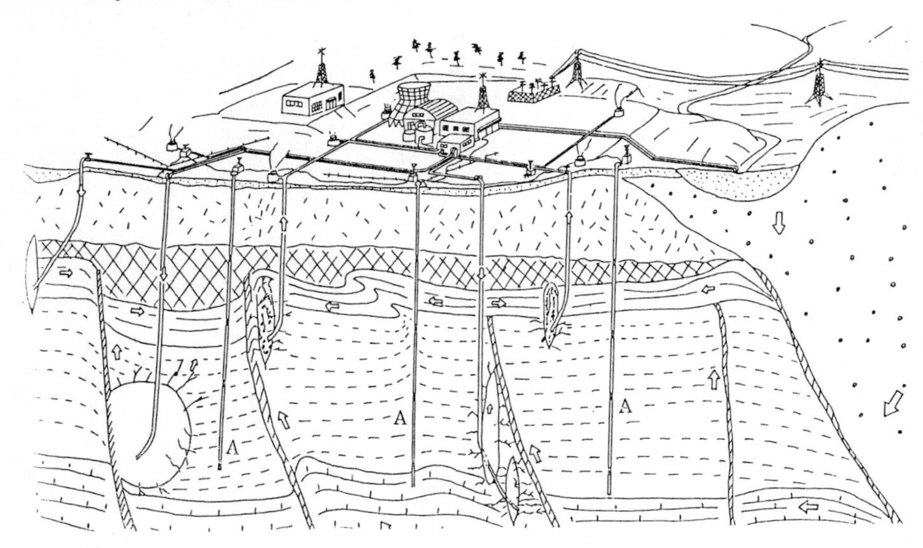

Fig. 3 A general concept on the function of cracks in a natural hydrothermal reservoir A:Crack monitoring well

The fracture mechanics approach is being employed to provide new insight into the fracture of rocks, especially, in designing fracture in a geothermal reservoir, and the evaluation of cracks in the earth crustal condition. Figure 4 shows a fracture evaluation procedure based on fracture mechanics, where a full understanding of three basic quantities, the fracture properties of rock mass under the crustal conditions, the location and size (geometry) of cracks, and the stress components including earth stress is prerequisite.

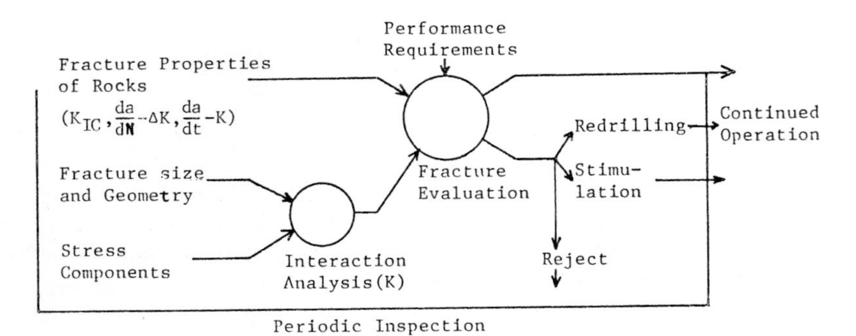

Fig. 4 Schematic representation of an idealized fracture control for geothermal reservoir

Growth Model of a Penny-Shaped Crack in Geothermal Reservoir [11]

Suppose that fracturing fluid is continuously injected from a wellbore into a fracture, and the fracture grows with respect to time. The fracture extends under the condition of the stress intensity factor K, being equal to the fracture toughness of the rock, K_c. As is shown in Figure 5, a vertical crack is considered. The radii of the crack and the inlet wellbore are denoted by R and R_o, respectively. The crack is maintained by a fluid pressure P where,

$$P(r, \theta) = P_o - \rho_f g r \cos\theta \qquad (R_o \le r \le R) \tag{1}$$

with P_o being the pressure at the wellbore, g the acceleration due to gravity and ρ_f the fluid density.

If the crack is absent, the compressive tectonic stress S (S>0) is acting on a vertical plane and can be divided into a constant stress (S_o) and a hydrostatic pressure. Hence,

$$S = S_o - K_a \rho_s g r \cos\theta \tag{2}$$

where ρ_s ($\rho_s > \rho_f$) is the density of the rock and K_a is the coefficient of active rock pressure. The stress intensity factor K is given by

$$K = \frac{\sqrt{2R}}{\pi} \left[P_o - S_o + \frac{2}{3} gR \left(K_a \rho_s - \rho_f \right) \cos\theta \right] \tag{3}$$

Usually $K_a \rho_s > \rho_f$, and stress intensity factor K is maximum at $\theta = 0°$ and minimum at $\theta = 180°$. Thus the fracture tends to extend upward with respect to time during injecting of the fluid.

The total mass of the fluid in the crack is expressed by,

$$Q = \frac{2\pi}{D} R^3 (P_o - S_o) \tag{4}$$

$$D = \frac{3\pi E}{8(1-\nu^2)\rho_f}, \qquad Q = \int_0^t q_o dt$$

where q_o is the injecting rate of the fluid (mass), E Young's modulus, and ν Poisson's ratio of the rock. It is assumed that the fracture criterion is given approximately by $\bar{K} = K_c$, where $\bar{K}$ is the stress intensity factor averaged along the crack tip circle, namely

$$\bar{K} = \frac{\sqrt{2R}}{\pi} (P_o - S_o) \tag{5}$$

(4) and (5) lead to,

$$Q = \frac{\sqrt{2}\pi^2}{D} R^{\frac{5}{2}} K_c \tag{6}$$

We can estimate the approximate radius of a crack from the injected total mass and density of the fluid, the fracture toughness, Young's modulus and Poisson's ratio of the rock, although further investigation is needed for establishing the fracture evaluation procedure of the geothermal reservoir more quantitatively.

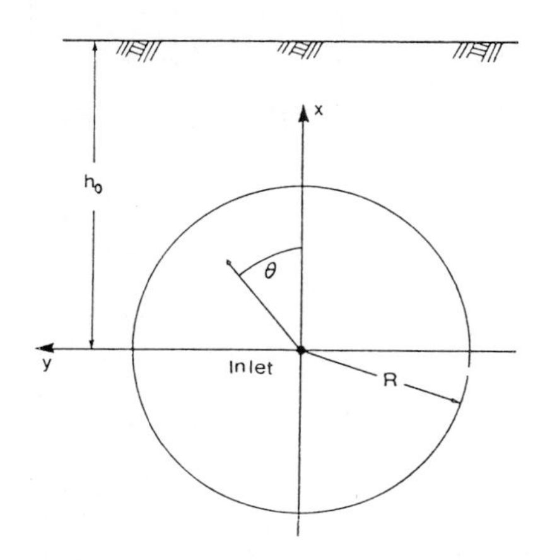

Fig. 5 Geometry and coordinate system for a penny-shaped crack with an inlet

Numerical Example

The SE-NE profile of the field crossing the center of the old caldera is shown in Fig. 6, together with the projection of the D-1 well passing through the plane from front to back in descending order (see also Figure 1). The location where the crack extension is expected from preceding drilling data, the circulation loss is marked with an arrow in the same figure. Since the increase of steam production had been observed in the preceding program in the fall of 1978, a fracture with some extension is thought to have been formed in the well. The formation there is composed of chert, with temperature 242°C. The confining pressure is estimated to be 18.5 MPa from the wellhead pressure-flow rate recorded during the fracturing. Thus, the representative value of fracture toughness for the wellbore chert is estimated as 0.33 MPa$\sqrt{m}$, taking the influence of temperature and confining pressure under a pressurized water circumstance into account. Details of the toughness evaluation will be described in another paper [14]. Young's modulus E and Poisson's ratio ν for chert were used as, $E=5.5\times10^4$ MPa, $\nu=0.2$ [15].

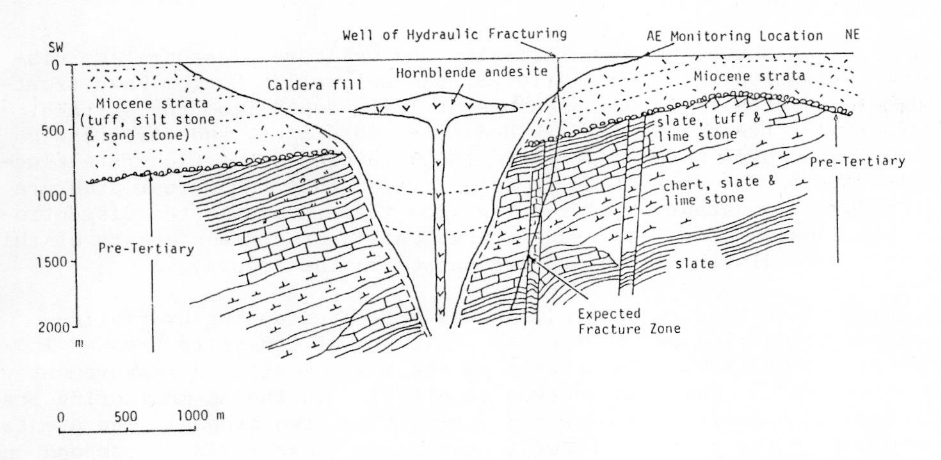

Fig. 6 A schematic view of sub-surface structure, fracturing well(D-1) and expected fracture zone in Nigorikawa field

Figure 7 shows the change of (P_O-S_O), the effective pressure at the wellbore, needed for crack propagation, versus crack radius R. Three values of fracture toughness are considered. The effective pressure (P_O-S_O) for a crack with large radius is very low compared with the water head pressure at the wellbore.

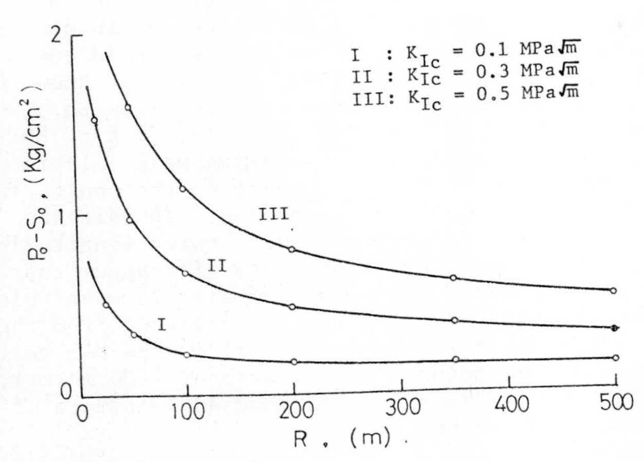

Fig. 7 The effective pressure needed for crack propagation in hydraulic fracturing, versus crack radius

Evaluation of a Fracture Extension

Figure 8 shows the flow rate-time and wellbore pressure-time diagram during hydraulic fracturing of the D-1 well. Bottom head treating pressure(BHTP) is the pressure at 1394m depth from the surface. A series of treatments was divided into two stages, and diverting agents were used between stages. Though pressure and flow rate fluctuate during the operation, it is estimated that the abrupt pressure drop during constant or increasing flow rate indicates the discontinuous extension of a subsurface crack. At least sixteen events of this type can be distinguished, as are marked in the figure.

In-situ AE measurement[16][17] was attempted during hydraulic fracturing carried out in the D-1 well, one of ten wells treated in the spring of 1980. The details of the instruments and measurement results are to appear in another paper[18]. In the AE monitoring station 43 AE signals were observed during these two stages. The events observed in the pressure-flowrate curve and those in AE correspond on the order of minutes, but not on the order of the seconds.

As shown in Fig. 7, the effective pressure (P_o-S_o) necessary for crack extension does not exceed 0.2 MPa for a crack larger than 20 meters. But, in Fig. 8, a pressure drop of the order of a few MPa is commonly observed. Probably a crack by hydraulic fracturing will discontinuously extend connecting large preexisting natural fractures scattered in the formations(Fig. 9). It is estimated that the volume change caused by the connection will be the reason of the abrupt wellhead pressure change observed in Fig. 8.

Since some part of the fracturing fluid fed from a wellhead will penetrate as a fluid loss into pores of formations or into secondary cracks formed by cooling rocks, the amount of lost fluid and the actual effective volume of the fracturing fluid is not known here. Figure 10 shows the change of crack radius derived from Eq. (6) versus the total fluid mass, where the cases of fluid loss 0 % to 80 % in the first stage of the D-1 well hydraulic fracturing were calculated. The change of the distance from an AE event source to the monitoring location is also shown in the figure for comparison. The distance was obtained from the difference of arrival time between longitudinal (P) and shear (S) waves for each event. The marks of scatter range are due to the influence of confining pressure on elastic wave velocity of the formations. The AE monitoring station deviated from the vertical of wellbore (-1394 m from the surface) in the D-1 well by about 30°, and the distance between them was about 1500 meters. By comparing these two data, it is estimated that 80 % or more of the injected fluid is lost in the fracturing.

Figure 11 also shows the change of crack radius versus the total mass of the fluid injected into the fracture when the fracture toughness changed in the range 0.1-0.5 MPa$\sqrt{m}$, assuming the fluid loss to be 80 %. Change of fracture toughness in this range may be due to the influence of high temperature pressurized water. The figures emphasize

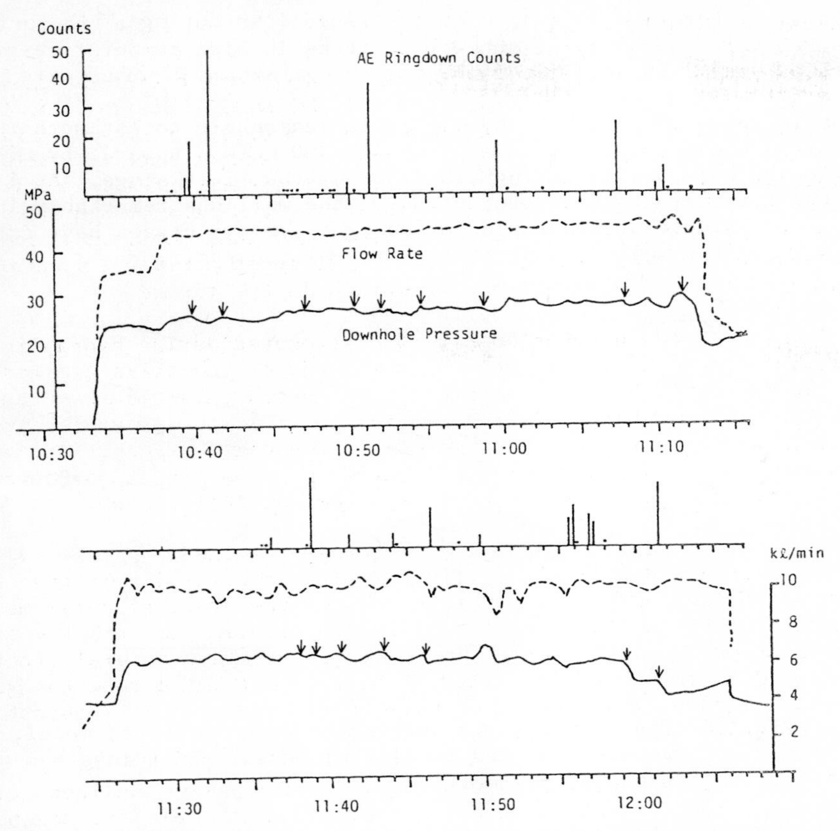

Fig. 8 Change of the flow rate and downhole pressure(BHTP) during hydraulic fracturing of D-1 well

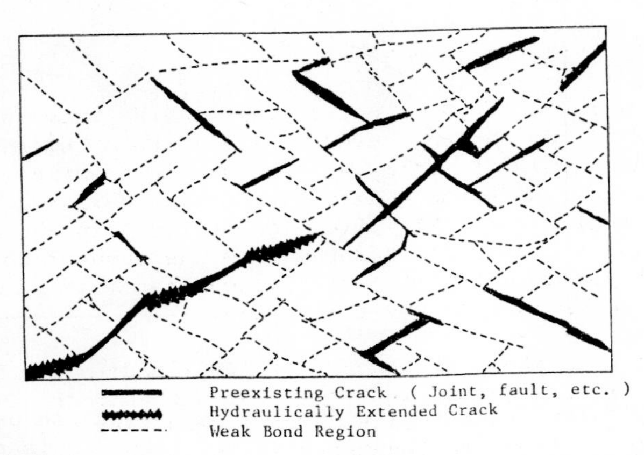

Fig. 9 Schematic view of crack extension in massive hydrofrac.

that an elaborate examination of the fluid loss during a fracturing and the fracture toughness of subsurface rocks in high temperature-pressurized water, is very important for an evaluation of crack extension.

As is shown in Figs. 10 and 11, it is reasonable to estimate the crack in the D-1 well to have grown 400-500 meters upward, by injecting about 400 tons of fracturing fluid during the first stage. As a result of the fracturing, steam production of the well has remarkably increased, indicating that fractures formed by these two stages have extended to connect the well to a promising natural reservoir.

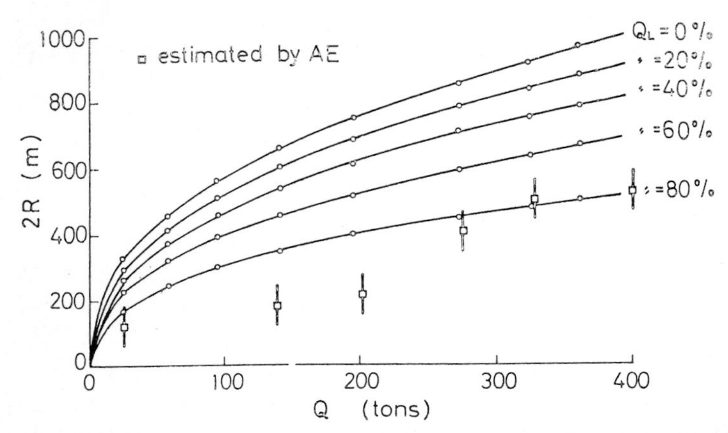

Fig. 10 The crack radius derived by a penny shaped model, versus injected total fluid mass. Fluid loss is taken as parameter

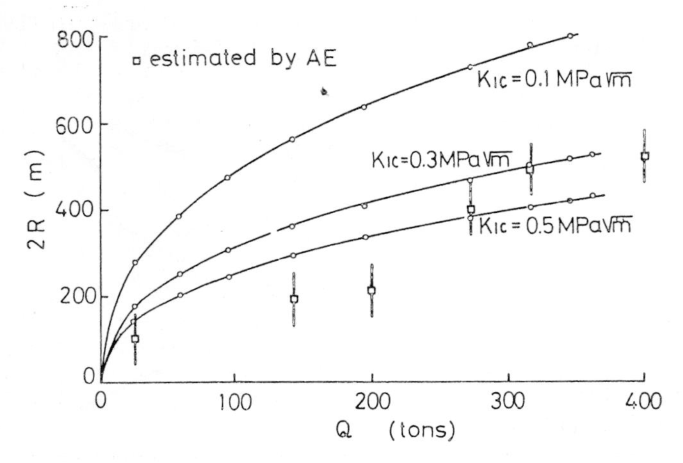

Fig. 11 The crack radius derived by a penny shaped model, versus injected total fluid mass. K_c is taken as parameter.

Among the ten treated wells, four wells (D-1, D-6, F-1a and F-6b) had been drilled into pre-Tertiary formations near the caldera wall and these responded to hydraulic fracturing by producing steam as well as by improving the injectivity. The six wells drilled in other areas excepting C-1a, responded only to an injecting test(Table 3). These results indicate that the natural fracture zone near the caldera wall is a potential reservoir and the connection of wells to the reservoir promises the improvement of steam or hot water productivity. Thus, the results of the fracturing experiment at Nigorikawa indicate that the elaborate preliminary examination of subsurface structure and the design of man-made fracture with the aid of fracture mechanics will greately increase the possibility of success of the stimulation, and will reduce total cost of geothermal energy development.

CONCLUDING REMARKS

The hydraulic fracturing for the stimulation of a geothermal reservoir performed at the Nigorikawa field is reviewed. In addition, the fracture extension in a representative well is evaluated by using a penny shaped crack growth model, where the radius of the fracture is derived from the amount of injected fluid and material constants of rocks under the assumption of an isotropic stress state. The result is compatible with other information obtained from AE measurement or geological structure.

The fact that the hydraulic fracturing was commercially useful in the stimulation of a hot water type geothermal reservoir will encourage future research development toward the further utilization of geothermal energy. The examination has just begun, and a main objective is the improvement of steam productivity at the present. Investigation of circulation mechanisms of reduced(reinjected) hot water is increasing in importance in order to lengthen the life of geothermal wells. Hydraulic fracturing will also serve in the design of injection systems, because hot water can spread into the ground through fractures with large surface area, while extracting heat from surrounding rocks. Extensive development of new technology in design, in monitoring and in controlling subsurface fractures is essential for the utilization of geothermal energy.

ACKNOWLEDGEMENT

The authors wish to express their thanks to Prof. M. Suzuki, Toyota Technological Institute and Prof. H. Abé, Department of Mechanical Engineering, Tohoku University for their encouragement throughout the course of the investigation. Thanks are also extended to Drs. H. Sekine, H. Niitsuma, K. Hayashi and T. Shoji, Tohoku University, for their helpful discussions. Financial support for a series of the work was provided by the Ministry of Education, Science and Culture under grand-in-aid 505514, 56045015 and 57045014.

REFERENCES

1) Katagiri, K., Ott, W. K. and Nutley, B. G., "Hydraulic Fracturing Aids Geothermal Field Development," World Oil, December (1980) pp. 75-88.

2) Hanold,R.J. and Morris,C.W.,"Induced Fractures···Well Stimulation Through Fracturing," Geothermal Resources Council's Spcial Report No. 12, Fractures in Geothermal Reservoirs, Honolulu, Aug. (1982) p. 1.

3) Bolton, R.S., "Fracturing in Wairakei Geothermal Field, New Zealand," Geothermal Resrouces Council's Special Report No. 12, Fractures in Geothermal Reservoirs, Honolulu, Aug. (1982) p. 85.

4) Stephen, P. D. and Allen, W. C., "Hydraulic Fracturing at the Baca Project, New Mexico," Geothermal Resources Council's Special Report No. 12, Fractures in Geothermal Resources, Honolulu, Aug. (1982) p. 139.

5) Abé, H., Takahashi, H. and Suyama, J., "Hydraulic Fracturing and Geothermal Energy Development in Japan," Proc. 1st Int. SME-AIME Fall Meeting (1982).

6) Suyama, J., "Present Status of Geothermal Development in Japan," Fourth Joint-Meeting MMIJ-AIME (1980) Tokyo, Technical Session-E-1, pp. 49-61.

7) Japan Geothermal Energy Association, "Geothermal Resources in Japan," Journal of the Japan Geothermal Energy Association, Special Issue No. 3 (1970) pp. 94-105.

8) Abé, H. and Takahashi, H., "Application of Fracture Mechanics to Geothermal Energy Extraction," Theoretical and Applied Mechanics, Vol. 29 (1981) p. 3.

9) Abé, H., Mura, T. and Keer, L. M., "Growth Rate of a Penny-Shaped Crack in Hydraulic Fracturing of Rocks," J. Geophys. Res., Vol. 81 (1976) pp. 5335-5340.

10) Abé, H., Keer, L. M. and Mura, T., "Growth Rate of a Penny-Shaped Crack in Hydraulic Fracturing of Rocks, 2," J. Geophys. Res., Vol. 81 (1976) pp. 6292-6298.

11) Abé, H., Keer, L. M. and Mura, T., "Theoretical Study of Hydraulically Fractured Penny-Shaped Cracks in Hot, Dry Rocks," International Journal for Numerical and Analytical Methods in Geomechanics," Vol. 3 (1979) pp. 79-96.

12) Hayashi, K. and Abé, H., "Opening of a Fault and Resulting Slip due to Injection of Fluid for the Extraction of Geothermal Heat," J. Geophys. Res., Vol. 87-B2 (1982) p. 1049.

13) Hashida, T., Yuda, S., Tamakawa, K. and Takahashi, H., "Determination of Fracture Toughness of Granitic Rock by Means of AE Technique," Proc. 6th International AE Symposium, NDI Japan, (1982) p. 78.

14) Wakabayashi, T., Hashida, T., Tamakawa, K., Takahashi, H. and Miyazaki, S., "Fracture Toughness Test of Granite in a Simulated Geothermal Reservoir Environment, Prep. of Annual Meeting of Geothermal Research Society of Japan, (1982) to be published.

15) Ogusa, S., Doboku Chishitsugaku (in Japanese), Asakura Book, Co., Ltd.

16) Nakatsuka, K., Niitsuma, H., Tamakawa, K., Takahashi, H., Abé, H. and Takanohashi, M., "*In-situ* Measurement of Extension of Hydraulically-Formed Fracture in Geothermal Well by Means of Acoustic Emission," Journal of Mining and Metallurgical Institute of Japan, Vol. 98 (1982) pp. 209-214.

17) Takahashi, H., Niitsuma, H., Tamakawa, K., Abé, H., Sato, R. and Suzuki, M., "Detection of Acoustic Emission during Hydraulic Fracture for Geothermal Energy Extraction," Proceedings of 5th AE Symposium, Tokyo, NDI Japan (1980) p. 443.

18) Niitsuma, H., Nakatsuka, K., Takahashi, H., Chubachi, N., Abé, H., Yokoyama, H. and Sato, R., "*In-situ* AE Measurement of Hydraulic Fracturing at Geothermal Field," Japan-U.S. Seminar on Hydraulic Fracturing and Geothermal Energy Development, Topic 3 (1982) p.227.

BREAKDOWN PRESSURE PREDICTION IN HYDRAULIC FRACTURING AT THE NIGORIKAWA GEOTHERMAL FIELD

by Ko Sato, Morihiko Takanohashi and Kunio Katagiri

Geologists and Geothermal general manager
Geothermal Development Division
Japan Metals and Chemicals Company Limited
Japan

ABSTRACT

Hydraulic fracturing stimulation in the Nigorikawa geothermal field made steam production and injectivity commercially feasible. A hydraulically induced fracture formation is estimated by a pressure drop, denoted as a breakdown. Breakdown pressure is provided by the use of an equation based on fracture mechanics and measurements of the fracture toughness, the instantaneous shut-in pressure and the length of the fracture. The values of the predicted breakdown pressure are nearly consistent with the ones of the treatment breakdown pressure after hydraulic fracturing treatment.

INTRODUCTION

The installed capacity of geothermal power stations in Japan has reached a total of 227.5 MWe (Fig. 1). A 50 MWe power station was put into operation in November, 1982, in the Nigorikawa (Mori) area by Dohnan Geothermal Energy Co., Ltd., a subsidiary of Japan Metals and Chemicals Co., Ltd. (JMC).

JMC has taken part in supplying steam for the Matsukawa, Kakkonda (the name of the power station constructed in the Takinoue area) and Mori (the name of the power station constructed in the Nigorikawa) geothermal power plants, and has drilled 29 production and 31 re-injection wells for three plants, up to now. To make wells commercial, sidetrack and hydraulic fracturing were performed for about 20 percent of all wells which did not intersect sufficient natural fractures. Table 1 shows JMC's hydraulic fracturing treatments and the fracturing tests conducted by the Project of the Government of Japan. For successful hydraulic fracturing, basic problems have been elucidated from studies of pump injection rate, pressure, required hydraulic horsepower, fluid properties, propping agents, retrievable packer and so forth. Normally a hydraulically induced fracture formation is estimated by a pressure drop, denoted as a breakdown. This paper deals with the breakdown pressure problem, that is, the comparison of predicted against measured breakdown pressure, by applying fracture mechanics.

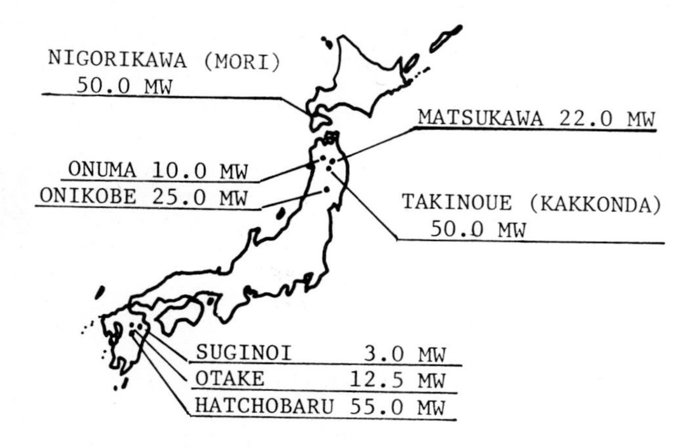

Fig. 1 JAPANESE GEOTHERMAL POWER STATIONS

Table 1 HYDRAULIC FRACTURING TREATMENTS AND TESTS IN JAPANESE GEOTHERMAL WELLS

JMC's hydraulic fracturing treatments		
Nigorikawa(Mori)		
Preliminary Treatment	spring 1978	3 wells
First Program	fall 1978	7 wells
Second Program	spring 1980	7 wells
Takinoue(Kakkonda)		
First Program	fall 1976	6 wells
Second Program	spring 1980	a well

Fracturing tests conducted by the Project of the Government	
Takinoue	1975 ～ 1976
Nigorikawa	1977
Yakedake	1978 ～
Hijiori	1981
Hatchobaru	1981
Sengan	1981 ～

METHOD TO PREDICT BREAKDOWN PRESSURE

During hydraulic fracturing treatments, breakdown pressure which is equal to the bottom hole treating pressure, is working to extend the subsurface fracture. To calculate wellhead pressure for any rate, it is necessary to determine the breakdown pressure, the hydrostatic pressure of the fluid, and the pipe friction pressure which is dependent upon pipe size, well depth and rheological properties of the fluids.

On the other hand, fracture mechanics states that a crack will advance whenever its stress intensity factor (K) reaches a critical value, fracture toughness (K_{1c}), assigning that the hydraulically created crack tip is in a state of a plane strain. When the pre-existing penny-shaped crack in the subsurface rock is under the condition of tectonic compressive stress (S), the fluid pressure in the crack (P) is written as follows :

$$P = S + (\pi / \sqrt{2R}) \cdot K \qquad (1)$$

where R is the radius of the penny-shaped crack. Therefore, breakdown pressure (Pb) is given by,

$$Pb \geqq S + (\pi / \sqrt{2R}) \cdot K_{1c} \qquad (2)$$

Pb can be solved to yield S, R and K_{1c}.

Tectonic Compressive Stress

If a hydraulic fracturing or injectivity test on a large scale was first performed, the tectonic compressive stress (S) can be obtained from the instantaneous shut-in pressure (Ps) of the pressure variation curve in time, because the minimum horizontal compressive stress is equal to Ps. In a hydraulic fracturing treatment for the first time, S will be estimated by the sum of the hydrostatic pressure and the pumping pressure during circulation losses or injectivity tests.

Fracture Toughness

Fracture toughness (K_{1c}) is determined by a fracture toughness test technique on rock samples. An acoustic emission technique monitored during the fracture toughness test is very useful for detecting the fracture toughness of rocks in the presence of not only airbut also pressurized water at elevated temperatures (1)(2).

Length of Fracture

If a hydraulic fracturing treatment was first performed, the length of the fracture (R) is obtained from the equations (2) and (3).

$$q_o = \frac{5\,\pi^{3/2}}{2\,D}\,K\,R^{3/2}\,\frac{dR}{dt} \qquad (3)$$

where q_o is the flow rate, D is the coefficient defined by Poisson's ratio, Young's modulus and density of water, and dR/dt is the velocity of crack propagation. On the contrary, difficulties for the prediction of R arise, if hydraulic fracturing treatment is conducted for the first time. However, the equations (3) and (5) suggest that R can be decided empirically by measuring the transmissibility around boreholes, lost circulation during drilling, and fracture toughness of rocks. Equations (4) and (5) led by Hele Shaw's flow which is an expression of the Navier-Stokes's law (the equation of motion of Hele Shaw's flow becomes $\partial P/\partial x = \mu \cdot \partial^2 V/\partial z^2$, where P is the fluid pressure, μ is the fluid viscosity and V is the fluid velocity) are as follows :

$$V = \frac{\bar{k}}{\mu} \cdot \frac{P_1 - P_2}{x} \qquad (4)$$

$$\bar{k} = \frac{m \quad ci \quad di^3}{12 \quad A \quad \alpha i} \qquad (5)$$

where $\bar{k}$ is the fracture permeability, m is the number of the fracture, ci is the height of the fracture, di is the width of the fracture, A is the unit area, and αi is the dimensionless length of the fracture.

The length of the fracture (αi) can be solved to yield $\bar{k}$, m, ci, di and A.

BREAKDOWN PRESSURE PREDICTION IN HYDRAULIC FRACTURING AT THE NIGORIKAWA GEOTHERMAL FIELD

Several wells were stimulated by hydraulic fracturing at the Nigorikawa geothermal field, Mori-machi in Hokkaido. Treatments were conducted in two separate projects with cooperation between Dohnan Geothermal Energy Co., Ltd. and Halliburton Manufacturing and Services Ltd. The details were discussed in separate papers (3)(4). Production and injection tests reveal that hydraulic fracturing stimulation at the Nigorikawa has been successful(Fig. 2 and Table 2). Stimulation makes the two steam production wells (D-1 and D-6) especially economical. All design for hydraulic fracturing of the first program was carried out by Halliburton Manufacturing and Services Ltd. JMC's engineers tried to anticipate the breakdown pressure using Equation (2) in the second program. It was decided to try four procedures.

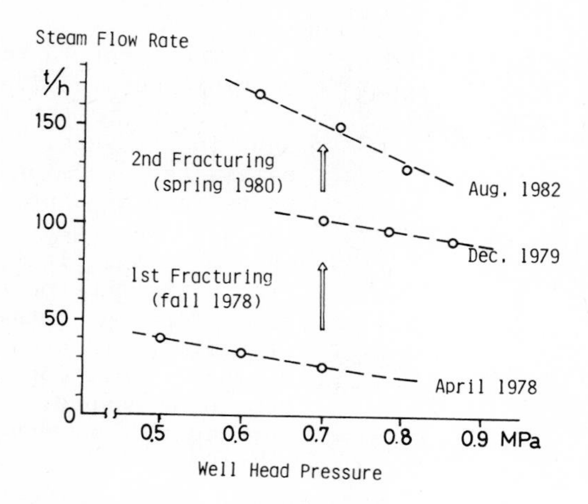

Fig. 2 CONFLUX PRODUCTION TESTS OF STEAM IN THE NIGORIKAWA FIELD (D-1 plus D-6)

Table 2 WELL INJECTIVITY TEST IN THE NIGORIKAWA FIELD

Well no.	Before Fracturing		After Fracturing		Fracturing Evaluation
	Flow Rate t/h	Pressure (MPa)	Flow Rate t/h	Pressure (MPa)	Up ↗
B-1	0	(3.55)	48	(3.55)	↗
B-2	69	(0.96)	100	(0.61)	↗
C-1a no.1	45	(0.51)	116	(0.41)	↗
no.2	100	(0.56)	92	(0.56)	
C-2	15	(1.52)	82	(1.32)	↗
D-2	23	(0.74)	48	(0.88)	
F-1a	28	(0.56)	57	(0.56)	↗
F-3c	Not Measured		110	(0.51)	
F-6b	0	(0.61)	64	(0.67)	↗

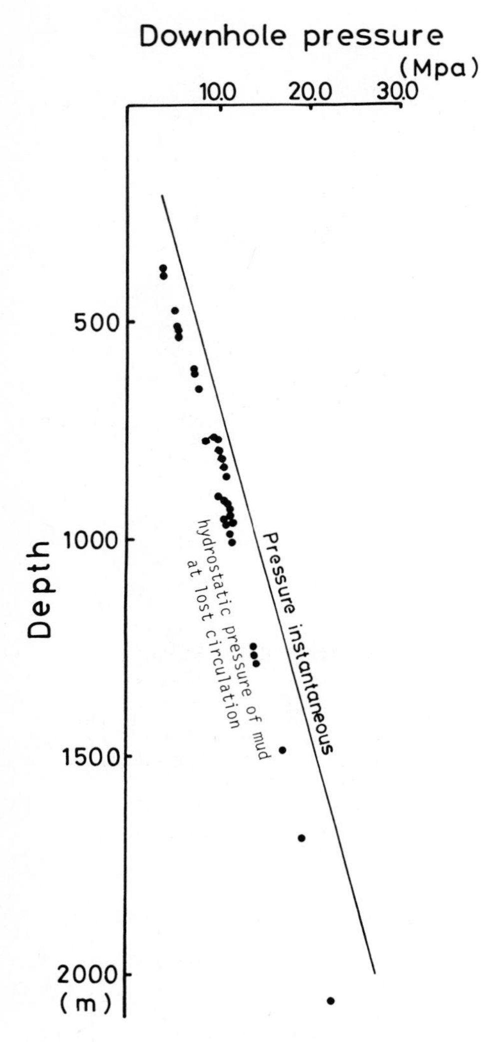

Fig. 3 RELATION BETWEEN PRESSURE INSTANTANEOUS AND HYDROSTATIC PRESSURE

Tectonic Compressive Stress

The first is to determine tectonic compressive stress, being equal to the instantaneous shut-in pressure, by two methods, namely the decision of the instantaneous shut-in pressure based on the pressure-time curves in the first hydraulic fracturing treatments (fall 1978) and the total pressure of hydrostatic pressure of mud during circulation losses and pumping pressure. Excellent agreement is observed between the two methods (Fig. 3).

Fracture Toughness

The second is to measure fracture toughness by fracture toughness tests on *in-situ* core samples. Experiments were excuted in which both notched three-point-bend configuration and notched cylindrical specimens were loaded to failure in the presence of pressurized water or elevated temperature (1)(2). The effect of increased temperature reduces K_{1AE} by about 30 percent in breakdown pressure. On the contrary, Schmidt and Huddle (5) pointed out that increased confining pressure raises fracture toughness in tests for jacketed specimens. It is reasonable for the confining pressure test to adopt the effect of the jacketed specimen test because hydraulic fracturing in the Nigorikawa field depends upon a large amount of high viscosity fluids, that is, gelling agents and sand.

Fig. 4 was based mainly on geological data and fracture toughness tests, making confining pressure and temperature corrections.

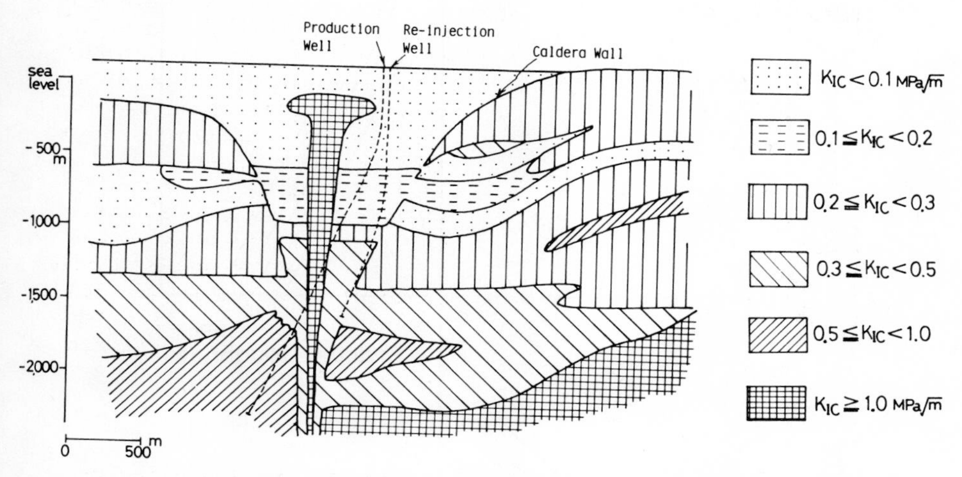

Fig. 4 PROFILE OF FRACTURE TOUGHNESS IN THE NIGORIKAWA GEOTHERMAL FIELD

Length of Fracture

The third is to decide the initial length of the fracture. Fracture toughness tests on *in-situ* core samples of the Nigorikawa field make clear that the fracture toughness is in proportion to the lost circulation during drilling. Fig. 5 represents the empirical relationship among fracture toughness, lost circulation factor, transmissibility, and dimensionless initial length of fracture (αi).

Lost circulation factor is newly defined by the following equation.

$$\sum_{i=1}^{n} \left\{ \frac{\text{Lost circulation volume (kl/h)} \times \text{Coefficient of lost circulation frequency}}{\text{Ratio of each rock facies to total rock facies}} \right\} \qquad (6)$$

where the coefficient of lost circulation frequency is divided into three degrees of 1 (seeping loss), 10 (partial loss) and 100 (complete loss), and n is the number of lost circulation.

Transmissibility which is defined by multiplication permeability by effective thickness of strata, is determined by production and injectivity tests of geothermal wells.

Equation (5) applied to a borehole, (Fig. 6), becomes

$$D \cdot \bar{k} = \frac{m \; ci \; di^3}{12 \; rw \; \alpha i} \qquad (7)$$

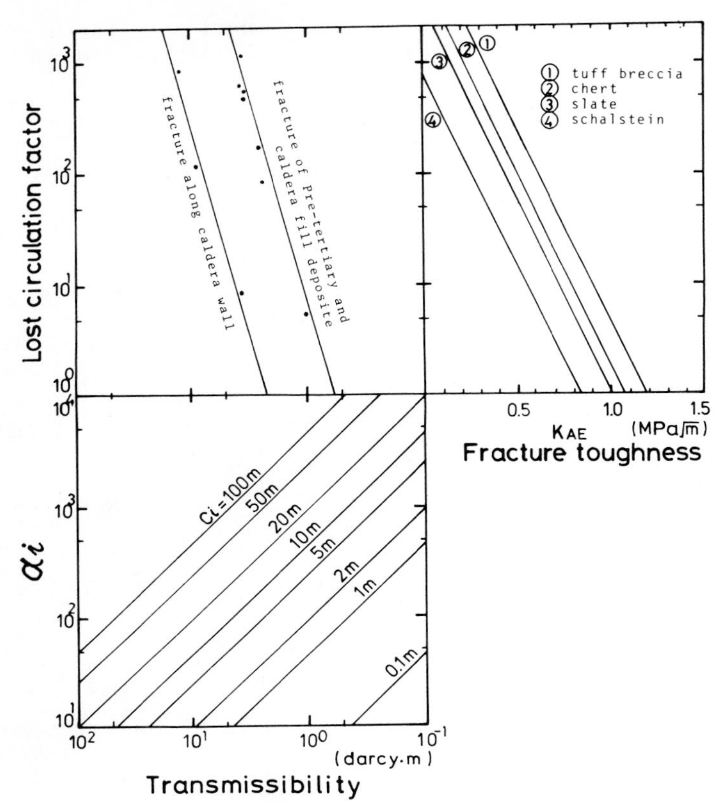

Fig. 5 CHART SHOWING DIMENSIONLESS INITIAL LENGTH OF FRACTURE (αi) DECIDED BY FRACTURE TOUGHNESS, LOST CIRCULATION FACTOR AND TRANSMISSIBILITY

where D is the effective thickness of strata (therefore $D \cdot \bar{k}$ is the transmissibility) and rw is the well bore radius.

The initial length of the fracture was calculated from Equations (7) and (8).

$$\alpha i = \frac{li}{xi} \qquad (8)$$

where xi is the length of the fracture and li is the maximum initial length of the fracture.

Breakdown Pressure Calculation

The last is to calculate the breakdown pressure before fracturing treatments in accordance with the above-mentioned procedures. In the Nigorikawa field, di measured from fissures of core samples

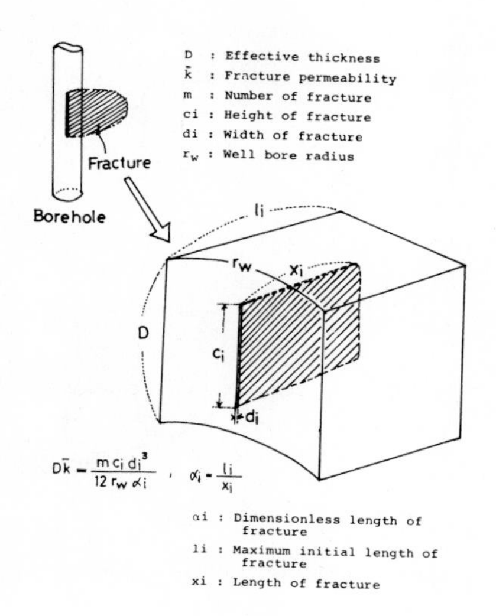

FIG. 6 DETERMINATION OF LENGTH OF FRACTURE APPLIED TO A BOREHOLE

avarages 0.05 cm, m is 1 and rw is 22 cm. The height of the fracture (ci) is estimated by geophysical logging and lost circulation. Fig. 7 shows an example of determination of the height of the fracture in F-1a well. Table 3 shows the results of calculation for 5 wells applicable to fracturing. In this case, the depth of the fracture was determined by the depth of a large amount of circulation losses right below the casing. The exterior boundary radius obtained from production or injectivity tests gives li ≒ 200 m.

The values for the predicted breakdown pressure are nearly consistent with the ones of treatment breakdown pressure after fracturing treatments.

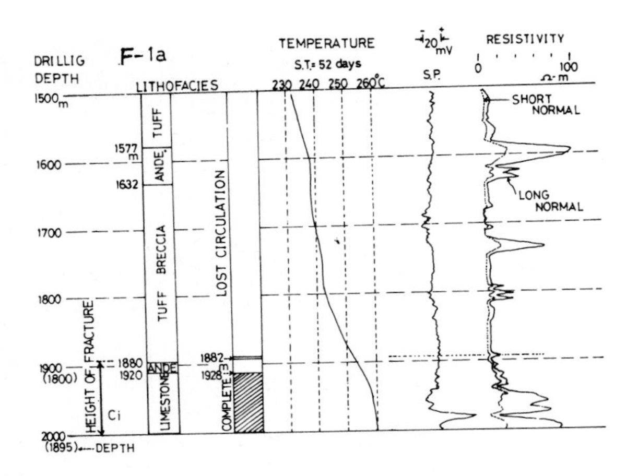

FIG. 7 DETERMINATION OF HEIGHT OF FRACTURE

Table 3 PREDICTED AGAINST MEASURED BREAKDOWN PRESSURE

Well no.	D-1	D-6	F-1a	F-3c	F-6b
Depth of fracture (m)	1935	1245	1788	744	1622
Rock facies	chert	chert	tuff breccia	schal-stein	tuff breccia
Pressure instantaneous (MPa)	19.11	17.25	24.21	10.78	22.44
Height of fracture (m)[ci]	10	20	100	50	100
Dimensionless initial length of fracture [αi]	145	315	330	1300	1300
Initial length of fracture (m) [xi=R]	1.38	0.63	0.47	0.15	0.15
Fracture toughness ($MPa\sqrt{m}$)	0.28	0.27	0.34	0.28	0.32
Predicted breakdown pressure (MPa)	19.63	18.01	25.31	12.39	24.28
Measured breakdown pressure (MPa)	23.87	22.84	27.33	21.74	23.63

CONCLUSIONS

Breakdown pressure during hydraulic fracturing treatments is predicted by the use of the equation based on fracture mechanics. Breakdown pressure can be solved to yield the tectonic compressive stress, the fracture toughness, and the length of the fracture. Tectonic compressive stress is estimated from the instantaneous shut-in pressure during fracturing or injectivity tests. Fracture toughness is determined by the fracture toughness tests depending upon confining pressure and temperature elevation in rock samples. Length of fracture is decided by the empirical relation charts among fracture toughness, lost circulation factor, transmissibility, and Hele Shaw's flow.

ACKNOWLEDGMENTS

The authors are deeply indebted to Professors Hiroyuki Abé and Hideaki Takahashi of the Department of Mechanical Engineering, Tohoku University, for their kind guidance concerning the present work. Thanks are also due to Associate Professor Hiroaki Niitsuma of the Department of Electrical Engineering, the same University, for his kind suggestion in the acoustic emission work.

REFERENCES

1. Yuda, S., "Evaluation of Fracture Toughness of Rocks by Acoustic Emission," 1981, Master Thesis, Tohoku University.

2. Abé, H., Takahashi, H., and Suyama, J., "Hydraulic Fracturing and Geothermal Energy Development in Japan," Sept. 1982, Mineral Resources in the Pacific Rim, Honolulu, Hawaii.

3. Katagiri, K., Ott, W. K., and Nutley, B. G., "Hydraulic Fracturing Aids Geothermal Field Development," World Oil, Vol. 191, No. 7, Dec. 1980, pp. 75-88.

4. Nakatsuka, K., Niitsuma, H., Tamakawa, K., Takahashi, H., Abé H., and Takanohashi, M., "*In-situ* Measurement of the Extension of Hydraulically-formed Fracture in Geothermal Well by Means of Acoustic Emission," Jour. Mining and Metallurgical Inst. Japan, Vol. 98, No. 1129, Mar. 1982, pp. 209-214.

5. Schmidt, R. A., and Huddle, C. W., "Effect of Confining Pressure on Fracture Toughness of Indiana Limestone," Inst. J. Rock Mech. Min. Sci. & Geomech. Abstr. Vol. 14, 1977, pp. 289-293.

THE INFLUENCE OF SUBSURFACE CONDITIONS ON HYDRAULICALLY INDUCED FRACTURES IN DEEP ROCKS

by Ahmed S. Abou-Sayed

President
Salt Lake City, Utah

Terra Tek International
Salt Lake City, Utah

ABSTRACT

A discussion is presented on the influence of subsurface conditions such as, stress, rock properties and pre-existing geologic structures on the shape and azimuth of hydraulically induced fractures in deep rocks. A systematic approach to optimum fracture design would require a combined field, laboratory and analytical effort to determine these conditions. Detailed field testing involves in situ stress measurements, fracture orientation determination and monitoring of flow rate and bottom hole pressure during the stimulation treatment. Laboratory tests would provide the mechanical and physical properties of the rock as well as its sensitivity to interaction with its pore fluid and the fracturing fluids used in the treatment. All information is to be synthesized into the fracture design utilizing computer simulators.

INTRODUCTION

The study presented by Fast, et al. [1] following their hydraulic fracturing experience at AMOCO's Wattenberg Field was one of the first papers to critically outline some of the basic important parameters controlling the success of massive hydraulic fracturing treatment (MHF) of underground strata for enhanced resources recovery.

The conclusions of Fast, et al. [1] from their experience at the Wattenberg Field are still applicable today. The general portion of those conclusions are as follows:

1. MHF is operational; however, careful pre-job planning is mandatory.

2. MHF is effective in increasing production rates and recoverable reserves when applied in appropriate areas.

3. As there are some extremely low permeability formations which may be noncandidates for MHF, pre-testing to determine product of permeability and formation height (kh) and reservoir pressure is vital.

4. Continued research and operational learning are still required to further develop MHF.

Since publication of Fast, et al. [1], the data and tests needed for stimulation design have become more clearly defined. Simonson, et al. [2], first defined the conditions needed for containment and which since have been detailed further by Cleary [3]. Ahmed, et al. [4] described tests to evaluate fracturing fluid damage to the reservoir matrix permeability and fracture conductivity. Meanwhile, fracturing fluid and proppant developments have greatly enhanced the opportunity of improving stimulation design.

Developments in hydraulic fracturing have come from advances in fracturing fluids, proppants, and improved understanding of the mechanisms controlling containment. Optimization of both the treatment size and fracturing technique depends on the individual well configuration as well as fracturing fluid and proppant selected. Pay sand quality (permeability and thickness), location of natural barriers to vertical fracture growth, location of aquifers and gross length of interval to be treated will govern the type and size of the treatment. The fracture treatment must be aimed at achieving a geometry as dictated by reservoir quality and controlling the fracture to meet this goal in view of the design constraints imposed by the reservoir formation and the formations surrounding the pay sand.

Containment: Several authors [5-9] have provided production increase estimates as a function of the reservoir parameters, fracture area in contact with the reservoir and fracture conductivity. McGuire and Sikora [5] eloquently summarized the parametric relations for production increase assuming pseudo-steady state; see Figure 1. Fracture geometry remains the elusive parameter in fracture design.

The first attempt at specifying parameters controlling fracture geometry was made by Simonson, et al. [2] Three factors were considered:

1. Elastic moduli contrast between the reservoir formation and bounding barriers.

2. In situ stress contract between the reservoir formation and bounding barriers.

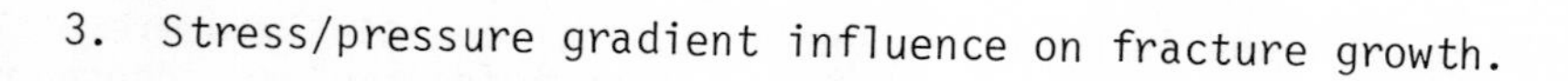
3. Stress/pressure gradient influence on fracture growth.

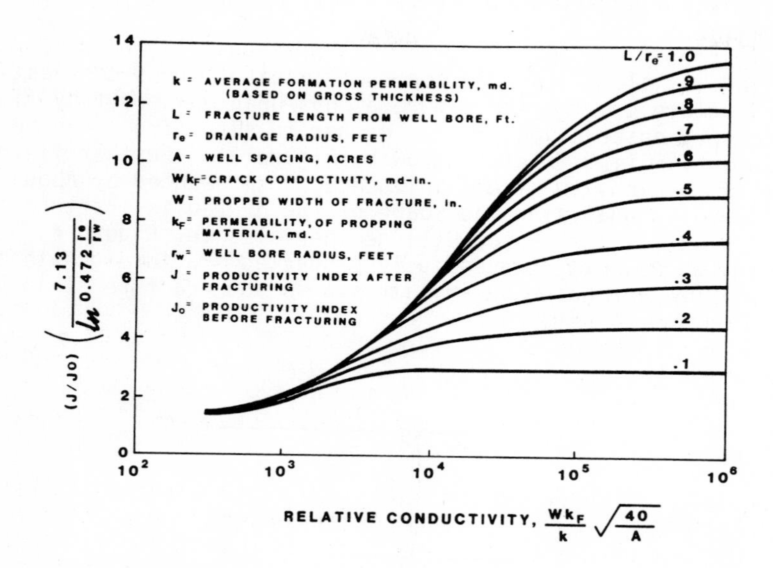

Figure 1. Productivity Increase from Fracturing Assuming Pseudo-Steady State (after McGuire and Sikora [5]).

Higher elastic moduli and higher in situ stress in the barriers were shown to retard fracture growth. Formations with higher moduli were considered to have the potential of being perfect barriers; see Figure 2a. As Cleary [3] has since shown the dimensions of the crack tip process zone prevents full development of a low stress intensity factor ahead of crack tip as it approaches the interface with the higher modulus formation. That is, the higher moduli formation will not be a barrier to fracture growth. This is supported by laboratory and field observations [10, 11]. Cleary [3] did conclude, however, that a higher moduli formation would present greatest resistance to continued fracture growth - still a matter of controversy [11]. AMOCO (Veatch [12]) has reported field observations to support Cleary's [3] conclusion. Veatch [12] reported "effective" fracture containment for barriers with moduli a factor of three higher than the pay formation.

The onset of slippage between the pay zone and the barrier can lead to crack fracture containment. Under these circumstances, a shear-like cracking occurs along the boundary, Figure 2b. The fracture may reinitiate, however, unless the frictional grip across the interface is too weak or the barrier formation too elastically compliant [3, 13].

Laboratory [11] and in-mine [14] observations (especially where a soft, thin shale layer exists at the roof) provide ample evidence of this behavior.

Simonson, et al. [2] prediction that in situ stress contrast would inhibit fracture growth (see Figure 3) has been borne out by field and laboratory observations [10]. Higher in situ stress can be very effective in limiting fracture growth (Figure 3). Further field evidence of the in situ stress influence was presented by Abou-Sayed et al. [15]. In analyzing the shape of the hydraulic fracture in massive granite at the Fenton Hill geothermal site, Figure 4, Abou-Sayed, et al. [15] successfully predicted fracture growth and final configuration based on in situ stress and fracture fluid pressure gradients.

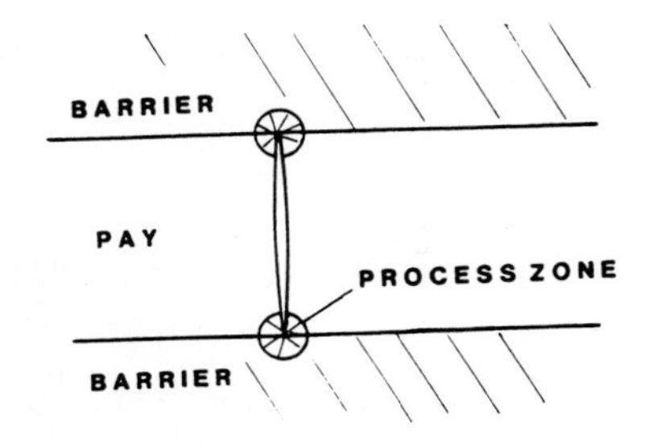

Figure 2a Uncontained Fracture

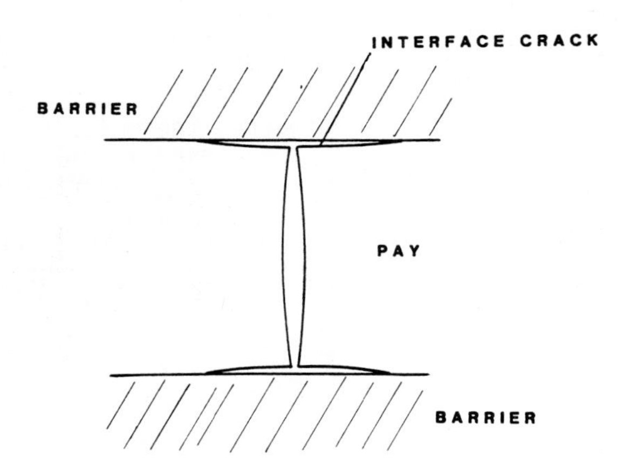

Figure 2b. Shear-like Cracking at the Interface (After Clifton [2]).

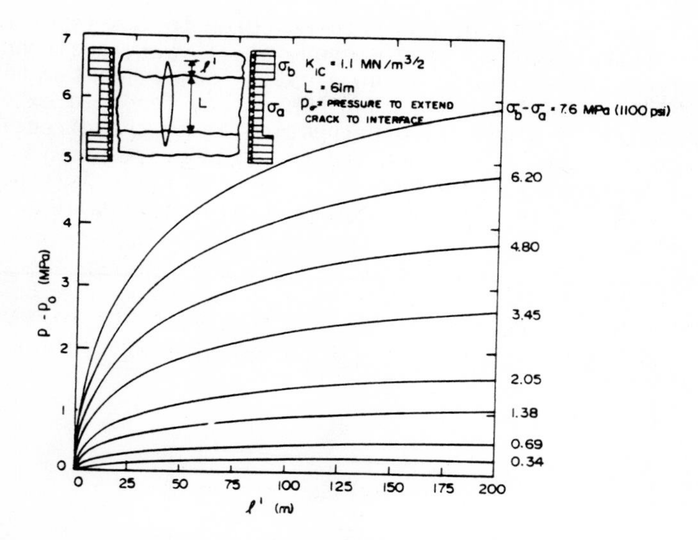

Figure 3. Plot of Excess Pressure $p - p_o$ versus Distance Advanced into the Region of High Stress. The In-Situ Stress Difference $\sigma_b - \sigma_a$ is a Parameter. (after Simonson et al. [2]).

Nolte and Smith [16] and Nolte [17] made a significant development to interpreting downhole pressure records assimilated during the fracture treatment. Based on the data interpretation they were able to assess the phase during which the fracture was confined within the pay zone and to quantify the pressure which leads to undesirable pressure increase (screen out) or undesirable height growth. With this interpretation, imporved designs can be achieved for future treatment in a given pay zone.

Sufficient insight of the dominant rock mechanics parameters has been gained over the last five years to initiate more comprehensive programs aimed at predicting fracture geometry. A number of efforts are underway; for example, Clifton and Abou-Sayed, [18, 19] Settari [20] Advani [21] and Cleary [22]. With input from field and laboratory measurements these will provide better design guidelines for fracturing reservoirs close to aquifers, limited entry methods for multiple sands (both continuous and lenticular), as well as lead to more economic treatments of lower permeability sands.

Fracture Direction:

Daneshy [23, 24] investigated fracture initiated at an arbitrary direction at the wellbore. Laboratory model simulation tests revealed that these fractures change their orientation as they extend away from the wellbore until they become perpendicular to the least compressive in situ principal stress. Abou-Sayed, et al. [25] analyzed the problem for fracture extension in an elastic, homogeneous material. They concluded that a sufficiently long open crack tends to extend in a direction which is more nearly perpendicular to the direction of minimum compressive stress than the existing crack provided the principal in situ stresses are unequal. This prediction is consistent with Daneshy's [23, 24] laboratory results and Warpinski, et al. [10] field tests.

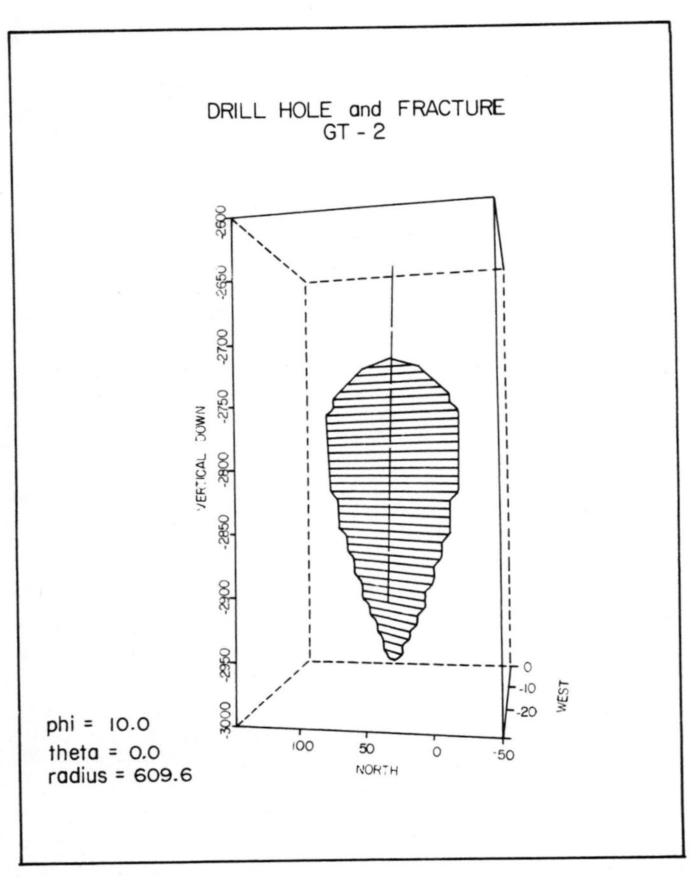

Figure 4. Three-Dimensional Perspective of the Fracture Originating in GT-2 (At 10° to the Fracture Surface at 2.81 Km).

Smith [26] has indicated the importance of knowing the fracture azimuth to reservoir planning in order to avoid direct interference between wells. Analysis of surface tiltmeters (resolution of 5 X 10^{-9} radians) arrayed at the surface surrounding the well to be stimulated has predicted fracture direction in excellent agreement with fracture azimuth verified by coring in a shallow well experiment [27]. These positive results have been confirmed in several successful tests.

In Situ Stress Determination in Wells:

Delineation of the effects of in situ stress contrasts on fracture containment requires the field measurement of the minimum horizontal in situ stress of the pay zone and adjacent layers. Small hydraulic fracturing tests are presently the most practical method with which to determine the state of stress acting at great depth within the earth.

The hydrofracturing technique for in situ stress determination has been used extensively in the past four years to measure the horizontal in situ stress field in several tight reservoirs. A total of more than fifteen wells (both open hole and cased hole) have been tested and measurements performed in up to four horizons in each well. Analysis of fracture containment and recommendations for pump schedules were deduced from the field measured in situ stress contrast. Furthermore, the availability of recently developed fracture geometry codes utilizing these measurements will enhance the cost/benefit aspects of in situ stress determination period to well fracturing.

If field measurements of in situ stress are not feasible, both 3-D sonic logs and one-dimensional strain tests may be utilized to infer the values of in situ stresses within the pay and surrounding formations. These techniques, however, should only be utilized where direct in situ measurement is impossible or where there is a need to accumulate a large data base. In the latter case, spot checks utilizing field tests of the predicted stress magnitude is strongly recommended.

Finally, determination of fracture azimuth is essential in planning well locations and fracture lengths in any large scale development. Two techniques are commonly utilized in the field, the earth tidal method and the tilt-meter method. The latter measures the amount of surface tilt during fracturing and is inappropriate for offshore application. The former technique utilizes the earth tide effects on changing downhole pressure in an inactive well to determine azimuth of high permeability channels (fracture) connected to the well. The technique is applicable offshore with ocean tide corrections. It has the advantage of postfracture analysis and requires only a wireline downhole pressure gage and surface data acquisition. It has been applied to reservoirs in the North American continent,Hanson and Owen [28].

EVALUATION OF IN SITU STRESS USING FIELD HYDRAULIC FRACTURING TECHNIQUE

At the present time the only technique which has been demonstrated to be viable for in situ stress determinations at depths greater than a few hundred meters is small scale hydraulic fracturing. The technique is well understood and frequently demonstrated for both cased and open operations,Haimson [29], Bredehoeft et al. [30], Zoback and Pollard [31], Voegele et al. [32]. Other techniques, utilizing strain relief or velocity birefringence, are proposed from time to time, but to date none of these techniques have been successfully performed in a deep hole. In addition, only the "mini" hydraulic fracturing technique is capable of being performed in a cased well, as usually required.

To determine the feasibility of using the small scale hydraulic fracturing technique as a means of in situ stress determination in cased wellbores, a series of laboratory tests,Voegel, et al. [32], were performed on specimens of hydrostone, loaded in polyaxial compression. API standard 7-inch O.D. 23 lb/ft casing was simulated using 28.6 millimeter X 26 millimeter steel tubing which was centered in the model before casting. This casing was perforated using a small right angle drive drill. It was found that the fractures always initiated through the perforations even when the perforations were oriented at 90 degrees to the maximum stress. It was also found that these fractures reoriented themselves as they extended and became normal to the minimum principal stress component.

By perforating the wellbore with a helical pattern there is a very good chance that one perforation will be oriented in or very near to the direction of the maximum horizontal stress and as such the fracture would initiate in the proper orientation. Results of tests performed with the helically perforated specimens show that this is indeed the case. For the most part the fractures were planar and oriented in the proper direction by the time they had propagated to several wellbore diameters into the rock mass.

For those cases where the preferred fracture orientation was vertical, it was found that the instantaneous shut in pressure (ISIP) was a reliable estimate of the minimum applied horizontal stress when analysis techniques based upon minimum reopening and shut in pressures were utilized. The estimation of the maximum horizontal stress was not so straightforward. It must be borne in mind that the actual presence of casing in a wellbore coupled with the cementing pressure history to which the casing has been subjected presents a marked deviation from the standard open hole situation upon which the interpretive theory is based. Nonetheless, it was found that for vertical fractures a reasonable estimate of the maximum horizontal stress could be estimated from a formulation incorporating the minimum reopening and shut in pressures. The laboratory findings have been confirmed by various field experiments conducted by Terra Tek where horizontal in situ stress has been measured in cased wells and in open holes in several proprietary wells within the North American Continent.

The general equation relating breakdown pressure to the in situ stress follows from Hubbert and Willis [33], Fairhurst [34], or

Abou-Sayed et al. [25], and is given in modified form by:

$$P_b = 3\sigma_{min} - \sigma_{max} - P_o + T(\ell, D, k) \quad (1)$$

where σ_{min}, σ_{max} are the minimum and maximum principal stresses perpendicular to the borehole, respectively, P_o is the fluid pore pressure in the rock, and P_b is the pressure required to initiate the hydraulic fracture breakdown pressure. The last term of the equation incorporates the hole size (D), the flaw size (ℓ), the formation permeability (k), and a measure of the tensile strength (T) of the rock. The above result assumes either an impermeable rockmass to the borehole, or injection rates high enough to minimize the effects of permeability.

The second and more important equation is:

$$P_s = \sigma_{min} \quad (2)$$

where P_s, the instantaneous shut in pressure (ISIP), is the quasi-static equilibrium pressure required to just keep the hydraulic fracture open without further propagation or closure. In the derivation of equations (1) and (2) it is assumed that one of the principal stresses aligns with the borehole. In this case, the problem reduces to the two-dimensional plane strain situation of a circular hole in an infinite plane. The fracture aligns with the borehole axis and leaves the borehole in the direction of σ_{max} so that the normal to the fracture is in the direction of σ_{min}. Recently Nolte [17], and Nolte and Smith [16], provided a procedure for estimation of the minimum in situ stress pressure decline after instantaneous shut in.

The mini-frac test follows procedures similar to those used in all hydraulic fracturing. In the open hole configuration the test section is isolated by a straddle packer; in the cased hole, as for this test, by a single positrieve packer and a retrievable bridge plug. Fracturing fluid is injected into the section sealed by the packers. Pressurization of the section is monitored by surface and downhole pressure gages. Pressure is raised slowly until the breakdown pressure (P_b) is reached and a fracture is formed in the rock surrounding the hole. Providing the flow rate remains approximately constant after breakdown, the pressure will drop to a constant level, known as the extension pressure, (P_f), as the fracture propagates into the formation. When the flow is stopped by shutting in the well, the entire system will come to an equilibrium where the in situ stress acting to close the fracture equals the fluid pressure; this equilibrium pressure is the instantaneous shut in pressure (P_s), and is approximately equal to the minimum transferred in situ stress (σ_{min}).

In an open hole configuration, fracture orientation is determined by an impression packer. A soft rubber membrane on the outside of the packer is extruded into the fracture trace by pressure, thereby forming a permanent trace of the fracture. A downhole compass correlated with a reference make on the outside of the packer provides absolute orientation of the fracture. In tests within cased holes the fracture orientation may be estimated from temperature surveys, Cooke [35],or from radioactive tracing methods.

Experimental results indicate that within a few minutes after break down (P_b), the fracture will be oriented perpendicular to the regional minimum stress, (σ_{min}). This effect has been demonstrated in several laboratories, including the large scale block tests performed by Terra Tek. Since the bulk of the fracture area following 1/2 to 2 minutes slow extension is oriented normal to σ_{min}, the instantaneous shut in pressure will give a good indication of σ_{min}. This minimum principal stress is the key parameter for optimal design of massive hydraulic fracturing of the well.

Several investigators have shown theoretically,Simonson et al. [2], Van Eckelen [36], Abou-Sayed [37],and experimentally,Warpinski et al. [38], Teufel and Clark [39]that the differences in the minimum horizontal in situ stresses between a productive formation and potential barriers are sufficient to inhibit the vertical growth of hydraulic fractures. The magnitude of this stress difference can be as low as 300-450 PSI (2-3 MPa),Warpinski et al. [38].

The way to optimize fracturing treatments and maximize the ratio of productive fracture area to volume of fluid pumped is to ensure that hydraulic fractures are effectively contained within the formation being treated. This requires that accurate determinations be made of the minimum in situ stresses acting in the producing formation and in potential barriers and to limit the bottomhole treating pressure to those values.

Errors in estimating the minimum stresses in either the producing formation or the potential barriers could result in overpressurization of the fracture and undesirable vertical migration of the fracture. However, the reliability of the instantaneous shut in pressure as an indicator of σ_{min} has been established through careful comparisons with overcoring measurements and carefully controlled laboratory block tests,Gronseth and Detournay [40], Haimson [41], Voegele et al. [32]. Numerous successful field applications demonstrate the value of these techniques.

MHF Orientation:

Hydraulic fractures, induced in an isotropic and homogeneous medium by fluid pressure confined in a small section of a borehole, tend to follow planes perpendicular to the least principal stress,Hubbert and Willis [33, 42]. These hydraulic fractures may be vertical, horizontal, or inclined to the axis of the borehole depending on the orientation and magnitude of the prevailing state of stress. For example, in deep sedimentary (tectonically relaxed) basins where the greatest principal stress is vertical and increases in magnitude with the weight of the overburden, and where the intermediate and least principal stresses are horizontal, the induced hydraulic fracture is vertical.

The orientation or azimuth of the intermediate and least principal stresses and, therefore, the direction of the vertical fracture in anisotropic and inhomogeneous rock is not readily established without additional geological, structural or mechanical evidence. In sedimentary basins characterized by normal faults, for instance, and typical of the Gulf Coast and Mid-Continent regions, the direction of the induced, vertical hydraulic fracture should parallel the strike of normal faults and associated natural fractures,Hubbert and Willis [33, 42], Hubbert [43], Bredehoeft et al. [30]. On the other hand, in basins where high-angle reverse faults (up-thrusts) are the dominant structural features, the prevailing stress state is such that the greatest and least principal stresses are horizontal and the intermediate principal stress is vertical,Min [44]. This stress condition is typical in many of the Rocky Mountain Basins,Stearns [45]. The direction of the least stress component is parallel to the strike of the high-angle reverse faults and the azimuth of the induced, vertical hydraulic fracture should, consequently, be perpendicular to the high-angle faults,Smith et al. [46].

The apparent influence of the regional structural features and associated stress states in sedimentary basins on the preferred direction of induced hydraulic fractures is reflected on small scales as well. Good statistical correlations have been established between the directional tendencies (anisotropies) of the tensile and shear strengths, sonic velocities and elastic moduli, and permeabilities measure on oriented cores and the preferred directions of the fabric elements or structural defects such as microfractures, fractures (joints), grain boundaries and dimensions of individual grains, McWilliams [47], Friedman and Logan [48], Friedman and Bur [49], Komar [50]. Similar statistical correlations have been demonstrated between the anisotropic mechanical properties and fabric of oriented cores and the preferred orientation of induced hydrualic fractures determined by impression-packer surveys in open holes,Overbey [51], and remote geophysical surveys of changes in the surface electric potential field and in the seismic and acoustic radiation field,Power et al. [52], Smith et al. [46].

The question of whether the strong directional properties of rock, particularly sedimentary rock, or the ambient stress state governs the preferred orientation of induced hydraulic fractures is at present difficult to answer. In regions that are presently tectonically relaxed and relatively free of seismic or earthquake activity, the prevailing stress state tends toward near hydrostatic or lithostatic conditions,Friedman and Heard [53]. A great number of fluid-injection tests in these areas have shown that the down hole pressure necessary to maintain or extend the induced hydraulic fracture (fracture gradient) averages about 0.8 of the overburden pressure and that the ratio of the vertical stress to the average horizontal stress is very close to unity, Hubbert [43], Haimson [54], Bredehoeft et al. [30]. Under these uniform stress conditions, anisotropy in the sedimentary rocks may well dominate the preferred orientation of induced hydraulic fractures, provided that the dominant anisotropy direction lies very near the principal plane containing both the greatest and intermediate stress components, Hubbert [43].

Based on these observations, the juxtaposition of the anisotropy and the prevailing stress field has been tentatively assumed to be the rule rather than the exception in many instances where hydraulic injection treatments have been performed. However, a thorough understanding of the stress field in the reservoir rock subject to MHF treatment and its variation in direction and magnitude vertically as well as horizontally is important in the design and containment of massive hydraulic fractures, Secor and Pollard [55], Simonson et al. [2]. The tendency of hydraulic-fracture wings to extend to significantly different lengths in opposite directions from the borehole, Power et al. [52], Smith et al. [46], attests to the importance of horizontal stress gradients. Unfavorable vertical stress gradients will prevent the containment of the hydraulic fracture within the pay zone.

Fracture Geometry:

The effects of in situ stress conditions, formation permeability, and rock properties on created fracture geometry are better illustrated using a numerical simulator. The simulator used here has been developed by Clifton and Abou-Sayed over the last three years. It has been designated as the 'HYFRAC' code. (See Figure 5 through Figure 8.)

The analysis underlying the development of HYFRAC has been described in a series of three papers. In the first paper the physical basis of the model is presented and the governing equations are derived. The fracture is assumed to open in a vertical plane and to be perpendicular to the minimum in situ compressive stress. The surrounding rock is assumed to be homogeneous, isotropic and linear elastic. The fracture is opened by the injection of an incompressible, non-Newtonian, powerlaw fluid. Leak-off of the frac fluid and the expansion of the crack opening are included in the fluid balance equation. Vertical migration of the fracture is modeled by including the effects of the weight density of the frac-fluid and the variation with depth of the minimum compressive in situ stress [4]. Linear elastic fracture mechanics is used to characterize the critical condition for crack advance at each position along the crack front.

CASE I

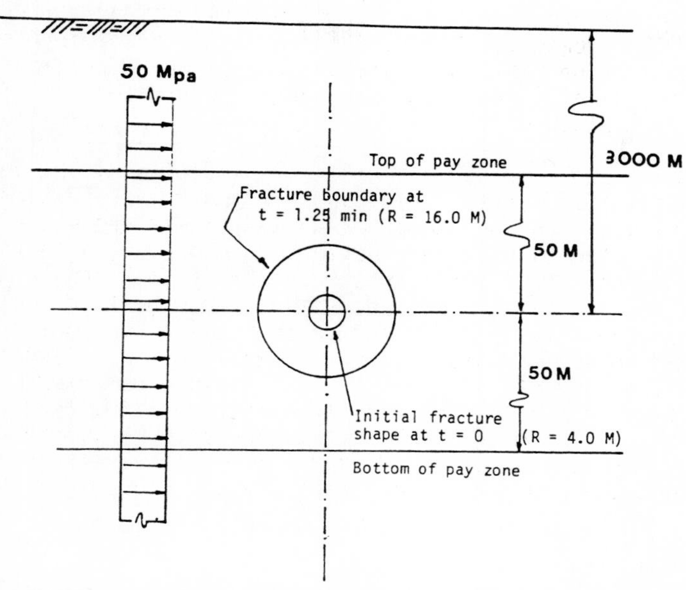

Figure 5. HYFRAC code output for uniform stress in the pay zone and adjacent barriers.

CASE II

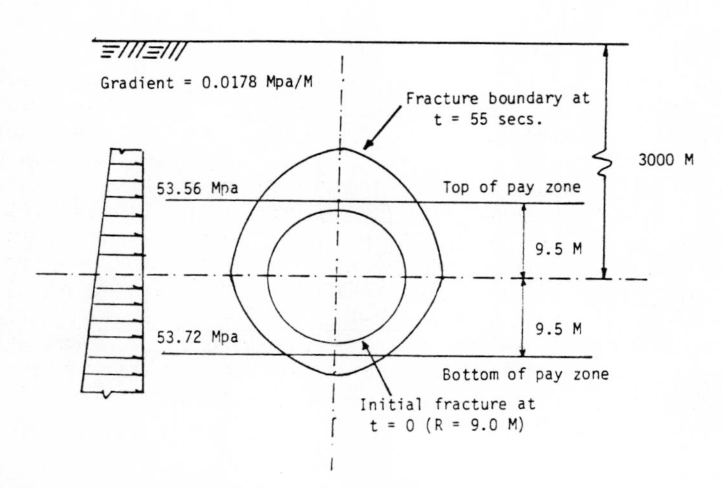

Figure 6. HYFRAC code output for stress gradient.

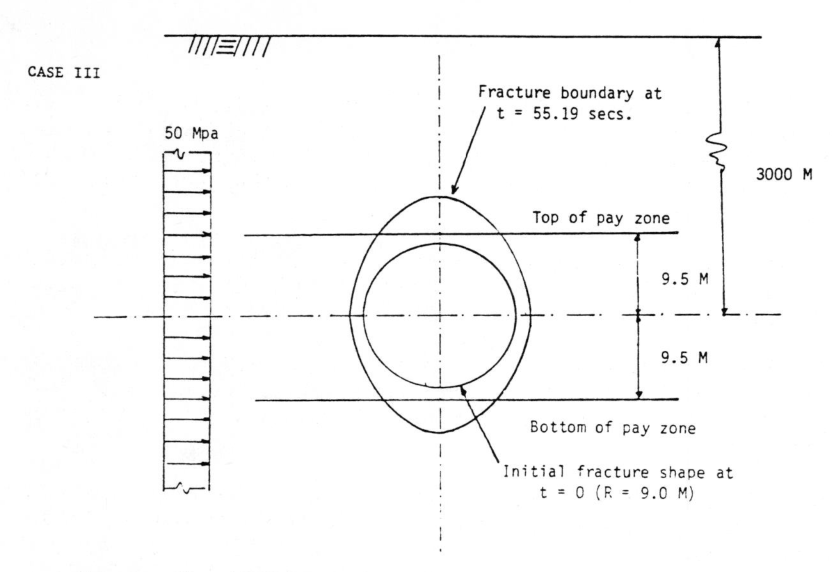

Figure 7. HYFRAC Code output for permeability barrier in overlaying and underlaying layers.

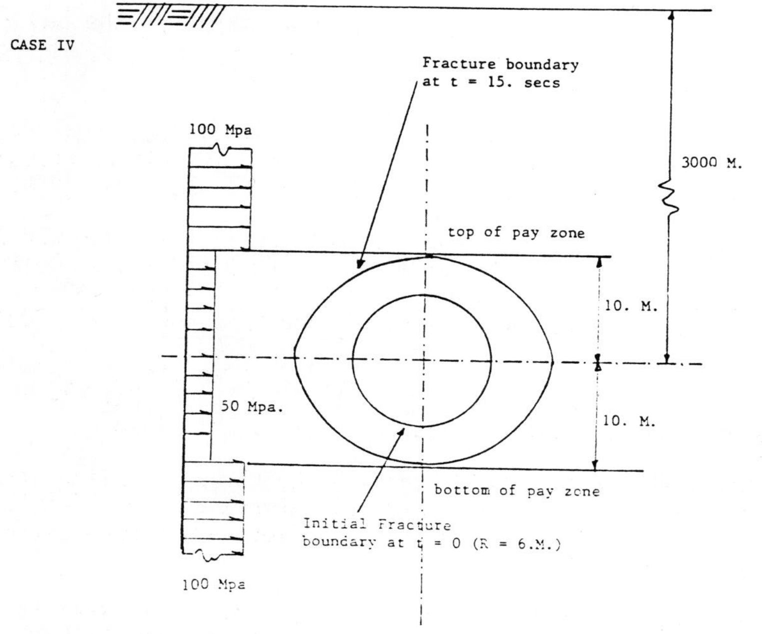

Figure 8. HYFRAC Code output for step contrast between overlaying and underlaying barriers.

REFERENCES

1. C.R. Fast, G.B. Holman and R.J. Covlin, "A Study of the Application of MHF to the Tight Muddy "J" Formation Wattenberg Field, Adams and Weld Counties, Colorado," Paper No. SPE 6524, 50th Annual Fall Meeting of the Society of Petroleum Engineers, Dallas Texas, Sept. 28 - Oct. 1, 1975.

2. E.R. Simonson, A.S. Abou-Sayed and R.J. Clifton, "Containment of Massive Hydraulic Fractures," Paper No. SPE 6089, 51st Annual Fall Technical Conference and Exhibition, New Orleans, Oct. 1976.

3. M.P. Cleary, "Primary Factors Governing Hydraulic Fractures in Heterogeneous Straticified Porous Formations," Paper No. 78-Pet-47, ASME Energy Technology Conference and Exhibition, Houston, Texas, November 1978.

4. U. Ahmed, A.S. Abou-Sayed and A.H. Jones, "Experimental Evaluation of Fracturing Fluid Interaction with Tight Reservoir Rocks and Propped Fractures," Paper No. SPE 7922, SPE/DOE Symposium on Low-Permeability Gas Reservoirs, Denver, Colorado, May 20-22, 1979.

5. W.J. McGuire and V.J. Sikora, "The Effect of Vertical Fractures on Well Productivity," Trans. AIME Vol. 219, pp. 401-403, 1960.

6. M. Prats, P. Hazebroek and W.R. Stickler, "Effect of Vertical Fractures on Reservoir Behavior - Compressible Fluid Case," Soc. of Pet. Eng. J., Vol. 2, pp. 87-94, June 1962.

7. J.M. Tinsley, J.R. Williams, Jr., R.L. Tiner and W.T. Malone, "Vertical Fracture Height - Its Effect on Steady-State Production Increase," J. of Pet. Tech., Vol. 21, pp. 633-638, May 1969.

8. R.G. Agarwal, R.D. Carter and C.B. Pollock, "Evaluation and Prediction of Performance of Low Permeability Gas Wells Stimulated by Massive Hydraulic Fracturing," Paper No. SPE 6838, 52nd Annual Fall Technical Conference and Exhibition, Houston, Texas, Oct 1978.

9. H. Cinco-L and F. Samaniego-V, "Transient Pressure Analysis for Fractured Wells," Paper No. SPE 7490, 53rd Annual Fall Technical Conference and Exhibition, Houston, Texas, Oct. 1978.

10. N.R. Warpinski, R.A. Schmidt and D.A. Northrop, "In Situ Stresses-The Predominant Influence on Hydraulic Fracture Containment," Paper No. SPE 8932, Unconventional Gas Recovery Symposium, Pittsburgh, PA, May 1980.

11. M.E. Hanson, G.D. Anderson and R.J. Shaffer, "Effects of Various Parameters on Hydraulic Fracturing Geometry," Paper No. SPE 8942, Unconventional Gas Recovery Symposium, Pittsburgh, PA, May 1980.

12. R. Veatch, "Massive Hydraulic Fracturing," Proceedings of the Geothermal Reservoir Well Stimulation Symposium, San Francisco, Feb. 1980 (Paper not submitted for publication).

13. R.J. Clifton, "Some Thoughts on Crack Growth in Hydraulic Fracturing," Terra Tek Report TR 76-82, Oct. 1974.

14. S.W. Lambert, M.A. Trevits, and P.F. Steidl, "Vertical Borehole Design and Completion Practices to Remove Methane Gas from Mineable Coalbeds," U.S. Department of Energy Report No. DOE/CMTC/TR-80/2, Aug. 1980.

15. A.S. Abou-Sayed, A.H. Jones and E.R. Simonson, "On the Stimulation of Geothermal Reservoirs by Downward Hydraulic Fracturing," Paper No. 77-Pet-81, ASME Energy Technology Conference and Exhibition, Houston, Texas, Sept. 1977.

16. K.G. Nolte and M.B. Smith, "Interpretation of Fracturing Pressures," Paper No. SPE 8297, 54th Annual Fall Technical Conference and Exhibition, Las Vegas, Nevada, Sept. 1979.

17. K.G. Nolte, "Determination of Fracture Parameters from Fracturing Pressure Decline," Paper No. SPE 8341, 54th Annual Fall Technical Conference and Exhibition, Las Vegas, Nevada, Sept. 1979.

18. R.J. Clifton and A.S. Abou-Sayed, "On the Computation of the Three-Dimensional Geometry of Hydraulic Fractures," Paper No. SPE 7943, SPE/DOE Symposium on Low Permeability Gas Reservoirs Denver, Colorado, May 1979.

19. R.J. Clifton and A.S. Abou-Sayed, "A Variational Approach to the Prediciton of the Three-Dimensional Geometry of Hydraulic Fractures," SPE Paper No. 9879, SPE/DOE Low Permeability Gas Reservoirs Symposium, May 1981.

20. A. Settari, "Simulation of Hydraulic Fracturing Process," SPE Paper No. 7693, Fifth SPE Symposium on Numerical Simulation, Denver, Colorado, 1979.

21. S.H. Advani, "Finite Element Model Simulations Associated with Hydraulic Fracturing," SPE Paper No. 8941, Unconventional Gas Recovery Symposium, Pittsburgh, PA, May 1980.

22. M.P. Cleary, "Comprehensive Design Formulae for Hydraulic Fracturing," SPE Paper No. 9259, 55th Annual Fall Technical Conference and Exhibition, Dallas, Texas, Sept. 1980.

23. A.A. Daneshy, "True and Apparent Direction of Hydraulic Fractures," Soc. of Pet. Eng. Jour., Vol. 12, pp. 149-163, 1971.

24. A.A. Daneshy, "A Study of Inclined Hydraulic Fractures," Soc. of Pet. Eng. Jour., Vol. 14, pp. 61-68, 1973.

25. A.S. Abou-Sayed, C.E. Brechtel and R.J. Clifton, "In Situ Stress Determination by Hydrofracturing: A Fracture Mechanics Report," J. of Geophys. Res., Vol. 83, No. 36, pp. 2851-2862, 1978.

26. M.E. Smith, "Effect of Fracture Azimuth on Production with Application to Wattenberg Gas Field," Paper No. SPE 8298, 54th Annual Fall Technical Conference and Exhibition, Las Vegas, Nevada, Sept. 1979.

27. M.B. Smith, J.M. Logan and D.M. Wood, "Fracture Azimuth - A Shallow Experiment," J. of Energy Resource Technology, Vol. 102, pp. 99-105, June 1980.

28. J.M. Hanson, and L.B. Owen, "Fracture Orientation Analysis by the Solid Earth Tidal Strain Method," SPE 11070, SPE Annual Technical Conference, Sept. 26-29, 1982, New Orleans, Louisiana.

29. B.C. Haimson, "Crustal Stress in the Continental United States as Derived from Hydrofracturing Tests," in J.C. Heacock, ed: The Earth's Crust, Geophys. Monograph Series, Vol. 20 AGU, 1978.

30. J.D. Bredehoeft, R.G. Wolff, W.S. Keys and E. Shuter, "Hydraulic Fracturing to Determine the Regional State of Tectonic Stress, Piceance Basin, Colorado", Geol. Soc. America Bull., Vol. 87, pp. 250-258.

31. M. Zoback and D. Pollard, "Hydraulic Fracture Propagation and Interpretation of Pressure Time Records for In-Situ Stress Determinations," 19th Symposium on Rock Mechanics, Stateline, Nevada, 1978.

32. M.D. Voegele, A.S. Abou-Sayed and A.H. Jones, "Optimization of Stimulation Design Through the Use of In Situ Stress Determination," Society of Petroleum Engineers, DOE. Low Permeability Symposium, Denver, SPE 10308, 1981.

33. M.K. Hubbert and D.G. Willis, "Mechanics of Hydraulic Fracturing," Am. Inst. Mining Engineers Trans., Vol. 210, pp. 153-168, 1957.

34. C. Fairhurst, "Methods of Determining In Situ Rock Stresses at Great Depths," Univ. of Minnesota, Contract No. DA-25-066-ENG-14, 765 for the Missouri River Division, U.S. Army Corps. of Eng 1967.

35. C. Cooke, "Radial Differential Temperature Logging - A New Tool for Dectecting and Treating Flow Behind Casing," SPE 7558, 1978.

36. H.A. Van Eckelen, "Hydraulic Fracture Geometry: Fracture Containment in Layered Formations," SPE 9261, Presented at the 55th Annual Fall Technical Conference and Exhibition of SPE, Dallas, Texas, Sept. 1980.

37. A.S. Abou-Sayed, U. Ahmed and A. Jones, "Systematic Approach to Massive Hydraulic Fracturing Treatment Design," Society of Petroleum Engineers, DOE, 1981.

38. N.R. Warpinski, J.A. Clark, R.A. Schmidt and C.W. Huddle, "Laboratory Investigation on the Effect of In Situ Stresses on Hydraulic Fracture Containment," SPE 9834, Presented at the 1981 SPE/DOE Low Permeability Symposium, Denver, Colorado, May 1981.

39. L.W. Teufel and J.A. Clark, "Hydraulic Fracture Propagation in Layered Rock: Experimental Studies of Fracture Containment," SPE 9878 Presented at the 1981 SPE/DOE Low Permeability Symposium, Denver, Colorado, May 1981.

40. J.M. Gronseth and E. Detourney, "Improved Stress Determination Procedures by Hydraulic Fracturing," Final Report for the U.S. Geological Survey, Menlo Park, California, Contract No. 14-08-0001-16768, 1979.

41. B.C. Haimson, "A Comparison of Four Sets of Stress Measurements at the Nevada Test Site," Proceedings, Workshop on Hydraulic Fracturing Stress Measurements, Monterey, California, 1981.

42. M.K. Hubbert and D.G. Willis, Mechanics of Hydraulic Fracturing; Am. Assoc. Petroleum Geologists Memoir 18, pp. 239-257, 1971.

43. M.K. Hubbert, "Natural and Induced Fracture Orientation," Am. Assoc. Petroleum Geologists Memoir 18, pp. 235-238, 1972.

44. K.D. Min, "Analytical and Petrofabric Studies of Experimental Faluted Drape-Folds in Layered Rock Specimens," Ph.D Dissertation, Texas A&M University, 1974.

45. D.W. Stearns, "Mechanism of Drape Folding in the Wyoming Province" 23rd Annual Field Conf. Guidebook, WY. Geol. Assoc., pp. 125-143, 1971.

46. M.B. Smith, G.B. Holman, C.R. Fast and R.J. Covlin, "The Azimuth of Deep, Penetrating Fractures in the Wattenberg Field," Paper SPE 6092, 51st Annual Fall Meeting SPE-AIME, New Orleans, 1976.

47. J.R. McWilliams, "The Role of Microstructure in the Physical Properties of Rock," Testing Techniques of Rock Mechanics, ASTM-STP 402, pp. 175-189, 1966.

48. M. Friedman and J.M. Logan, "Influence of Residual Elastic Strain on the Orientation of Experimental Fractures in Three Quartzose Sandstones," Jour. Geophys. Res., Vol. 75, No. 2, pp. 387-405, 1970.

49. M. Friedman and T.R. Bur, "Investigations of the Relations among Residual Strain, Fabric, Fracture and Ultrasonic Attenuation and Velocity in Rocks," Int. J. Rock Mech. Min. Sci. & Geomech. Abstr., Vol. 11, pp. 221-234, 1974.

50. G.A. Komar, W.K. Overbey and R.J. Watts, "Prediction of Fracture Orientation from Oriented Cores and Aerial Photos in Sand Draw Field," Wyoming; MERC/TPR-76/4, 12 pp., 1976.

51. W.K. Overbey, "Effect of In Situ Stress on Induced Fractures," Devonian Shale Production and Potential, Proc. 7th Appalachian Petroleum Geol. Symp., Morgantown, pp. 182-211, 1976.

52. D.V. Power, C.L. Schuster, R. Hay and J. Twombly, "Detection of Hydraulic Fracture Orientation and Dimensions in Cased Wells," Paper SPE 6526, 50th Annual Fall Meeting SPE-AIME, Dallas, 1975.

53. M. Friedman and H.C. Heard, "Principal Stress Ratios in Cretaceous Limestones from Texas Gulf Coast," Am. Assoc. Petroleum Geologist Bull., Vol. 58, No. 1, pp. 71-78, 1974.

54. B.C. Haimson, "The Hydraulic Fracturing Technique for Stress Measurement," Preprint, ISRM Symp. Advances in Stress Measurement, Sydney, Australia, 1976.

55. D.T.J. Secor and D.D. Pollard, "On the Stability of Open Hydraulic Fractures in the Earth's Crust," Geophys. Res. Letters, Vol. 2, No. 11, pp. 51-513, 1974.

EVALUATION OF ARTIFICIAL FRACTURE VOLUME BASED ON VENTED WATER AT YAKEDAKE FIELD

by Hirohide HAYAMIZU and Hideo KOBAYASHI

Chief of Mining Machine and Explosives Section, National Research Institute for Pollution and Resources
Tsukuba, JAPAN

Mining Machine and Explosives Section
National Research Institute for Pollution and Resources
Tsukuba, JAPAN

ABSTRACT

An analysis on fluid flow while artificial fractures created by hydraulic pressure are squeezing, has been carried out to evaluate the fracture volume based on vented fluid measured at the well-head. Assuming that the flow rate from the fracture into the well is approximately equal to the vented fluid per unit time, fracture volumes have been estimated with the vented fluid measured after three hydraulic fracturing operations conducted at YAKEDAKE test site, 1981. In the case where closed fractures were created and/or fluid loss through passes from fracturing well to other investigation wells were not so much, the flow rate of vented fluid calculated from fracture volume was comparable with the measured vented fluid.

INTRODUCTION

It may be the general consideration that artificial fractures created in hot dry rocks by hydraulic fracturing play a principal role in extracting heat energy from hot dry rocks, because these fractures act as a heat exchanger between rocks and fluid and also artificial geothermal reservoirs. Therefore, to know the characteristics such as the configuration and size of fractures during the developing stage of artificial fluid circulation system in the formation, is very important for evaluating the heat extraction efficiency.

Considerable techniques have been developed to evaluate fracture size and configuration, and some typical ones available now are based on a) source location by means of acoustic emission which occurs when rocks are broken, b) stress wave propagation characteristics across the fracture and c) fluid volume injected during hydraulic operation. Only one of these techniques is not sufficient to evaluate fracture size and configuration, so synthetic interpretation of fracture characteristics is usually conducted based on data from different kinds of techniques.

A topic that we are going to present in this paper concerns the evaluation of fracture volume based on vented fluid measured at the well-head after a hydraulic fracturing operation. Evaluation of fracture volume by this technique has been carried out using the field data measured at YAKEDAKE test site.

CONCEPT OF PRESSURE BALANCE IN FRACTURE AND SURROUNDING ROCKS

First, let us consider the evaluation technique which depends on the measured value of injected fluid throughout the hydraulic fracturing. During pressurization, some of the injected fluid may permeate into the surrounding formation which requires other procedures to estimate the volume of permeated fluid into the formation through the fracture surface and the wall of the well. Therefore it is also necessary to know the mechanism of fluid permeation into rocks and several characteristics of rocks which are influenced by fluid permeation.

Now, if we open a valve at the well-head just after the hydraulic fracturing treatment, the fluid held in the fracture is pushed out by squeezing of the fracture due to formation stresses. In this case, fluid pressure in the fracture is in equilibium with the formation stress perpendicular to the fracture plane. Because the pore pressure in the formation is proportional to the hydraulic pressure maintained in the fracture, i.e. the formation stress perpendicular to the fracture, therefore the fluid pressure in the fracture coincides with the permeated fluid pressure in the formation. Through these considerations described above, it can be thought that the fluid flow into the fracture from the surrounding formation is very small while the fracture is squeezing. If we can eliminate back-flow of the permeated fluid in the formation and the compressibility of the well, the total flow out of the well, i.e. the vented fluid, is approximately equal to the fracture volume.

Measurements of the total vented fluid after opening the valve at the well-head, therefore, may give us one value of fracture volume, and this evaluation can be performed more easily than the evaluation based on the injected fluid volume because further estimation of fluid volume permeated into the formation is not necessary. But it may require much time over several hours to perform complete data sampling of the vented fluid. An attempt to clarify the fluid flow mechanism while the fracture is squeezing has been carried out to be able to predict fracture volume using data of vented fluid measured in only a few minutes.

FLUID FLOW ANALYSIS WHILE FRACTURE IS SQUEEZING

In general, the orientation of an artificial fracture created in the formation by hydraulic pressure is considered to be perpendicular to the maximum stress direction (throughout this report, compressive stresses are indicated by minus values), and the axis of a well is not usually parallel to the principal stress directions. Therefore, it is very rare that the fracture plane is perpendicular to the axis of a well.

Considering the recent tendency of drilling schemes for geothermal wells to drill wells perpendicular to the fracture plane by an inclined drilling technique, and for simplicity to build a model for analysis, the orientation of fracture is assumed to be perpendicular to the axis of the well.

The fluid flow and fracture model design is shown in Fig.1. Fluid flow rate from the fracture into the well corresponds to the squeezing rate of fracture volume. Therefore we have the next equation.

$$-\frac{dQ}{dt} = 2 \pi r B v \qquad (1)$$

where
- Q : fracture volume
- B : fracture width
- v : velocity of fluid flow from fracture into well
- r : radius of well

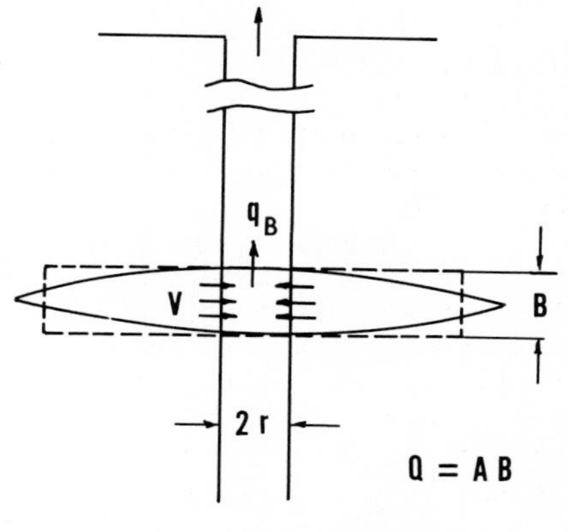

Fig.1 Model of fracture and fluid flow

Let A = Q/B and substitute it into Eq.1.

$$\frac{dQ}{Q} = -\frac{2 \pi r}{A} dt = - S\, dt \qquad (2)$$

where $S = 2 \pi r v / A$

If we integrate Eq.2 with respect to t, considering that v and A are functions of t, the following equations are given.

$$Q = Q(t_o) \exp \left\{ - \int_{t_o}^{t} S \, dt \right\} \tag{3}$$

As is obvious from Eq.3, $Q(t_o)$ represents the fracture volume at $t=t_o$.

Now take the differential of Eq.3. Then, fluid flow rate from the fracture into the well, $-dQ/dt$, is given by the next equation.

$$- \frac{dQ}{dt} = Q(t_o) \ S(t) \ \exp \left\{ - \int_{t_o}^{t} S \, dt \right\} \tag{4}$$

If q_B denotes $-dQ/dt$, the fluid flow rate from the fracture into the well at $t=t_o$ is expressed by the following equation.

$$q_B(t_o) = Q(t_o) \ S(t_o) \tag{5}$$

Then, take a very short interval, $\Delta t = t_1 - t_o$. When Δt approaches zero, an approximation of function S is given by

$$S \fallingdotseq S(t_o) + \left| \frac{dS}{dt} \right|_{t=t_o} \xi \ \Delta t \tag{6}$$

where $\xi = (t - t_o) \ / \ \Delta t$

If we substitute Eq.6 into the logarithms taken from Eq.4, we have the next equation.

$$\log q_B = \log \left[Q(t_o) \left\{ S(t_o) + \left| \frac{dS}{dt} \right|_{t=t_o} \xi \ \Delta t \right\} \right]$$

$$+ \left[- \int_{t_o}^{t} \left\{ S(t_o) + \left| \frac{dS}{dt} \right|_{t=t_o} \xi \ \Delta t \right\} dt \right] \tag{7}$$

Assuming that we can neglect terms of Eq.7 which involve Δt, the zero order approximation of $\log q_B$ is expressed by Eq.8 in the region of $t_o \leqq t \leqq t_1$.

$$\log q_B = \log \left\{ Q(t_o) \ S(t_o) \right\} - S(t_o) \ (t - t_o) \tag{8}$$

Differentiation of Eq.8 with respect to t, gives

$$\frac{d}{dt} \log q_B = - S(t_o) \tag{9}$$

Supposing that the vented fluid from the well per unit time, q_S, is equal to the value of the flow rate from the fracture into the well, q_B, the fracture volume, $Q(t_o)$, at $t=t_o$ is obtained using the measured vented fluid, $q_S(t_o)$, and Eq.5 and Eq.9.

But it is expected that several parameters connected to fluid flow in the well such as gravity, flow velocity, and pressure affect the measured value of q_S at $x = h$, and q_S may take a slightly different value from that of q_B at $x = 0$. Additional flow analysis in the well has been performed to make clear the differences between q_S and q_B.

The following results were obtained for the relation between q_S and q_B. Details for deriving the resulting equations will be presented in the Appendix.

$$q_S - q_B = A \exp\left(- \int_0^t \phi \, dx \right) \tag{10}$$

where
$$\phi = \frac{a^2 \kappa}{\rho h q_S} + \frac{1}{q_S} \frac{d}{dt} q_S \tag{11}$$

$$A = q_S(0) - \sqrt{q_S(0) - \frac{a^2}{\rho} p_S'(0)} \tag{12}$$

$$p_S'(0) = p_B(0) - \rho g h \tag{13}$$

Using the measured value, q_S, which indicates the vented fluid volume per unit time from the well, other quantities such as ϕ, A and $p_S'(0)$ could be fixed. Then the fluid flow rate from the fracture into the well, q_B, could be calculated. So, the fracture volume, $Q(t_o)$, at $t=t_o$ is estimated by Eq.5, Eq.9 and $q_S(0)$.

A numerical analysis conducted with the equations listed above showed that the estimated fracture volume, Q, includes a large aberration; therefore, a further try to improve the accuracy of the approximation has been carried out.

Because $- dQ/dt$ is denoted by q_B, Eq.4 can be changed to

$$\log q_B = \log Q(t_o) + \log S - \int_{t_o}^{t} S \, dt \tag{14}$$

Differentiation of Eq.14 with respect to t gives us the following equation.

$$\frac{d}{dt} \log q_B = \frac{1}{S} \frac{ds}{dt} - S \tag{15}$$

The left side of Eq.15 is calculated based on the measured vented fluid q_S. If we denote

$$-\alpha = \frac{d}{dt} \log q_B$$

Eq.15 can be formulated as

$$-\alpha = \frac{1}{S} \frac{ds}{dt} - S \tag{16}$$

The zero order approximation of S at $t=t_o$ is indicated in $\alpha(t_o)$; therefore, the first order approximation of S at $t=t_o$ is given by the following equation.

$$S_1(t_o) = \frac{1}{\alpha(t_o)} \frac{d\,\alpha(t_o)}{dt} + \alpha(t_o) \qquad (17)$$

Further, the second order approximation of S can be calculated by substituting $S_1(t_o)$ from Eq.17 into Eq.16. But when we calculate dS/dt, a round-off error occurs and $S_2(t_o)$ does not reach a good approximation. So, we selected $S_2(t_o)$ which is shown by the next equation, as a kind of approximation value.

$$S_2(t_o) = \frac{S_o(t_o) + S_1(t_o)}{2} = \frac{\alpha(t_o) + S_1(t_o)}{2} \qquad (18)$$

Using S obtained from Eq.18 and Eq.5, the fracture volume at a certain time, Q, can be calculated and this approximation offers a better estimate of the fracture volume than that from α and/or S_1.

The analysis on the fracture volume based on the vented fluid showed that the fracture volume at $t=t_o$ is easily evaluated by measuring the flow rate of the vented fluid in a very short time which includes $t=t_o$.

In this section, we have investigated the deducing technique from the measured value, q_S, to q_B. But an evaluation of fracture volume using field data which will be presented in the next section will be carried out with the assumption that q_B can be replaced by the measured vented fluid q_S.

EVALUATION OF FRACTURE VOLUME BASED ON FIELD DATA (AT YAKEDAKE)

In 1981, hydraulic fracturing field tests were carried out at the west side of Mt.YAKEDAKE by contractors of HDR Project,MITI. Object of the tests was to obtain basic information on hydraulic fracturing and for fracture mapping by the AE technique. Five wells of depth of about 300 m are used for the tests. One well of 160 mm in diameter is located at the center of the test site and hydraulic fracturing operations are conducted with this well. Four other wells of about 60 mm in diameter are about 40 m away from the central well and used as investigation wells. Six fracturing tests are performed at depths ranging from 228 to 299 m of the central well with a LINES Production-Injection Packer and a Halliburton HT-400 Pump.

An impeller-type flow meter was arranged in the drainage line to measure the rate of vented fluid after the hydraulic operation. Because of insufficient capacity of the flow meter, we could not open the valve completely for about the first 5 minutes of measurement, and only 3 series of data from 6 hydraulic fracturing tests were obtained as shown in Fig.2. It is obvious from Fig.2 that the rate of vented fluid of No.5-81 is greater than that of the others, and the fracture volume generated by the No.5-81 test might be bigger than the others.

Approximations for each curve were made with power functions using data at times ranging from about 5 to 40 minutes after opening the well-head. By integration of these approximated equations, fracture volumes created by three operations were estimated as 1.019 m^3(No.2-81), 1.703^3m (No.3-81) and 3.363 m^3(No.5-81) with the assuption that $q_S = q_B$.

Fluid injection rates during these 3 tests are shown in Fig.3. Total amounts of injected fluid also shown in Fig.3 are several times greater than the fracture volume estimated by integration of the vented fluid curves. While hydraulic operations of No.3-81 and No.5-81 were continued, fluid was drawn from the investigation wells. From Fig.4, it is clear that artificial fluid passes were created between

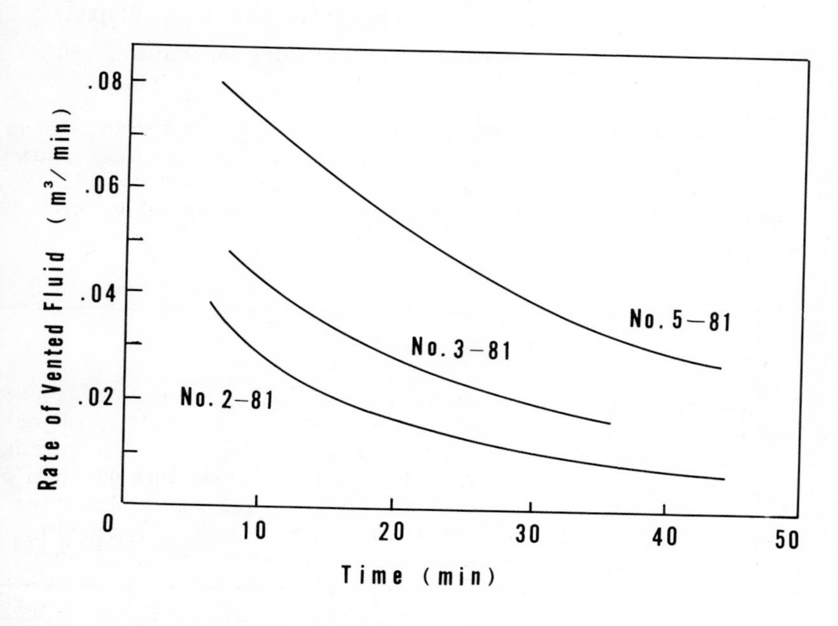

Fig.2 Vented fluid measured after hydraulic fracturing operations conducted at YAKEDAKE test site

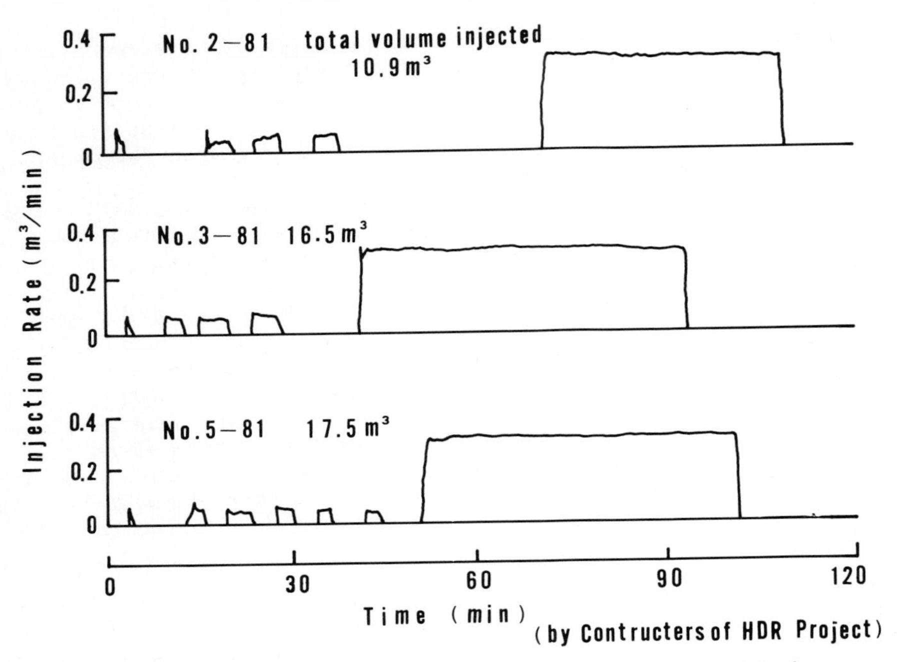

Fig.3 Fluid injection rate of hydraulic fracturing operation

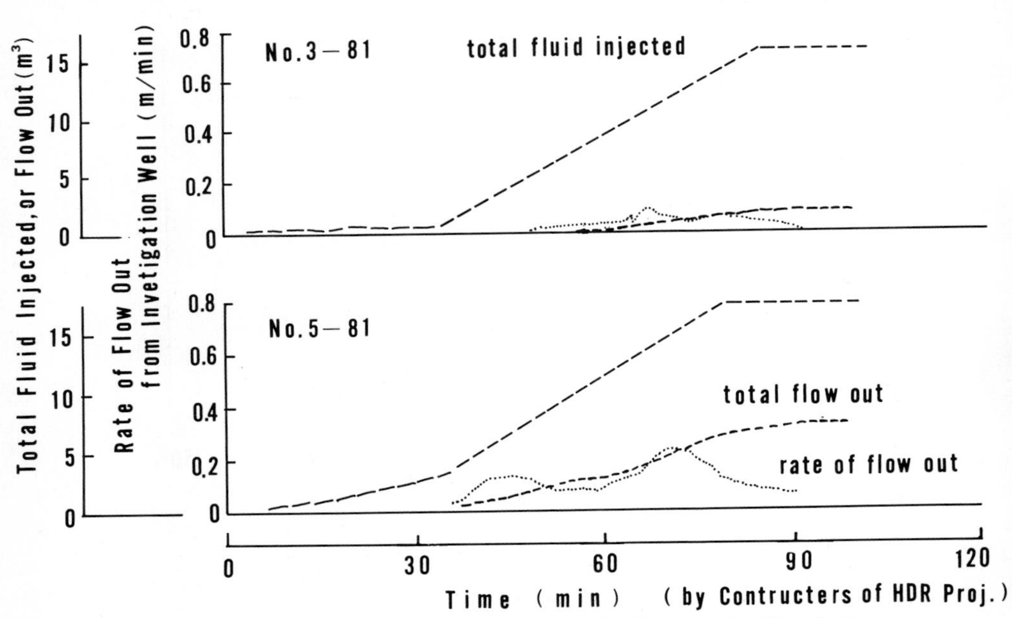

Fig.4 Relation between injected fluid and flow out fluid during hydraulic fracturing operations

the fracturing well and the surrounding investigation wells, which resulted in difficulty for fracture volume evaluation based on vented fluid.

Now, let us begin to evaluate fracture volume based on field data of vented fluid measured in a very short time. First, take logarithms of the vented fluid rate every 2 minutes from the measured data shown in Fig.2. Differential coefficient of these logarithmic values, $\alpha(t_o)$, can easily be obtained using two logarithmic values of q_s at times $t=t_o - 2$ and $t=t_o + 2$, and $d\alpha(t_o)/dt$ can also be calculated from $\alpha(t_o - 2)$ and $\alpha(t_o + 2)$. Then, a first order approximation of $S_1(t_o)$ at every 2 minutes is given by Eq.17, and $S_2(t_o)$ is obtained by Eq.18. Substituting $S_2(t_o)$ and $q_s(t_o)$ into Eq.5, the fracture volume every 2 minutes after opening the valve can be estimated.

Fig.5 shows estimated fracture volumes based on vented fluid data every 2 minutes after opening the valve for each hydraulic fracturing test. If we take logarithms of the vented fluid data at very short intervals, we may be able to estimate the fracture volume at any time after opening the valve.

A supplementary check was performed to see if the estimated fracture volume was comparable with the measured data of vented fluid. From the estimated fracture volume shown in Fig.5, vented fluid volumes are calculated back which are shown as dotted lines in Fig.6. Solid line in Fig.6 indicates measured value of vented fluid. In the case of No.2

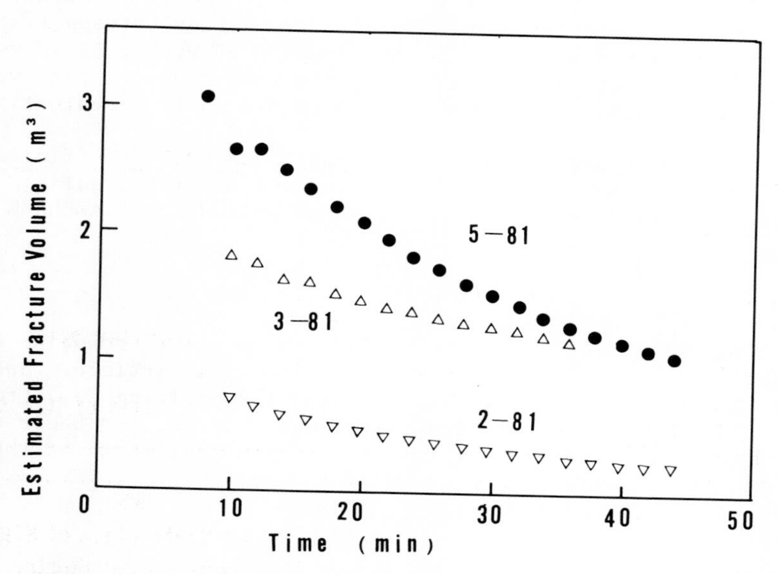

Fig.5 Estimated fracture volume based on vented fluid measured after hydraulic fracturing operations

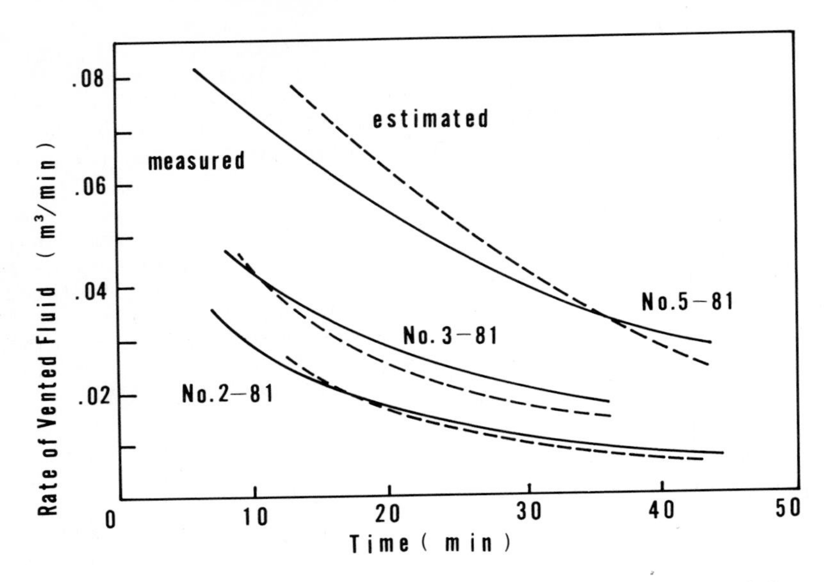

Fig.6 Rate of vented fluid calculated back from estimated fracture volume and measured vented fluid rate

and No.3 tests, it is clear that fairly good evaluations were made, but in the case of No.5, vented fluid values calculated back were not in good agreement with measured values. Several influences such as 1) slightly poor measured data due to lack of flowmeter capacity, 2) fluid flow paths generated by hydraulic fracturing between fracture well and investigation wells and 3) application of vented fluid measured at well-head, q_S, instead of flow rate from fracture into well, q_B, are considered causes which result in errors in this evaluation.

In October, 1982, measurements of vented fluid with a sufficient capacity flow meter are going to be carried out after hydraulic fracturing which will be scheduled for a new well at the YAKEDAKE field.

ACKNOWLEDGMENTS

The authors gratefully acknowledge Sunshine Project Promotion Headquarters, MITI, for their permission to publish this article. Special thanks are extended to S.Suga, Mitsui Mining and Smelting Co., Ltd., Y.Ogata, H. Karasawa, K.Morita and T.Yamaguchi, NRIPR, for their cooperation and valuable contribution in conducting measurements at the YAKEDAKE field.

REFERENCE

Contractors of HDR Project, The report "Feasibility Study on the Power Generation System from Hot Dry Rock", March, 1982, which was addressed to Sunshine Project Promotion Headquarters.

APPENDIX : Relation between q_S and q_B

Considering a fluid flow model shown in Fig.7, fluid pressure in the well is expressed by

$$\frac{\partial p}{\partial t} = -\kappa \frac{\partial q}{\partial x} \qquad \text{(A-1)}$$

where
p : pressure in well
q : flow rate of fluid
κ : pressure necessary to expand the unit volume of the well, i.e. the compressibility of the well

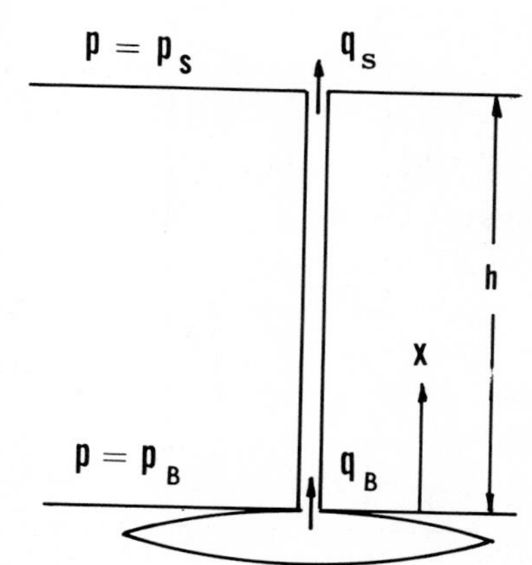

Fig.7 Fluid flow model in well and its coordinate

If we integrate Eq.1 from 0 to h with respect to x, we have

$$\frac{\partial}{\partial t} \int_0^h p \, dx = -\kappa (q_S - q_B). \qquad \text{(A-2)}$$

Rearrangement of Eq.A-2 with assumption, $p(x) = \frac{p_B}{h}(h - x)$, gives us

$$\frac{\partial}{\partial t} \left(\frac{1}{2} p_B h \right) = -\kappa (q_S - q_B). \qquad \text{(A-3)}$$

Applying the law of conservation of momentum to fluid flow in the well, we will able to get following equation.

$$\frac{\partial}{\partial t} \rho g = -\frac{\partial}{\partial x} \left(\frac{\rho q^2}{a} \right) - \frac{\mu}{4r} \left(\frac{q}{a} \right)^2 - a\rho g - a \frac{\partial p}{\partial x}$$

where
a : sectional area of well
r : radius of well
ρ : density of fluid μ : coefficient of friction

Integration of this equation from 0 to h with respect to x is

$$\rho \frac{\partial}{\partial t} \left(\int_0^h q \, dx \right) = - \frac{\rho}{a} \left(q_S^2 - q_B^2 \right) - \frac{\mu}{4ra^2} \int_0^h q^2 dx - a\rho gh + ap_B \, . \tag{A-4}$$

If we can neglect the left side and the second term of the right side of Eq.A-4, the following equation is given as for the zero order approximation of Eq.A-4.

$$p_B = \frac{\rho}{a^2} \left(q_S^2 - q_B^2 \right) + \rho gh \tag{A-5}$$

Substituting p_B of Eq.A-5 into Eq.A-3, we have

$$\frac{1}{2} \frac{\rho h}{a^2} \frac{d}{dt} \left\{ \left(q_S - q_B \right) \left(q_S + q_B \right) \right\} = -\kappa \left(q_S - q_B \right).$$

Now, if we consider $(q_S + q_B) = 2 q_S$, then we have the following equation.

$$\frac{\rho h}{a^2} \frac{d}{dt} \left(q_S - q_B \right) q_S = -\kappa \left(q_S - q_B \right)$$

$$\frac{\rho h}{a^2} q_S \frac{d}{dt} \left(q_S - q_B \right) + \frac{\rho h}{a^2} \left(q_S - q_B \right) \frac{d}{dt} q_S = -\kappa \left(q_S - q_B \right)$$

$$\frac{d}{dt} \left(q_S - q_B \right) = - \left(\frac{a^2 \kappa}{\rho h q_S} + \frac{1}{q_S} \frac{d}{dt} q_S \right) \left(q_S - q_B \right)$$

$$\text{Let} \quad \frac{a^2 \kappa}{\rho h q_S} \quad \frac{1}{q_S} \frac{d}{dt} q_S = \phi; \tag{A-6}$$

$$\text{then we have} \quad q_S - q_B = A \exp \left(- \int_0^t \phi \, dt \right) \tag{A-7}$$

where A : arbitrary constant.

From Eq.A-5, q_{BO} at t=0 is given as

$$q_{BO} = \sqrt{q_{SO}^2 - \frac{a^2}{\rho} p_{BO} + \frac{a^2}{\rho} \rho gh} = \sqrt{q_{SO}^2 - \frac{a^2}{\rho} p'_{SO}} \tag{A-8}$$

where $p'_{S0} = (p_{B0} - \rho gh)$, which represents the pressure at the well-head at time of t=0, that means the time just before opening the valve. Substituting q_{B0} denoted by Eq.A-8 into Eq.A-7 gives us the next equation.

$$A = q_{S0} - \sqrt{q_{S0}^2 - \frac{a^2}{\rho} p'_{S0}} \tag{A-9}$$

Then, we have reached the resulting equation with which calculation of flow rate, q_B, from the fracture into the well is made.

$$q_B = q_S - \left(q_{S0} - \sqrt{q_{S0}^2 - \frac{a^2}{\rho} p'_{S0}} \right) \exp \left(-\int_0^t \phi \, dt \right) \tag{A-10}$$

Propagation of Hydraulic Fracture and Its Conductivity in Layered Media

A.A. Daneshy

Halliburton Services
Duncan, Oklahoma

ABSTRACT

Hydraulic fracture propagation in layered media is controlled by the magnitude of the least insitu principle stress and the mechanical properties of each medium, with the least principle stress being dominant.

The most effective mechanism for fracture containment is slippage and the accompanied blunting of the fracture tip.

In the absence of slippage, fractures tend to mainly extend in zones with the lowest least insitu principle stress, although they will also extend into other adjacent formations.

When fracture conductivity is achieved thru etching of the fracture face, only those parts of the fracture chemically suitable for etching can contribute to production.

If fracture conductivity is obtained by the use of proppant, then distribution of this proppant is also governed by the stresses and mechanical properties of each formation.

INTRODUCTION

In most industrial applications hydraulic fractures encounter different formations with different mechanical and physical properties and under different insitu stresses. The location of the fracture is important for the production of reservoir fluid or extraction of thermal energy.

Hydraulic fractures are initiated by applying fluid pressure inside the borehole beyond the tolerance of the rock. Once a fracture is initiated, continuous injection of the fluid under sufficient pressure will extend the fracture away from the borehole.

The effectiveness of a hydraulic fracture depends on its conductivity and length after the treatment. The conductivity is achieved by either a chemical reaction between fracturing fluid and the formation rock (e.g., HCl in carbonates) or by adding solid particles (proppant) to the fracturing fluid. The propping agent, if used in sufficient quantities, will keep the fracture open and give it conductivity.

Most hydraulic fractures are assumed to be vertical and are likely to encounter several formations. Since every one of these formations does not contribute to production, it is important to know where the fracture is and which parts of it are conductive.

This paper is a qualitative attempt towards analysis of the vertical growth of the fracture and factors which influence it. The discussion is divided into two parts; vertical fracture extension and proppant distribution.

The term fracture is used in a general sense to indicate a crack created by fluid pressure. In special cases when fracture conductivity is achieved by chemical etching of the formation, the term fracture acidizing is used to describe the process.

VERTICAL FRACTURE EXTENSION

Suppose a hydraulic fracture is initiated from a line source AB in the wellbore, Fig. 1. AB can be either an open-hole segment or the interval between top and bottom perforations accepting fluid in a cased hole. As this fracture extends away from the well, it will propagate upward, downward, and laterally. The pattern of fracture propagation in an isotropic homogeneous brittle material has been shown[1] to be elliptical, with the major axis along the wellbore. In layered formations the vertical extension of the fracture will bring it into contact with other formations which are likely to have different mechanical and physical properties.

It has been shown[2] that hydraulic fractures propagate perpendicular to the least insitu principle stress. The fracturing fluid pressure, in a single formation, should be larger in magnitude than the least insitu principle stress.

In layered formations the least insitu principle stress may be different in each formation. Therefore, in addition to differences in physical and mechanical properties, one has to also consider the influence of stresses.

We will designate the formation where fracture has initiated by the number (1). The formations above will be designated successively by even numbers (2), (4), (6) etc. Similarly, odd numbers (3), (5), (7) etc will successively designate formations below. The letters E, ν, γ, and σ will designate Young's modulus, Poisson's ratio, effective surface energy and the least insitu principle stress. When referring to properties of a specific formation, these letters will be subscripted according to formation number. Thus, for example, σ_3 will designate the least insitu stress in formation (3).

Vertical or horizontal growth of the fracture are influenced by changes in mechanical or physical properties, variations in the least insitu principle stress, or by blunting of the fracture tip. Of these, blunting of the tip has the most drastic effect on growth and will be discussed first.

Blunting of Fracture Tip

An effective mechanism for preventing the growth of a fracture is blunting of its tip. This will drastically reduce the tensile stress concentration at the fracture tip, thus stopping the propagation.

Blunting of the fracture tip can be caused by several mechanisms;

1. Slippage

   If the hydraulic fracture intersects a natural plane of weakness (e.g., a bedding plane, interface between adjacent formations, etc.) then slippage can occur along the plane of weakness (Fig. 2-a). This will blunt the fracture tip and prevent its local growth. Slippage is aided if the following conditions exist:

   a. Plane of weakness is smooth at the vicinity of the fracture. Roughness of the face of the plane of weakness can cause initiation of a second displaced fracture such as shown in Fig. 2-b.

   b. The fluid pressure needed to shear the rock along the plane of weakness is less than the pressure needed to extend the fracture. The shear strength (frictional resistance) of the plane of weakness increases with increasing the stress normal to the plane, which in turn increases with depth. Therefore, this kind of slippage is more likely to happen at shallower depths.

2. Natural Fractures

   If the hydraulic fracture intersects an open natural fracture, its growth will halt momentarily. The fracture will soon continue its propagation by either surrounding the natural fracture, or, re-initiation somewhere at the tip of the natural fracture. In the latter case, the extension of the fracture on the other side of the natural fracture will be displaced as shown in Fig. 2-c, at the vicinity of natural fracture. The effect of the open natural fracture will be directly related to its size and the frequency of its occurrence. In severely fractured formations, the hydraulic fracture may not have a planar shape. Instead, it can have a zigzag pattern resulting from repeated dislocation of the fracture due to the natural fractures. On a large scale, the orientation of the fracture will still be perpendicular to the least insitu principle stress. If the natural fracture is large (comparable to the hydraulic fracture) then it can act as a local barrier to fracture growth.

If the blunting of the fracture tip is localized (such as caused by natural fractures) then the overall effect on the fracture propagation will be minor. Blunting of a regional scale (which may occur along bedding planes) will have a much more pronounced effect on fracture growth. It could totally prevent fracture growth into an adjacent formation. In order for a fracture to stay totally contained within a formation, slippage needs to take place at the interfaces above and below.

Pressure Requirements for Fracture Extension

If a hydraulic fracture is not blunted at its tip, it will always have the potential for growth into adjacent zones.

The fluid pressure, p, needed for fracture extension in a single formation depends on fracture shape and is given by

$$p - \sigma = \sqrt{\frac{2E\gamma}{\pi L(1-\nu^2)}} \qquad \text{Griffith Crack} \qquad (1)$$

$$p - \sigma = \sqrt{\frac{\pi E\gamma}{2L(1-\nu^2)}} \qquad \text{Penny-shaped Crack} \qquad (2)$$

$$p - \sigma = \sqrt{\frac{3E\gamma}{2(1-\nu^2)} \cdot \frac{(L^2+H^2)\ [E(k)]^2}{L[2(L^2+H^2)E(k)-L^2K(k)]}} \qquad \text{Elliptical Crack} \qquad (3)$$

where:
- L = Fracture Extent
- H = Height in Eliptical Fracture
- $K(k), E(k)$ = Complete elliptic integrals of the first and second type
- k = Parameter of the elliptic integral

These equations show that the pressure p needed for fracture propagation increases as σ, E and γ increase. Of these parameters σ is the most dominant. Young's modulus E, and γ have similar, less prominent, effect. The value of γ is variable even in a given rock formation and depends on local anistropy. The value applicable for the above equations is the γ at the fracture tip, which is constantly moving due to fracture extension. Of course γ varies amongst different rock types. Friedman et al[3] list values of γ varying between $0.9 * 10^4$ to $8.8 * 10^4$ ergs/cm^2 for different limestones and sandstones. Hudson et al[4] give the value of $2.0 * 10^4$ ergs/cm^2 for granite.

The value of Young's modulus E applicable to the above equations is the regional value. Although variations in E amongst different rocks is about the same as γ, the regional nature of dependence makes it more dominant. For a fracture propagating in two different formations, one is likely to see slower propagation rate in the formation with the larger E.

For a formation with $E = 60.9 * 10^9$Pa ($10 * 10^6$ psi), $\nu = 0.2$, $\gamma = 2 * 10^4$ ergs/cm^2 (0.114 lb in/in^2), and for a fracture extent of 50 feet, (p-σ) will be about 245 KPa (35.5 psi) for a Griffith crack, 384.4 KPa (55.8 psi) for a penny-shaped crack, and a value between these two for an elliptical crack. Changing E or γ will change the value of p-σ, but its magnitude will be about the same. On the other hand, σ can potentially change an order of magnitude more. Therefore p in different formations is substantially more influenced by σ than the mechanical properties. In general, changes in σ in the same order of magnitude as those created by mechanical properties can be considered insignificant. The reason is that equal changes could result from variations in E and γ , and the latter are usually ignored because they are not measureable under insitu conditions. If E and γ are known from laboratory testing (for E or γ) or logs (for E) and σ is known from mini-frac jobs in adjacent formations then one can compute the overall effect of the combination for various geometry fractures. When the variations within adjacent zones are small (in the order of 700 KPa or 100 psi or less) then the hydraulic fracture can be assumed to be propagating in an essentially isotropic medium, which will eventually give it a circular shape. When the differences are larger then one can add the influence of all the variables into one and assume that variable to be σ. In this manner, the problem will be reduced to that of determining the effect of varying insitu stresses on the fracture propagation.

In the above analysis the influence of ν on p was intentionally left out. First of all, as Eqs. 1 - 3 show, ν does not have a significant effect on p. Although attempts have been made to relate σ to ν, these are not known to have proven themselves useful for field use. In addition, the value of ν measured in the lab depends on the stress level and its proximity to ultimate strength. Since the underground stresses are generally not known, a unique and reliable value of ν cannot be established for describing the underground conditions.

In the mathematical analysis of hydraulic fracturing one usually assumes that mechanical properties and insitu stresses are independent of each other, although a dependent variation can also be easily formulated. Logically, one would expect a high σ associated with increasing depth, higher E (assuming nearly equal geologic displacements in adjacent formations), higher temperature or higher coefficient of thermal expansion. The increase in σ with depth is well established from drilling and fracturing data. However, the actual influence of E or thermal properties have not been sufficiently studied to yield conclusive results.

Effect of σ on Fracture Extension

Suppose a hydraulic fracture is extending in a formation called (1) with the least insitu principle stress, σ_1. Formations (2) and (3) are adjacent to (1), and (4) and (5) are adjacent to (2) and (3), respectively, Fig. 3. We will assume $\sigma_1<\sigma_2<\sigma_3$ and that all of them have the same orientation and are parallel to the planes of interfaces, which are all in turn parallel to each other.

A hydraulic fracture is initiated from a line source AB in the wellbore, Fig. 3. The initial pattern of fracture propagation will be elliptical until the hydraulic fracture reaches the interface between (1) and (2), and (1) and (3). The hydraulic fracture will initially propagate in (1) only, as shown by curves a-b in Fig. 3. Since the hydraulic fracture is not restricted in its vertical growth, the fluid pressure needed for its extension will exceed the value predicted by Eq. 3. The increase in fluid pressure will also increase the stresses at the fracture tips adjacent to the interfaces, and these stresses continue to increase as fluid pressure increases due to further fracture growth in (1). Eventually, this will result in either slippage in the interface or facture growth into formation (2). The difference between σ_1 and σ_2 is very important in this regard. As the value of $\sigma_1 - \sigma_2$ increases, so also does the chance of slippage. Should slippage take place, the fracture will not grow from (1) into (2).

Suppose the fracture breaks into (2). (Actually, this would be the more likely event, especially at greater depths.) The fracture propagation rate in (2) will be lower than in (1) because $\sigma_2 > \sigma_1$, with the difference increasing as $\sigma_2 - \sigma_1$ increases. The fracture will propagate along paths somewhat similar to curves c, d, and e in Fig. 3.

The hydraulic fracture can break from (1) and (2) with a fluid pressure p less than σ_2. Suppose A_1 and A_2 denote the fracture surface areas in (1) and (2), and, γ_1 and γ_2, the effective surface energies of (1) and (2). The force F_1, exerted by σ_1, and A_1 is

$$F_1 = \sigma_1 A_1$$

Similarly $$F_2 = \sigma_2 A_2$$

The force F applied by the fracturing fluid on its surface area (and thus creating the fracture) is

$$F = (A_1 + A_2)p$$

The first requirement for fracture extension is that the force exerted by the fluid exceed the resistance supplied by σ_1 and σ_2. This means

$$F > F_1 + F_2$$

or

$$p > \frac{\sigma_1 A_1 + \sigma_2 A_2}{A_1 + A_2} \tag{4}$$

In addition to overcoming the force exerted by σ_1 and σ_2, the fluid pressure p has to provide the energy needed for creating the fracture surface area, as well as the strain energy of the fracture. The surface energy needed for the fracture is

$$2A_1\gamma_1 + 2A_2\gamma_2$$

If V_f denotes the created fracture volume

$$V_f\ \Delta p_e = 2A_1\gamma_1 + 2A_2\gamma_2$$

gives the excess pressure, Δp_e, needed to provide the surface energy. The fracture volume can be computed by

$$V_f = A_1\bar{\omega}_1 + A_2\bar{\omega}_2$$

where ω_1 and ω_2 are average fracture widths in (1) and (2). Thus,

$$\Delta p_2 = \frac{2A_1\gamma_1 + 2A\ 2\gamma_2}{A_1\bar{\omega} + A_2\bar{\omega}_2}$$

gives the excess pressure needed for creation of the fracture surface.

The fracturing fluid pressure should also provide the change in the strain energy of the system, it therefore follows that

$$p > \frac{\sigma_1 A_1 + \sigma_2 A_2}{A_1 + A_2} + \frac{2A_1\gamma_1 + 2A_2\gamma_2}{A_1\bar{\omega}_1 + A_2\bar{\omega}_2} \tag{5}$$

Since hydraulic fractures follow the path of least resistance, therefore, the growth of the fracture (2) is always just to the extent that would be necessary to accomodate its growth in (1). Equation 5 also shows that for a given p, as σ_2 increases, one needs a smaller A_2 to satisfy the inequality. This means that as $\sigma_2 - \sigma_1$ increases, A_2

becomes smaller. The same is true for $\gamma_2 - \gamma_1$ although to a much lesser extent.

The above discussion relates the fluid pressure needed for fracture extension to its areal extent in two formations. It does not establish a relationship between A_1, A_2, and formation properties. Such a relationship has not yet been developed in the literature known to this author. However, certain qualitative comments could be made on the subject

1. As $\sigma_2 - \sigma_1$ increases, the fracture extends longer in (1) before it breaks into (2). Furthermore A_1/A_2 is larger for larger $\sigma_2 - \sigma_1$.

2. As the thickness of formation (1) decreases, the hydraulic fracture extension into formation (2) occurs at shorter fracture lengths. This is because the restrictive influence of (2) is felt earlier. The fluid pressure increase due to this restriction will also occur earlier. The fracture will break into (2) at a higher fluid pressure.

3. For a fixed fracture length, as the thickness of formation (1) decreases, the fracture width increases. This increased width is the result of higher fluid pressures.

4. The relative thicknesses of formations (1) and (2) determine whether p is more influenced by σ_1 or σ_2. If (1) is much thicker than (2), then p will be mostly governed by σ_1. If (2) is much thicker, then σ_2 will be the dominant factor.

The propagation of the fracture in (1) and (2) will extend it to the interface between (2) and (4). Whether the fracture will break into (3) during this period depends on the difference between σ_3 and σ_2. If this difference is small, then the fracture can break into (3) before it reaches the upper part of (2). Otherwise, it will stay in (1) and (2). The restrictive effect of (3) will make the fracture behave as if the source AB in Fig. 3 is moving upward (but still contained in [1]). The next vertical extension of the fracture will be into formation (3). Once again, this breakage can take place at a pressure $p<\sigma_3$.

Most of the discussion on fracturing two adjacent formations can be extended into three or more formations. Generally, when $\sigma_2,\sigma_3 > \sigma_1$ the fluid pressure will have an upward trend. The lateral extent of the fracture will be larger than its vertical extent. The ratio of the lateral to vertical extents will increase as the difference between σ_1 and the largest least principle stress encountered by the fracture in the adjacent layers increases.

Suppose $\sigma_1 > \sigma_2$ or σ_3. Hydraulic fracture will seek and extend in the zone with the lowest σ. The pattern of fracture propagation will change to make it appear like it has initiated in the zone with the lower σ. All of the above discussion will hold true if one replaces formation (1) with the formation having the lowest σ. If the lower σ is below the perforated zone, hydraulic fracture will take a pear shape. If it is above, it will look like an upside down pear.

The changes in σ can sometimes be detected from the measurement of the instanteous shut-in pressure. If this pressure is much higher (several hundred psi) at the end of the job compared to the beginning, it would imply fracture extension into higher stress zones. If it is lower, it would imply major fracture growth out of zone. The latter case (which occurs rarely in industrial treatments)is more definitive. To take full advantage of the instantaneous shut-in pressure, one needs to record its value very early during fluid injection and before fracture has had time to grow out of zone. The best practice is to record this pressure as soon as the pressure is stablized and use a very small initial injection rate before shut-in.

Fracture Width in Layered Formations

The only parts of the fracture which can contribute to production are those with sufficient conductivity (defined as fracture width times its permeability). In fracture acidizing this conductivity is obtained by etching the formation rock (usually carbonates) with an acid. The only part of the fracture etched is where the formation is chemically suitable. Even if a fracture grows out of the productive perforated zone into adjacent formations, if these other formations are non-reactive (such as shales) once the fracturing fluid pressure is released the fracture in the non-reactive zones will close. Consequently, these zones will not contribute to production. Furthermore, if the fracture had grown thru and beyond the non-reactive zones, these zones will act as a barrier to communication of the two parts of the fracture. Therefore, in fracture acidizing, the productive part of the fracture is usually limited to the reactive formation in direct contact with perforations or open-hole.

If fracture conductivity is obtained by the use of a proppant, then distribution of the proppant and its amount in each zone becomes important. Both these factors are greatly influenced by the width of the fracture in each formation.

The width equations for a single fracture in an isotropic medium are:

$$\omega = \frac{4(1-\nu^2)\ \Delta p}{E}\ L \qquad \text{Griffith Crack} \qquad (6)$$

$$\omega = \frac{4(1-\nu^2)\,\Delta p}{\pi E} \quad L \qquad \text{Penny-shaped Crack} \tag{7}$$

$$\omega = \frac{4(1-\nu^2)\,\Delta p}{E} \cdot \frac{L}{E(k)} \qquad \text{Elliptical Crack} \tag{8}$$

where $\Delta p = p - \sigma$

In all these equations the fracture width increases as Δp increases, or E decreases. While these equations in their exact form are not applicable to fractures generated in layered media, it is reasonable to assume that the effect of Δp and E will be the same.

The fluid pressure inside a hydraulic fracture increases from its tip towards the wellbore. This is due to frictional pressure losses due to fracturing fluid flow. Assuming that the fracturing fluid behavior is Newtonian with viscosity μ, then the pressure difference due to flow is given by

$$\Delta p = 12\mu \int_0^L \frac{q}{h\omega^3}\, d\xi \tag{9}$$

where integration is performed along the path of fluid flow described by ξ, inside the fracture. The fluid is flowing at the rate of $q(\xi)$ in a channel with a rectangular cross section separated by fracture width $\omega(\xi)$ and $h(\xi)$ such that ω, h, and ξ will be mutally perpendicular at each point. Since fluid pressure has the same magnitude along the fracture tip, therefore Δp has the same magnitude along each flow path. Equations 6 to 8 indicate that ω will have smaller values in formations with largers σ's. Equation 9 shows a much stronger dependence on ω than q. Therefore, even though as ω becomes smaller q also decreases, the magnitude of the integrand rapidly increases with decreasing ω. To obtain equal Δp's with different ω's one will need a shorter flow length along the path with smaller ω's. These in turn occur where σ's are larger; the smallest ω corresponding to largest σ. Therefore, as the hydraulic fracture propagates into formations with higher σ, the extent of fracture growth in these formations substantially decreases.

The variations of ω influence proppant distribution inside the fracture. First of all, proppant cannot enter those parts of the fracture where ω is smaller than proppant diameter. Such locations are likely to occur close to fracture tip and also adjacent to zones with larger σ's. If the proppant bridges inside the fracture, it will begin to impede fluid flow thru it with a resulting additional Δp, which in turn will reduce fracture growth along its path.

In cases where slippage has caused creation of an offset fracture, the slippage plane will have very narrow width and may not be able to accept proppant. Therefore, the proppant may not be able to move to the other side of the slippage plane.

The amount of proppant inside the fracture is proportional to the original concentration of the proppant in fracturing fluid, amount of fluid leak-off (which tends to increase proppant concentration) and fracture width. In narrow parts of the fracture there will be less proppant. The smaller width together with smaller quantities of proppant result in substantially decreased fracture flow capacity. This means that assuming the adjacent formations have similar permeability, reservoir fluid pressure and viscosity, then the major part of the production will come from the zone where the combination of σ and E (small σ and E) is most favorable. When formation permeabilities are not equal in adjacent zones then the dominant permeability for reservoir fluid flow is that of the dominantly fractured zone.

Concluding Remarks

Much of the available literature on fracture growth in layered media contains a qualitative discussion of the subject. The reason is an absence of the following;

1. How is the pressure needed for fracture breakthrough related to formation properties and stresses.

2. Variations of fracture width in layered media.

3. Computation of the strain energy for a fracture in layered media.

4. Three dimensional fracturing criterion in an isotropic medium.

Any major breakthrough in understanding hydraulic fracture propagation will await an answer to at least one of the above questions. This will require a concerted effort by universities, major research organizations and industry. Of these, the theoretical work is perhaps best suited for universities, while industry can fit the results for everyday use. Any theoretical development will need to be experimentally verified. This can be done by either universities, industry, or major research laboratories.

REFERENCES

[1]Daneshy, A. A., "Three Dimensional Propagation of Hydraulic Fractures Extending from Open Holes", Applications of Rock Mechanics, Proceedings of ASCE, 15th Symposium on Rock Mechanics, E.R. Haskins, Jr. (ed.)

[2]Haimson, B.C., and Fairhurt, C., "Initiation and Extension of Hydraulic Fractures in Rocks" SPEJ., Sept., 1967, pp. 310-318.

[3]Friedman, M., Handin, J., and Alani, G., "Fracture Surface Energy of Rocks," Int. J. Rock Mech. Min. Sci., Vol 9, pp 757 - 766, 1972.

[4]Hudson, J.A., Hardy, M.P., and Fairhurst, C., "The Failure of Rock Beams, Part II - Experimental Studies", Int. J. Rock Mech. Min. Sci. Vol 10, pp. 69-82., 1973.

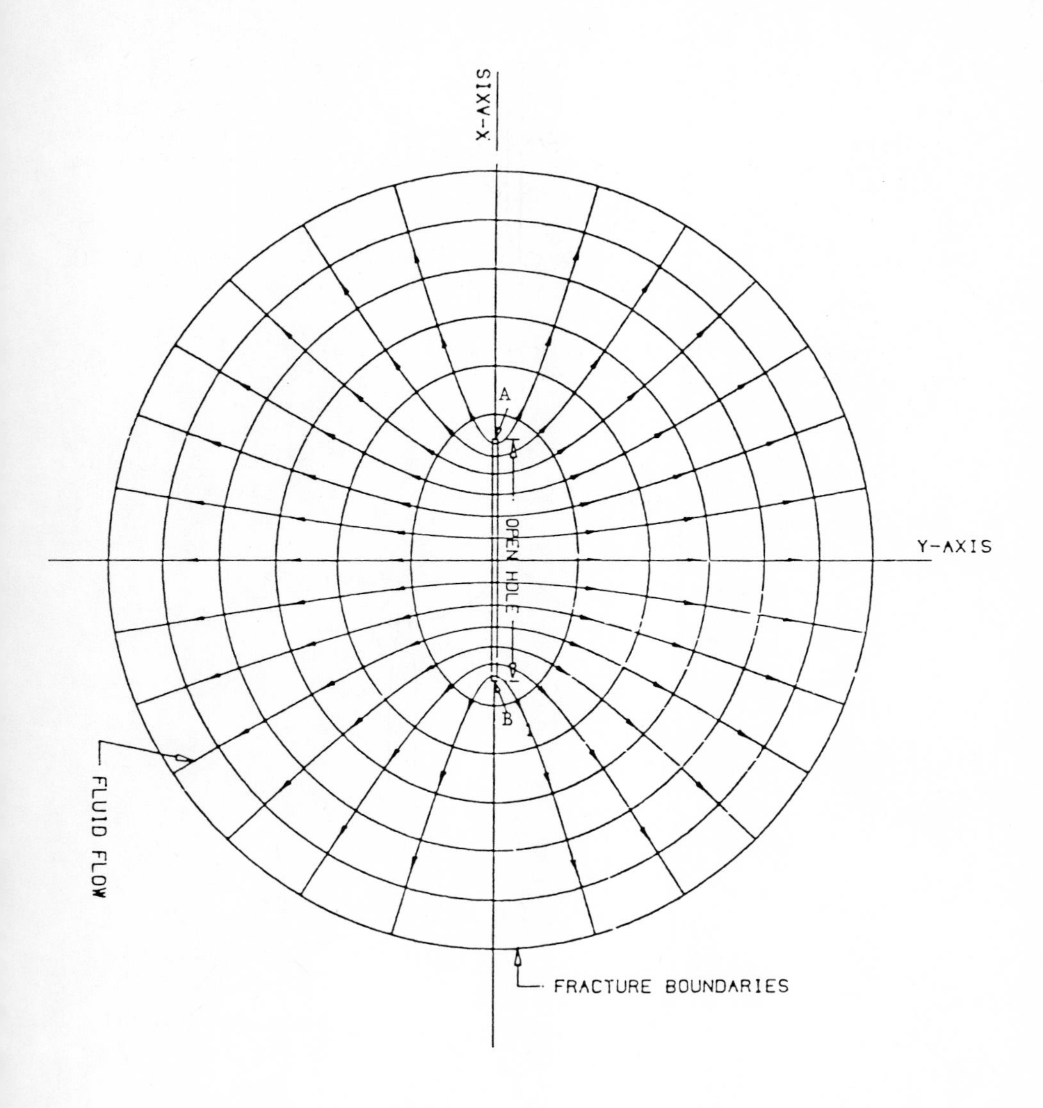

Fig. 1. Fracture Boundaries and the Fluid Flow Path During Hydraulic Fracture Extension

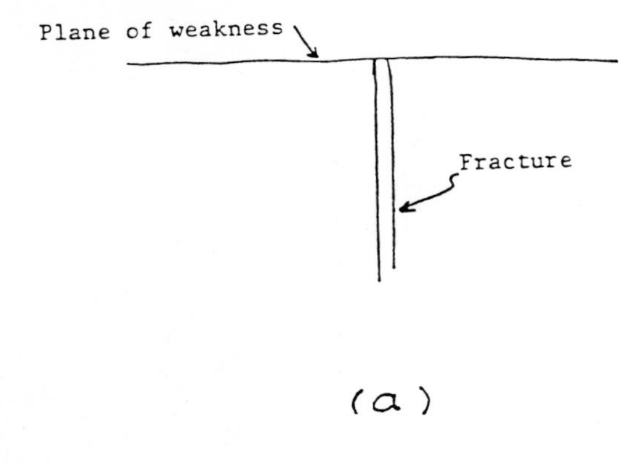

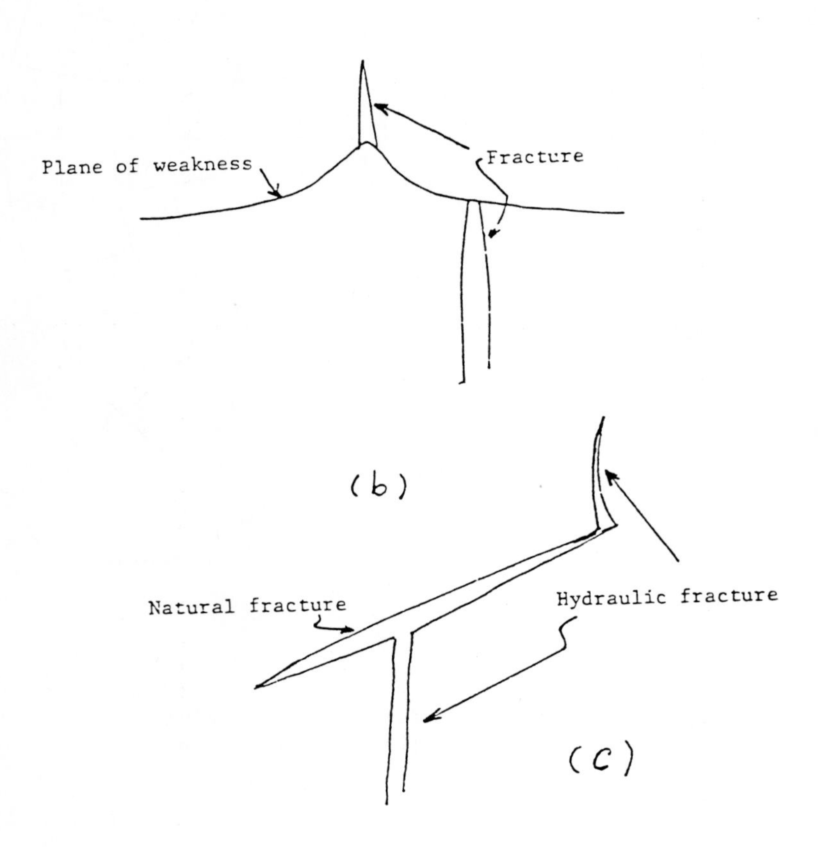

Fig. 2. Blunting of Fracture Tip

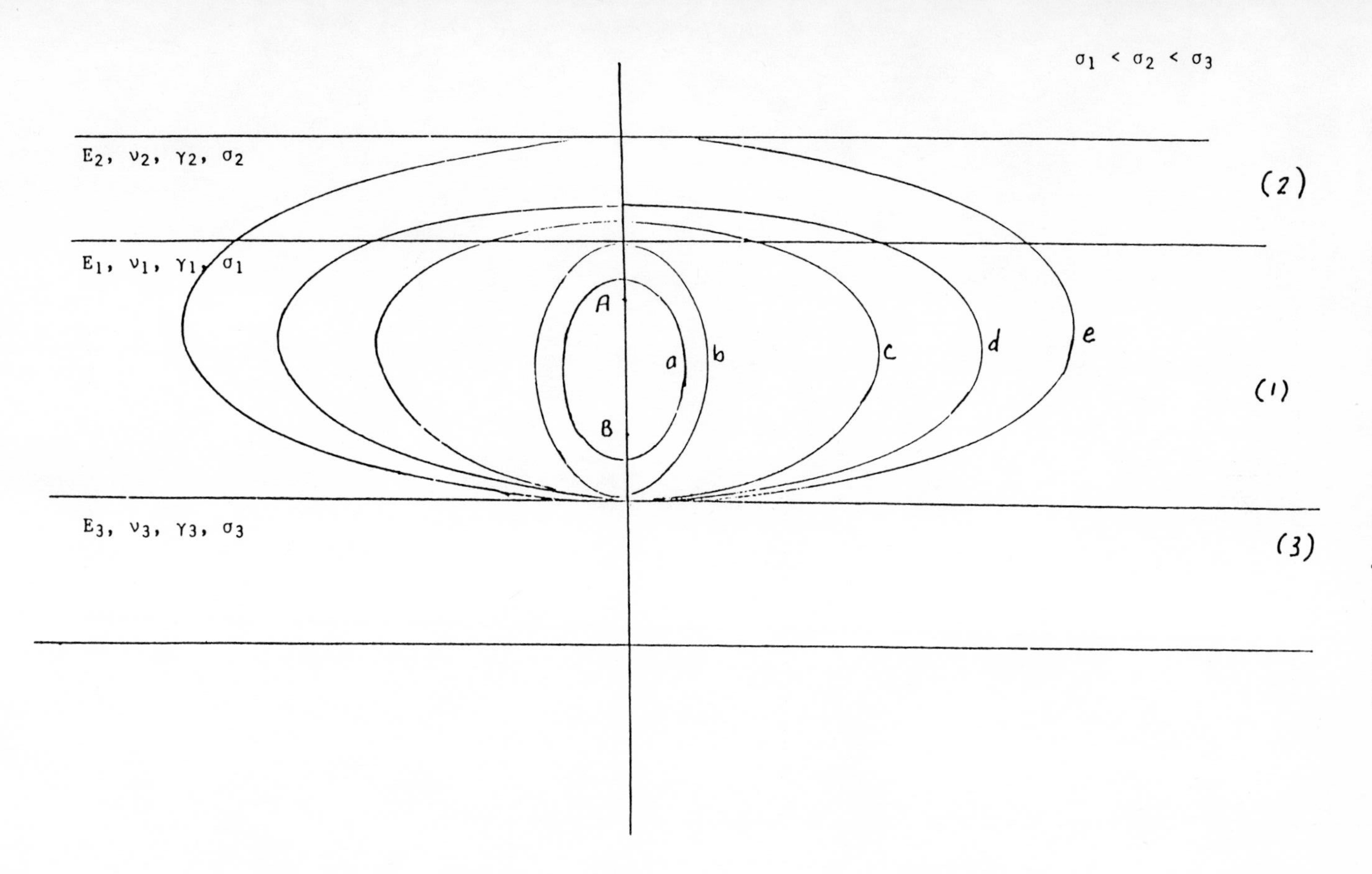

Fig. 3. Propagation of the Fracture in Layered Formations

HYDRAULIC FRACTURING IN GEOLOGICAL PROCESSES: A REVIEW

by Tetsuya Shoji and Sukune Takenouchi

Faculty of Engineering
The University of Tokyo
Hongo, Tokyo 113

ABSTRACT

In the Earth's crust there occur many emplacement bodies and fissure-filling columns, such as salt plugs, igneous stocks, sheet intrusions, clastic dikes, and mineral veins. Since they were formed with relation to the activity of some fluid materials, such as plastic salt, magma, water, aqueous solution, they may be analogs of hydraulic fractures. With the increase of pore-pressure, rocks become weak and also brittle even under a relatively high confining pressure. This implies that the pore-pressure affects the fracture pattern as well as the strength of rocks.

INTRODUCTION

There are many fissures ranging from several microns to several hundred meters in the Earth's crust. Most of them are filled with certain materials. A vein consists of minerals precipitated from an aqueous solution. A sheet intrusion of igneous rocks, such as dikes and sills, is composed of minerals solidified from a magma. A clastic dike is filled by clastic fragments such as mud, sand, and breccia. These materials are regarded to have been essentially derived from certain fluids, such as magma, hydrothermal solution, and muddy water. Diapirs of salt or igneous rock have pierced also the country rocks as liquid or liquid-like phases.

Some fractures are formed by tectonic stresses. They are called either faults along which the blocks on both sides have displaced each other, or extensional cracks along which no displacement is recognized. They correspond to the shear fracture and the longitudinal splitting in rock mechanics, respectively. The fracture pattern indicates the directions of the principal axes of the causative tectonic stress in the area concerned. Usually, the analysis of the fracture pattern has been carried out on the basis of the results of uniaxial or triaxial dry compression tests. In the last two and half decades, however, the effect of fluid pressure has been taken into account on the discussion of the formation of faults and extensional cracks. With pore-pressure, the fracturing of rock is controlled by the effective stresses. This concept was introduced in the study of soil by Terzaghi (45), and its importance with regard to the fracturing of rocks now has been well established by Hubbert and his co-workers (16, 17).

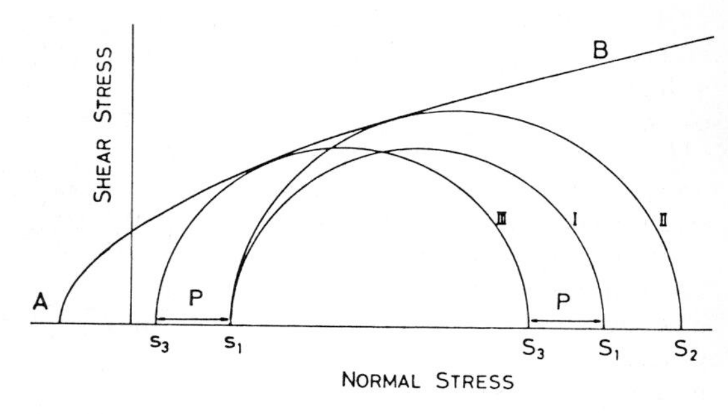

Fig. 1. Diagram showing the effect of pore-pressure on failure (see text).

Fig. 1 illustrates how pore-pressure affects failure based on the Griffith theory. Let us consider the case where the maximum and minimum principal stresses (σ_{max}, σ_{min}) are S_1 and s_1, respectively. If the pore-pressure is zero, the Mohr circle (Curve I) lies inside the Mohr envelope (Curve A-B). In this case, the rock does not break down. If the confining pressure (σ_{min}) is constant, failure will take place at $\sigma_{max} = S_2$ (Curve II). Meanwhile, as the pore-pressure (P) increases, the Mohr circle moves to the left. At $\sigma_{max} = S_3 = S_1 - P$ and $\sigma_{min} = s_3 = s_1 - P$ (Curve III), it touches the envelope and failure takes place. Consequently, the pore-pressure plays a role in assisting the compressive stress by which the rock is broken, rather than leading directly to failure.

The concept of effective stress can explain the effect of pore-pressure on the strength from the macroscopic point of view, but cannot explain the initiation and extension of fracture from a cavity filled by compressed fluid. Many authors (13, 24, 32) have analyzed theoretically the stress condition near the cavity. According to them, the extensional failure extends to the direction normal to the minimum principal stress. The mechanics here is hydraulic fracturing. In hydraulic fracturing, the crack is opened by the internal fluid pressure overcoming the confining pressure.

It is suggested that the previously mentioned fractures were formed under the pore-pressure condition, or by hydraulic fracturing, because their formation is closely related to a fluid or fluid-like materials. Accordingly, it is important to review the formation mechanics of fractures from the viewpoint of the effective stress and hydraulic fracturing.

DIAPIRS AND STOCKS

Diapirism is a process by which the earth materials from deeper levels pierce upward to shallower levels. The term was formerly applied only to intrusive evaporites, but was later used to describe the intrusion of igneous bodies. Fig. 2 shows a schematic representation of a salt plug.

The concept of buoyancy is proposed in the mechanics of salt dome intrusion, that is an unbalance in the geostatic load of overburden. When salt is heated above 200°C, it becomes soft and plastic, and behaves hydrodynamically (10). Accordingly, when sedimentary salt is buried at a depth greater than 8000 m, it moves laterally to places of lower overburden pressure, where doming or piercement occurs. Diapirism may be induced, therefore, either by (i) differences in thickness of overburden, (ii) lateral differences in the density of overburden, or (iii) topographical irregularities in the surface of the salt deposits (8). The effect of buoyancy increases gradually with the height of the intrusive column and flow continues as long as the salt remains plastic and is under hydraulic pressure.

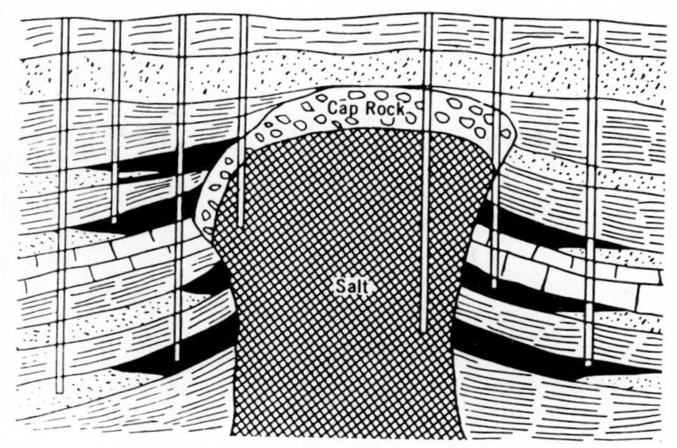

Fig. 2. Diagram showing salt plug, caprock, and upturned strata (21).

When the salt in the deeper part of a sedimentary basin reaches a temperature of a critical value, it becomes plastic. Consequently,

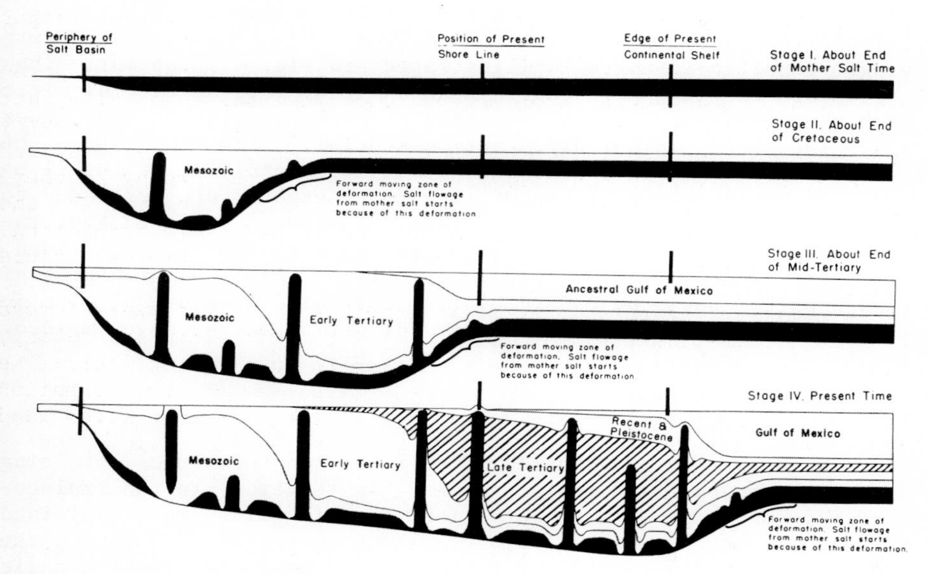

Fig. 3. Rise of Gulf Coast salt domes through geologic time (10).

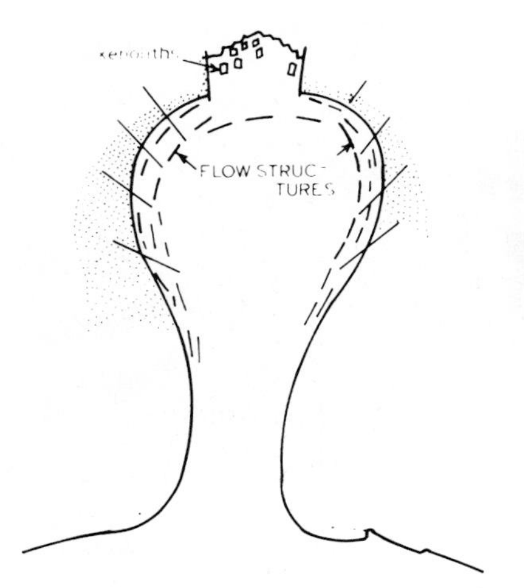

Fig. 4. Schematic representation of an igneous diapir (8).

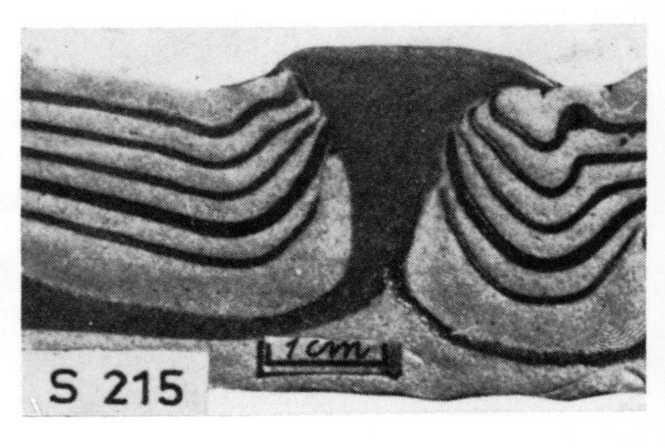

Fig. 5. Cross section of a model of silicone dome which has risen through alternating layers of putty and modelling clay (darker and thinner sheets) (39).

the intrusion of salt domes would be progressively younger from the deeper parts of a subsiding basin towards the margin. As shown in Fig. 3, downwarping of the Gulf Coast area occurred in stages and moved progressively seaward (14).

Gussow (10) maintains that the emplacement of igneous masses such as volcanic plugs, granite batholith, kimberlite pipes, carbonatites, and serpentinite bodies, and of such intrusive masses as mud volcanoes, shale diapirs, is similar to that of salt piercement. In all cases, the prime motivating force for intrusion is the weight differential of the overburden, or geostatic load (9). The idealized form of single granitic diapirs is illustrated in Fig. 4. The intrusions are composed of the upper cupola zone and the intermediate diapiric zone which rise from the abyssal zone of emplacement occupying lower levels. The form is not different essentially from the shape of the salt dome. Ramberg (39) conducted the model analyses of the rises of salt and magma due to buoyant forces. The experiments succeeded to reproduce the diapirism (Fig. 5). Figs. 2, 4 and 5 suggest that the self-inducing material did not rise concordantly with the overburden, but penetrated it forcibly. It is postulated that the spreading of the upper part should be caused by the hydraulic pressure of plastic or liquid material. This implies that diapirism may be an example of hydraulic fracturing.

IGNEOUS SHEETS AND DIKES

Many authors have discussed the formation mechanics of ring dikes, cone sheets, and radial dikes, in relation to the emplacement of igneous magma. The formation of ring dikes was explained by the subsidence of magma (2), or cauldron subsidence following the eruption of magma (4). The cone sheets were explained to be tensile fractures (2), or shear fractures caused by the rising magma (40).

The formation of radial dikes was explained by the wedging effect of intruded magma (3). Phillips (34) discussed the mechanics of the formation of these intrusions, and concluded as follows: (i) ring dikes are not hydraulic fractures, but deveolop on shear fractures formed in the overlying rocks when a magma subsides, and in a certain situation the mechanical effect of the fluid occupying fractures might bring about the hydraulic extension of the fractures; (ii) cone sheets occupy shear fractures formed as the result of dynamic stress arising from the rapid expansion of a magma undergoing retrograde boiling; and (iii) radial dikes are simple hydraulic tension fractures formed during the upwelling of magmas.

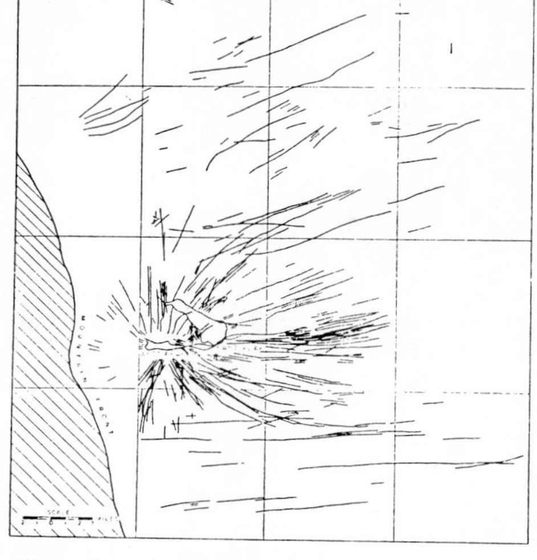

Fig. 6. System of dikes surrounding the Spanish Peaks (26).

As illustrated in Fig. 6, the dike pattern of the Spanish Peaks area, Colorado, shows the peculiar shape (26). The pattern of the dikes is clearly symmetrical about a line running N75°E. Most dikes originating from West Spanish Peak tend to curve eastward. The westward-trending dikes are much shorter than those trending eastward. On the basis of Anderson's hypothesis (3), Ode (32) explained this dike pattern by the superposition of hydrostatic pressure caused by a magma on a previously existing regional stress system.

The dikes resulted from the intrusion of a magma along the extensional fractures which were normal to the minimum principal stress, namely parallel to the maximum principal stress. This implies that the average strike of dikes indicates the directions of the maximum and minimum principal stresses, where the intermediated principal stress is gravitational (30). The method to obtain the orientation of stress is the same as the hydrofracturing stress meassuring method (11, 12), although usually the value of stress is not estimated in geology. Where the stress field is uniform, radial dikes are formed by the localized magmatic pressure. On the other

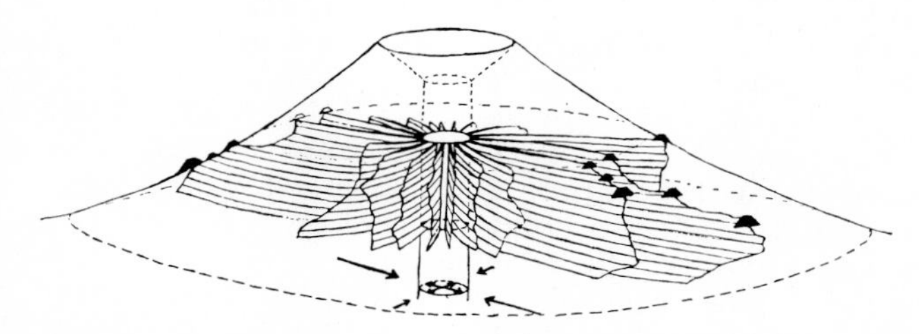

Fig. 7. Diagram showing the radial dikes and flank volcanoes of a polygenetic volcano under differential horizontal stresses (29).

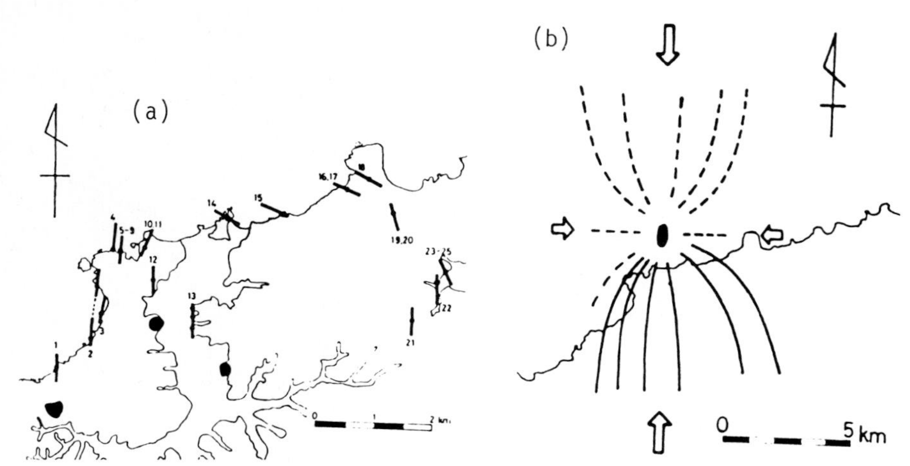

Fig. 8. The dike swarm in Tango Peninsula, Kyoto (23). (a) Distribution of dikes. (b) Schematic model showing a central crater and regional stress field.

hand, when the stress differential exists, radial dikes tend to curve and run paarallel to the maximum principal stress beyond some distance from the central conduit. This is just the same penomenon in the hydrofracturing mechanics as the extension of the previously existing crack tends to curve to the direction perpendicular to the direction of minimum compressive stress (1).

According to Nakamura (28, 29), flank volcanoes of some groups are distributed along a specific line indicating a dike within the polygenetic volcanic edifice (Fig. 7). Consequently, volcanic features as well as dike swarms suggest their validity as indicators of tectonic stress regimes (28-31).

Following the above-mentioned reasoning, Kobayashi (23) inferred the regional stress field and the conduit of magma in the northern Tango penninsula, Kyoto. The average direction of a dike swarm is N-S in this area. The dikes occurring on the western coast, however, strike N-S to N20°E, while those on the eastern coast strike N60°W (Fig. 8a). This pattern of the dike swarm indicates that the direction of the maximum principal stress of the regional stress field is in N-S, and that a central conduit exists offshore (Fig. 8b).

Pollard and his co-workers (35-38, 42) revealed valuable data important to the interpretation of dike system evolution, and discussed the intrusion mechanics of dikes and sills on the basis of hydraulic fracturing. Acoding to Pollard (36), natural sheet intrusions may be good analogs for hydraulic fractures. A similarity between the processes of hydraulic fracturing and sheet intrusion is that both fractures are opened by the internal fluid pressure. On the contrary, an obvious dissimilarity is the temperature difference between fluids and rocks. Temperature differences (= fluid temperature - rock temperature) up to 1000°C may be expected for igneous intrusions. On the other hand, temperature differences over 100°C for

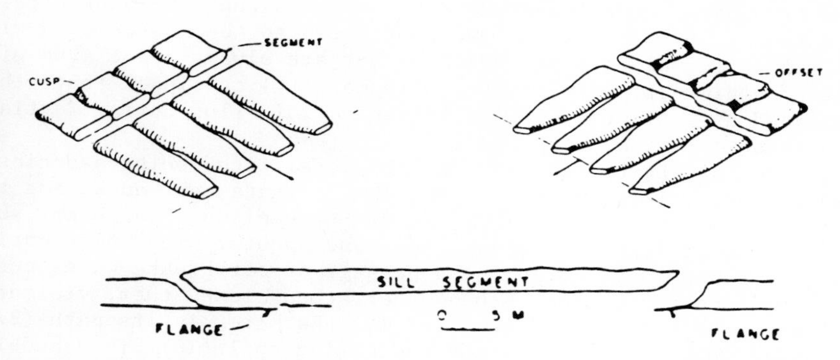

Fig. 9. Idealized segments of sheet intrusion illustrating typical pattern in cross section and geometric features resulting in coalescence (36).

hydraulic fracturing are probably rare. In some hydraulic fracturing, especially for geothermal fields, the fluid is even cooler than the host rock.

The thickness-length ratio of sheet intrusions commonly ranges between 1/100 and 1/1000 (36). It is expected that a planar crack in a homogeneous and isotropic material shows a symmetrical ellipse under uniform internal pressure and uniform regional stress (44). The form in cross section of a dike from the Spanish Peak, however, is asymmetical and similar to an elongated teardrop. This form suggests a linear gradient in the regional stress and magma pressure (37).

Usually, dikes and sills are composed of many separate segments (36). Vertical dike segments are often arranged in echelons. Segments of horizontal sills leave a cusp-shaped indentation along the path of coalescence, or leave a step-like offset where they coalesce (Fig. 9). Pollard (36) concluded that the discontinuous segments, offsets, and echelon arrangements are geometric features related fundamentally to fractures which open as they propagate and that, therefore, they should occur commonly in hydraulic fractures.

CLASTIC DIKES

Clastic dikes are tabular intrusive bodies of clastic materials transecting the bedding of sedimentary formations. They are named breccia dikes, pebble dikes, sandstone dikes, shale dikes, siltstone dikes, and claystone dikes in accordance with the size of the fragments. According to Hayashi (15), clastic dikes resulted from various diastrophisms, such as earthquakes, landslides, and so on.

Sedimentary volcanism characterizes the Guydan Formation of South Texas (7). Evidence for sedimentary volcanism includes relict mud-volcano vents along deep-seated faults and fractures, mud flow deposits, erratic igneous blocks and boulders, limestone blocks, orthoquartzite blocks, serpentinite masses, and numerous structurally controlled clastic dikes. All the dikes fill the fractures which probably were channels of extrusion. Several sandstone dikes are considered to have been injected from below to their present stratigraphic level. Clastic dikes are not known to be associated with the shallow-seated diapiritic mudlumps, but are always associated with deep-seated sedimentary volcanism. Freeman (7) concluded that they were emplaced as intrusive masses by rapid fluid flow during Tertiary diastrophic activity.

Sandblows have been observed in many localities, being associated with earthquakes, for instance in the 1964 Niigata earthquake and the 1964 Great Alaska earthquake. Different mechanisms have been suggested to account for the behavior of sand spouts. The alternative tension and compression during the passage of earthquake waves opens fissures and sucks down the groundwater, then closes them, violently forcing out the water with sand or mud that lie in its path (27). An earthquake shaking liquefies the underlying soil (16). The sandblow extrusion results from the high pressure temporarily confined in underlying aquifers (46).

Scott and Zuckerman (41) carried out a series of laboratory experiments on the mechanics of liquefaction and the generation of sandblows and spouts. Their conclusions are as follows: (i) the presence of a relatively fine-grained layer of soil overlying the liquefiable zone is necessary for the formation of sandblow; (ii) the excess water produced by liquefaction makes its way to the surface by a process of cavity formation and then channel formation in the upper layer; and (iii) the cavities begin to form at the base of the upper layer of the relatively fine-grained soil, and propagate to the surface. This appears to be a kind of hydraulic fracturing.

Some clastic columns are cylindrical in the horizontal section, and thus they are called breccia pipes or pebble pipes. Breccia and pebble columns are frequently associated with epigenetic ore deposits. Especially, they are important for the interpretation of the genesis of porphyry-type ore deposits. According to Bryner (5) who discussed the origin of breccia and pebble columns, (i) collapse might result from withdrawal of magma, slackening of intrusive activity, melting of rock by gases compressed by advancing magma, by distension of the crust caused by doming, by solution of limestone by groundwater, by shrinkage due to the corrosion by gaseous or hydrothermal fluids and by shrinkage due to the cooling of an intrusive magma, and (ii) intruded fragments generally are impelled by magma or by gaseous or aqueous fluids that arise from the magma or are convected by magmatic heat.

VEINS

Many hydrothermal mineral deposits are associated with normal faults, and frequently they are characterized by brecciation, although distinct relationships are not necessarily found among the amount of brreciation, the size of mineral deposits, and the amount of displacement of a fault. Phillips (33) states as follows: (i) normal faults develop as a consequence of an increase in the magnitude of the differential stress above the critical limit under the prevailing pore water condition; (ii) the accumulation of a body of hydrothermal solution on the fault zones under pressure greater than the pore water pressure, results in the extension of the faults by hydraulic fracturing, and (iii) the abrupt drop in the pressure of the hydrothermal solution when fracturing occurs, causes the bursting apart of the rock into which the hydrothermal solution has permeated under high pressure, thus forming angular breccias.

The mineralized zone of the Pueblo Viejo oxide gold deposit, Dominica, is roughly funnel-shaped, and contains numerous, narrow, closely spaced, irregular veins that can be mined by bulk excavating methods (22). Though there is no reference to the displacement of these veinlets, they seem to be of extensional cracks similar to the porphyry-copper type. It is thought by Kesler and others (22) that the vein network was formed by hydraulic fracturing of carbonaceous sediments caused by fluid pressures as the hot spring system developed. This is suggested to account for the upper extensive part of the funnel-shaped geothermal zone at Wairakei, where overpressure formed when the fluid moving upward was ponded and expanded laterally beneath an impermeable zone (6).

According to Pollard (36), sheet intrusions are composed of many segments being arranged in an echelon pattern, or leaving cusps or offsets. Similar figures are not frequently found for hydrothermal veins. Fig. 10 shows some cross sections through segmented quartz veins at the Ohtani scheelite deposit located 20 km to the west of Kyoto. The vein system is composed predominantly of tension fractures striking N20°E and shear fractures striking N40°E (19).

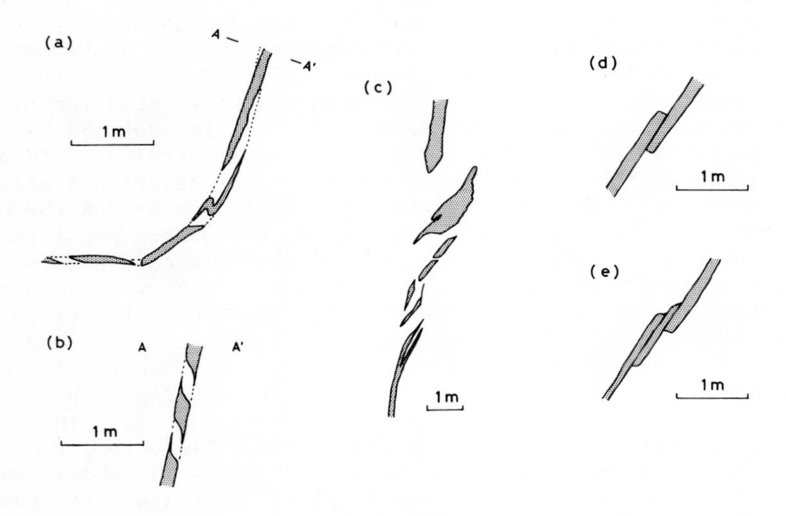

Fig. 10. Ceiling sketches showing quartz veins in Ohtani scheelite deposit, Kyoto. (a) No. 12 vein at 50 m level. (b) Cross Section at line A-A' of (a). (c) No. 13 vein at 50 m level. (d) No. 3 vein at -150 m level. (e) No. 3 vein at -150 m level.

VEIN SYSTEMS IN THE AKENOBE DEPOSIT (43)

The Akenobe Cu-Pb-Zn-W-Sn deposit is located 100 km to the northwest of Osaka. The mining field consists of three areas. The Daiju area is west of the Akenobe fault. The Daisen-Sekiei area and the Kanagidani area are north and south of the 25-go fault, respectively.

The gallery system in the Sekiei area (Fig. 11a) consists predominantly of three directions striking N10-20°W, N40-50°W and N70-80°W, excluding cross-cuts. These correspond to the dominant directions of the vein system. Accordingly, it is thought that the vein system in this area consists of extensional cracks striking N40-50°W and a conjugate set of shear fractures striking N10-20°W and N70-80°W, which were formed in the stress field where the maximum and minimum principal stresses were directed toward NW-SE and NE-SW, respectively. On the other hand, the gallery pattern of the Chiemon vein group in the Daiju area (Fig. 11b) is very complicated, as compared with the Sekiei area. The underground map does not indicate the predominant direction of the vein system. This is explained by the fact that there are many branch veins.

The relationships between veins and their wall rocks are different in accordance with the varieties of the gallery patterns. A distinct boundary is observed along the veins which belong to the Sekiei type of the simple pattern. Fig. 12a is an example. This indicates that the failure took place along a single plane. On the contrary, the boundary between a vein and its wall rock is not well defined for the complicated gallery pattern, because the disseminated ore occurs along the veins which belong to the Chiemon type (Fig. 12b). The existence of disseminated ore suggests that the failure took place not only along a single plane but also in its neighbor where a number of micro-fractures were formed.

The two types of vein systems seem to have resulted from the difference in fracture mechanics. The simple pattern seems to correspond to brittle failure and the complicated one to ductile deformation. In the former case, the dominant directions of brittle fractures coincide with the maximum principal stress and the maximum shear planes. Consequently, many micro-fractures were not formed near the fracture zone. This implies that the veins concentrate in the three directions, and that the boundary of a vein is distinct. On the other hand, in the latter case, the greatest strain in the ductile deformation is due to many small fractures formed in the material (25). The directions of micro-fractures scatter in ductile deformation, as compared with brittle fracturing. These micro-fractures might be observed as branch veins or disseminated ore.

Why did brittle failure take place in the Sekiei area, and ductile deformation occur near the Chiemon vein? Fig. 13 shows schematic stress-strain curves for rocks at various confining pressures and pore-pressures. Under a certain confining pressure, brittle fracture occurs with the increase of strength and the increase in permanent strain before the fracture. The stress-strain curve is completely different when the confining pressure goes over a certain value. A great strain occurs without loss of strength. In other words, there is a transition from the typical brittle behavior to fully ductile

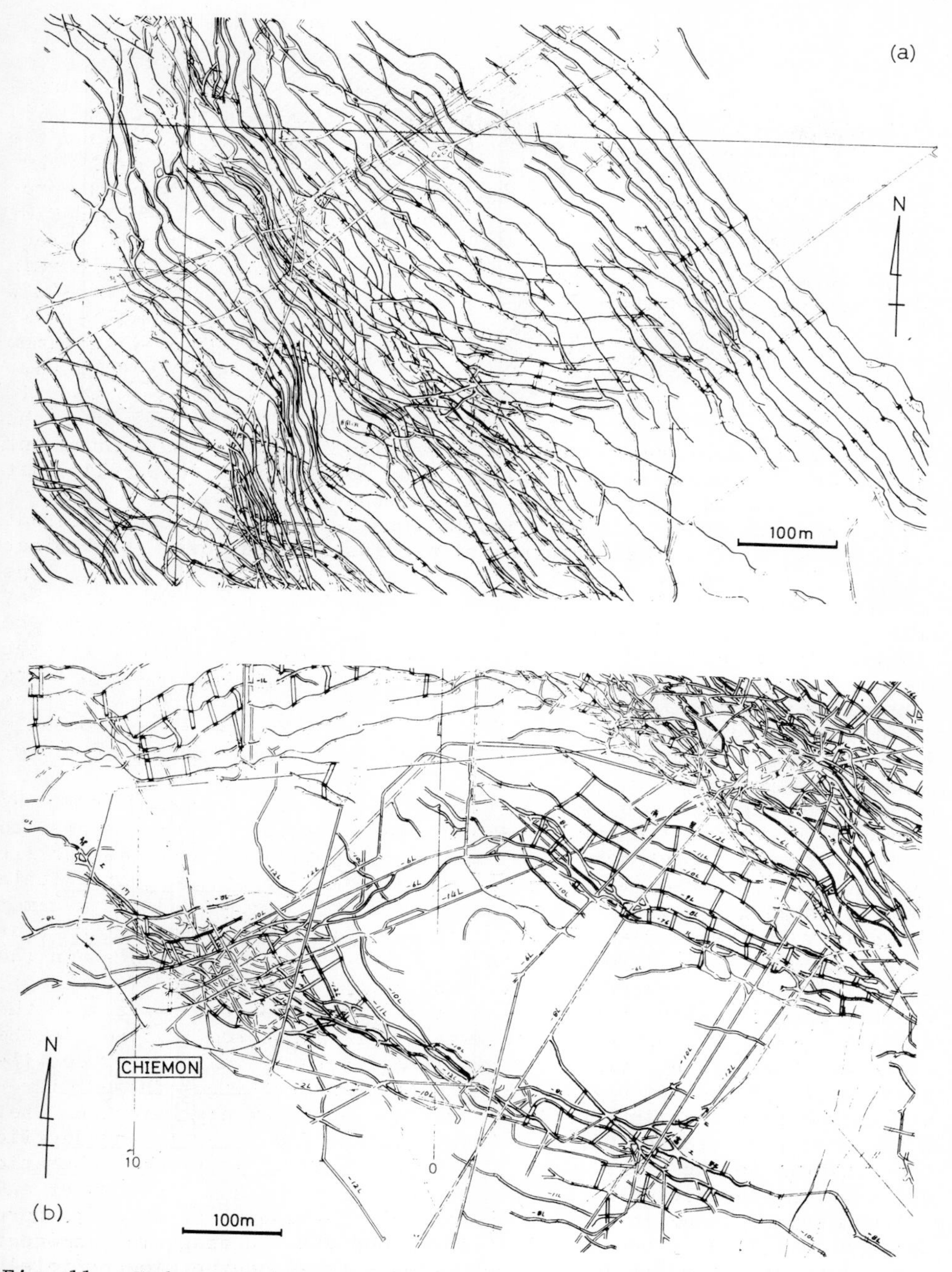

Fig. 11. Underground maps of (a) a part of the Sekiei area, and (b) a part of the Kaiju area (43).

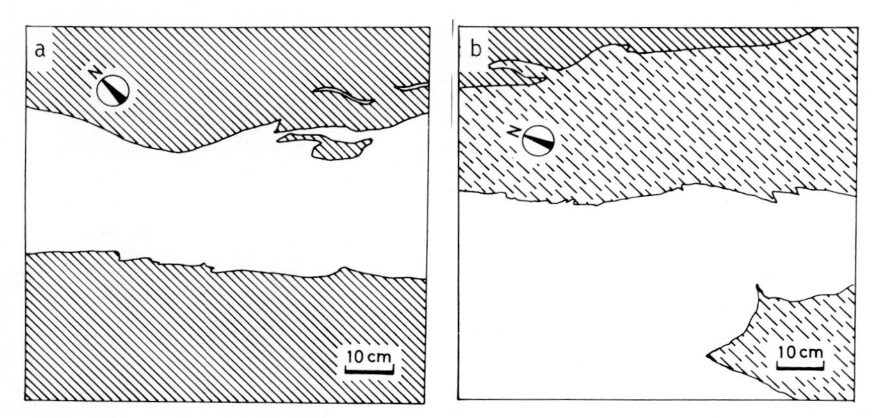

Fig. 12. Underground sketches of shieling showing the contact between veins (white) and country rocks (parallel solid lines) (43). (a) Shirogane 3-go vein (-6 L): distinct contact. (b) Chiemon 7-go vein (sublevel of -11 L): accompanied with disseminated ore (broken lines).

behavior. This is called the brittle-ductile transition. The transition is controlled by the effective confining pressure (20). If the actual confining pressure is constant, the curve for high confining pressure corresponds to the curve for low pore-pressure, and vice versa (Fig. 13). Since geology does not differ for the Sekiei area and the Daiju area, the confining pressures appear also to have been similar in both areas. Consequently, if the pore-pressure in the Sekiei area was higher than that in the Daiju area, brittle failure took place in the Sekiei area, while ductile deformation was dominant in the Daiju area. This is one of the interpretations. Two other interpretations are proposed. The one concerns the mechanical

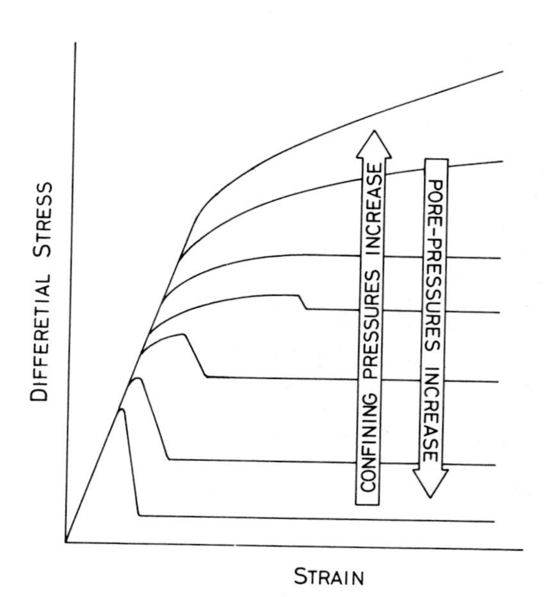

Fig. 13. Schematic diagram showing the relation between brittle-ductile transition and confining pressure or pore-pressure.

properties of the country rocks. This is inferred from the fact that the veins of the simple pattern are localized in the pumpellyite-actinolite facies zone, and that the veins of the complicated pattern occur in the prehnite-pumpellyite facies zone. The other concerns the temperature. Generally, the rock becomes ductile with the increase of temperature. We cannot, however, advance the discussion, because of little information on temperature during fracture formation.

CONCLUDING REMARKS

In the Earth's crust, there occur many emplacement masses and intrusive bodies whose constituents were derived from fluid or fluid-like phases, such as magma, aqueous solutions, and so on. Some of them appear to have simply filled previously existing fissures or cavities, and others may have spread the country rock in the manner of the fluid of hydraulic fracturing. Unfortunately, we cannot, however, affirm geological hydrofracturing, because of little fractographical information on hydraulic fracture. Pollard (36) has represented many characteristic features of sheet intrusions. Are they sufficient evidence of hydraulic fracturing? Is there any other evidence? The concept of effective stress implies that the values themselves of pore-pressure and each principal stress are not important for failure as long as the condition is represented by Curve III in Fig. 1. Fractographical differences may be found, however, between the conditions where both pore- and confining pressures are high, and where both are low. To answer these questions, we must await the advance in the fractographical research of hydraulic fracturing and pore-pressure effects.

The pressure differential between a pump and the edge of a crack is usual in actual hydraulic fracturing. This implies that the extension of failure leads to a temporary drop in pore-pressure and the adiabatic expansion of the pore-fluid. This change of the physico-chemical condition may lead to phase transition, such as boiling of the aqueous solution and degassing from the magma, and then may accelerate the crystallization (or solidification) of constituents of dikes or veins which are forming from fluid materials. It seems to be sufficiently significant to investigate the relation between crystallization and such features of sheet intrusions as described by Pollard (36), or such features of veins as shown in Fig. 10.

ACKNOWLEDGEMENTS: We wish to thank Professor Yuichi Nishimatsu, and Associate Professor Kazuaki Nakamura, the University of Tokyo, and Dr. Hitoshi Koide, Geological Survey of Japan, for critical reading of the manuscript and many valuable suggestions.

REFERENCES

1. Abou-Sayed, A.S., Brechtell, C.E., and Clifton, R.J., "In Situ Stress Determination by Hydrofracturing: A Fracture Mechanics Approach," J. Geophy. Res., Vol. 83, No. B6, 1978, pp. 2851-2862.
2. Anderson, E.M., "The dynamics of the Formation of Cone Sheets, Ring-dykes and Caldron-subsidences," Proc. Roy. Soc. Edinburgh, Vol. 56, 1936, pp. 128-157.

3. Anderson, E.M., The Dynamics of Faulting and Dike Formation, 2nd ed., Oliver and Boyd, London, 1951, in (32).
4. Billings, M.P., "Mechanics of Igneous Intrusion in New Hampshire," Am. J. Sci., Vol. 243A (Daly Vol.), 1945, pp.40-68.
5. Bryner, Leonid, "Breccia and Pebble Columns Associated with Epigenetic Ore Deposits," Econ. Geol., Vol. 56, 1961, pp. 488-508.
6. Elder, A. J., The Bowels of the Earth, Oxford University Press, London, 1976, in (8).
7. Freeman, P.S., "Exposed Middle Tertiary Mud Diapirs and Related Fractures in South Texas," Diapirism and Diapirs, Jules Braunstein, and G.D. O'Brien, eds., Am. Assoc. Petroleum Geol., Mem. 8, 1968, pp. 162-182.
8. Fyfe, W.S., Price, N.J., and Thompson, A.B., Fluid in the Earth's Crust, Elsevier Scientific Publishing Company, Amsterdam, 1978.
9. Gussow, W.C., "Energy Source of Intrusive Masses," Trans. Roy. Soc. Canada, Series III, Section III, Vol. 56, 1962, pp.1-19.
10. Gussow, W.C., "Salt Diapirism: Importance of Temperature, and Energy Source of Emplacement," Diapirism and Diapirs, Jules Braunstein, and G. D. O'Brien, eds., Am. Assoc. Petroleum Geol., Mem. 8, 1968, pp. 16-52.
11. Haimson, B.C., "A Simple Method for Estimating In Situ Stresses at Great Depths," Field Testing and Instrumentation of Rock, ASTM STP 554, Am. Soc. Testing and Materials, 1974, pp. 156-182.
12. Haimson, B.C., "The Hydrofracturing Stress Measuring Method and Recent Field Results," Inter. J. Rock Mech. Min. Sci. & Geomech. Abstr., Vol. 15, 1978, pp. 167-178.
13. Haimson, B. and, Fairhurst, C., "Initiation and Extension of Hydraulic Fractures in Rocks," Soc. of Petroleum Eng. J., Vol. 7, 1967, pp. 310-318.
14. Hanna, M.A., "Salt Domes: Favorite Home for Oil," Oil & Gas J., Vol. 57, No. 6, 1959, pp. 138-142.
15. Hayashi, Tadaichi, "Clastic Dikes in Japan (I)," Jap. J. Geol. Geogr., Vol. 37, No. 1, 1966, pp. 1-20.
16. Housner, George W., "The Mechanism of Sandblows," Bull. Seis. Soc. Am., Vol. 48, 1958, pp. 155-161.
17. Hubbert, M.K., and Rubey, W.W., "Role of Fluid Pressure in the Mechanics of Overthrust Faulting. I. Mechanics of Fluid-Filled Porous Solid and its Application to Overthrust Faulting," Bull. Geol. Soc. Am., Vol. 70, 1959, pp. 115-166.
18. Hubbert, M.K., and Willis, D.G., "Mechanics of Hydraulic Fracturing," Trans. Soc. Petroleum Eng. AIME., Vol. 210, 1957, pp. 153-168.
19. Imai, Hideki, Kim, Moon Yung, and Fujiki, Yoshinori, "Geologic Structure and Mineralization of the Hypothermal or Pegmatitic Tungsten Vein-Type Deposits at the Ohtani and Kaneuchi Mines, Kyoto Prefecture, Japan," Mining Geol., Vol. 22, 1972, pp. 371-381.
20. Jaeger, J.C., and Cook, N.G.W., Fundamentals of Rock Mechanics, Chapman and Hall, London, 1976.
21. Jensen M.L., and Bateman, A.M., Economic Mineral Deposits, 3rd ed., Revised Printing, John Wiley & Sons, New York, 1981.

22. Kesler, S.E., Russell, N., Seaward, M., Rivera, J., McCurdy, K., Cumming, G.L., and Sutter, J.F., "Geology and Geochemistry of Sulfide Mineralization Underlying the Pueblo Viejo Gold-Silver Oxide Deposit, Dominican Republic," Econ. Geol., Vol. 76, 1981, pp. 1096-1117.
23. Kobayashi, Yoji, "Late Neogene Dike Swarms and Regional Tectonic Stress Field in the Inner Belt of Southwest Japan," Bull. Volcanol. Soc. Japan, 2nd Series, Vol. 24, No. 3, 1979, pp. 153-168 (in Japanese with English abstract).
24. Koide, Hitoshi, and Bhattacharji, S., "Formation of Fractures around Magmatic Intrusions and their Role in Ore Localization," Econ. Geol., Vol. 70, 1975, pp. 781-799.
25. Koide, Hitoshi, and Hoshino, Kazuo, "Development of Microfractures in Experimentally Deformed Rocks (Preliminary Report)," J. Seis. Soc. Japan, 2nd Series, Vol. 20, 1967, pp. 85-97 (in Japanese with English Abstract).
26. Knopf, A., "Igneous Geology of the Spanish Peaks Region, Colorado," Bull. Geol. Soc. Am., Vol. 47, pp. 1727-1784.
27. Macelwane, J. B., When the Earth Quakes, Bruce Publishing Company, Milwaukee, 1947, in (41).
28. Nakamura, Kazuaki, "Arrangement of Parasitic Cones as a Possible Key to Regional Stress Field," Bull. Volcanol. Soc. Japan, 2nd Series, Vol. 14, No. 1, 1969, pp. 8-20 (Japanese with English abstract).
29. Nakamura Kazuaki, "Volcano Structure and Possible Mechanical Correlation between Volcanic Eruptions and Earthquakes," Bull. Volcanol. Soc. Japan, 2nd Series, Vol. 20, 1975, pp. 229-240 (in Japanese with English Abstract).
30. Nakamura, Kazuaki, "Volcanoes as Possible Indicators of Tectonic Stress Orientation - Principle and Proposal," J. Volcanol. Geoth. Res., Vol. 2, 1977, pp. 1-16.
31. Nakamura, Kazuaki, Jacob, K.H., and Davies, J.H., "Volcanoes as Possible Indicators of Tectonic Stress Orientation - Aleutians and Alaska," Pageoph, Vol. 115, 1977, pp. 87-112.
32. Odé, Helmer, "Mechanical Analysis of the Dike Pattern of the Spanish Peak Area, Colorado," Bull. Geol. Soc. Am., Vol. 68, 1957, pp. 567-576.
33. Phillips, W.J., "Hydraulic Fracturing and Mineralization," J. Geol. Soc. London, Vol. 128, 1972, pp. 337-359.
34. Phillips, W.J., "The Dynamic Emplacement of Cone Sheets," Tectonophys., Vol. 24, 1974, pp. 69-84.
35. Pollard, D.D., "On the Form and Stability of Open Hydraulic Fractures in the Earth's Crust," Geophys. Res. Let., Vol. 3, No. 9, 1976, pp. 513-516.
36. Pollard, D.D., "Forms of Hydraulic Fractures as Deduced from Field Studies of Sheet Intrusions," Preprint-Proc. 19th U.S. Symp. Rock Mech., Stateline, Nevada, 1978, pp. 1-9.
37. Pollard, D.D., and Muller, O.H., "The Effect of Gradients in Regional Stress and Magma Pressure on the Form of Sheet Intrusions in Cross Section," J. Geophy. Res., Vol. 81, No. 5, 1976, pp. 975-984.
38. Pollard, D.D., Muller, O.H., and Dockstader, D.R., "The Form and

Growth of Fingered Sheet Intrusions," Geol. Soc. Am. Bull., Vol. 86, 1975, pp. 351-363.

39. Ramberg, Hans, Gravity, Deformation and the Earth's Crust, Academic Press, London, 1967; 2nd ed., 1981.

40. Robson, G.R., and Barr, K.G., "The Effect of Stress on Faulting and Minor Intrusions in the Vicinity of a Magma Body," Bull. Volcanol., Tome 27, 1964, pp. 315-330.

41. Scott, R.F., and Zuckerman, K.A., "Sandblows and Liquefaction," The Great Alaska Earthquake of 1964, Engineering, U.S. Nat. Acad. Sci., Wash., D.C., 1973, pp. 179-189.

42. Secor, D.T., Jr., and Pollard, D.D., "On the Stability of Open Hydraulic Fractures in the Earth's Crust,", Geophys. Res. Let., Vol. 2, No. 11, 1975, pp. 510-513.

43. Shoji, Tetsuya, "Vein Systems Interpreted from the viewpoint of Brittle and Ductile Behavior of Rocks: A Case Study of the Akenobe Poymetallic Deposit, Southwestern Japan," Mining Geol., Vol. 32, No. 1, 1982, pp. 47-54 (in Japanese with English abstract).

44. Sneddon, I.N., "The Distribution of Stress in the Neighborhood of a Crack in an Elastic Solid," Proc. Roy. Soc. London, Ser. A, Vol. 187, 1946, pp. 229-260.

45. Terzaghi, K. Van, Theoretical Soil Mechanics, Wiley, New York, 1943.

46. Waller, Roger M., "Effects of the March 1964 Alaska Earthquake on the Hydrology of the Anchorage Area, Alaska," U.S. Geol. Surv. Prof. Paper 544-B, 1966.

A METHOD OF ESTIMATING UNDERGROUND TEMPERATURE FROM A SOLUTION OF ITERATIVE LEAST-SQUARES INTERPRETATION OF SCHLUMBERGER'S RESISTIVITY SOUNDING CURVES IN GEOTHERMAL FIELDS

Seibe ONODERA

Prof. of Exploration Geophysics
Department of Mining
Faculty of Engineering
Kyushu University

ABSTRACT

An empirical formula for estimating underground temperature is presented. This method uses a solution of the iterative least-squares interpretation of Schlumberger's resistivity sounding curves.

If an attainable depth of resistivity sounding is satisfied, then the temperature calculated by the formula represents that of a fracture-type geothermal reservoir. If not, it gives only the low temperature of the underground.

INTRODUCTION

In order to find geoelectrical indications of a geothermal reservoir by resistivity sounding curves and to evaluate the result of interpretation of RS (resistivity sounding) data, a comparison was made between the results of interpretation of RS curves and geothermal well data. As a result, the well data correlating with the solution was really a bottom hole temperature in the course of drilling.

On the other hand, the application of the iterative least-squares interpretation method, based on the assumption of multi-layered structure, to Schlumberger's RS curves observed at some stations in the Hatchobaru geothermal field produced good results, and the geothermal resistivity interpretation system (installed by the Grant-in-Aid for Scientific Research (A) in 1980) displayed the step resistivity distribution.

These play an important part in the present investigation.

The object of this paper is to present a method for estimating the underground temperature (temperature of a fracture-type geothermal reservoir, if possible,) at a depth corresponding to the deepest interface between resistivity layers.

AREA OF INVESTIGATION

Figure 1 shows the location of wells and the arrangement of resistivity lines at the Hatchobaru field. After careful

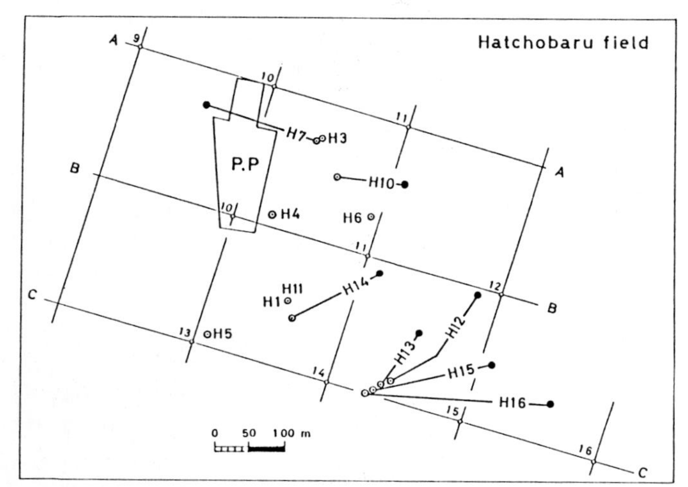

Fig. 1. Location of wells and arrangement of resistivity lines at the Hatchobaru geothermal field

examinations of the field RS curves, the interpretations were limited to only several RS curves observed at stations, A-10, B-10, B-11, B-12, C-14, C-15 and C-16. Also, steam production wells, H-1, H-4, H-6, H-7, H-10, H-12, H-13, H-15 and H-16 were taken into consideration.

INTERPRETATION OF RS CURVES

In general, the iterative least-squares interpretation produces a number of solutions with the precondition of convergence. These solutions can be grouped in accordance with the accuracy range and depend upon initial values.

For example, the solution, that is, the result of the interpretation of the RS curve observed at station, A-10 is shown in Figure 2 with the highest accuracy of curve fitting. This graph was produced by the geothermal resistivity interpretation system (graphic display apparatus connected to a computer).

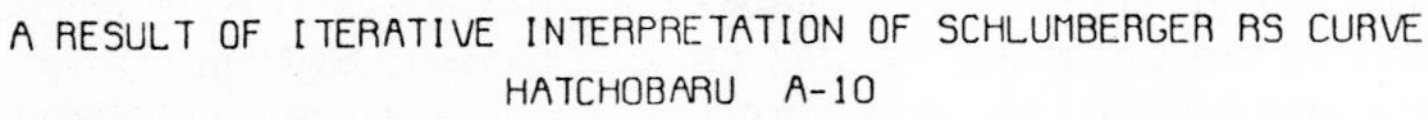

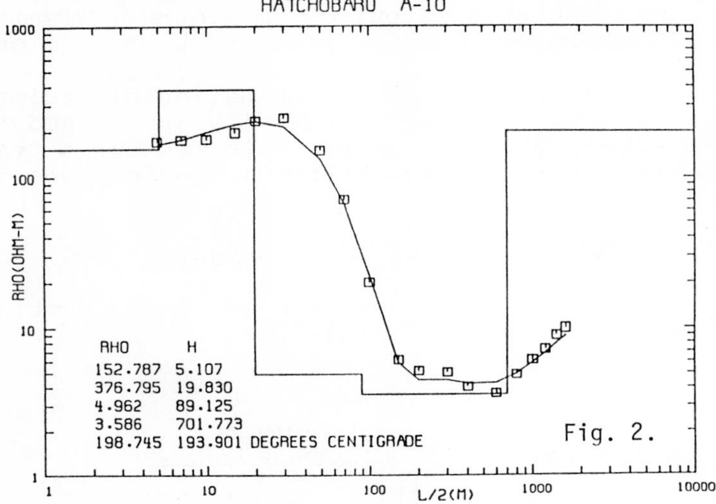

Fig. 2.

The abscissa represents the half electrode separation of the Schlumberger array, denoted by L/2, and the ordinate shows resistivity in ohm-m. Regular square marks show observed apparent resistivities corresponding to the selected electrode separations, and a solid curve represents the theoretical RS curve obtained by automatic curve fitting. The solution of iterative least-squares interpretation of RS curves is given by a step resistivity distribution, whose numerical values of resistivity in ohm-m for successive layers and of depth in m of interfaces for the assumed multi-layered structure are plotted below Labels, RHO and H, respectively, in the left-hand lower corner of Figure 2.

The RS curve observed at A-10 shows the type 7(1221) of 5-layer, and gives the characteristic form in a geothermal field.

Note resistivity values of ρ_4=3.586 ohm-m and ρ_5=198.745 ohm-m for the 4-th and 5-th resistivity layers, respectively, and the depth of h_4=701.773 m for their 4-th interface. Then the 4-th resistivity layer can be correlated with the strongest altered formation, and the 5-th resistivity layer is correlated with the fracture-type geothermal reservoir, in which cracks probably exist. In general, the target of drilling depth is expressible as 701.8xα. Since the value of α in the Hatchobaru field is statistically given by 1.13, the depth of drilling target becomes 793 m approximately.

Therefore the result of the interpretation of the RS curve at A-10 indicates the existence of a fracture-type geothermal reservoir at a depth of order 800 m, and a circular area with radius of about 100 m around the same station is a suitable position for drilling a production well.

From this point of view, the combination of resistivity values for the 4-th and 5-th resistivity layers and the depth of their interface, together with the curve type, is defined as "Geothermal necessary condition" or "Geothermal requirement" in this paper.

COMPARISON OF SOLUTION AND TEMPERATURE LOG

The comparison of the temperature log in production well, H-7, and the solution for A-10 are shown in Figure 3.

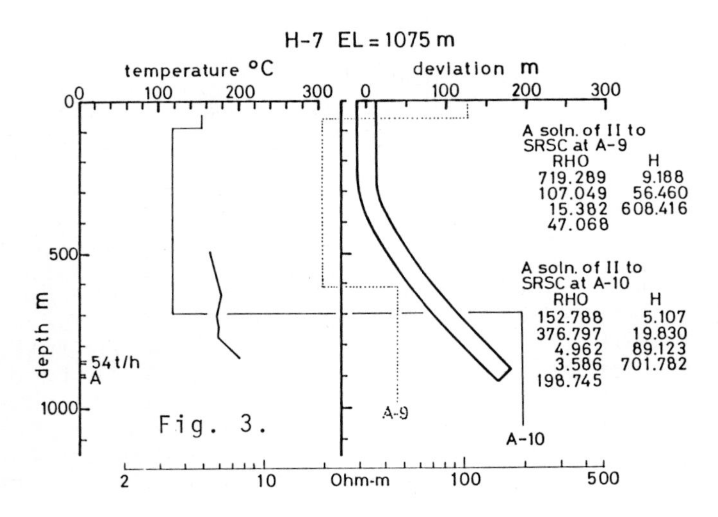

Fig. 3. Correlation of the temperature log in production well, H-7 with the interpreted results of RS curves observed at stations, A-10 and A-9 in the Hatchobaru geothermal field

The temperature in the well at a depth corresponding to the 4-th interface (702 m) is 175 ^{o}C. (Thus the difference of the observed and estimated temperatures is 18.9 ^{o}C, which is the maximum among the case studies.)

A METHOD OF ESTIMATING UNDERGROUND TEMPERATURE

Tables 1 and 2 show solutions and temperatures in vari-

Table 1. Solutions of the iterative least-squares interpretation of resistivity sounding curves observed at Hatchobaru geothermal field

STATION	TYPE CURVE	SOLUTIONS	VALUES OF F
A-09*	4-layer 7(221)	RHO(719.289, 107.049, 15.382, 47.068) H(9.188, 56.460,608.416)	3.06
A-10	5-layer 7(1221)	RHO(152.787, 376.795, 4.962, 3.586,198.745) H(5.107, 19.830, 89.125,701.773)	55.42
A-11*	4-layer 7(221)	RHO(327.935, 12.016, 7.172, 58.934) H(6.455, 45.457,1021.131)	8.22
B-10	4-layer 7(221)	RHO(460.064, 10.136, 6.066, 59.760) H(22.847, 86.853,754.050)	9.85
B-11	4-layer 7(221)	RHO(265.945, 89.475, 4.616, 30.095) H(6.490, 135.884,512.065)	6.52
B-12	6-layer 15(12221)	RHO(480.083, 617.182,585.169, 26.122, 1.952,198.745) H(4.040, 26.409, 93.320,227.259,694.375)	101.82
C-13*	4-layer 7(221)	RHO(160.819, 25.091, 8.025,105.889) H(8.182, 595.530,996.791)	13.19
C-14	4-layer 7(221)	RHO(398.346, 148.835, 5.054,115.635) H(4.857, 143.436,544.696)	22.88
C-15	5-layer 7(1221)	RHO(474.792,1391.752,119.623, 4.577,234.537) H(4.918, 55.065,168.159,702.721)	51.24
C-15	5-layer 7(1221)	RHO(474.792,1391.752,119.623, 4.577,520.678) H(4.918, 55.065,168.159,702.721)	113.76
C-16	5-layer 15(2221)	RHO(9673.617,2440.377,122.436, 1.258, 50.000) H(26.177, 86.951,221.719,871.718)	39.74
C-16	5-layer 15(2221)	RHO(9673.617,2440.377,122.436, 1.258,101.117) H(26.117, 86.951,221.719,1000.000)	80.38

*Schlumberger RS curves at stations with star are omitted from this temperature estimation.

Table 2. Observed temperature in various wells at depth corresponding to interface between altered formation and fracture-type geothermal reservoir at Hatchobaru geothermal field

WELL NO.	t,°C	di,m	rs	dr,m	ERROR	REMARKS
H-1	105	754m	(B-10)	1000	-24.6%	
	70	512m	(B-11)	1000	-48.8%	
H-4	95	754m	(B-10)	1000	-24.6%	
H-6	65	512m	(B-11)	1250	-59.0%	
H-7	175	702m	(A-10)	850	-17.4%	
H-10	90	512m	(B-11)	650	-21.2%	
H-12	220	694m	(B-12)	810	-14.3%	Good result
H-13	230	694m	(B-12)	703	- 1.3%	Very good
	155	545m	(C-14)	703	-22.4%	
	230	703	(C-15-2)	703	0.0%	Very good
H-15	190	703	(C-15-1)	1000	-29.7%	
H-16	195	872	(C-16-1)	1000	-12.8%	Good result
	220	1000	(C-16-2)	1000	0.0%	Very good

t,°C: temperature in well at a depth corresponding to the interface between altered formation and fracture-type geothermal reservoir
di,m: depth corresponding to the interface between altered formation and fracture-type geothermal reservoir, which is estimated from the solution
rs : resistivity sounding curve observed at station, ( )
dr,m: depth of fracture-type geothermal reservoir estimated from temperature log.

ous wells at depths corresponding to interfaces, hi.

Suppose that the temperature of the bottom hole for each well in the course of drilling represents the temperature of the underground formation involved, and that a resistivity structure was obtained by the interpretation of a resistivity sounding curve.

Let F be the ratio of resistivity values of the lowest layer and its over layer, let h be the depth of interface of the two adjacent layers, and let t in ^{0}C be the temperature in the well at a depth corresponding to the interface involved.

The numerical values of F and h necessary to evaluate the underground temperature of the fracture-type geothermal reservoir are tabulated in the fourth column of Table 1, and the third column of Tables 1 and 2, respectively.

If the values of the temperature given in the second column of Table 2 are plotted as abscissas and those of F as ordinates, then Figure 4 is obtained.

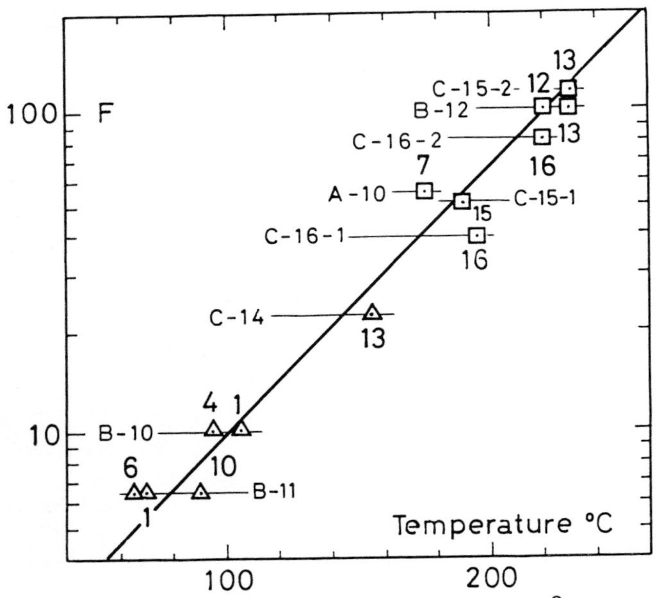

Fig. 4. Relation between F and Temperature in ^{0}C in wells at Hatchobaru geothermal field

It is seen that the relation between t and F can be roughly expressed by an equation of a straight line. Thus the underground temperature, t ^{o}C corresponding to h is expressed by an empirical formula of the form

$$t = 50.15 \cdot \ell nF + \frac{\partial \theta}{\partial h} \cdot h - 28.5 \quad ^{o}C \quad \ldots\ldots\ldots\ldots \quad (1)$$

where $\partial\theta/\partial h$ is the normal geothermal gradient, which is given by 3 ^{o}C/100 m.

As for the underground temperature calculated by the formula, in fact, there are estimations having close relation to the temperature of a fracture-type geothermal reservoir in the Hatchobaru field (A-10 and H-7, B-12 and H-12 & H-13, C-15-2 and H-13, C-16-1 & C-16-2 and H-16) and are not so (B-10 & B-11 and H-1, B-10 and H-4, B-11 and H-6, B-11 and H-10, C-14 and H-13, C-15-1 and H-15) as shown in Table 2 and Figures 5, 6, 7, 8, 9, 10, 11 and 12.

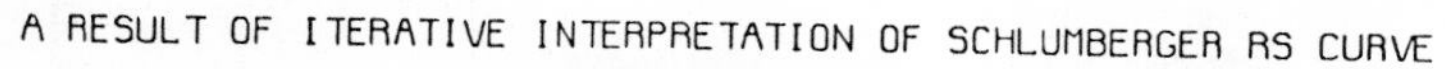

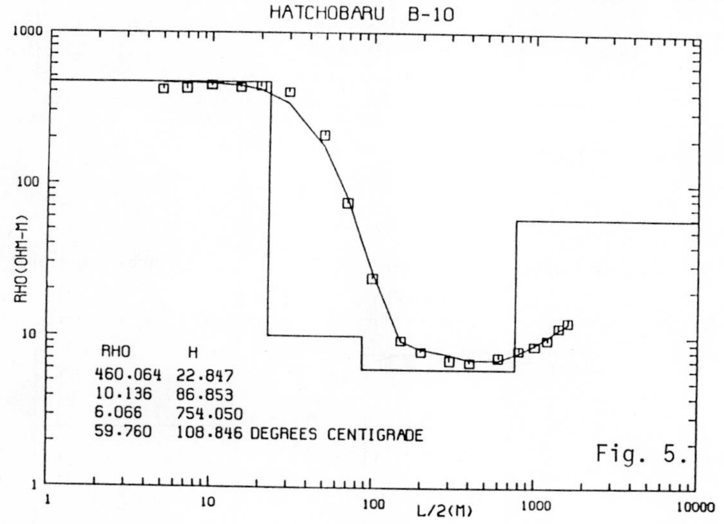

Fig. 5.

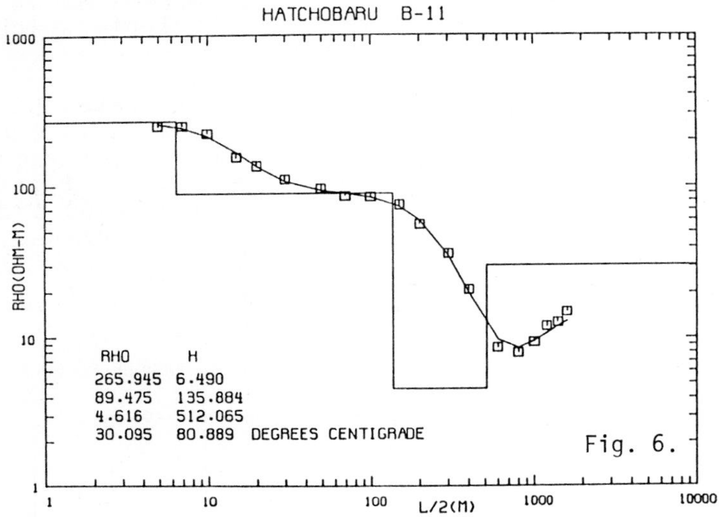
A RESULT OF ITERATIVE INTERPRETATION OF SCHLUMBERGER RS CURVE
HATCHOBARU B-11
RHO(OHM-M)
L/2(M)
1000
100
10
1
1
10
100
1000
10000
RHO H
265.945 6.490
89.475 135.884
4.616 512.065
30.095 80.889 DEGREES CENTIGRADE

Fig. 6.

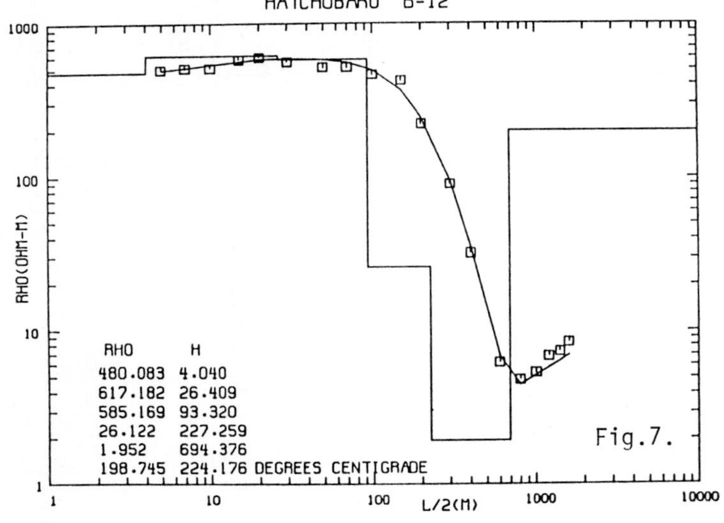
A RESULT OF ITERATIVE INTERPRETATION OF SCHLUMBERGER RS CURVE
HATCHOBARU B-12
RHO(OHM-M)
L/2(M)
1000
100
10
1
1
10
100
1000
10000
RHO H
480.083 4.040
617.182 26.409
585.169 93.320
26.122 227.259
1.952 694.376
198.745 224.176 DEGREES CENTIGRADE

Fig.7.

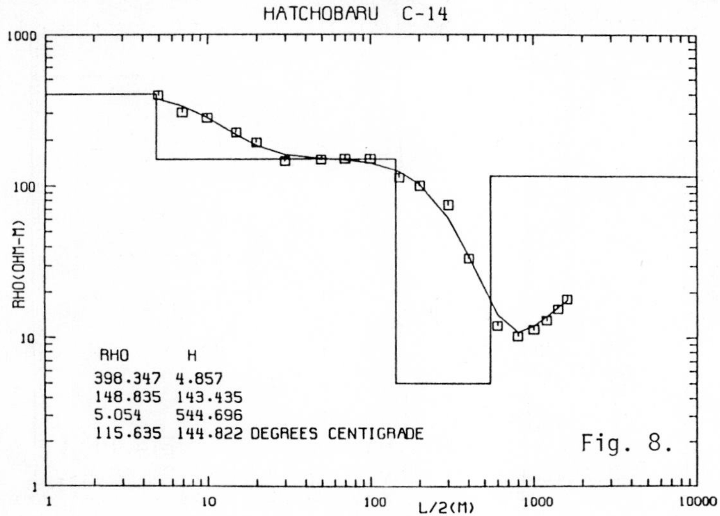
A RESULT OF ITERATIVE INTERPRETATION OF SCHLUMBERGER RS CURVE
HATCHOBARU C-14
1000
100
10
1
RHO(OHM-M)
1
10
100
1000
10000
L/2(M)
RHO H
398.347 4.857
148.835 143.435
5.054 544.696
115.635 144.822 DEGREES CENTIGRADE

Fig. 8.

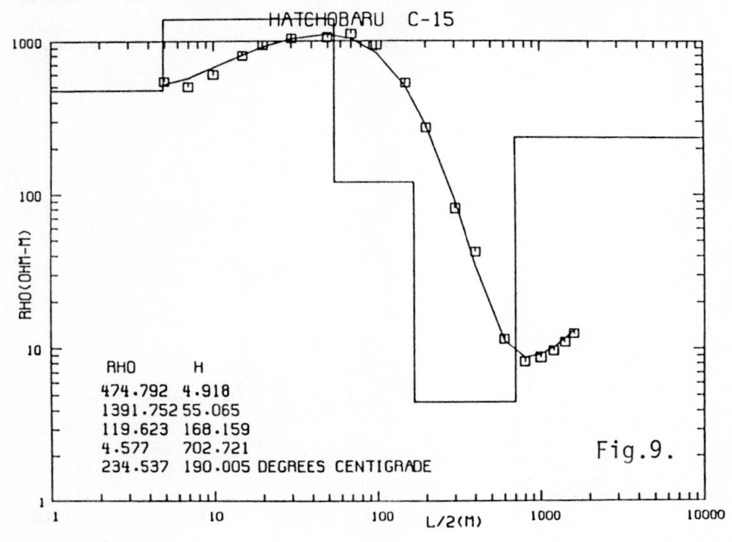
A RESULT OF ITERATIVE INTERPRETATION OF SCHLUMBERGER RS CURVE
HATCHOBARU C-15
1000
100
10
1
RHO(OHM-M)
1
10
100
1000
10000
L/2(M)
RHO H
474.792 4.918
1391.752 55.065
119.623 168.159
4.577 702.721
234.537 190.005 DEGREES CENTIGRADE

Fig.9.

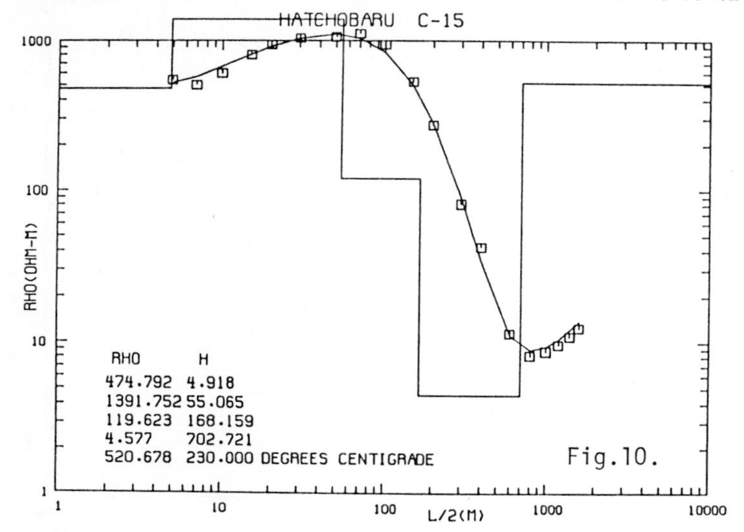

Fig.10.

A RESULT OF ITERATIVE INTERPRETATION OF SCHLUMBERGER RS CURVE

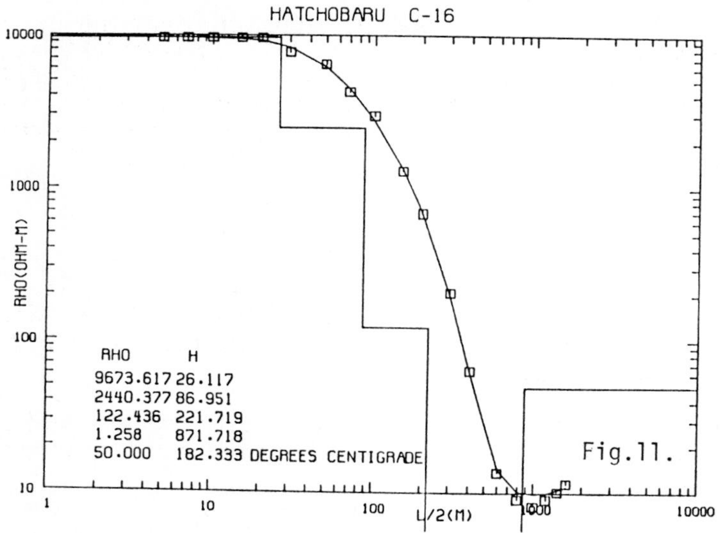

Fig.11.

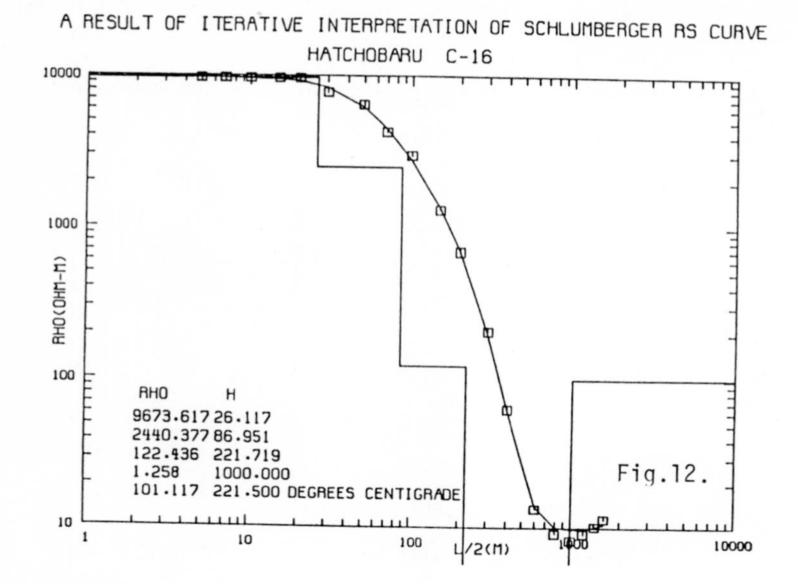

From the attainable depth or the resolving power of resistivity sounding, the effect of topography on the resistivity sounding curve, etc., we can know whether the computation represents the temperature of a fracture-type geothermal reservoir or not.

Such an evaluation of underground temperature points out a "Geothermal sufficient condition" in resistivity exploration for geothermal energy resources. If the calculated temperature exceeds 200 ^{o}C, then it provides a geothermal sufficient condition for exploiting geothermal energy resources to generate electricity, so far as the temperature is concerned.

APPLICATION

The application of equation (1) to estimate the temperrature of a fracture-type geothermal reservoir is also valid for the permeable geothermal reservior in El Tatio, Chile as shown in Figures 13, 14 and 15.

In this case, only the 2nd level reflects on the RS curve, because of the resolving power of Schlumberger's resistivity sounding.

Summarizing the results, for the 2nd level, it is estimated to be 213 ^{o}C on the average for the range from 176 ^{o}C to 243 ^{o}C. In practice, the observed temperature for the same level is 228 ^{o}C at a depth ranging from 480 m to 530 m

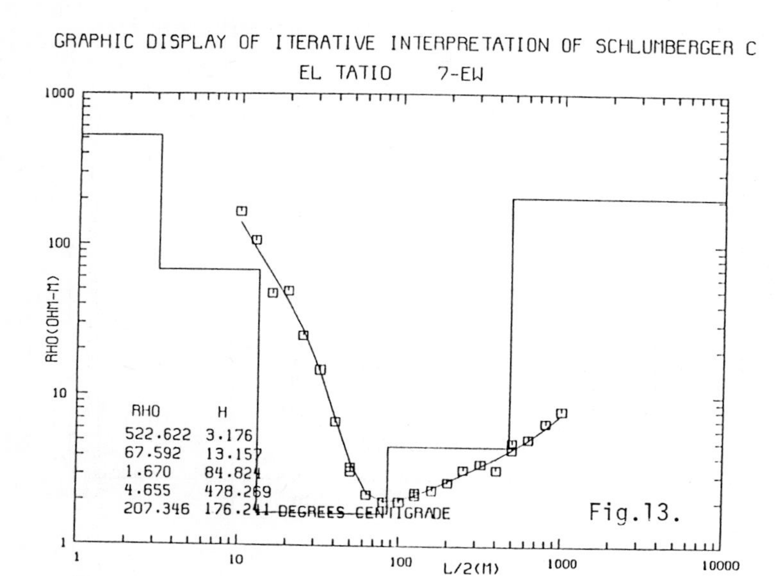

Fig.13.

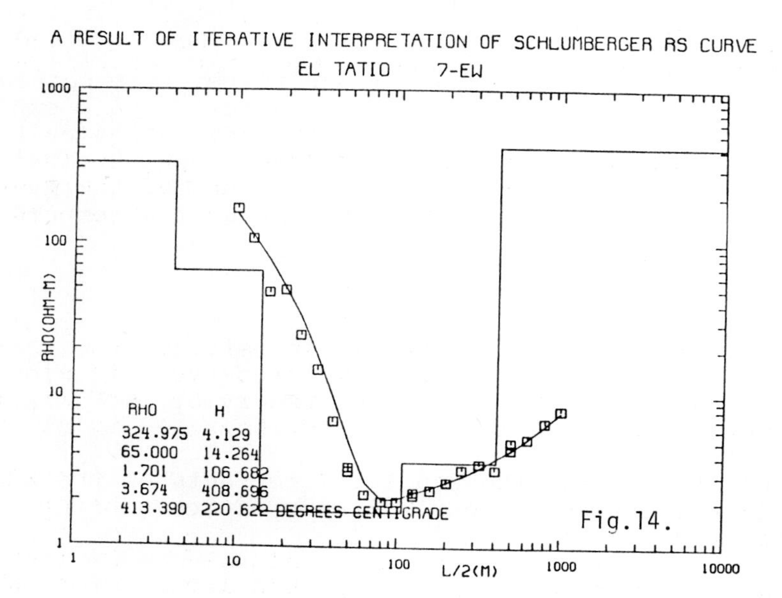

Fig.14.

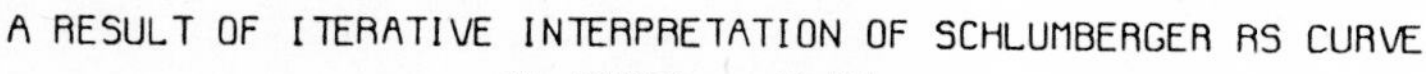

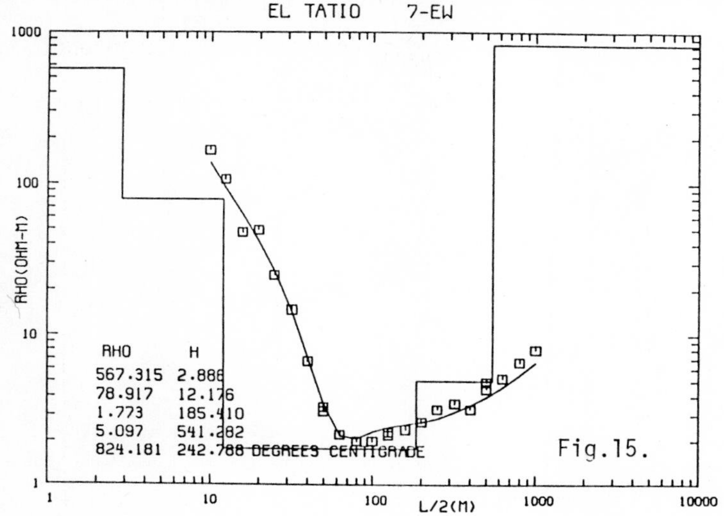

Fig.15.

CONCLUSION

An empirical formula for computing underground temperature was derived from the solution of the iterative least-squares interpretation of Schlumberger's resistivity sounding curves observed at the Hatchobaru geothermal field, and a method of estimating underground temperature was presented.

When the resistivity sounding curve provides a sufficient attainable depth to the fracture-type geothermal reservoir, the estimate gives the temperature of the geothermal reservoir. If not, it shows only the underground temperature.

ACKNOWLEDGMENT

This study was conducted by the use of the Grant-in-Aid for Energy Special Research (1) in 1981. Here the author is very glad to express his heartiest thanks to the Ministry of Education for the Grant-in-Aid for Energy Special Research (1).

Further, I would like to extend my sincere appreciation to the Kyushu Electric Power Co., Inc. for the offer of interesting well data.

Finally, the writer is greatly indebted to Professor S. Nemat-Nasser for help in preparing this article for publication.

MEASUREMENTS OF IN-SITU STRESS, FRACTURE DISTRIBUTION, PERMEABILITY, AND SONIC VELOCITY

by Mark D. Zoback

U.S. Geological Survey
Menlo Park, California 94025

ABSTRACT

Recent in-situ measurements attempting to characterize fractured crystalline rock are reported. The parameters we have chosen to study, in-situ stress, macroscopic fracture distribution, seismic velocity, and bulk permeability are all quite important for commercial development of geothermal energy.

IN-SITU STRESS

Understanding the state of stress is important for geothermal energy production in several ways; it controls the orientation, vertical extent, and propagation pressure of production-related hydraulic fractures, and in a fractured media, the stress field can cause permeability anisotropy which affects the optimal pattern of production and injection wells in a geothermal field.

Stress Magnitudes

Figure 1 presents in-situ stress data as a function of depth from a well (XTLR) drilled in granitic rock near the San Andreas Fault in Southern California (from Zoback et al., 1980). The stress measurements were made with the hydraulic fracturing technique-procedures and interpretation methods are discussed by Hickman and Zoback (1982). Figure 1 illustrates two important points about the change in stress magnitudes with depth that we have observed in crystalline rock in several areas.

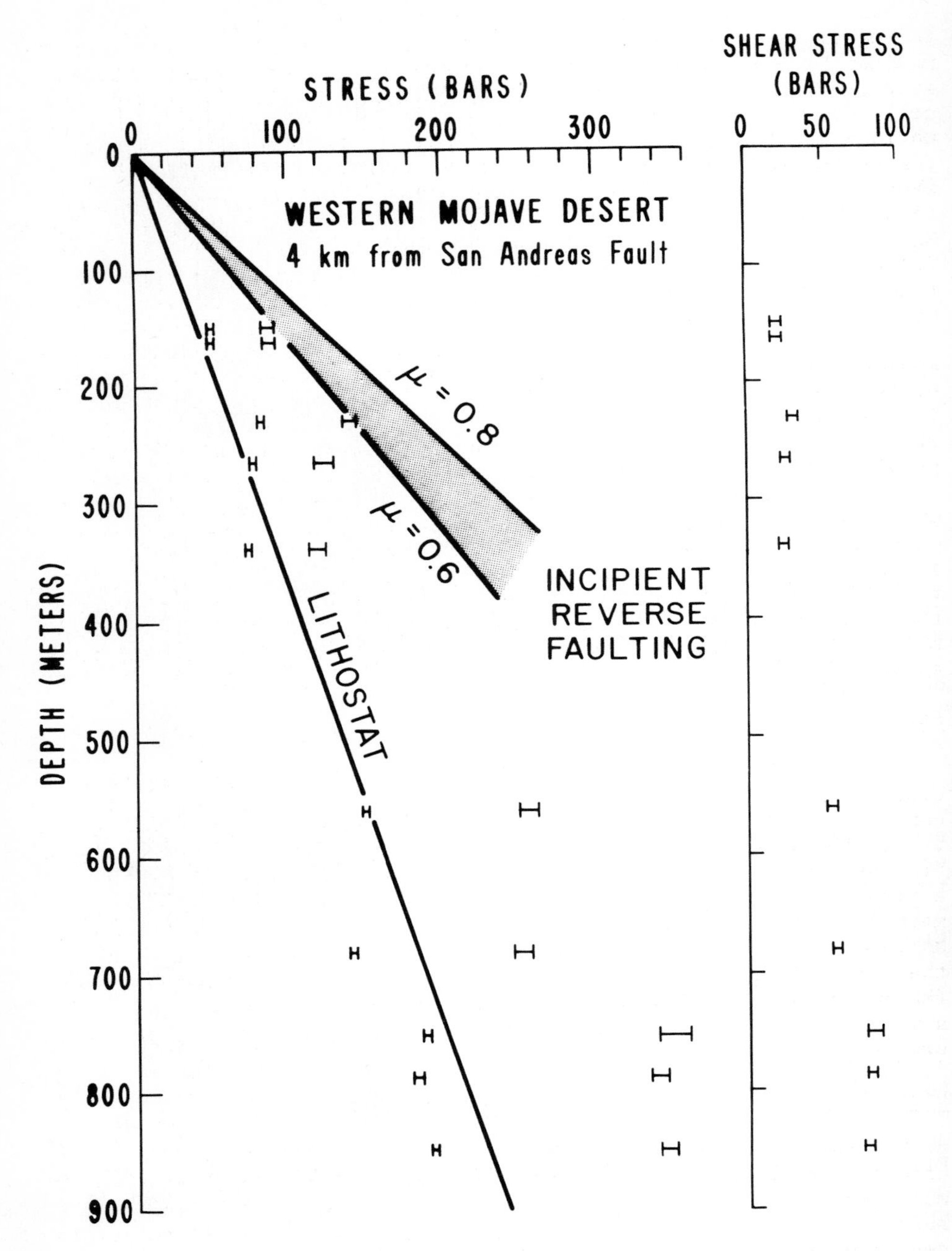

Fig. 1. The horizontal principal stresses (and maximum horizontal shear stress) measured in well XTLR near the San Andreas fault in Southern California.

First, due to the low magnitude of the vertical stress, S_V, at relatively shallow depths (< 200-300 m), the relative magnitude of the maximum horizontal stress, S_H, and vertical stress is limited by the frictional strength of rock. This is because in highly faulted rock at least one plane will be optimally oriented for reverse type fault motion. It is straightforward to show that the critical level of S_H for reverse-type faulting is $S_H^* = (S_V - P_p)((\mu^2 + 1)^{1/2} + \mu)^2 + P_p$, where μ is the coefficient of friction and P_p is the pore pressure (see Zoback and Hickman, 1982). Figure 1 shows that at shallow depth S_H is limited by frictional strength for reasonable coefficients of friction. We have noted this at various sites around the U.S., and Brace and Kohlstedt (1980) note a similar phenomena for sites in South Africa and Canada.

The second interesting feature of Figure 1 that is typical of many measurement sites is that stress is observed to change discontinuously with depth. That is, although stress is generally observed to increase with depth in our wells (see also McGarr and Gay, 1978), the stresses change in a step-like manner with depth. In another well in granitic rock in Southern California (located 35 km from the well shown in Figure 1), the stresses are actually observed to decrease over a limited depth range. The discontinuous changes in stress with depth can have important effects on the vertical propagation of massive hydraulic fractures because their growth will be controlled by the difference between the internal fluid pressure in the fracture and the magnitude of the least principal stress. Thus, through knowledge of the stress state of control of pumping pressures, the vertical extent of hydraulic fractures can sometimes be controlled.

Stress Orientations

Figure 2 shows a compilation of *in-situ* stress data in the western U.S. (from Zoback and Zoback, 1980). The direction of least-principal compressive stress is shown in Figure 2 determined in a variety of ways. The importance of this work to this symposium is that hydraulic fracture orientations determined at the wellbore generally indicate a direction of maximum compression that is consistent with other methods (such as earthquake focal mechanisms and dike orientations). Moreover, Figure 2 demonstrates that tectonic stresses seem to have the same relative magnitudes at depths of a few hundred meters, or a few km, as at the mid-crustal depths sampled by the earthquakes. So, a variety of information can be used to predict hydraulic fracture orientation.

Distribution of Macroscopic Fractures at Depth

Fractures are very important in geothermal fields because of their effect on permeability. To study the distribution and orientation of macroscopic fractures at depth, comprehensive borehole televiewer surveys have been conducted in many of our wells. The televiewer is described in detail by Zemanek et al. (1970) and the results of our fracture studies have been described in detail by Seeburger and Zoback (1982).

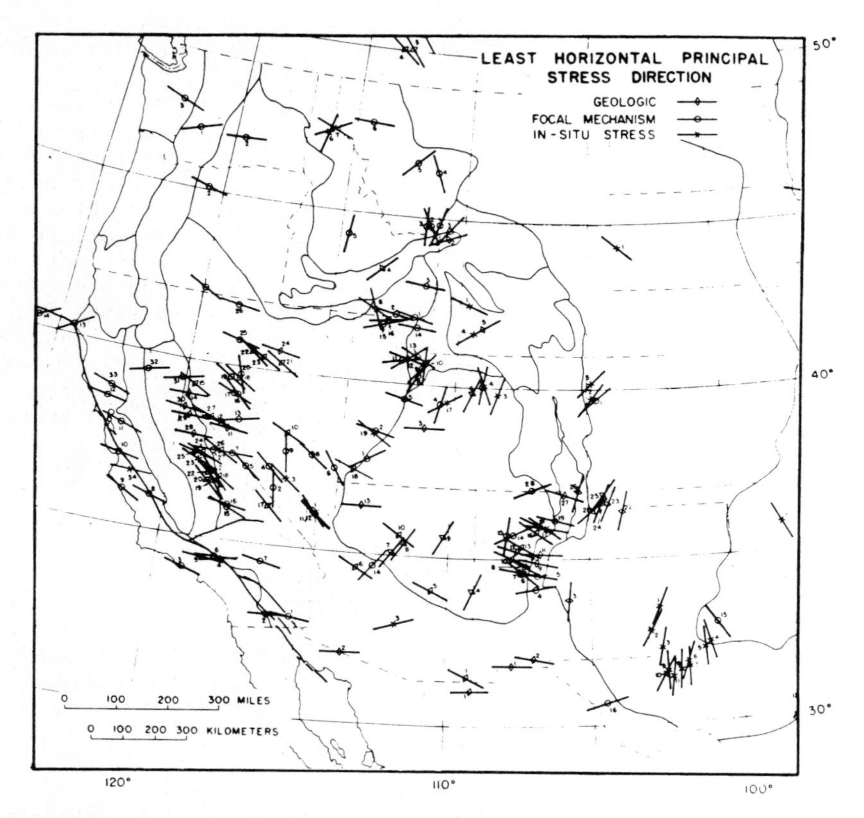

Fig. 2. Direction of least horizontal principal stress in the western U.S. (from Zoback and Zoback, 1980). Symbol represents the type of stress-field indicator. The number refers to description of each point in the original work.

Figure 3 shows an example of borehole televiewer data. The dark sinusoids in the figure indicate planar fractures intersecting the borehole. The brightness of a trace is the amplitude of an ultrasonic pulse reflected off the borehole wall. The pulse is emitted by a rotating transducer synchronized with a flux-gate magnetometer. Thus, the strike and dip of the fractures can be straightforwardly determined.

Figure 4 shows the density of fractures encountered in the XTLR well. It is typical of all the wells we have investigated in that the fracture density decreases only moderately with depth. This is completely unlike the anticipated rapid decrease of fracture density with depth suggested by Snow (1968).

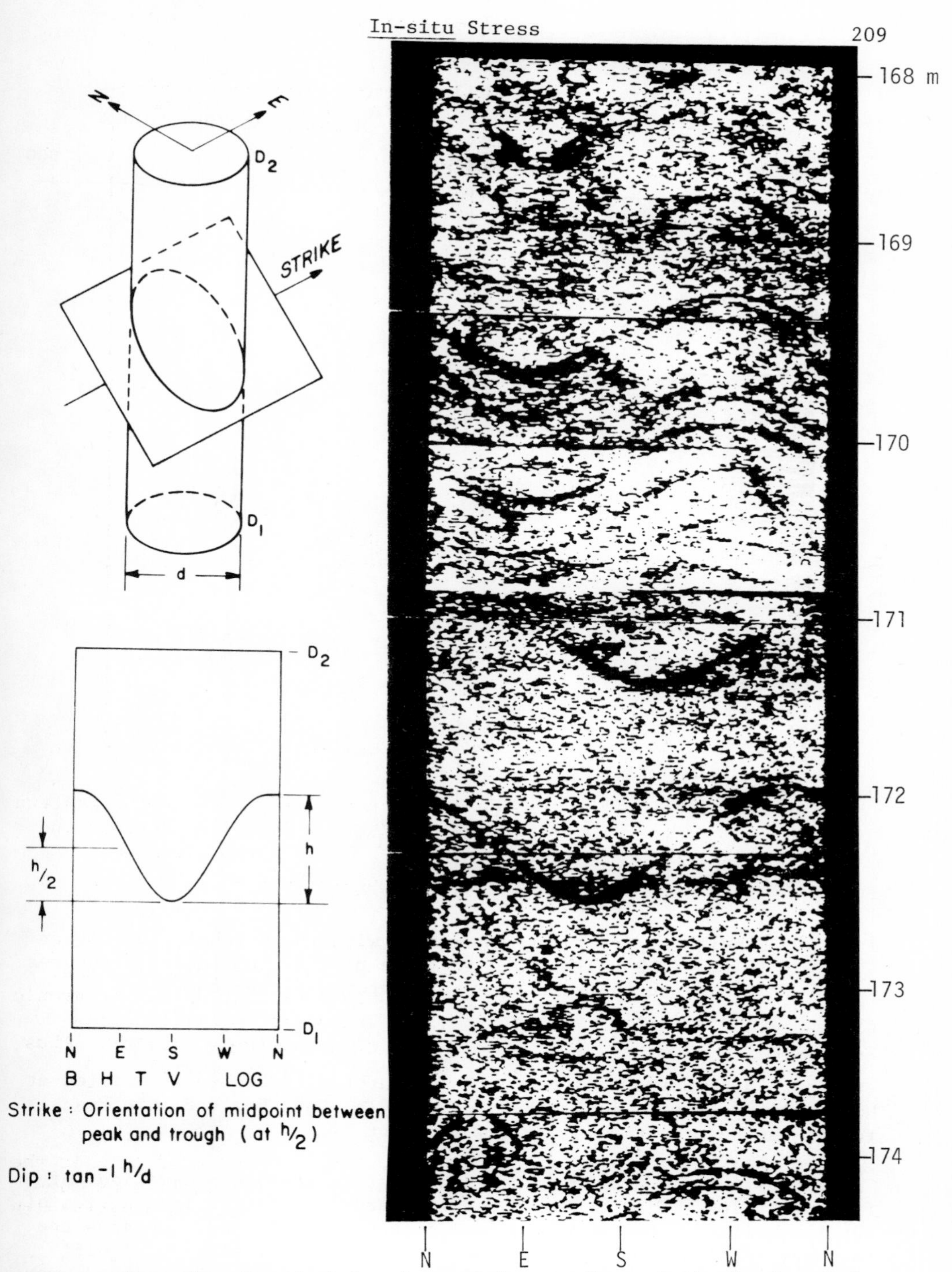

Fig. 3. Example of method for determining strike and dip for borehole televiewer data. Borehole televiewer record for 6 meter section of well in fractured granitic rock.

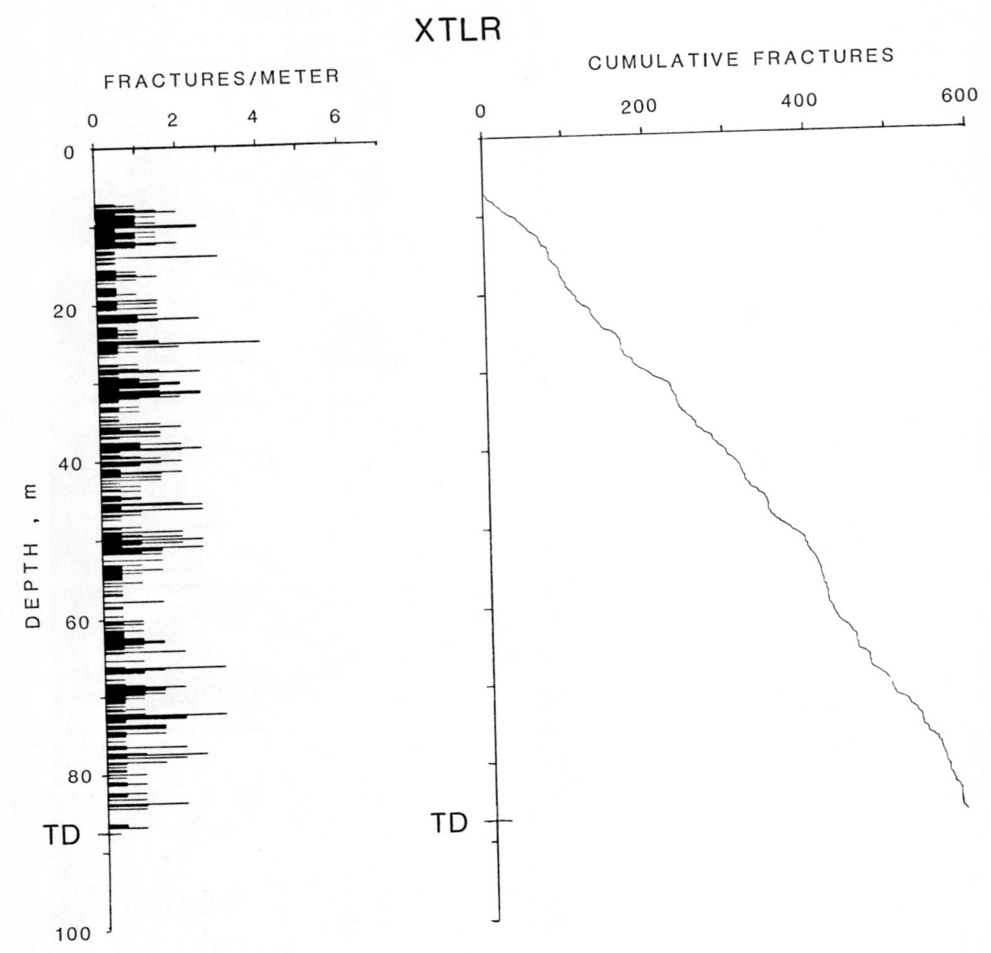

Fig. 4. Density of fractures observable with the borehole televiewer in the XTLR well. Note from the plot of cumulative fractures with depth that the rate of fractures decreases very gradually with depth.

Figure 5 compares the fractures encountered in two wells drilled at Monticello Reservoir in South Carolina. It shows how variable fracturing can be at sites only a few km apart.

Figures 6 and 7 are lower hemisphere stereographic projections of fracture poles. Figure 6 compares fracture orientation in a series of shallow wells in the western Mojave desert, and Figure 7 compares the two Monticello wells. Although at every site statistically significant fracture concentrations could be found (contours >3σ in the lower part of each figure), the fracture orientations are extremely variable from site to site and in no case could preferential fracture orientations be predicted from knowledge of the present or past regional stress field.

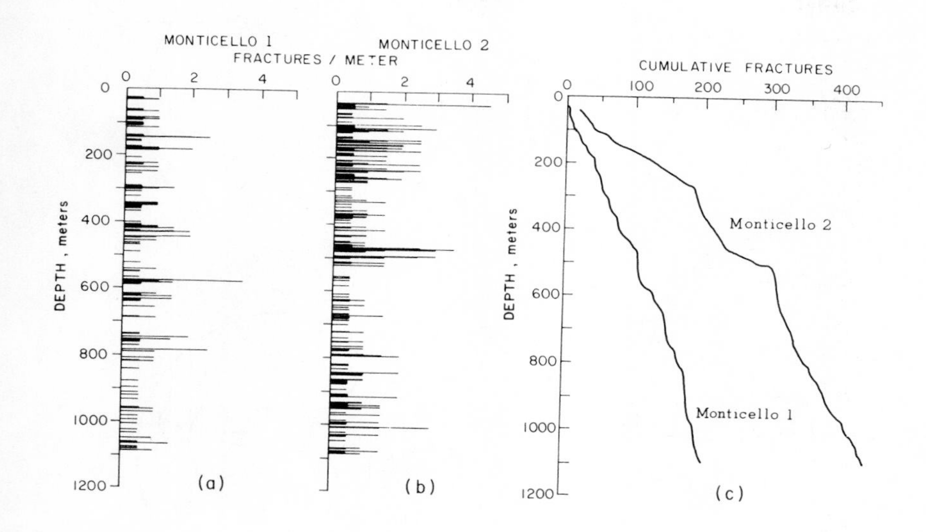

Fig. 5. Comparison of fracture density in the two wells drilled at Monticello Reservoir.

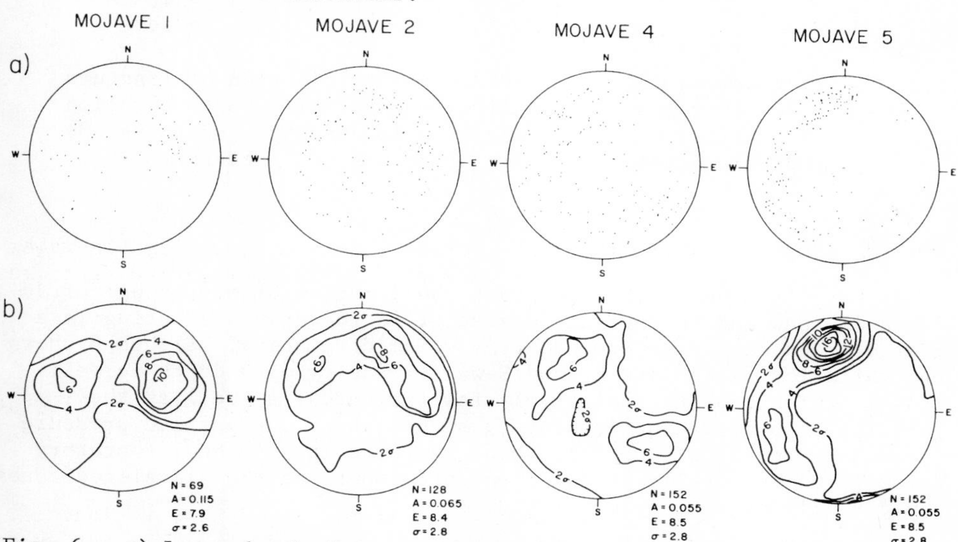

Fig. 6. a) Lower hemisphere stereographic projection of fracture poles in wells in the western Mojave Desert. b) Statistical contouring of pole density. Contours >3σ indicate significant concentrations of fractures.

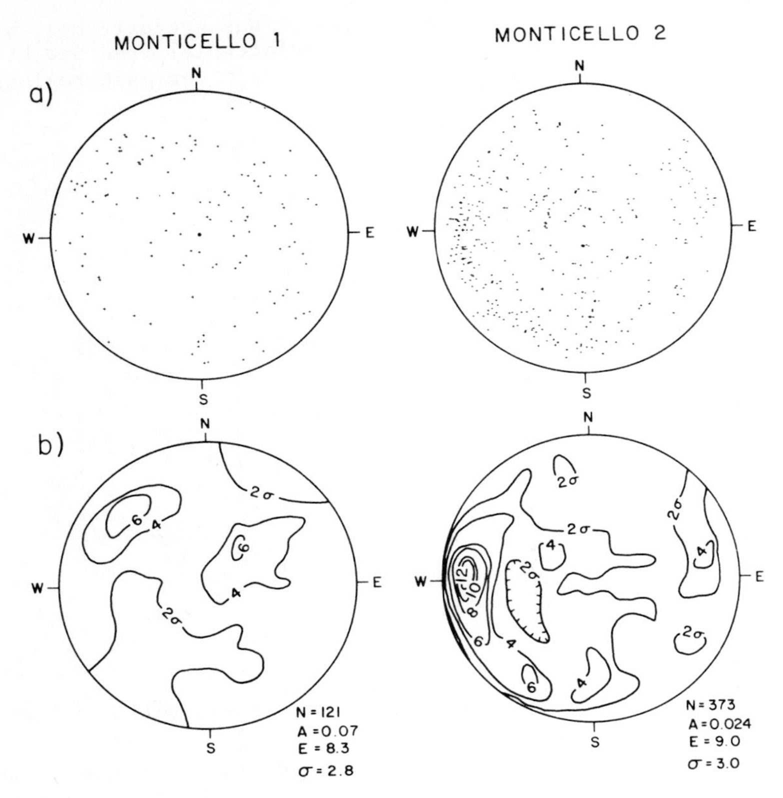

Fig. 7. a) Lower hemisphere stereographic projection of fracture poles in wells near Monticello Reservoir, South Carolina. b) Statistical contouring of pole density. Contours >3σ indicate significant concentrations of fractures.

Seismic Velocity Measurements

It has often been suggested that the location of macroscopic fractures in rock and the overall degree of macroscopic fracturing in a body of rock can be determined from seismic velocity. Figure 8 shows the variation of *in-situ* P- and S-wave velocity with depth in well Mont-1 (which contains relatively few fractures) as well as the velocity of a saturated laboratory sample under appropriate pressure (from Moos and Zoback, 1982). Note that the *in-situ* and laboratory velocity are generally in excellent agreement and that localized zones could easily have been identified.

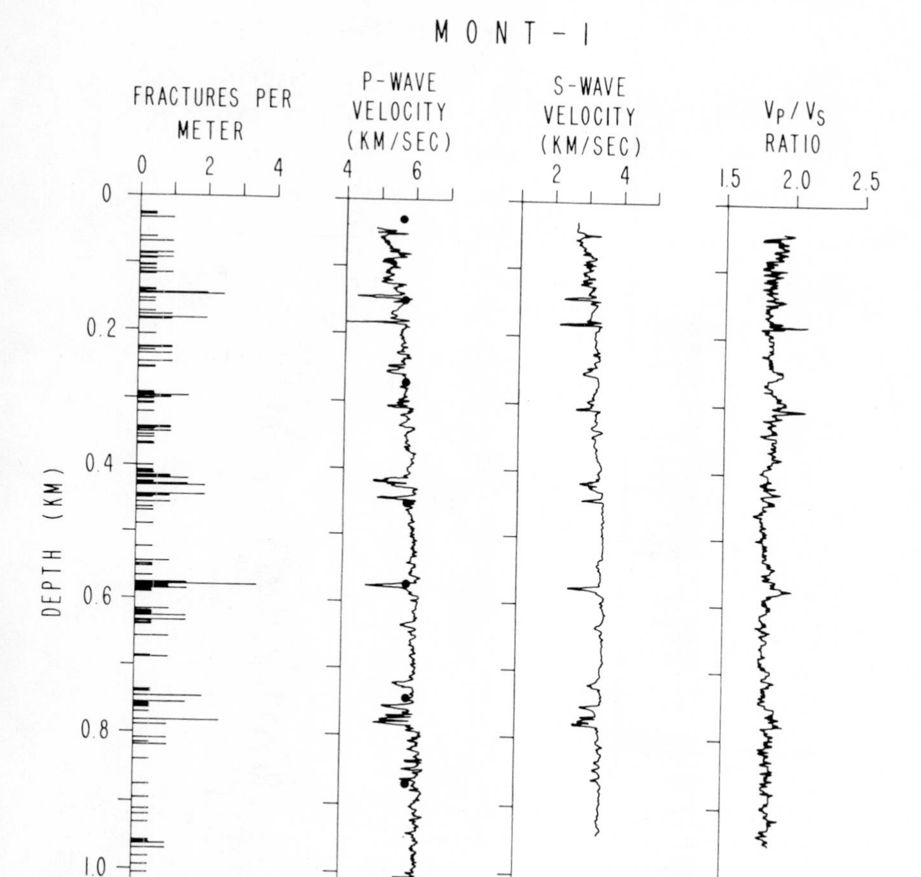

Fig. 8. P- and S-wave seismic velocity, observable macroscopic fractures, and V_p/V_s as a function of depth in Monticello.

Figure 9 presents the velocity data from the XTLR well (which contains far more fractures than Mont-1) and there is a marked discrepancy between the *in-situ* velocity and that of the lab sample under pressure. The *in-situ* velocity has been lowered remarkably by the numerous macroscopic fractures and distinct fracture zones are hard to pick out. Although it is easy to say with certainty that a granite with velocities of 3-4 km/sec is highly fractured, one cannot simply relate velocity to fracture density because of the importance of composition and microcrack density.

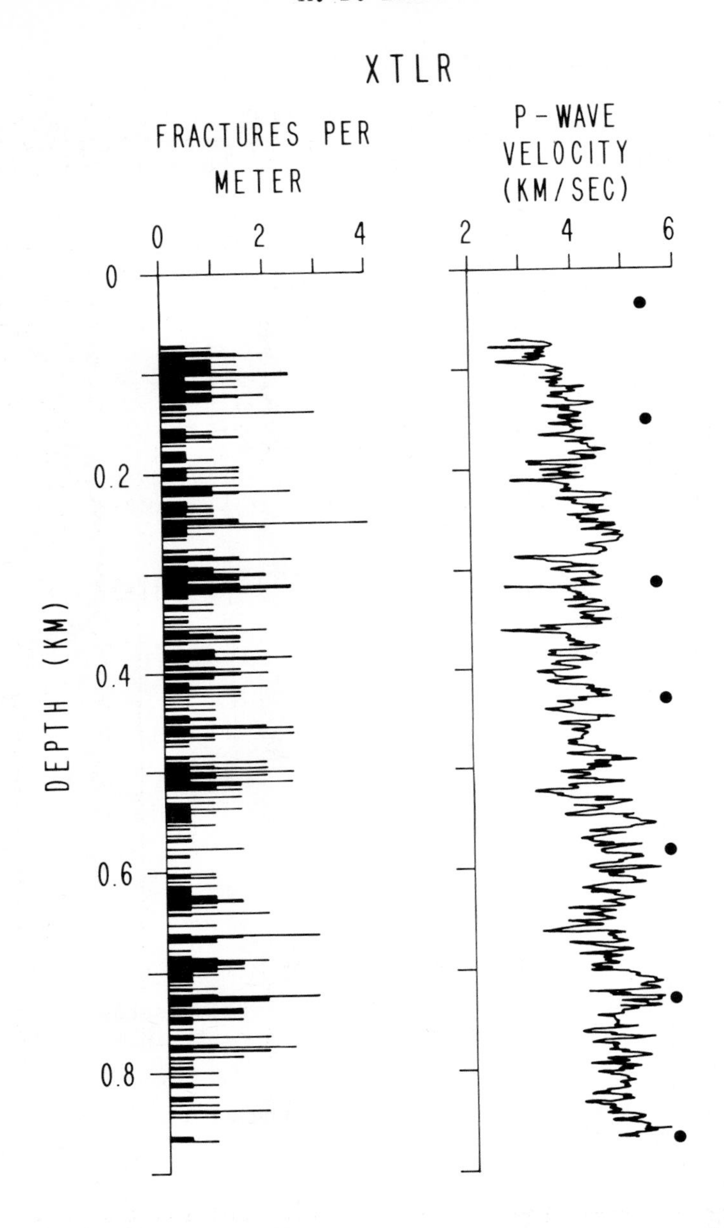

Fig. 9. P-wave seismic velocity and observable macroscopic fractures as a function of depth in well XTLR.

Bulk Permeability Measurements

Although bulk permeability is of considerable importance, it is relatively difficult to measure in-situ. It would be very useful to know if bulk permeability was simply related to macrofracture density in fractured crystalline rock.

We have measured the bulk permeability in a number of wells drilled in fractured crystalline rock using relatively standard pulse-test techniques (Bredehoeft and Papadapolus, 1980). As shown in Figure 10, in the XTLR well there appears to be a simple correlation between fracture spacing and bulk permeability (the dashed straight line is drawn in Figure 10 knowing that a data point should lie near the origin). Thus, in this area one could probably roughly estimate bulk permeability from knowledge of fracture spacing even when the fracture spacing is fairly low. For example, if in the vicinity of XTLR well the fracture density was only 1 fracture per 10 meters, the bulk permeability would probably be a few tenths of a millidarcy.

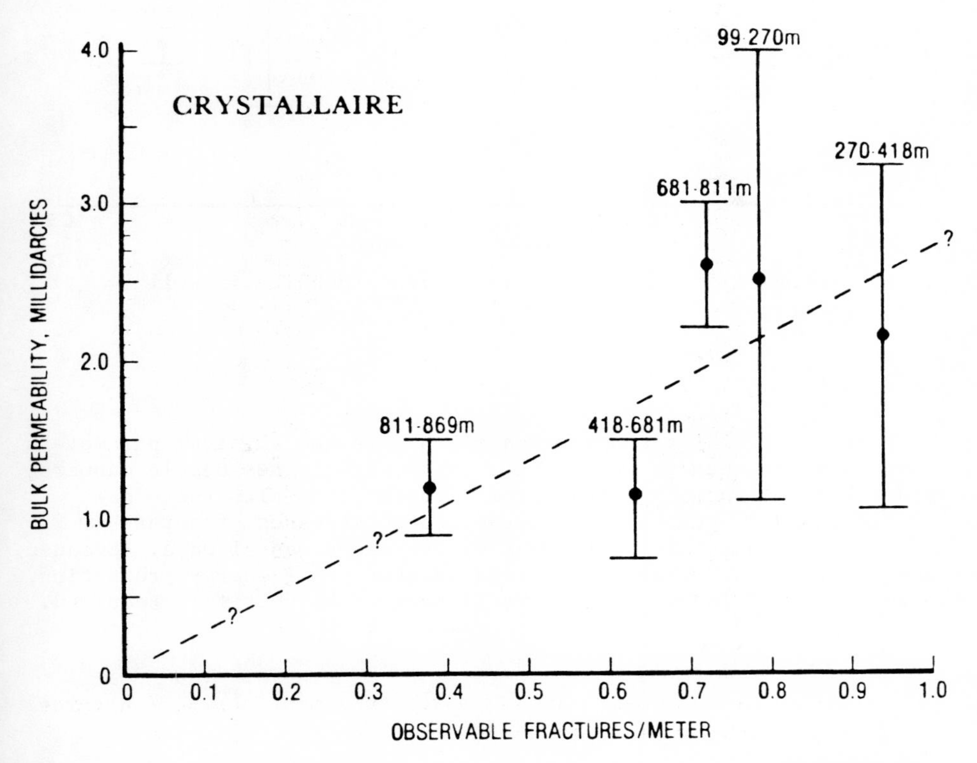

Fig. 10. Permeability as a function fracture density in well XTLR.

In the Mont-2 well (Figure 11), it is not possible to simply relate fracture density to permeability because of the marked effect of remineralization in the fractures. In areas such as this it is necessary to measure bulk permeability as no simple dependence of permeability on fracture density exists.

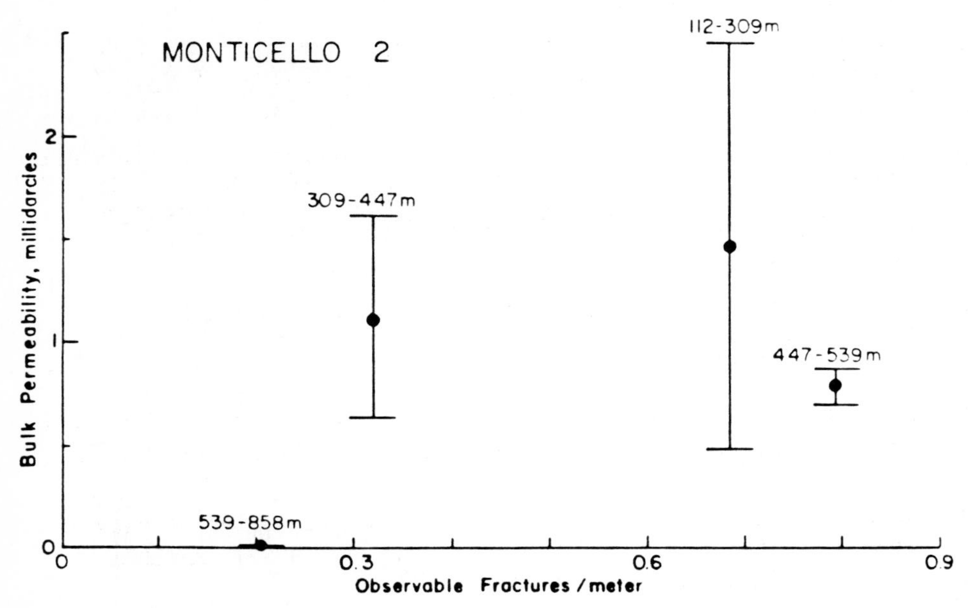

Fig. 11. Permeability as a function fracture density in well Monticello 2.

Summary

We have found that although at a given site the physical parameters we have measured seem to interrelate in physically reasonable manners, considerable variance can occur from site to site. It, therefore, seems difficult to generalize about expected stresses, fracture patterns, or permeabilities without direct experimental data. Because of the importance of these parameters on geothermal energy production, extensive in-situ investigations would seem to be routinely required.

REFERENCES

1. Brace, W.F., and Kohlstedt, D.L., "Limits on Lithospheric Stress Imposed by Laboratory Experiments," Journal of Geophysical Research, Vol. 82, No. B11, 1980, pp. 6248-6252.

2. Bredehoeft, J.D., and Papadapulos, S.S., "A Method for Determining the Hydraulic Properties of Tight Formations," Water Resources Research, Vol. 16, 1980, pp. 233-238.

3. Hickman, S.H., and Zoback, M.D., "The Interpretation of Hydraulic Fracturing Pressure-Time Data for In-Situ Stress Determination," in Proc. Workshop on Hydraulic Fracturing Stress Measurements, U.S. National Commission on Rock Mechanics, Washington, D.C., in press, 1982.

4. McGarr, A.F., and Gay, N.C., "State of Stress in the Earth's Crust," Annual Review of Earth and Planetary Sciences, Vol. 6, 1978, pp. 405-436.

5. Moos, D., and Zoback, M.D., "In-situ Studies of Seismic Velocity in Fractured Crystalline Rocks," Journal of Geophysical Research, in press, 1982.

6. Seeburger, D.A., and Zoback, M.D., "The Distribution of Natural Fractures and Joints at Depth in Crystalline Rock," Journal of Geophysical Research, in press, 1982.

7. Zemanek, J., et al., "Formation Evaluation by Inspection with the Borehole Televiewer," Geophysics, Vol. 35, 1970, pp. 254-269.

8. Zoback, M.D., Healy, J.H., and Roller, J.C., "Preliminary Stress Measurements in Central California Using the Hydraulic Fracturing Technique," Pure and Applied Geophysics, Vol. 115, 1977, pp. 135-152.

9. Zoback, M.D., Tsukahara, H., and Hickman, S.H., "Stress Measurements at Depth in the vicinity of the San Andreas Fault: Implications for the Magnitude of Shear Stress at Depth," Journal of Geophysical Research, Vol. 85, 1980, pp. 6157-6173.

10. Zoback, M.L., and Zoback, M.D., "State of Stress in the Conterminous United States," Journal of Geophysical Research, Vol. 85, 1980, pp. 6113-6156.

11. Zoback, M.D., and Hickman, S.H., "In-situ Study of the Physical Mechanism Controlling Induced Seismicity at Monticello Reservoir, South Carolina, Journal of Geophysical Research, Vol. 87, 1982, pp. 6959-6974.

MEASUREMENT OF TECTONIC STRESS BY HYDRAULIC FRACTURING IN JAPAN

Hisao Ito

Geological Survey of Japan
Tsukuba, Japan

ABSTRACT

Using the hydraulic fracturing technique, we have made a series of in-situ stress measurements in wells drilled in the Sengan geothermal area. The wells are all 250 m deep and 158 mm in diameter. The stratigraphy of the well is tight dacite and rhyolite welded tuff of the late Miocene-Pleistocene Periods. From the results of the well logging, ultrasonic borehole televiewer, and the observation of the core samples, we have selected five intervals for each well. Vertical stress was calculated as overburden pressure. Horizontal principal stress was calculated from the pressure records. The impression packer and ultrasonic borehole televiewer have been used to detect the orientation of the fractures. In many cases, impression packer results show clearly two vertical fractures (approximately 180 degrees apart). Calculated stress results show that the maximum principal stress is the maximum horizontal stress, the intermediate principal stress is the minimum horizontal stress, and the minimum principal stress is the vertical stress. Direction of the maximum horizontal stress is NW to SE. Absolute value of the stress shows that the state of stress is very compressive. Absolute value and the direction of the stress by our hydraulic fracturing stress measurements are in harmony with the tectonic stress. This shows that Tamagawa welded tuff is hard and tight enough to maintain tectonic stress and act as a cap rock of the underlying geothermal reservoir. Since we expect that a less compressive stress field is dominant in large scale geothermal areas, hydraulic fracturing stress measurements in deeper wells are necessary to understand the Sengan geothermal area.

INTRODUCTION

Hydraulic fracturing experiments in geothermal areas have been done such as the Hot Dry Rock Project or the geothermal well stimulation (Katagiri et al., 1980)[1]. In-situ stress measurements in geothermal fields are also important for several reasons. Nakamura and Uyeda (1980)[2] pointed out that large scale volcanic and geothermal areas exist in less compressive state of stress region. Voegele et al., (1981)[3] insist that in-situ stress determination is essential to the optimization of stimulation design. Induced seismicity in a geothermal field is closely related to the state of stress and the pore pressure. More practically, in-situ stress determination will be useful to determine the re-injection site and re-injection pressure.

We have been conducting various kinds of geophysical, geological, and geochemical research in the Sengan geothermal area, Tohoku Japan, to understand geothermal reservoir properties and to confirm the effectiveness of prospecting techniques for deep geothermal resources. In one study, we have been making a series of hydraulic fracturing stress measurements to evaluate the state of stress in the Sengan geothermal field.

SITE DESCRIPTION

The location of the Sengan geothermal area is shown in Fig. 1. The wells are all 250 m deep and about 158 mm in diameter. The stratigraphy of the well is tight dacite and rhyolite welded tuff of the late

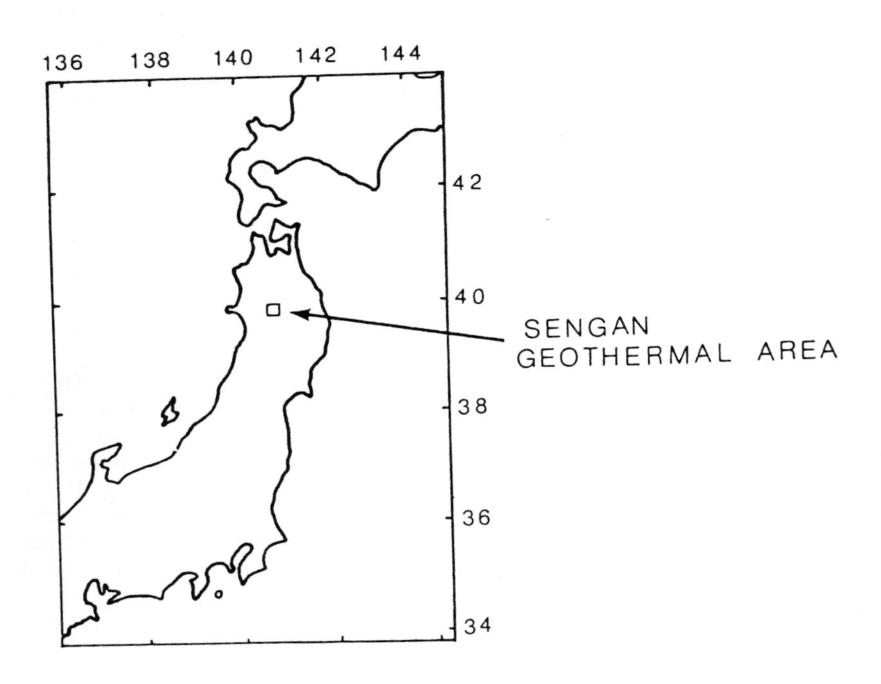

Fig. 1. Location of the Sengan geothermal area.

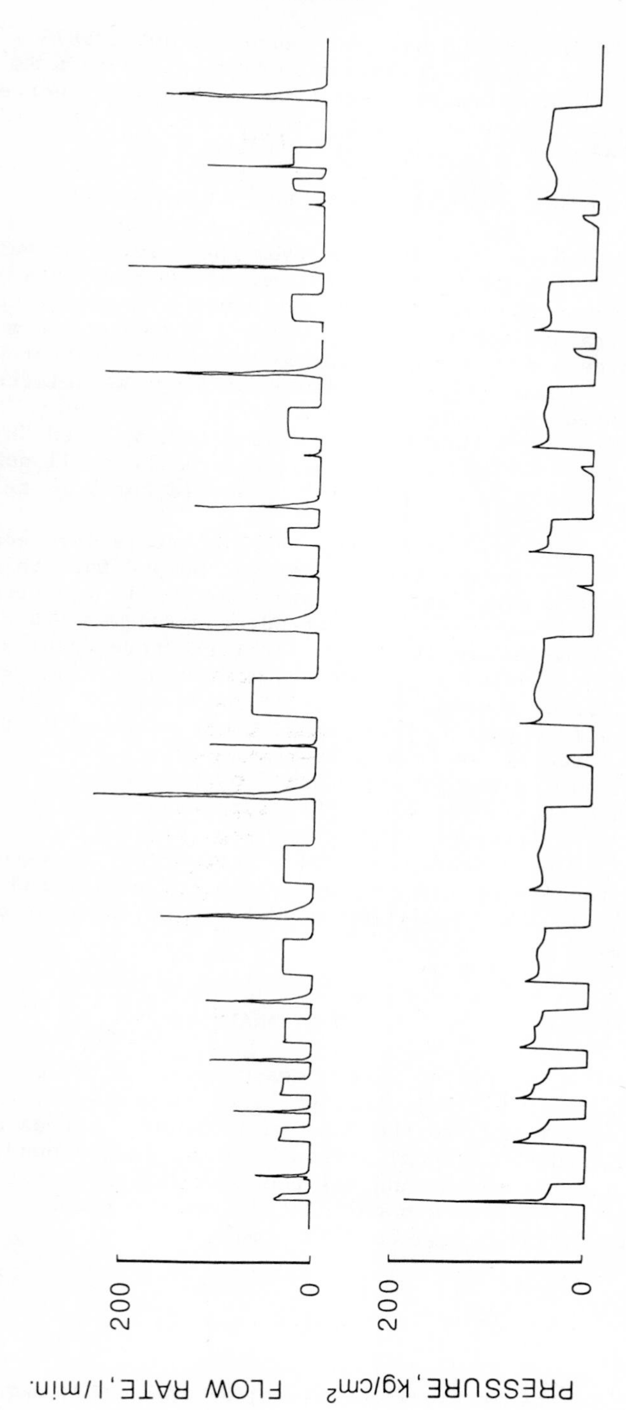

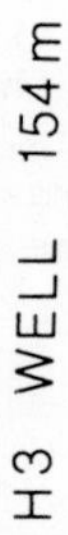

Fig. 2. Pressure and flow rate records in the H-3 well.

Miocene-Pleistocene Periods (Tamanyu and Suto, 1978)[4], from near the surface to the bottom of the well. This welded tuff is widely distributed in this area and acts as a cap rock of the geothermal reservoir underneath.

HYDRAULIC FRACTURING

Test intervals were selected from the results of well logging data, ultrasonic borehole televiewer (Zemanek et al., 1969)[5] data, and the observation of the core samples. We conducted sonic, caliper and precise temperature logging from the surface to the 250 m depth. Although core samples and ultrasonic televiewer showed fractures intersecting the well, which are probably due to cooling, we selected four or five fracture free test zones for each well.

Test sites were then cleaned with water. We used "hydrajet" to clean the borehole wall by injecting water from the small nozzles. After the cleaning procedure, we repeated ultrasonic borehole televiewer logging for the test zones.

Approximately two meter long test intervals were sealed by LYNES straddle rubber packers, and water was pumped into this interval until tensile fracture was induced. Once the tensile fracture was induced and the breakdown pressure was recorded, pumping was stopped.

In order to record the reopening (fracture opening) pressure, a series of pressurization, shut-in, and bleed cycles were repeated (Fig. 2). The pumping time was 1-3 minutes for each cycle. Flow rate was carefully controlled to be as steady as possible using a HALLIBURTON T-10 three piston pump. In most cases, flow rate was about 40 liters/minute. Water pressure and flow rate were recorded at the surface with a strain-gauge type pressure transducer and an electromagnetic flow meter, respectively. By simple integrating circuits, the total amount of injected water was calculated from flow rate records.

After the pressurization, an impression packer and an ultrasonic borehole televiewer were used to detect the orientation of the induced fractures.

DATA ANALYSIS

Hydraulic fracturing stress measurement is based on the assumption that (1) one of the principal stresses is vertical, (2) the fracture propagates parallel to the maximum horizontal stress and perpendicular to the minimum horizontal stress, (3) the rock around the well is homogeneous and isotropic, and (4) the rock deforms elastically.

Horizontal stress was calculated from the pressure records. Hubbert and Willis (1957)[6] derived the formula

$$P_B = 3S_h - S_H - P_p + T \tag{1}$$

where P_B is the breakdown pressure, S_h is the minimum horizontal stress, S_H is the maximum horizontal stress, P_p is pore pressure, and T is tensile strength.

Minimum horizontal stress S_h is determined from the shut-in pressure measured in later pressurization cycles,

$$S_h = P_s \tag{2}$$

where P_s is the shut-in pressure.

To determine S_h from equation (1), we have to know T and P_p. We assumed pore pressure P_p is equal to hydrostatic pressure. Bredehoeft et al., (1976)[7] and Zoback et al., (1980)[8] showed that the fracture opening pressure can be used to calculate S_H when the tensile strength T is not known, by the following equation

$$P_F = 3S_h - S_H - P_P \tag{3}$$

where P_F is the fracture opening pressure attained in later pressurization cycles (Fig. 3). Equation (3) is more favorable than equation

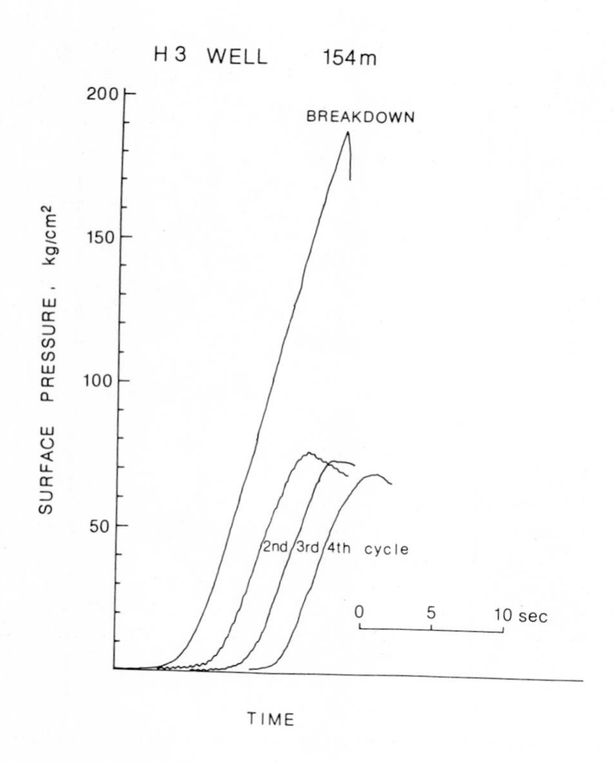

Fig. 3. Pressure records of the beginning of pressurization cycles.

(1), because (1) breakdown pressure is pumping rate dependent, and (2) T is not always equal to the tensile strength measured for core samples in the laboratory.

The shut-in pressure and fracture opening pressure are determined for each pressurization cycle where the pressure versus time curve begins to depart from linearity. We determined the final shut-in pressure and the fracture opening pressure from the stable values obtained in later pressurization cycles (Fig. 4) after Zoback et al., (1980).[8]

Vertical stress was calculated as overburden pressure with the core sample density of 2.45 gm/cm^3.

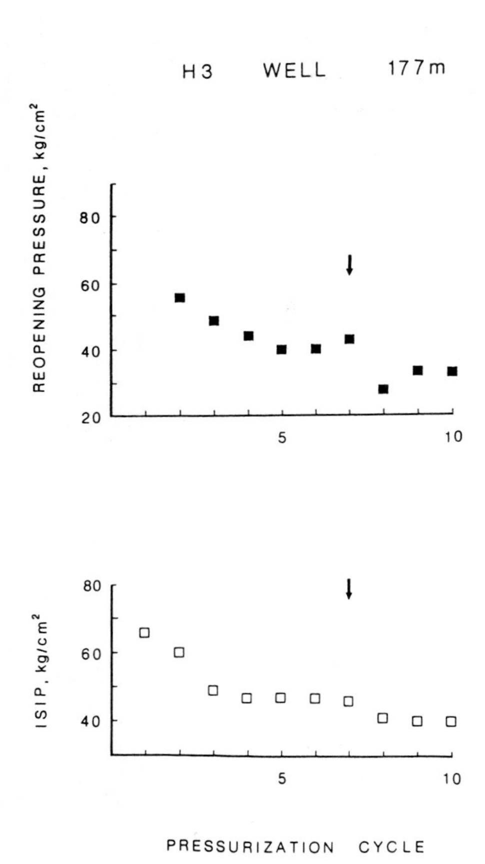

Fig. 4. Fracture reopening pressure P_F and shut-in pressure (ISIP) are shown. At the pressurization cycle shown by the arrow, the flow rate was increased to about 60 liters/minute.

RESULTS

In most experiments, a clear breakdown was observed. Judging from this, in addition to the facts (1) in many cases, impression packer results show clear vertical fractures which are almost 180 degrees apart from each other, and (2) we recorded a considerable amount of flow back after bleed-out, we believe that new fractures were induced by pumping.

The stress results of the present hydraulic fracturing experiments are summarized in Table 1. For greater detail, refer to Ito et al., (1982).[9] In Fig. 5 and Fig. 6, the maximum horizontal stress (open square) and the minimum horizontal stress (closed square), together with the lithostatic pressure (straight line), are plotted as functions of depth from the surface.

Both the maximum and minimum horizontal stresses increase linearly with depth. In all experiments, vertical stress is the least principal stress, minimum horizontal stress is the intermediate stress, and maximum horizontal stress is the maximum principal stress.

When the least principal stress is vertical, a vertical fracture will be induced at the borehole wall if the inflatable straddle packer is used (Haimson and Fairhurst, 1970,[10] Roegiers et al., 1973[11]).

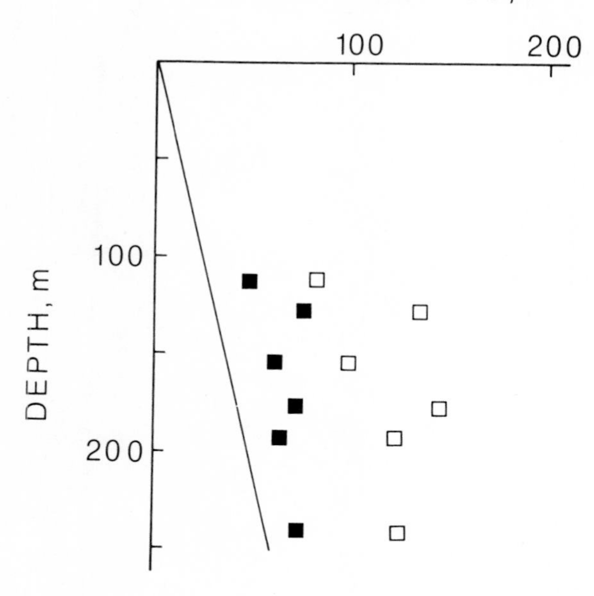

Fig. 5. Calculated principal stresses for the H-3 well.

As it propagates, the fracture turns into a horizontal plane, and two shut-in pressures will be recorded: one is for vertical fracture and is equal to the minimum horizontal stress, and the other is for horizontal fracture and is equal to the vertical stress. We have tried to detect this second shut-in pressure. As is shown in Fig. 4, typically the shut-in pressure decreases in later pressurization cycles, but the value is still higher than the theoretical vertical stress. This is probably because the fracture reaches preexisting fractures or joints, as it propagates, before it tends to horizontal fracture completely.

Maximum and minimum horizontal stresses are represented as linear functions of depth with the data of the H-1, H-2, H-3 and H-6 wells, as follows:

$$S_h = 112 + 0.16h \tag{4}$$

$$S_H = 223 + 0.36h \tag{5}$$

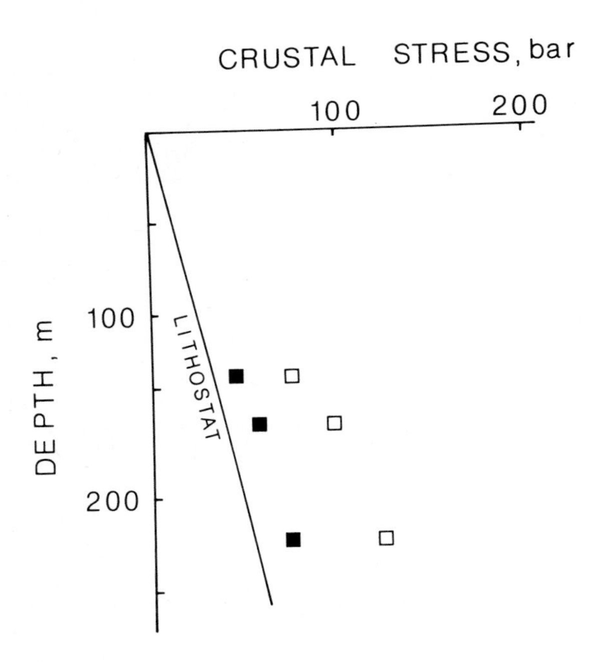

Fig. 6. Calculated principal stresses for the H-6 well.

Table 1. Summary of Stress Measurements

Well	Depth (m)	Hydraulic Fracturing Data			Principal Stress			
		Breakdown Pressure (bars)	Shut-in Pressure (bars)	Fracture Opening Pressure (bars)	Minimum Horizontal Stress (bars)	Maximum Horizontal Stress (bars)	Vertical Stress (bars)	Tensile Strength (bars)
H-1	122	92	40	49	51	81	29	43
	133	92	37	49	49	73	32	43
	201	137	100	82	118	234	48	55
	217	142	47	43	67	174	52	99
	245	---	95	72	117	232	59	--
H-2	127	60	25	28	37	59	31	32
	156	83	46	40	60	111	38	43
	206	92	30	47	61	49	50	45
H-3	111	118	37	40	48	82	27	78
	128	164	64	69	77	136	31	95
	154	184	51	57	66	111	37	127
	177	82	56	39	73	146	43	43
	192	118	47	37	66	123	46	81
	239	126	51	51	74	125	58	75
H-6	137	68	31	31	44	75	33	37
	156	74	39	41	56	97	38	33
	223	94	66	51	73	124	54	43

where S_h , S_H are in bars, and h is depth from sea level in meters.

The azimuth of the maximum horizontal stress coincides with the orientation of the hydraulically induced fracture at the well wall. We used the ultrasonic borehole televiewer and impression packer to detect the orientation of the induced fracture. Unfortunately the ultrasonic borehole televiewer failed to detect vertical fracture, but the impression packer successfully obtained the fracture orientation in many cases.

The results of the azimuth data are summarized in Fig. 7. The azimuth is southeast-northwest. Neither impression packer nor ultrasonic borehole televiewer showed induced vertical fracture in the H-6 well.

Although we expected detectable temperature anomaly around the induced fracture (Ito et al., 1982)[9], we did not find any temperature change even with very precise temperature logging.

We also installed several seismometers on the exposed rock at the surface, several hundred meters away from the well head. However, we did not detect any seismic signal associated with hydraulic fracturing.

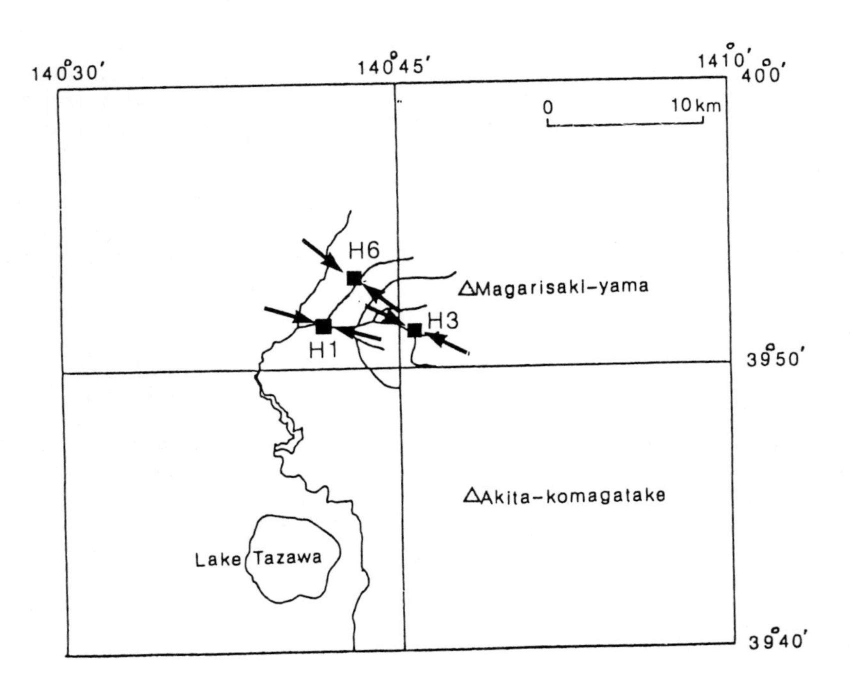

Fig. 7. The azimuth of the maximum horizontal stress.

DISCUSSION

Haimson (1980)[12] insists hydraulic fracturing stress measurements in very shallow wells are affected by the topography. He showed that the direction of the maximum horizontal stress is parallel to the slope axis of the surface topography. Nakajima (1982)[13] points out, based on three-dimensional finite element calculation, that the surface topography affects the stress field up to the depth of two or three times the value of the topographical variation. From this point of view, present hydraulic fracturing experiments are not deep enough to overcome the surface topography effect. Our experiments, however, were conducted under different topographical conditions, and both the stress value and azimuth of maximum horizontal stress are consistent with each other.

It is well known that the focal mechanism of shallow earthquakes in the Tohoku region is a low angle thrust faulting with east-west compression. Geodetic measurements also show east-west compression in the Tohoku area. Our present stress data by hydraulic fracturing are consistent with earthquake focal mechanisms and geodetic data.

In the Takinoue geothermal area, about 10 km east of the present hydraulic fracturing experimental site, very shallow earthquakes of one or two kilometers focal depth have been observed. The seismicity and focal mechanism changed during 50 MW geothermal power plant operation and hot water re-injection. This phenomenon will be interpreted with in-situ stress data and pore pressure data.

In the present data analysis, we tentatively determined the shut-in pressure where the pressure versus time curve begins to depart from linearity. However, in some cases, the pressure decreases so gradually after shut-in that it is difficult to determine the shut-in pressure in this manner. A more reliable method to determine the shut-in pressure in these cases is needed.

The present results show that the Tamagawa welded tuff, in which hydraulic fracturing was conducted, is hard and tight enough to maintain tectonic stress and acts as a cap rock of the geothermal reservoir.

CONCLUSION

We obtained the stress data by hydraulic fracturing in the Sengan area, Tohoku Japan. The results are:

1) The greatest principal stress is the maximum horizontal stress, the intermediate principal stress is the minimum horizontal stress, and the least principal stress is the vertical stress. This stress state causes reverse faulting.
2) The azimuth of the maximum horizontal stress is northwest to southeast.
3) The present results are in harmony with tectonic stress and consistent with earthquake focal mechanisms.

ACKNOWLEDGEMENT

The author is grateful to the members of the New Energy Development Organization for their kindness to give us the opportunity to conduct hydraulic fracturing stress measurements in their wells.

REFERENCES

1. Katagiri, K., Ott, W. K., and Nutley, B. G., "Hydraulic Fracturing aids Geothermal Field Development", World Oil, December 1980, pp. 75-88.

2. Nakamura, K., and Uyeda, S., "Stress Gradient in Ark Back-Ark Regions and Plate Subduction", J. Geophys. Res., Vol. 85, 1980, pp. 6419-6428.

3. Voegele, M. D., Abou-Sayed, A. S., and Jones, A. H., "Optimization of Stimulation Design through the Use of In-situ Stress Determination", SPE 10308, 1981.

4. Tamanyu, S., and Suto, S., "Stratigraphy and Geochronology of Tamagawa Welded Tuff in the Western Part of Hachimantai, Akita Prefecture", Bull. Geol. Surv. Japan, Vol. 29, 1978, pp. 159-169.

5. Zemanek, J., Caldwell, R. L., Glenn, E. E., Holcom, S. V., Norton, L. J., and Strauss, A. J. O., "The Borehole Televiewer - A New Logging Concept for Fracture Location and Other Types of Borehole Inspection", J. Pet. Tech., Vol. 21, 1969, pp. 762-774.

6. Hubbert, M. K., and Willis, D. G., "Mechanics of Hydraulic Fracturing", J. Pet. Tech., Vol. 9, 1957, pp. 153-168.

7. Bredehoeft, J. D., Wolff, R. G., Keys, W. S., and Shutter, E., "Hydraulic Fracturing to determine the Regional In Situ Stress Field, Piceance Basin, Colorado", Geol. Soc. Am. Bull., Vol. 87, 1976, pp. 250-258.

8. Zoback, M. D., Tsukahara, H., and Hickman, S., "Stress Measurements at Depth in the Vicinity of the San Andreas Fault: Implication for the Magnitude of Shear Stress at Depth", J. Geophys. Res., Vol. 85, 1980, 6157-6173.

9. Ito, H., Ishido, T., Matsubayashi, O., and Sugihara, M., "Hydraulic Fracturing Stress Measurements in the Sengan Geothermal Area", The Report to Sunshine Project Promotion Headquarters,1982, in press.

10. Haimson, B. C., and Fairhurst, C., "In-situ Stress Determination at Great Depth by means of Hydraulic Fracturing", Rock Mechanics-Theory and Practice,Proceedings, 11th Symposium on Rock Mechanics, Berkeley, 1969, Society of Mining Engineers of AIME, New York,

1970, pp. 559-584.

11. Roegiers, J. C., Kudo, O., and Fairhurst, C., "Influence of the Type of Packer on the Stress Distribution around a Borehole and the Resulting Fracture Orientation", J. Jap. Am. Pet. Tech., Vol. 38, 1973, pp. 209-222.

12. Haimson, B. C., "Near-Surface and Deep Hydrofracturing Stress Measurements in the Waterloo Quartzite", Int. J. Rock Mech. Min. Sci. & Geomech. Abstr., Vol. 17, 1980, pp. 81-88.

13. Nakajima, T., " Topography and the Tectonic Stress", Abstr. Seism. Soc. Japan, (in Japanese), 1982, pp.114.

ROCK MASS BEHAVIOUR COUPLED WITH PRESSURIZED WATER FLOWING THROUGH THE FRACTURE

by Yoshiaki Mizuta and Charles Fairhurst

Associate Professor, Yamaguchi University
Ube, Yamaguchi, JAPAN

Professor, University of Minnesota
Minneapolis, Minnesota, USA

1. Introduction

Recently, the hydraulic fracturing technique has been examined as a potential method for the determination of *in-situ* stress at depth as well as for improving the efficiency of geothermal energy extraction.

For the determination of *in-situ* stresses, hydraulic fracturing has an advantage in that it can be done without over-coring.

Despite the increasing use of hydraulic fracturing, however, there are uncertainties associated with the interpretation of the resulting data.

The authers carried out laboratory hydraulic fracturing of impermeable rock in order to investigate rock fracture extension under conditions in which the principal stresses were inclined to the axis of the pressurized borehole. They found from the experiment that two typical fracture patterns were produced, which were able to occur at any borehole inclination and that re-opening of one type of fracture, called here a "longitudinal" fracture, using low fluid injection rates leads to a stable flow established, whereby fluid flows into the borehole beyond the sealing elements [1,2].

In consideration of such behaviours of rock and fluid obtained through the experiment and coupled stress-flow analysis, they developed an improved method for three-dimensional stress determination by hydraulic fracturing. Moreover, they examined theoretically the applicability of the method.

On the other hand, the project of heat extraction from hot dry rock has created a demand for the exact evaluation of output and life in generation of electricity obtained through heat exchange at the fracture produced by hydraulic fracturing.

In order to investigate the problem, it is also neccesary to take into account rock and fluid behaviours revealed through coupled rock mechanics and hydromechanics.

First, this report describes experimental results and the concept of coupled stress-flow analysis with some calculation examples. Second, an outline of a new method for stress determination is introduced and the applicability of the method is discussed. Finally, modelling of the coupled stress-flow analysis to be incorporated in the problem of heat extraction from hot dry rock is presented and a calculated example is shown.

2. Information from laboratory experiment

Experiments were carried out on relatively large specimens of brittle, competent and impermeable rock containing vertical or inclined boreholes. These specimens were fractured hydraulically by means of a miniature straddle packer under biaxial loading, and the fracture was re-opened by cycling the pressure at different flow rates.

Significant results obtained:

1) For any borehole inclination, either of two kinds of typical patterns of fracture will be produced.

2) The first type of fracture is longitudinal, parallel to the borehole axis, and will be produced when the differences among principal stresses are small or when borehole inclination is slight, fluid injection is rapid. It is also apt to be produced when initial packer pressure is low and clearance between the straddle packer and the borehole wall is slight. The normal stress component perpendicular to the borehole axis, p_y, can be estimated

from a pressure-time record obtained from a re-opening test on this type of fracture using a relatively low flow rate, because the stable flow pressure p_{st} is scarcely affected by the other stress components.

3) The second type of fracture is transverse to the borehole and perpendicular to the minimum principal stress p_3. The magnitude of p_3 can be determined from the corresponding pressure-time record.

Induced rock fracture extension patterns are summarized in Fig.1, and measured relationship between p_{st}/p_y ratio and flow rate is summarized by the symbols in Fig.2.

MAGNITUDE OF PRESSURES APPLIED TO SIDES OF TEST CUBES (MPa) Note. $p_3 = 0$.	OIL FLOW RATE (nm^3/s)	FRACTURE PATTERN a			a + b	a'		b'	b	
$p_1 = 13.78$, $p_2 = 6.89$	210									AI-36
	105			AI-9	AI-18	A-0 AII-18	AII-9	AII-27 AII-27	AII-36	A-0 AI-27
	59.3	A-0								
	26.2				A-0					
$p_1 = 8.27$, $p_2 = 6.89$	26.2	BI-9	BI-18	B-0		BII-18	BII-9			BI-27 BI-36 BII-27 BII-36
$p_1 = 6.89$, $p_2 = 3.45$	26.2	C-0 CI-18	CI-36 CII-18 CII-36							

Note Designation AI-9 indicates stress state AI with 9° borehole inclination

Fig.1 Relationship between induced fracture pattern, applied stresses and fluid flow rate obtained from fracturing tests

In Fig.1, the letter A corresponds to maximum principal stress p_1 = 13.78 MPa, intermediate principal stress p_2 = 6.89 MPa, minimum principal stress p_3 = 0, the letter B corresponds to p_1 = 8.27 MPa, p_2 = 6.89 MPa, p_3 = 0, the letter C corresponds to p_1 = 6.89 MPa, p_2 = 3.44 MPa, p_3 = 0, and the numeral I indicates that the borehole axis is perpendicular to p_2, the numeral II indicates that the borehole axis is perpendicular to p_1.

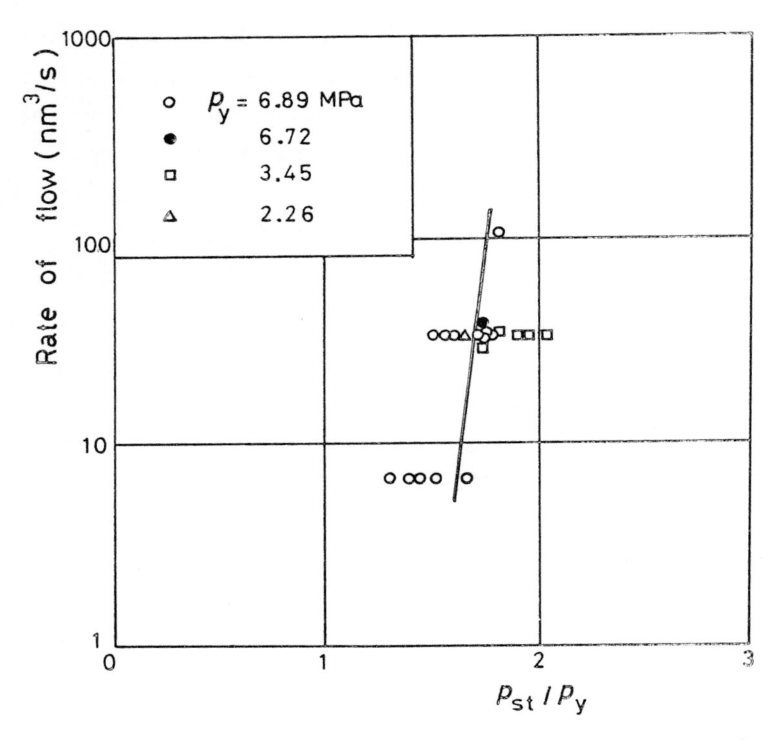

Fig.2 Relationship between p_{st}/p_y ratio and fluid flow rate obtained from re-opening tests for longitudinal fractures.

In Fig.2, the solid line is the calculated relationship between p_{st}/p_y ratio and the flow rate, as is described later.

The results obtained mean that even if a longitudinal fracture is produced, the direction of one of principal stresses is not always parallel to the borehole axis.

Also, it is presumed that even when longitudinal fracture pattern (a') is produced, p_y can be determined by using a shorter straddle packer at re-opening than at fracturing.

3. Coupled stress-flow analysis for longitudinal fracture

The numerical model employed two-dimensional flow analysis coupled with three-dimensional stress analysis, both analyses being by the finite element method. The finite element meshes based on the experimental configurations are shown in Fig.3. Each brick element shown in the figure is divided into six tetrahedrons in computation.

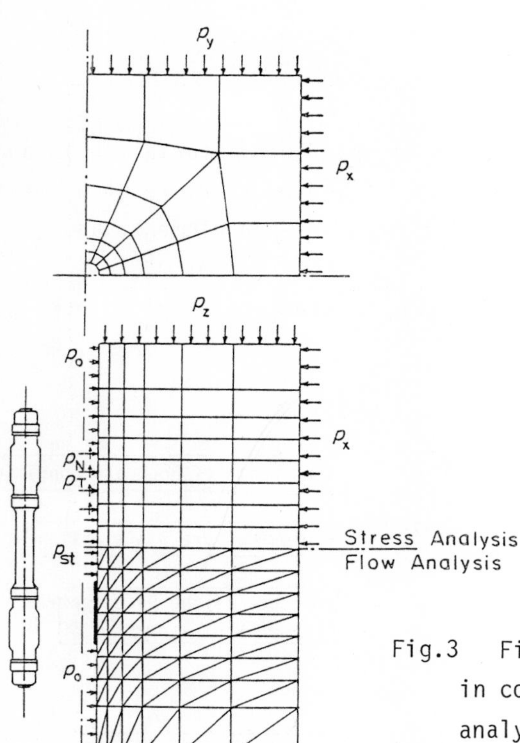

Fig.3 Finite element meshes used in coupled stress and flow analyses for longitudinal fractures.

The solution technique employed an iterative loop to ensure full coupling between the stable flow and the resulting stress field. Input parameters were fluid pressure, p_{st} at inlet and p_0 at outlet. For a given input pressure then, a given crack surface and aperture will maintain stable flow, so that both flow rate and crack re-opening surface result from computations by assuming that : 1) rock is an elastic, homogeneous and isotropic material, and 2) rock is impermeable. No attempt has been made to optimize the crack shape which was assumed to be perpendicular to the direction of p_y.

The following expression of the diffusion equation can be derived, based on the validity of the parallel plates cubic law for a rock fracture, as shown by Detournay [3].

$$ke[\frac{\partial}{\partial x}(\frac{\partial p}{\partial x}) + \frac{\partial}{\partial z}(\frac{\partial p}{\partial z})] + \frac{Q}{\Delta} = 0 \quad (1)$$

$$k = \frac{1}{\mu}\frac{e^2}{12} \quad (2)$$

where μ is the absolute viscosity of the fluid material, e is an average element thickness corresponding to an aperture of a re-opening crack, Q is the total discharge of each element, and Δ is the area of each element.

Flow charts depicting the main computational logic blocks are shown in Figure 4.

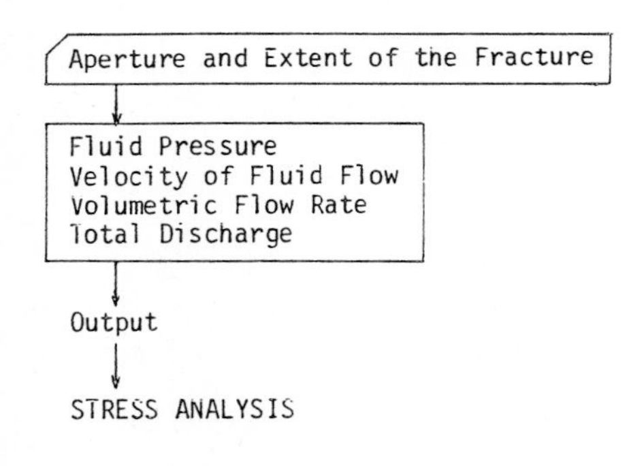

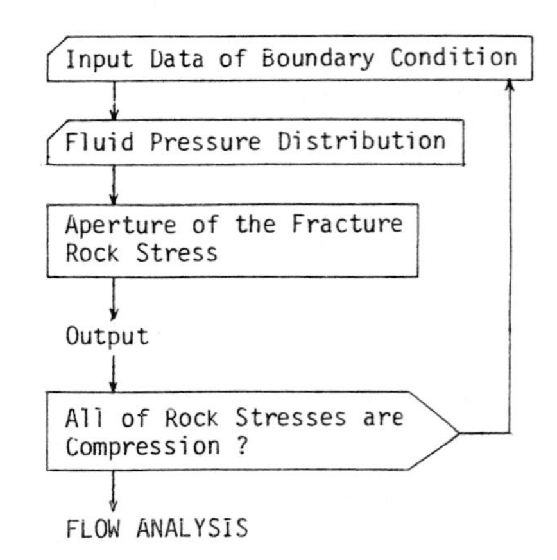

Fig.4 Flow chart of coupled stress and flow analysis.

Two examples of numerical results obtained from analyses on the experimental configurations are shown in Figure 5. For the flow analysis, the value of μ was assumed to be 40 mPa·s. For the stress analysis, the values of Young's modulus E and Poisson's ratio ν of the rock were assumed to be 70.7 GPa and 0.245, respectively. The

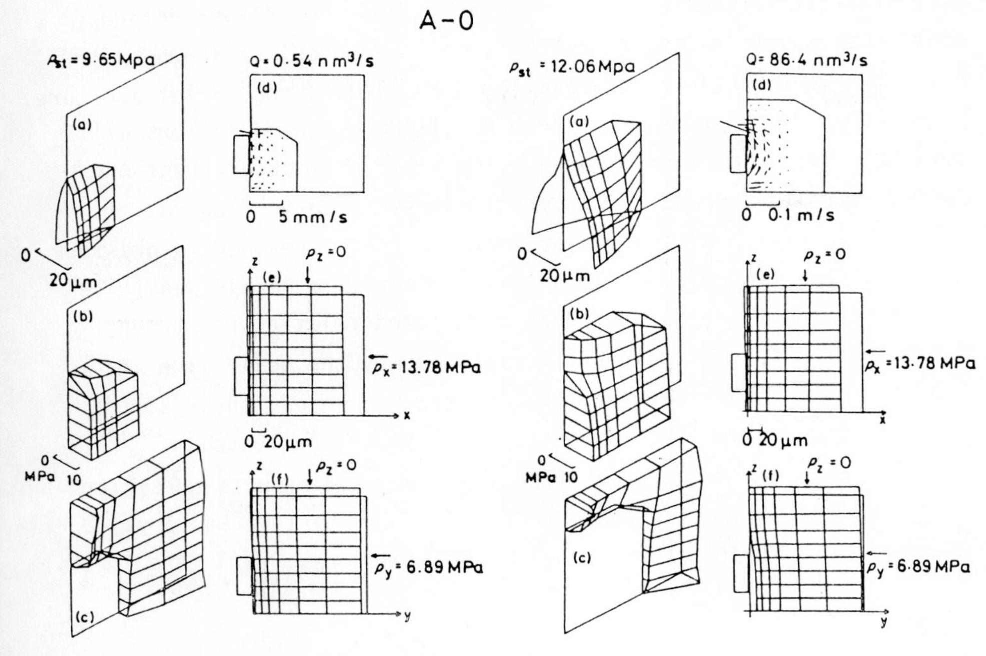

Fig.5 Examples of numerical results computed for the longitudinal fracture with applied stress state A.

figure graphically presents the results of fluid injection in terms of :

(a) Crack opening displacements
(b) Fluid pressure distribution in the crack
(c) Normal stress distribution on the continuation of the crack plane in the rock
(d) Fluid velocity field within the crack
(e) Rock specimen deformation in the X-Z plane
(f) Rock specimen deformation in the Y-Z plane

In the figure, the Z axis is taken to coincide with the borehole axis, and the Y axis coincides with the direction perpendicular to the fracture. The actual area of the opening fracture must be larger than the calculated result since, in this analysis, the existence of pore area and fluid pressure in the closed fracture was neglected. Similarly the actual crack opening and total discharge must be slightly greater, while the effective rock stress acting across the closed fracture must be less. However, as shown by the figure, flow rate is almost totally controlled by the outlet aperture. Conversely, for a given flow rate, equilibrium considerations will lead to a given aperture through which fluid will escape past the packers into the borehole. A very large increase in flow rate is required to effect a significant increase in the re-opening pressure, implying that the crack opening displacement is a rather sensitive measure of local pressures. The calculated relationship between p_{st}/p_y ratio and flow rate are represented by the solid line in Fig.2. In the analyses for the inclined boreholes, shear stress components among applied stress components were neglected.

It can be seen that numerical analysis gave results consistent with experiment when the concept of hydraulic aperture was applied to configurations involving longitudinal fractures.

4. Improved procedures for stress determination by hydraulic fracturing

4.1 Three-dimensional stress determination from hydraulic fracturing data

Consider a coordinate system $OX_iY_iZ_i$ chosen such that OZ_i is the borehole axis, and OY_i is perpendicular to both the borehole axis and the direction of minimum principal stress p_3, as shown in Figure 6. Consider another coordinate system $Ox_iy_iZ_i$ chosen such that Ox_i lies in the X_i-Y_i plane and is included in the longitudinal fracture. The orientation of $OX_iY_iZ_i$ with respect to $Op_1p_2p_3$ can be expressed in terms of its direction cosines which are the trigonometric functions of two angles, α_i and β_i. The orientation of $Ox_iy_iZ_i$ with respect to $OX_iY_iZ_i$ can be expressed in terms of its direction cosines which are trigonometric functions of the angle θ_i. Those three angles are defined as :

α_i = the angle between OZ_i and Op_3

β_i = the angle between OY_i and Op_2

θ_i = the angle between Ox_i and OX_i

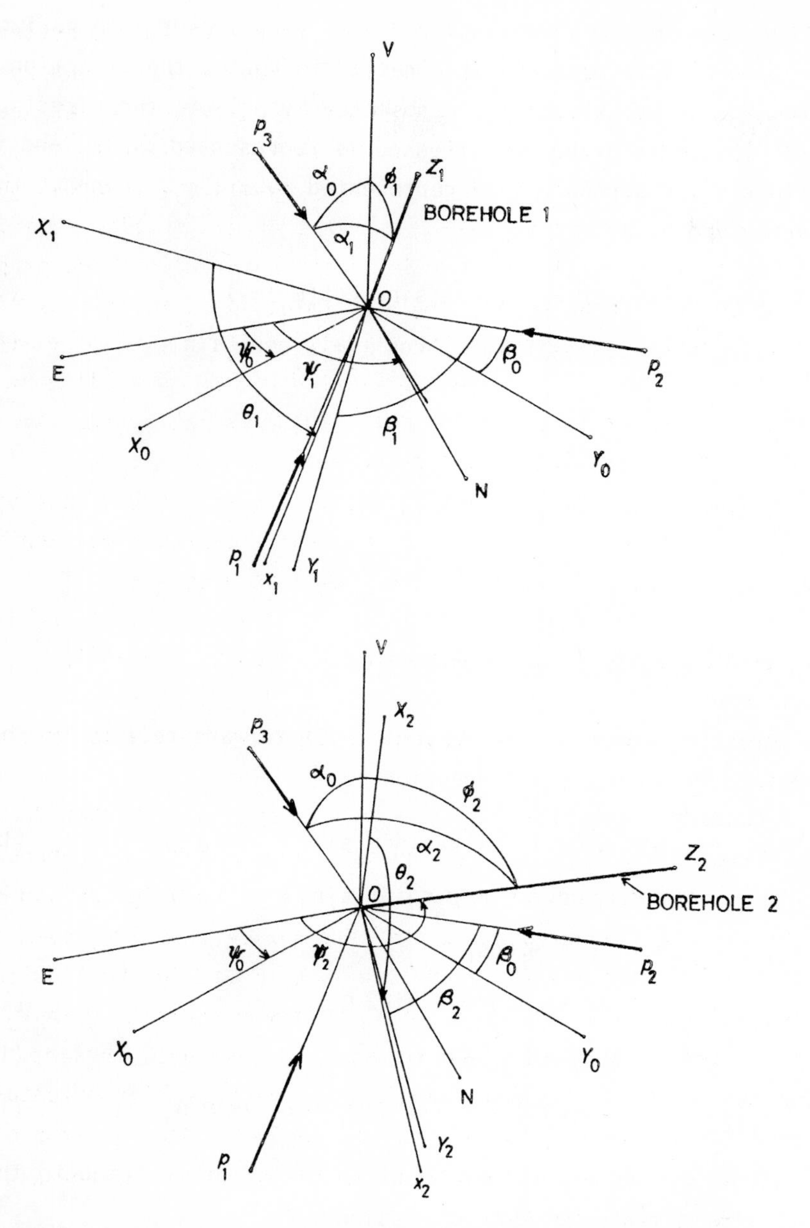

Fig.6 Orientation of two longitudinal fractures (lies in the x-Z plane) induced in the borehole 1 and borehole 2 with respect to OXYZ and the *in-situ* principal stresses.

Consider that OZ_i (=OV) is vertical (OX_i and OY_i are horizontal) OE is directed to East, ON is directed to North, the direction of minimum principal stress is represented by α_0, ψ_0, the direction of one of the other principal stresses is represented by β_0, and the inclination of borehole i is represented by ϕ_i, ψ_i. Then, α_i and β_i are given by :

$$\cos\alpha_i = \cos\alpha_0\cos\phi_i + \sin\alpha_0\sin\phi_i\cos(\psi_i-\psi_0) \tag{3}$$

$$\cos\beta_i = -L_i\cos\alpha_0\sin\beta_0 + M_i\cos\beta_0 + N_i\sin\alpha\,\sin\beta_0 \tag{4}$$

where

$$L_i = \pm m_i\cos\alpha_0/\sin\alpha_i \tag{5}$$

$$M_i = (\cos\phi_i - \cos\alpha_0\cos\alpha_i)/(\sin\alpha_0\sin\alpha_i) \tag{6}$$

$$N_i = \pm m_i\sin\alpha_0/\sin\alpha_i \tag{7}$$

$$m_i = \pm(\sin^2\phi_i - l_i^2)^{\frac{1}{2}} \tag{8}$$

$$l_i = (\cos\alpha_i - \cos\alpha_0\cos\phi_i)/\sin\alpha_0 \tag{9}$$

The stress components with respect to OX-Y-Z are related to the *in-situ* stresses by the expression :

$$p_{Zi} = p_1\sin^2\alpha_i\cos^2\beta_i + p_2\sin^2\alpha_i\sin^2\beta_i + p_3\cos^2\alpha_i \tag{10}$$

$$p_{Xi} = p_1\cos^2\alpha_i\cos^2\beta_i + p_2\cos^2\alpha_i\sin^2\beta_i + p_3\sin^2\alpha_i \tag{11}$$

$$p_{Yi} = p_1\sin^2\beta_i + p_2\cos^2\beta_i \tag{12}$$

$$p_{XiYi} = 0.5(p_1 - p_2)\cos\alpha_i\sin2\beta_i \tag{13}$$

$$p_{YiZi} = -0.5(p_1 - p_2)\sin\alpha_i\sin2\beta_i \tag{14}$$

$$p_{ZiXi} = -0.5(p_1\cos^2\beta_i - p_2\sin^2\beta_i + p_3)\sin2\alpha_i \tag{15}$$

The stress components with respect to Ox-y-Z are related to the above stress tensor by the expression :

$$p_{xi} = p_{X_i}\cos^2\theta_i + p_{Y_i}\sin^2\theta_i + p_{X_iY_i}\sin2\theta_i \tag{16}$$

$$p_{yi} = p_{X_i}\sin^2\theta_i + p_{Y_i}\cos^2\theta_i - p_{X_iY_i}\sin2\theta_i \tag{17}$$

$$p_{xiyi} = -0.5(p_{X_i} - p_{Y_i})\sin2\theta_i + p_{X_iY_i}\cos2\theta_i \tag{18}$$

Now it is presumed that if the values of the overburden pressure and the tensile strength of rock T are estimated, the factor η which gives p_{st}/p_y ratio is determined from laboratory experiment or coupled stress-flow analysis, and the values of breakdown pressure p_{bi}, p_{sti}, ϕ_i, θ_i, ψ_i and shut-in pressure p_s ($=p_3$) are measured from hydraulic fracturing test, then the values of p_1, p_2 and β_0 can be calculated by using a method for unconstrained optimization problems [4]. The object function to be minimized is expressed by :

$$W = \sum_{i=1}^{n}[(p_{yi} - p_{sti}/\eta)^2 + p_{xiyi}^2 + (\sigma_{\theta i} + T)^2] + (p_V - \gamma h)^2 \tag{19}$$

where

$$\sigma_{\theta i} = -p_{xi} + 3p_{yi} - p_{bi} \tag{20}$$

Hence, six stress components with respect to OENV are determined by the expression :

$$p_E = p_{X_0}\cos^2\psi_0 + p_{Y_0}\sin^2\psi_0 - p_{X_0Y_0}\sin2\psi_0 \tag{21}$$

$$p_N = p_{X_0}\sin^2\psi_0 + p_{Y_0}\cos^2\psi_0 + p_{X_0Y_0}\sin2\psi_0 \tag{22}$$

$$p_V = p_{Z_0} \tag{23}$$

$$p_{EN} = 0.5(p_{X_0} - p_{Y_0})\sin2\psi_0 + p_{X_0Y_0}\cos2\psi_0 \tag{24}$$

$$p_{NV} = p_{Y_0Z_0}\cos\psi_0 - p_{Z_0X_0}\sin\psi_0 \tag{25}$$

$$p_{VE} = p_{Y_0Z_0}\sin\psi_0 + p_{Z_0X_0}\cos\psi_0 \tag{26}$$

4.2 Examination for applicability of the method

Consider an example where longitudinal fractures are produced in two boreholes with different inclinations and a transverse fracture is produced in one of these boreholes. Assuming that

$\phi_1 = 0$, $\alpha_0 = \alpha_1 = \alpha_2 = \psi_0 = \phi_2 = 30°$, $T = 10$ MPa, $\gamma h = 7.35$ MPa, $\eta = 1.5$, $p_s = 5$ MPa, $p_{b1} = 21.33$ MPa, $p_{b2} = 14.55$ MPa, $p_{st1} = 13.15$ MPa, $p_{st2} = 9.06$ MPa are measured, stress states in the rock will be calculated by minimizing the value w of equation (19).

The stress states calculated and expressed by six stress components are shown in column one in Table 1.

Table 1. Calculated stresses in the rock and their errors

Six stress components	Calculated stresses (MPa)	Errors due to ±10% or ±5° errors in every measured values (MPa)
p_E	10.49	-0.68 0.22
p_N	12.16	0.42 -0.03
p_V	7.35	0.99 -0.76
p_{EN}	1.33	-0.73 0.67
p_{NV}	1.34	0.37 -0.19
p_{VE}	-3.93	0.07 -0.35

The errors in the calculated stress components caused by ±10% or ±5° errors in every measured value are shown in column two in Table 1. It is found from Table 1 that three-dimensional stresses may be given with some accuracy by the new method.

Optimization calculations were carried out by using a mini-computer,Hewlett Packard 85, and took 7 minutes per problem.

For the purpose of design of underground construction, rock stress determination by hydraulic fracturing can be done economically by using a simple apparatus as is shown in Fig.7, while great expense is required in a conventional stress determination procedure.

Rock stress determinations using this apparatus are carried out

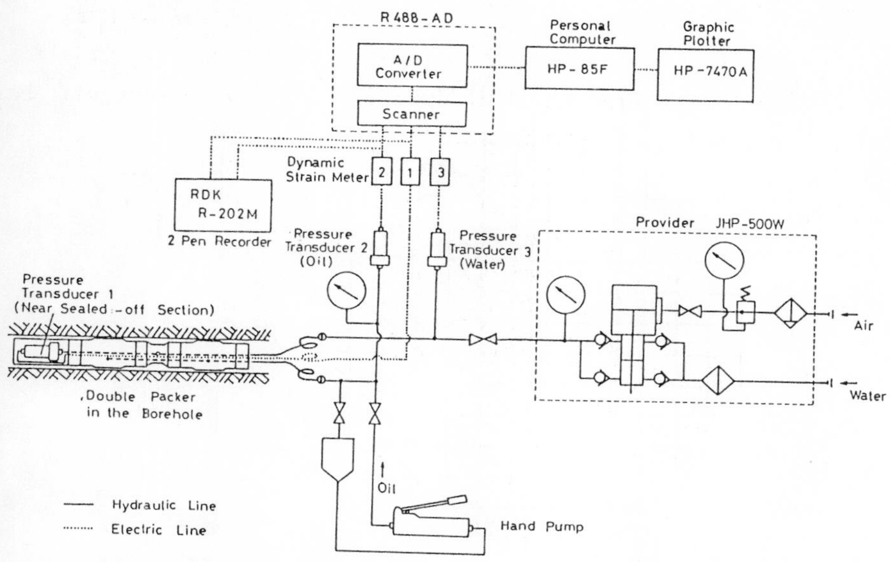

Fig.7 Outline of the hydraulic fracturing apparatus for rock stress determination with the purpose of design of underground construction.

at the underground repository for low-level radioactive waste disposal, which is under construction in the Sanpo Mine in Okayama Prefecture.

5. Coupled stress-flow analysis for transverse fracture

In order to investigate exact thermal energy extracted from hot dry rock, both rock and fluid behaviour must be incorporated into the analysis of the problem.

Consider a fracture plane intersecting two boreholes, one an inlet and the other an outlet for fluid flowing in the fracture. The numerical model employed two-dimensional flow analysis by the finite element method coupled with three-dimensional stress analysis by the boundary element method.

The boundary element squares [5], and the finite element triangles for assumed configuration are shown in Fig.8.

As an example, the numerical results obtained from the coupled stress-flow analysis are shown in Fig.9. For the flow analysis,

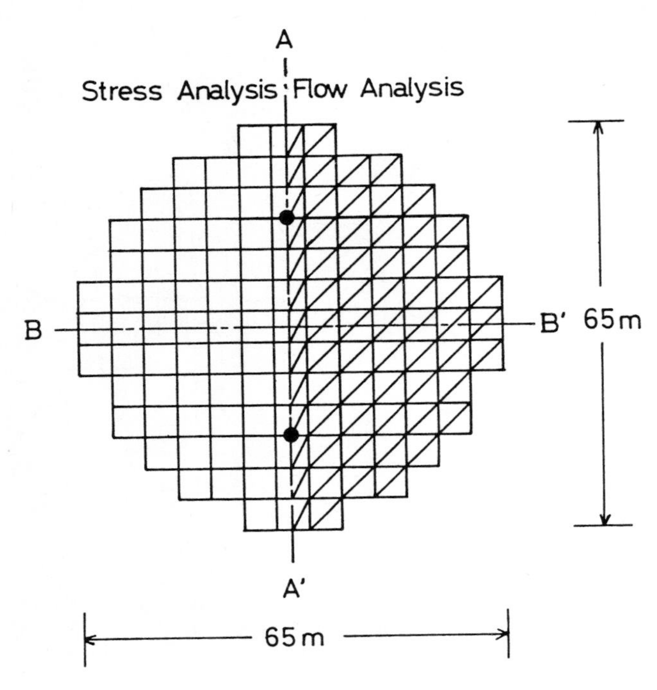

Fig.8 Boundary element and finite element meshes for assumed configuration.

the value of μ was assumed to be 1.002 mPa·s which is the value for water. For the stress analysis, it was assumed that the values of E, ν, p_1, p_2 and p_3 are 30.0 GPa, 0.25, 15.0 MPa, 11.25 MPa and 7.5 MPa, respectively, and the directions of p_1, p_2 and p_3 lie on B-B', A-A', and the vertical, respectively. The figure graphically presents the results of fluid injection in terms of :

(a) Crack opening displacement at section A-A'

(b) Crack opening displacement at section B-B'

(c) Fluid pressure distribution in the crack at section A-A'

(d) Volumetric fluid flow rate field within the crack

In Figure 9 (c), the solid line is obtained from the stress analysis and the dashed line is from the flow analysis. It can be seen from the figure that the volumetric rate of flow within the area with small aperture is almost negligible.

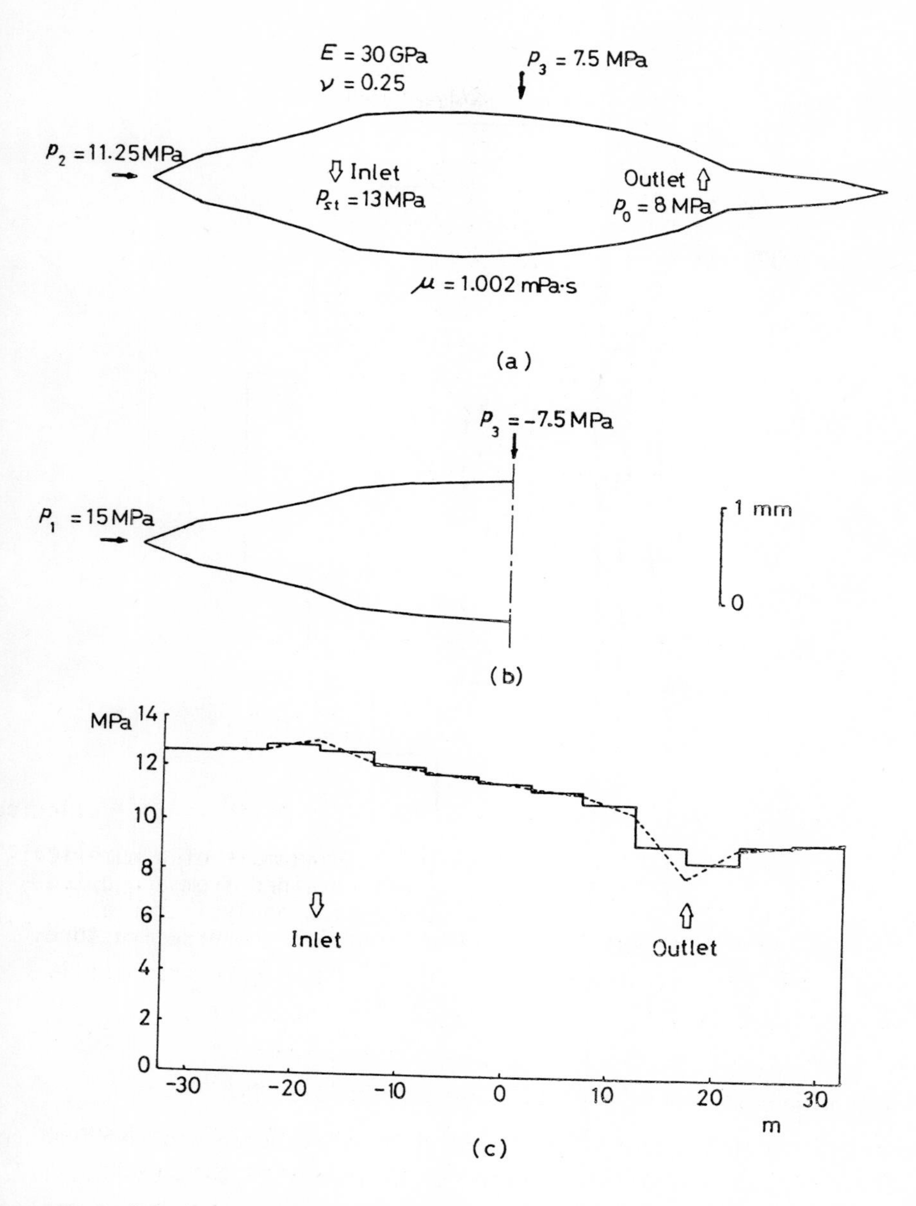

Fig.9 (a),(b),(c) An example of a numerical result obtained from coupled stress flow analysis for a horizontal transverse fracture.

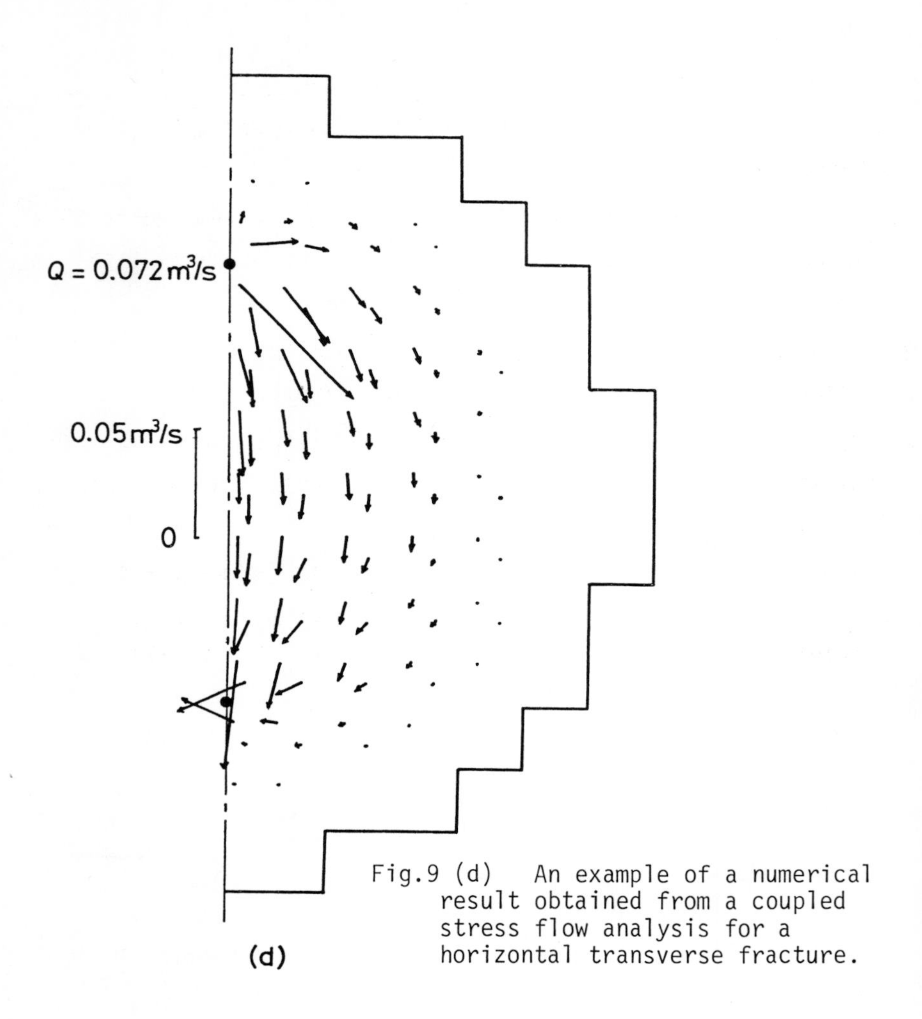

Fig.9 (d) An example of a numerical result obtained from a coupled stress flow analysis for a horizontal transverse fracture.

Acknowledgements

Financial support for a part of this research was received from the U.S. Geological Survey under grant number USID-14-08-0001-17775.

The authors would like to thank Mr. H.Kobayashi, Dr. J.Santich, Mr. E.Detournay, Prof. W.D.Lacabanne and Mr. M.Nakura for their invaluable help in the work.

References

1) Y.Mizuta & H.Kobayashi, An Analytical and Experimental Study of Inclined Hydraulic Fracture, 13th Canadian Rock Mech. Symposium, I, 1980

2) Y.Mizuta et al, Improved Stress Determination Procedures by Hydraulic Fracturing, Technical Report to U.S. Geological Survey, USDI-14-08-0001-17775, 1980

3) E.Detournay, The Hydraulic Conductivity of Closed Rock Fracture, 13th Canadian Rock Mech. Symposium,III, 1980

4) Kowalik J. & Osborne M.R. Methods for unconstrained optimization problems. American Elsevier Publishing Company, Inc. (1968).

5) B.H.G.Brady and J.W.Bray, The Boundary Element Method for Elastic Analysis of Tabular Orebody Extraction, Assuming Complete Plane Strain, Int.J.Rock Mech., 15, 29-37, 1978

AN ASSESSMENT OF THE FACTORS AFFECTING HYDRAULIC FRACTURE CONTAINMENT IN LAYERED ROCK: OBSERVATIONS FROM A MINEBACK EXPERIMENT

Lawrence W. Teufel and Norman R. Warpinski
Sandia National Laboratories
Albuquerque, New Mexico 87185, U.S.A.

ABSTRACT

In situ experiments, which were accessible for direct observation by mineback, have been conducted to determine the effect that geologic discontinuities, elastic properties, and *in situ* stress differences have on hydraulic fracture propagation and the resultant overall fracture geometry. Vertical fractures were observed to terminate only in regions of high minimum horizontal *in situ* stress. Fracture growth into a higher (by a factor of 5 to 15) elastic modulus region was preferred to propagation into a region of higher (by a factor of 2) stress. Geologic discontinuities did not arrest the lateral or vertical propagation of hydraulic fractures. However, hydraulic fracture growth was affected by discontinuities, because fracture-fluid leakoff occurred along intersecting discontinuities which the hydraulic fractures crossed, and thereby reduced the total extent of the hydraulic fractures. The results of the mineback experiment clearly indicate that differences in the minimum horizontal *in situ* stress between the reservoir rock and adjacent layers is the most critical factor affecting hydraulic fracture containment.

INTRODUCTION

Massive hydraulic fracturing is the most promising technique for stimulation of low-permeability gas reservoirs at the present time. This technique is at least an order of magnitude scale-up from conventional hydraulic fracture technology and it is designed to create long penetrating fractures which contact large areas of the reservoir. However, the results to date have often been disappointing and the general applicability of these treatments for unconventional gas resources is uncertain.

Although there are several possible causes for the lack of success, one of the most likely reasons is the failure of the fracture to contact a sufficiently large area of the reservoir due to unfavorable vertical propagation out of the reservoir into formations lying above and below the producing zone. When a treatment is designed, the height of the fracture is the parameter about which the least is known, *a priori*, yet this influences all aspects of the design (1). Therefore, it is extremely important to recognize and understand the mechanisms which may influence the height of a fracture by restricting vertical fracture propagation (containment). Of course, these same features would be available to influence lateral propagation also.

Several parameters have been suggested as being important for hydraulic fracture containment. A difference in elastic modulus between the reservoir rock and the barrier rock is often singled out as a primary mechanism controlling containment. In their work on composite materials, Cook and Erdogan (2) calculated the stress intensity factor for a two-dimensional crack approaching an interface between two materials with different elastic moduli. Simonson *et al.* (3) applied these results to hydraulic fracturing and observed that since the stress intensity factor, K, at the tip approaches zero as a fracture in a lower modulus material propagates toward a higher modulus material, the fracture will tend to be arrested. Conversely, for a fracture propagating in a higher modulus material toward a lower modulus material, they observed that K becomes large as the interface is approached, and the fracture should accelerate through the interface.

Daneshy (4) conducted laboratory experiments and found that differences in rock properties were insufficient to stop fracture growth at an interface. He suggested that barriers may only reduce vertical fracture growth rather than prevent it. Daneshy further suggested that fracture containment may be more a result of the nature of the interface itself rather than any difference in material properties, but he thought that this would be most often the case at shallow depths where the bonding is likely to be weaker. Teufel (5) and Anderson (6) performed laboratory experiments to study crack growth near both bonded and unbonded interfaces. They showed that, for unbonded surfaces, the stress normal to the interface (thus the friction along the interface and the interfacial shear strength) was the determining parameter for crack arrest or continued propagation. In addition to layer interfaces, other geologic discontinuities, such as faults and natural fractures, would similarly effect the propagation of hydraulic fractures (7).

Perkins and Kern (1) suggested that *in situ* stress differences in different rock strata may limit fracture height. Simonson *et al.* (3) also studied the effect that stress variations have on fracture propagation. They showed that a layer with greater *in situ* stress would provide an effective barrier because of the increase in the fracturing pressure necessary to continue propagating a fracture in this layer. The effectiveness of the barrier would, of course, depend on the difference in stress between the two layers. They also showed that the upward or downward migration of the fractures can be influenced by the hydrostatic gradient of the fracture-fluid relative to the vertical gradient of the minimum horizontal *in situ* stress.

While all of these theoretical studies and laboratory experiments offer insight to the problem of containment, it is clear that such idealized results are not easily applied directly. This paper presents the results of realistic *in situ* experiments which have been conducted in an existing tunnel complex at the U. S. Department of Energy's Nevada Test Site to examine hydraulic fracture behavior under many different conditions. These facilities are ideal for hydraulic fracturing experiments because they provide an *in situ* medium with the appropriate boundary conditions (*in situ* stresses, no free surfaces) yet still allow for detailed examination of the created fractures and geological features through mineback (physical excavation of the rock to observe the fracture directly). A detailed physical description can be obtained through photography and mapping, and this can be correlated with measured geologic properties, *in situ* stress distributions, fluid behavior, and the operational parameters of the test. Although the various volcanic tuffs in which these fractures are propagated are not the sandstones and shales usually encountered in gas reservoirs, proper application of rock mechanics principles allows the extrapolation of these results to gas well conditions.

EXPERIMENT

The hydraulic fracture experiments were conducted in a tunnel complex mined into volcanic tuffs that comprise Rainer Mesa at the Nevada Test Site. At tunnel level there is approximately 420 m of overburden which provides a realistic *in situ* stress distribution. The experiments were conducted in ash-fall tuffs, which are soft, low modulus, high porosity, low permeability tuffs that allow for easy excavation with a continuous mining machine. Overlying the ash-fall tuff is an ash-flow tuff which is much denser, higher modulus and lower porosity than the ash-fall tuff. The ash-flow tuff grades upward from an unwelded basal ash-flow tuff into a densely welded ash-flow tuff. The elastic properties of the three volcanic tuff layers are shown in Figure 1. The elastic modulus of the ash-flow tuff is a factor of 5 to 15 greater than the ash-fall tuff. Poisson's ratio of the ash-flow tuff is 2 to 3 times lower than the ash-fall tuff. The permeability of the tuffs are very low, ranging from 10^{-17} to 10^{-18} m^2. Consequently, fluid leak-off problems during fracture propagation are minimal.

The experiments utilize the immediate contact between the ash-fall tuff and basal ash-flow tuff as a material-property interface. As shown in Figure 2, inclined boreholes are cored either below or through the interface from an existing nearby tunnel. The holes are 0.1 m in diameter and are uncased. The core is examined and several unfractured intervals in each hole are chosen for hydraulic fracturing. These zones are usually 3 to 5 m apart.

For fracturing, a mobile pump system is brought into the tunnel and situated near the collar of the hole. The pump system is capable of pumping dyed water at a maximum flow rate of 0.15 m^3/sec at 35 MPa. Straddle packers are inserted in the open hole, centered over the zone and inflated. The length of the remaining open hole zone is approximately 1.5 m. The length of the packer elements is 1.7 m. Packers are typically inflated to 5 to 10 MPa.

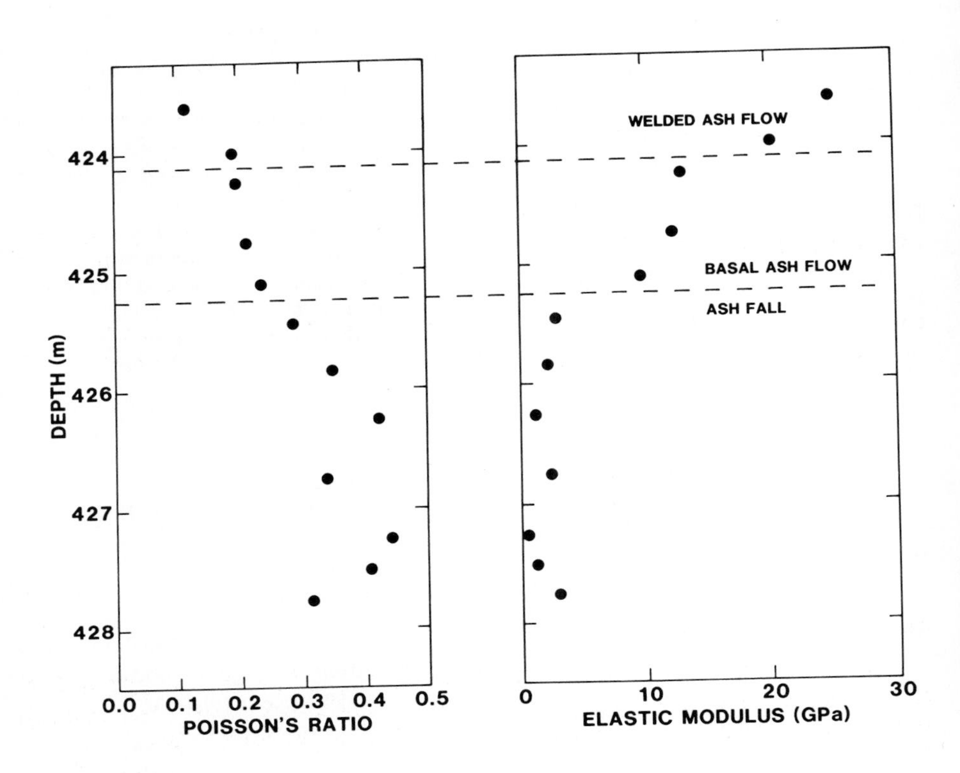

Figure 1. Elastic properties of volcanic tuff.

The fracturing procedure is as follows. After the packers are situated and inflated, the fracturing begins with a very small breakdown at maximum flow rate to determine the breakdown pressure and the instantaneous-shut-in-pressure (ISIP). This usually requires about 0.6 m^3 of dyed water, just sufficient an amount for the formation to break down and the fracturing pressure to stabilize. In this way the fracture has propagated beyond the borehole effects, yet the fracture is still small enough so that the ISIP, which is equivalent to the minimum principal *in situ* stress, is measured over a small area. After the pressure decays to a small value, another 0.2 to 0.6 m^3 of fluid are injected into the fracture to obtain a repeat value of the ISIP. Finally, when the pressure again decays to a low value, the remaining volume to be pumped is injected into the formation at the prescribed flow rate. These experiments are conducted with volumes of fluid from 5 m^3 to 15 m^3. Afterwards, the packers are moved to the next zone and the procedure is repeated.

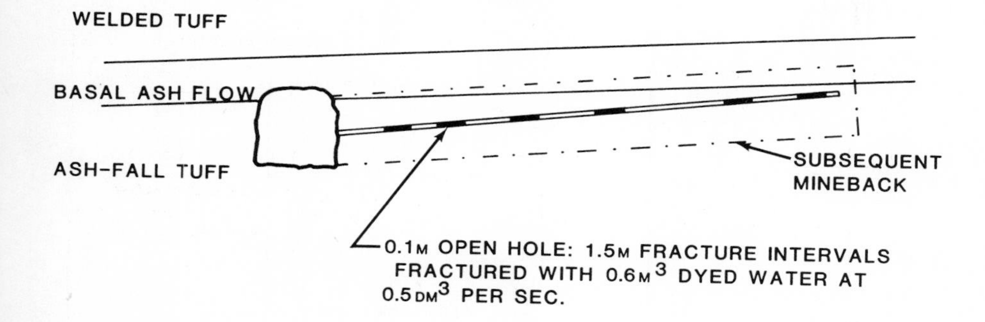

Figure 2. Schematic of experiment.

RESULTS

Mineback Observations

Hydraulic fractures were initiated in two inclined boreholes, CFE-1 and CFE-2, at different stratigraphic horizons in the ash-fall tuff. The instantaneous-shut-in-pressure obtained after the initial breakdown of each fracture zone was used to establish the vertical distribution of the magnitude of the minimum principal *in situ* stress below the ash-flow tuff/ash-fall tuff interface. The magnitude of the minimum principal *in situ* stress varied from 2.2 MPa immediately below the interface to 8 MPa at 2 m below the interface (Figure 3). Multiple ISIP stress determinations within the same stratigraphic horizon showed differences of less than 0.4 MPa.

After completion of all of the hydraulic fractures in the CFE-1 and CFE-2 holes, a mineback was conducted of each hole to observe directly the behavior of hydraulic fractures near a major material-property interface and in regions of varying in situ stress. During mineback it was found that each fracture initiated nearly perpendicular or at high angles to the borehole, usually near one of the packers, and propagated essentially in a vertical plane which consistently had a trend of N20°E to N30°E (Figure 4).

In the CFE-1 drift, all of the fractures preferentially propagated horizontally outwards and upwards through the interface into the higher modulus ash-flow tuff (Figure 5). The orientation of the fractures did not change during propagation across the interface, but remained in a vertical plane. There were, however, significant changes in the width of the fractures in regions of differing modulus with the larger width found in the low modulus region. The upper extent of these fractures was always several meters into the densely welded ash-flow tuff. The higher modulus welded tuff was clearly not a containment barrier to fracture propagation as predicted by current fracture models (3).

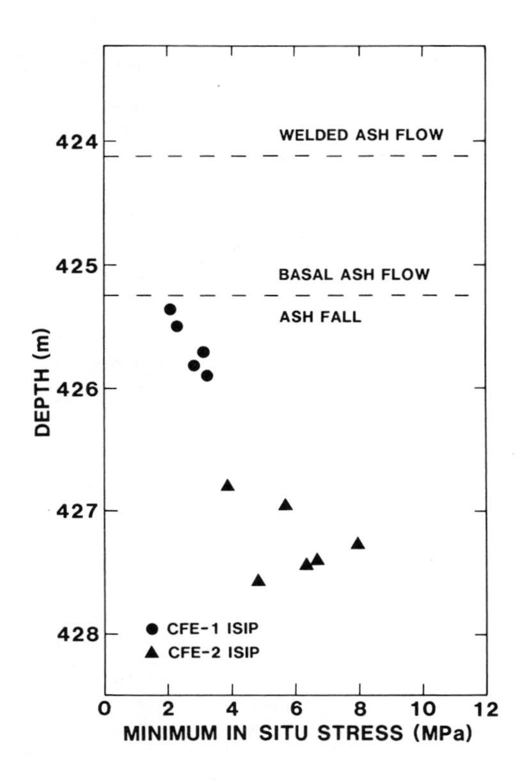

Figure 3. Plot of minimum principal *in situ* stress in ash-fall tuff.

In sharp contrast to the uncontained upward propagation of the CFE-1 fractures, downward propagation in the ash-fall tuff was severely restricted. All of the hydraulic fractures were arrested at the same stratigraphic horizon about 1 m below the interface (Figure 4). Near the location of the fracture termination, the fractures were generally impregnated with more dye than at any other location. This indicates that the fractures probably propagated to this point at an early time and remained inflated during pumping even though fracture propagation had terminated in that direction. Fluid continued to leak off, resulting in the heavy dye residue along the fracture surface at the termination point. The cause of this behavior will be discussed in the next section.

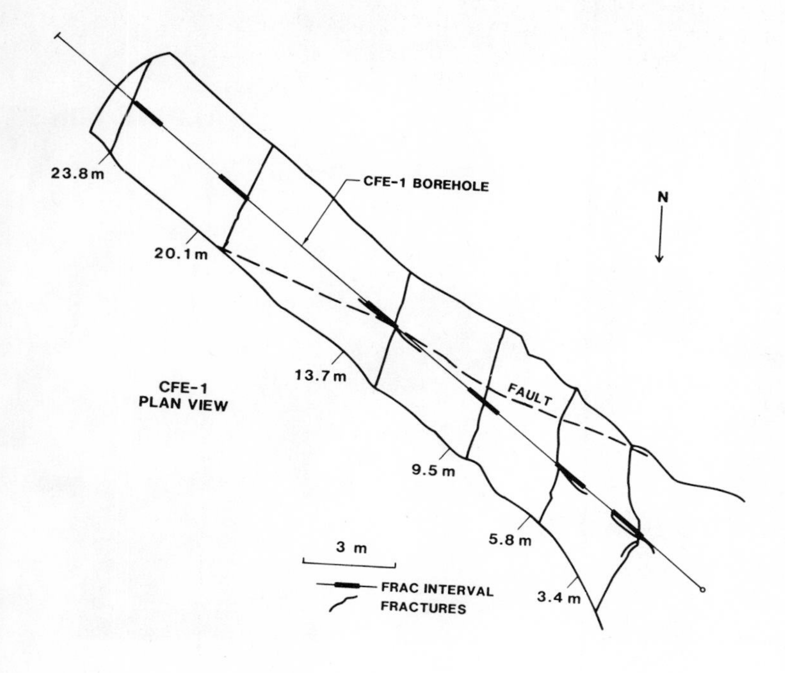

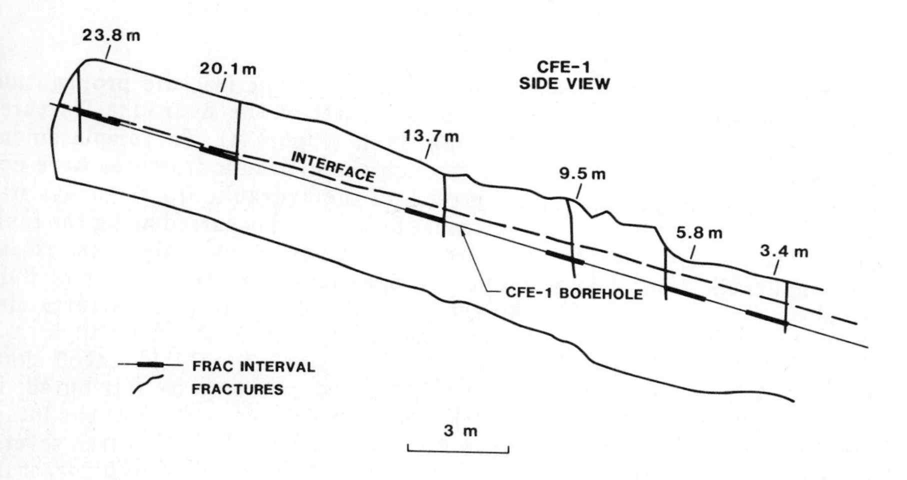

Figure 4. Plan view and side view schematic of CFE-1 drift showing lateral and vertical extent of hydraulic fractures.

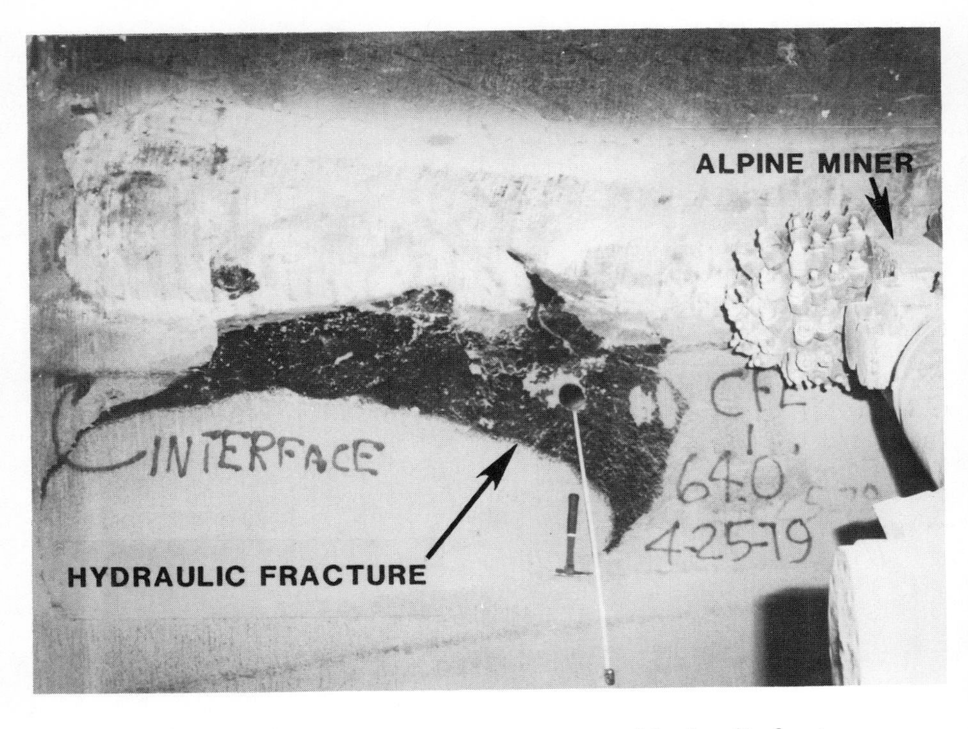

Figure 5. Photograph of partially exposed hydraulic fracture which is nearly parallel to the mineback face.

The effect of geologic discontinuities on hydraulic fracture propagation was also observed during the mineback. Several of the hydraulic fractures intersected and crossed a small normal fault (Figure 4). The displacement along this fault was about 30 cm. Although the hydraulic fractures were not arrested by this discontinuity, the growth of the hydraulic fractures was still significantly affected by the fault, because fluid leakoff occurred along the fault plane after the hydraulic fracture crossed it. Consequently, the total extent of the hydraulic fracture was reduced proportionally to the volume of fluid lost along the fault plane. In the ash-flow tuffs the hydraulic fractures also intersected and crossed natural extension fractures. As with the small normal fault, fluid leakoff also occurred along these discontinuities. Fluid leakoff along the normal fault and the natural extension fracures may be attributed, in part, to the higher permeability of these discontinuities relative to the intact rock. However, in many cases, heavy dye residue of the fluid occurred several meters away from the point of intersection where the hydraulic fracture crossed the fault or the natural fractures, suggesting that these discontinuities were inflated during pumping and hydraulic fracture growth. Consequently, in highly fractured regions the hydraulic fracture was not a single plane, but a zone with mutiple strands.

Stress Variations and Fracture Containment

A possible explanation of why the lower termination of all the CFE-1 hydraulic fractures occurred abruptly about 1 m below the basal ash-flow/ash-fall tuff interface was that the minimum *in situ* stress was significantly greater in that horizon than in overlying horizons where the fractures had been initiated. Previous ISIP data clearly showed that high stress regions can occur in the ash-fall tuff (Figure 3). In order to obtain additional data on the vertical distribution of the magnitude of the minimum principal *in situ* stress, particularly in the region of fracture termination, overcoring stress measurements were made in the CFE-1 drift at five stratigraphic horizons (Figure 6). In both of the ash-flow tuff layers there was one overcoring horizon, approximately 0.5 m above the lower interface of each of these layers. In the ash-fall tuff there were three overcoring horizons at 0.2 m, 0.6 m, and 1.0 m below the interface with the basal ash-flow tuff. Two of these horizons corresponded to horizons where hydraulic fractures had been initiated and thus a direct comparison could be made between stresses calculated from overcoring strain-relief measurements and ISIP data. The lowest horizon was the horizon of fracture termination.

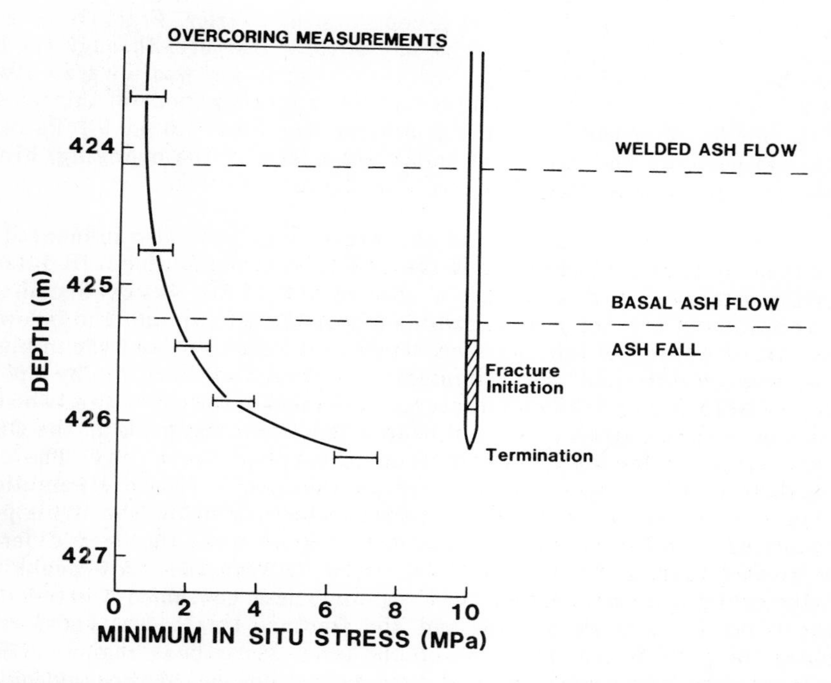

Figure 6. Plot of minimum principal horizontal *in situ* stress and vertical extent of hydraulic fractures in the CFE-1 drift.

All of the overcoring stress determinations were made with the well-documented U.S. Bureau of Mines (USBM) overcoring technique (8). Elastic moduli for overcore samples were determined with a biaxial loading device and the USBM three-component borehole deformation gage (9). The overcoring measurements were at drill-hole depths of 9.4 to 10.2 m, which is outside the zone of significant influence of the mineback drift. As such, the overcoring stress data should be representative of free-field *in situ* stresses. The drill-holes were oriented parallel to the general azimuth of the vertical hydraulic fractures, which is the orientation of either the maximum or intermediate principal *in situ* stress, and had an orientation of N25°E. Because the drill-holes were aligned with either the maximum or intermediate principal stress direction, an accurate determination of the minimum principal *in situ* stress could be made using only one drill-hole, instead of the usual three drill-holes at different orientations, which is required for determining the principal stress magnitudes. Three overcoring stress measurements were made at each horizon.

At the abrupt lower termination of all the CFE-1 hydraulic fractures, 1 m below the basal ash-flow/ash-fall tuff interface, the magnitude of the minimum principal *in situ* stress was 6.8 ± 0.6 MPa, which was a factor of 2 greater than the two overlying horizons where these fractures were initiated (Figure 6). The 3 MPa increase in the minimum *in situ* stress was sufficient to arrest the fractures and act as a containment barrier. Fracture pressure was apparently not sufficient to drive any of these fractures through the high stress horizon. Instead, fracture growth after initiation was always outward and upward into the higher modulus and lower stress regions of the ash-flow tuff. In the ash-flow tuff layers the minimum stress was 1.0 ±0.5 MPa in the basal ash-flow tuff and decreased to 0.6 ±0.4 MPa in the overlying, higher-modulus, welded ash-flow tuff.

Variations in the minimum *in situ* stress magnitude also influenced the propagation of fractures initiated in the CFE-2 hole. Combining ISIP data and overcoring measurements provides a detailed plot of the vertical distribution of the minimum *in situ* stress magnitude (Figure 7). At 1 m and 2 m below the basal ash-flow/ash-fall tuff interface, stress peaks occurred and the minimum stress was greater than in the immediate surrounding horizons by up to a factor of 3. In the CFE-2 hole, fractures were initiated between the two stress peaks, in the lower stress peak, and below the lower stress peak. In the CFE-1 hole, fractures were initiated only above the upper stress peak. The effect of these two stress peaks on fracture propagation is shown schematically in Figure 8. Fractures which were initiated above or below the stress peaks propagated away from the high *in situ* stress regions when the stress difference was greater than 2 MPa. Fractures initiated between the stress peaks were totally contained vertically in that region. These contained fractures had rectangular fracture geometries and the fracture length was considerably greater than the height. Fractures in the lower stress peak region (ISIPs of fractures were 6.5 to 8.0 MPa) had uncontained downward propagation, but upward propagation was contained at the upper stress peak after the fractures had propagated through the intervening lower stress region. These results demonstrate that hydraulic fracture growth can be contained by a region of high stress. In this particular experiment with small fractures and 420 m of overburden, containment occurred only when the resulting stress difference ahead of the propagating fracture wass greater than 2 MPa.

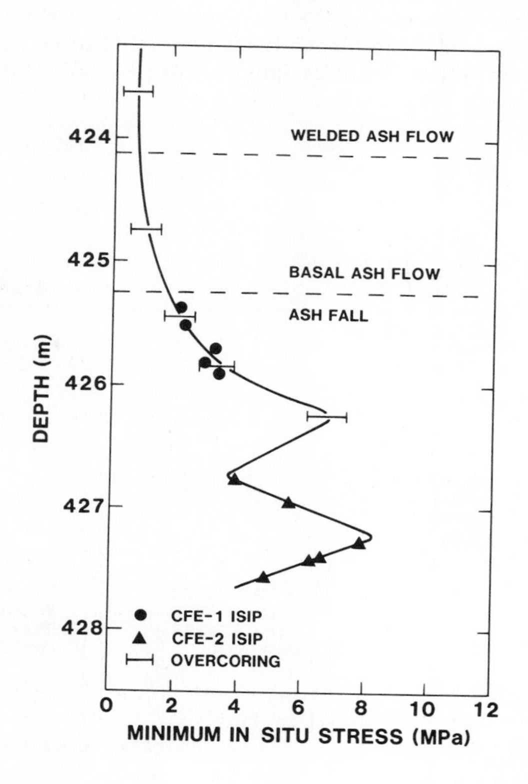

Figure 7. Plot of minimum principal horizontal *in situ* stress from hydraulic fracture ISIP data and overcoring measurements.

DISCUSSION

Hydraulic fractures were initiated below a material property interface in a region with large variations in the minimum principal *in situ* stress magnitude. Each fracture was injected with 5 to 10 m^3 of dyed water. The location of fracture initiation varied from 0.15 m below the basal ash-flow/ash-fall tuff interface (higher modulus material above) to 2.3 m below the interface. None of the fractures had their growth restricted in any apparent way by thc higher modulus layer (factor of 5 to 15 difference). These results are consistent with those of a previous large scale experiment (10,11) which was conducted near the same interface. Laboratory experiments also show that fractures will not be contained by a change in properties across an interface (5, 7). These results do not agree with the present two-dimensional analyses of crack behavior at an interface which would predict containment (3). The failure of these analyses apparently lies in modeling the behavior at the interface using only the stress intensity factor at the tip of the crack for the failure criterion. In order to properly model the mechanisms operating in this problem, it may be necessary to link a more realistic failure criterion with the stress analysis of a propagating crack (12).

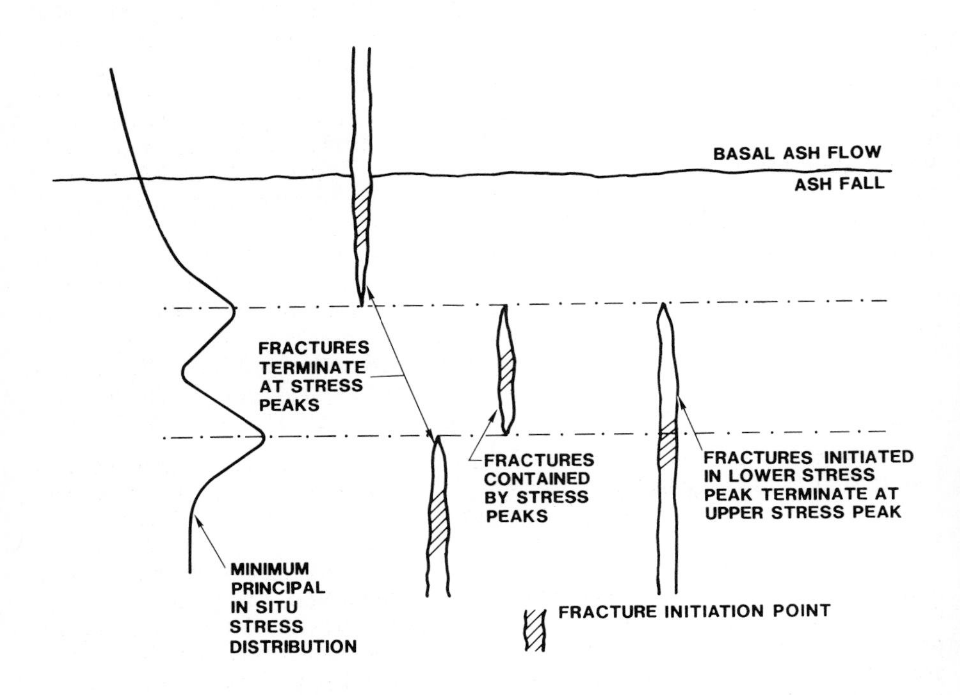

Figure 8. Schematic of mineback results showing effect of stress peaks on fracture containment.

The ISIP obtained during breakdown of each fracture test and the overcoring measurements were used to establish the vertical distribution of the minimum principal *in situ* stress magnitude (Figure 7). At two horizons (0.2 m and 0.6 m below the basal ash-flow/ash-fall interface) the ISIP data and overcoring measurements overlapped and were in good agreement. In the volcanic tuff the minimum stress was neither constant nor a linear gradient with depth, but exhibited large variations in magnitude. In the ash-fall tuff, at 1 m and 2 m below the interface, the minimum horizontal stress increased by a factor of 2 to 3 (2 to 4 MPa) relative to the surrounding horizons. These variations in stress magnitude had a significant and dominant influence on the propagation and overall geometry of the hydraulic fractures (Figure 8). Fractures were arrested whenever the they propagated from a low stress to a high stress region where the difference in the minimum horizontal *in situ* stress was greater than 2 MPa. Fractures initiated in a low stress layer that was bounded by high stress layers were completely contained vertically and had rectangular fracture geometries.

The high stress layers provide an effective containment barrier because fracture propagation into these layers requires an additional increase in fracture pressure (3). If the fracture pressure does not increase sufficiently propagation terminates at the high stress boundary but continues in other directions. The effectiveness of the stress barrier depends on the difference in the minimum *in situ* stress between the bounding layer and the layer with the propagating fracture. In the small scale hydraulic fracture experiments in volcanic tuff, at a depth of 420 m, the critical stress difference was 2 to 3 MPa. Laboratory experiments have shown that the critical stress difference to contain a hydraulic fracture ranges from 2 to 5 MPa, depending on the boundary conditions and lithologies used in the experiments (7, 13). For massive hydraulic fracture treatments in deep wells, the critical stress difference for containment would be higher, because in these large scale fractures the pumping pressure required to extend the fracture is usually at least 5 MPa greater than the minimum principal *in situ* stress. We are presently conducting a field experiment in low-permeability, gas-bearing sandstones with bounding shales to evaluate the magnitude of the critical stress difference required for massive hydraulic fracture containment in this type of reservoir at a depth to 2500 m.

The cause of the *in situ* stress variations found in this experiment is thought to be a combination of gravitational loading applied to the layered rock mass, which had varying elastic properties and boundary relief at the sides of the mesa. For an unconfined, layered rock mass subjected to gravitational loading, the magnitude of the horizontal stresses in individual layers increases with increasing depth. The relative difference in the horizontal stress from one layer to another will be a function of the relative difference in elastic properties of the layers (7). An increase in the horizontal stress will occur in going from a layer with a high elastic modulus and low Poisson's ratio into a layer with a low elastic modulus and high Poisson's ratio. In this experiment the minimum horizontal stress was lowest in the ash-flow tuff layers which had a higher modulus and lower Poisson's ratio than the ash-fall tuff (Figure 1). Stress peaks occurred in horizons of the ash-fall tuff where the elastic modulus was the lowest and Poisson's ratio was the highest. This indicates that differences in elastic properties may be important, not as a containment barrier *per se,* but because of the manner in which elastic property variations influence the vertical distribution of the minimum horizontal *in situ* stress magnitude. It is expected that the most significant changes in *in situ* stress will occur in the vicinity of a change in rock properties, typically due to layering.

The results of this study have important implications to the design of massive hydraulic fracture treatments for stimulation and enhanced recovery of low-permeability gas reservoirs. The predominant influence of differences in the minimum horizontal *in situ* stress on hydraulic fracture containment clearly indicates that an estimate of the height of potential fractures, prior to stimulation, can be made by determining the *in situ* stress state in the reservoir-fracture interval as well as in the bounding layers. This will enable selection, *a priori,* of zones that are bounded by high stress layers and that will most likely provide economic production due to favorable growth and containment conditions.

This study also indicates that geologic discontinuities, such as faults, fractures, and bedding planes, may also significantly affect the overall geometry of massive hydraulic fractures. Although geologic discontinuities did not arrest the lateral or vertical propagation of hydraulic fractures observed in the mineback experiment, fluid leakoff did occur along intersecting discontinuities that the hydraulic fractures crossed, suggesting that these discontinuities were inflated during pumping and hydraulic fracture growth. Accordingly, for massive hydraulic fracture treatments in gas reservoirs which have an abundance of natural fractures, fluid leakoff along these fractures could significantly reduce the total extent of the fracture. Futhermore, in massive hydraulic fracture treatments pumping pressures are usually maintained well above the minimum *in situ* stress in order to extend these large fractures. Consequently, inflation of intersecting natural fractures which the hydraulic fracture crosses would be enhanced, thus increasing the fluid leakoff along these discontinuities, and futher reducing the total extent of the hydraulic fracture and potential recovery of the reservoir.

CONCLUSIONS

Mineback experiments have provided insight into the mechanisms which are responsible for controlling fracture geometry. It has been demonstrated that a factor of 5 difference in elastic modulus between adjacent layers is insufficient to arrest a fracture at an interface, as predicted by present fracture models. Geologic discontinuities also did not arrest hydraulic fracture propagation, but did reduce the extent of the fracture as a consequence of fluid leakoff. In massive hydraulic fracture treatments the presence of natural fractures could significantly reduce the size of the hydraulic fracture and thereby, reduce the potential recovery of the reservoir. These experiments have shown that the minimum principal *in situ* stress is the predominant influence on fracture containment. Fractures were contained whenever there was a sufficiently large increase in stress in the bounding rock layers. In shallow, small-scale, hydraulic fractures in layered volcanic tuff, the critical stress difference was 2 to 3 MPa. For massive hydraulic fracture treatments in deep wells the critical stress difference would be higher because of higher pumping pressures in these large scale treatments. The mineback results indicate that an assessment of the potential success of fracture treatments of low-permeability gas reservoirs, in which containment is desired, can be made by determining the *in situ* stress in the fracture interval and bounding layers.

ACKNOWLEDGEMENTS

This work performed at Sandia National Laboratories supported by the U. S. Department of Energy under Contract Number DE-ACO4-76-DP00789.

REFERENCES

1. Perkins, T. K. and Kern, L. R. (1961). *Widths of hydraulic fractures:* Jour. Pet. Tech. (13), 937-949.

2. Cook, T. S. and Erdogan, F. (1972). *Stresses in bonded materials with a crack perpendicular to the interface:* Int'l. Jour. Eng. Sci. (10), 677-697.

3. Simonson, E. R., Abou-Sayed, A. S., and Clifton, R. J. (1978). *Containment of massive hydraulic fractures:* Soc. Pet. Eng. Jour. (18), 27-32.

4. Daneshy, A. A. (1978). *Hydraulic fracture propagation in layered formations:* Soc. Pet. Eng. Jour. (18), 33-41.

5. Teufel, L. W. (1979). *An experimental study of hydraulic fracture propagation in layered rock:* Ph.D. Dissertation, Texas A&M University, College Station, TX.

6. Anderson, G. D. (1981). *Effects of friction on hydraulic fracture growth near unbonded interfaces in rocks:* Soc. Pet. Eng. Jour., 21, 21-29.

7. Teufel, L. W. and Clark, J. A. (1982). *Hydraulic fracture propagation in layered rock: experimental studies of fracture containment:* Soc. Pet. Eng. Jour., Paper No. 9878, in press.

8. Hooker, V. E. and Bickel, D. L. (1974). *Overcoring equipment and techniques used in rock stress determination:* U.S. Bureau of Mines Inform. Circ. 8618, 32p.

9. Fitzpatrick, J. (1962). *Biaxial device for determining the elasticity of stress-relief cores:* U.S. Bureau of Mines Report 6128, 13p.

10. Warpinski, N. R., Northrop, D. A., and Schmidt, R. A. (1978). *Direct observation of hydraulic fractures: behavior at a formation interface:* Sandia National Laboratories Report, SAND78-1935.

11. Warpinski, N. R., Schmidt, R. A., and Northrop, D. A. (1982). *In situ stresses: the predominant influence on hydraulic fracture containment:* Jour. Pet. Tech., (34), 653-664.

12. Schmidt, R. A. (1980). *A microcrack model and its significance to hydraulic fracturing and fracture toughness testing:* in Proceedings of the 21st U.S. Symp. Rock Mech., Rolla, MO, 581-590.

13. Warpinski, N. R., Clark, J. A., Schmidt, R. A., and Huddle, C. W. (1982). *Laboratory investigation on the effect of in situ stresses on hydraulic fracture containment:* Soc. Pet. Eng. Jour., 22, 333-340.

Theory and Application of Hydraulic Fracturing Technology

by Michael P. Cleary
Associate Professor of Mechanical Engineering
Massachusetts Institute of Technology
Cambridge, Mass. 02139

ABSTRACT

In our industry-sponsored project (M.I.T. UFRAC), we are conducting a combined comprehensive theoretical and laboratory investigation of fracturing schemes which hold promise for providing or improving access to underground reserves of energy and other natural resources. The main focus has been on quasi-static methods, particularly on ramifications of the hydraulic fracturing technique and enhancing complements such as thermal cracking or induction of high pore pressures (e.g., due to expansion tendencies of highly energetic trapped second phases). However, consideration is also being given to various controlled explosive methodologies (e.g., as preparation for hydrafrac) and to possibilities for constructively dispersing electrical or chemical energy (e.g., with pumped fluids).

The project has proceeded on four fronts:

1. Identification of mechanisms (e.g., mechanical, thermal, chemical) which have potential for creating suitable fracture patterns in underground reservoirs or other target rock.
2. Detailed analytical and numerical modelling of fracture evolution, aimed at incorporating within efficient computer programs each potentially important field phenomenon, at least phenomenologically.
3. Experimental verification of analytical predictions at all levels of sophistication, from pocket calculator to mainframe programs, using both laboratory and field data.
4. Provision of insight and readily intelligible formulae (plus computer routines) for immediate application toward improvement of current operations in the field and development of new technology.

This paper will concentrate on the particular technology of hydraulic fracturing, which has emerged as a central component in most practical endeavors, complemented by the many other techniques (such as heating, acidising and explosive breakdown). This is receiving primary attention in our new RESOURCE EXTRACTION LABORATORY at MIT.

1. INTRODUCTION

The need to generate underground fractures with suitable extent, orientation and distribution is a problem well appreciated by the community of engineers concerned with extraction of energy and natural resources from pay zones which must be exploited under increasingly adverse conditions of inaccessibility, low transmissivity and tightening regulation. Research in this general area has mushroomed over the past decade especially, both among industrial concerns and in government-sponsored laboratories; fewer university groups have been involved, perhaps because the problems were not considered to be sufficiently fundamental in nature. However, there is now a growing recognition that some very basic issues have to be resolved quickly if there is to be proper guidance for and understanding of expensive field trials which have yielded all too little insight in the past.

A complete survey of the literature and industrial activity in this area would probably be superfluous for most readers of this report, so we limit the discussion to a few brief remarks on foregoing work. Adequate amount of review may be found in our published papers (Cleary, 1978, 1979, 1980), but a cross-section of relevant papers may be found cited at the end. These reveal, among other things, the great variety of uses that have been found for the technique of hydraulically breaking down a formation by producing excess pressure in a borehole and then continuing to pump as the fracture propagates; other mechanical, thermal and chemical fracturing processes (many common in nature) are also beginning to find technological application in the resource extraction area. As well, it seems that (perhaps predictably) limited success has attached to the use of (chemical or nuclear) explosives for stimulation purposes; some attention has been paid to coupling of controlled detonations or deflagrations with hydrafracing. The primary foci of our work (in the project MIT UFRAC) are, indeed, these twin topics of quasi-static fracturing and suitable methods for creating regimes of dynamic crack growth, whenever these might enhance the success-ratio of stimulation efforts or other reservoir-access activities; we concentrate our discussion here on hydraulic fracturing, central to most practically viable techniques.

The literature on "hydrafracing" abounds with (often inconsistent) assumptions which yield operative models for design of field operations. Two primary groups of working idealisations have evolved among industrial groups (see figure I.1), the one (called PKN after its authors) assuming a vertically-closed laterally-similar elliptical crack opening (Perkins et al, 1961; Nordgren, 1972) while the other (called CGDD after its authors), also vertically bounded, allows enough slippage at interfaces that the crack is vertically-similar (e.g. Christianovich, 1957; Geertsma and De Klerk, 1969; Daneshy, 1971; Daneshy, 1973); obviously, neither quite fits the real situation where the crack may partially blunt -- or, especially, may even spread vertically appreciably further than assumed. The latter is much more pessimistic about effective lateral length achieved for a given volume of fluid pumped; CGDD may therefore give a better length, in relation to a model which realistically allows vertical extension of the fracture, but such a more realistic model (e.g. P3DH in Section 2B) is typically closer to the PKN assumption for crack-width! We have performed computations (e.g. based on formulae in Cleary, 1980) for a much broader group of assumptions about shape, formation and fluid parameters; we substantially conclude that predictions typically made for the extent of fractures generated by conventional operational procedures in the field are often too optimistic -- a conclusion that many of our industrial colleagues have apparently reached as a result of various correlation studies in the field.

Although one finds astute empirical realisation (e.g. Pugh, 1978, White et al. 1981) of procedures which might enhance the tendency for containment of fracturing within the stratum of interest ("pay zone"), there seems to have been only limited recognition of the mechanics and materials features which govern this aspect. Indeed, some insights have been displayed in the identification of elasticity barriers (Simonson, 1978), in-situ stress contrast (Abou-Sayed, 1977) and other mechanisms (Daneshy, 1978) of retardation or acceleration of crack growth across interfaces between strata, but some confusion of the issues seems to prevail(e.g. Warpinski et al., 1980); some clarifications and more detailed consideration (Cleary, 1978, 1980) shows that these aspirations are usually over-optimistic, unless some attempt is made to render the formation response more amenable to operation of such mechanisms. We have established a number of possibilities (Cleary, SPE 9260, 1980) and we have detailed various methods whereby this can be achieved (Cleary et al., 1981). It is perhaps interesting that these and numerous other deductions of ours (Cleary, 1980) are based on calculations that we might classify as "back-of-the-envelope"; one tends to miss such phenomena in large computer analyses (e.g. using finite elements), when the mechanisms involved are not model-intrinsic.

However, we do not advocate over-simplification of the fluid-solid structural interaction, which is at the heart of the hydrafrac methodology: indeed, we quickly felt the need for more precise quantification of our estimates for crack shapes and extent at successive increments of time after initiation. To achieve this detailed simulation, while preserving the correct character of discontinuities on crack surfaces and slippy interfaces, we gradually developed a preference for the effective and very insightful method of force and dipole distributions, from which there arises only the need to solve a set of surface integral equations (Cleary, 1978, 1980) on each of the surface sub-domains which is active at that instant. The development of a general purpose computer scheme, based on this formulation, for simulation of fracture evolution under arbitrary driving stress distributions (e.g. internal hydraulic pressure), has become one of our major undertakings in this project.

That care is needed, to avoid over-simplification of relevant structural calculations, is also well illustrated by the interpretations that have been assigned to critical pressures measured in mini-frac efforts to deduce tectonic stress conditions (Haimson et al, 1970, 1975, Voegele and Jones, 1980): the problem here is that simple solutions (e.g. for a round hole in isotropically linear elastic material) tend to be employed, to the exclusion of important nonlinear, anisotropic and mechanistic effects. Considerable rationalisation has since been offered (e.g. Abou-Sayed et al, 1978; Zoback et al, 1978) and we have given a thorough discussion of the matter (Cleary, 1979). Simple laboratory experiments (Haimson et al, 1970; Medlin et al, 1976; Cleary, 1979; Morita et al., 1981) quickly illustrate the extent of deviation from the idealisation, and they emphasize the importance of mechanistic considerations.

We are now convinced that mini-frac has enormous potential both for initial breakdown of the well-bore (hence avoiding detrimental initially unstable growth of subsequent massive fractures) and for determination of reservoir parameters (especially in-situ stress and material deformation/transport properties, hitherto unrealised); however, our lab work shows that great care is needed to achieve reliable interpretation of (pressure and flow) measurements made. In this respect, laboratory simulation is indispensable: it may not rigorously represent the field environment but any analytical model of worth should certainly be capable of predicting sample response under these controlled conditions. This feeling has prompted us to pursue the development of a laboratory facility which we can use to test the reliability of our theoretical calculations: our aim is to make this effort increasingly sophisticated, quite feasibly to the point where we eventually have both analytical and experimental analogues of reasonably realistic field circumstances. In beginning this work, we noted that previous lab arrangements did not really take proper account of scaling laws which must govern any reduced model of field operations: great attention was typically focused on use of feed-back control, stiff fluids and other artifacts to produce stable crack growth, whereas the field operation is innately stable. By using suitably viscous fluids, and modulus/toughness of the sample being fractured,properly scaled to the reduced length, and especially by recognising the dominant stabilising influence of the confining stress on the exterior, we have been able (Papadopoulos et al, 1980, 1981, 1982; Crichton, 1980) to generate innately stable quasi-static spreading of fractures easily observable visually or deductively: these may well be the first true laboratory simulations of the hydrafrac operation and we now plan to carry the capability through a series of stages --

until we have exploited its full potential. Thus, a second major part of our project (MIT UFRAC) is to design various test configurations and material microstructures, which will provide invaluable insight into the character of fracture evolution within increasingly complex (e.g. porous, stratified, naturally fractured) structural formations of the kind produced by geological processes.

Conjugate to all this analysis and experimentation has been the desire to evaluate what response is actually elicited in the real-life circumstances of field operations, which are being performed everyday and could serve as the best test of all. Toward this, numerous monitoring methodologies have been proposed (Wood et al, 1976; Keck et al, 1978; Swolfs, 1977; Roegiers et al, 1974; Pierce et al, 1974; Power et al, 1975; Greenfield et al, 1977; Komar et al, 1973, 1975) and some have proven out to some extent in the field (Smith et al., 1978). However, there are a number of readily appreciated limitations on the applicability of all these techniques; we have been seeking alternative more comprehensive schemes, based on more reliable signals and measuring tools (preferably downhole), and a number of concepts have been developed. Related to this is our desire to work with opaque porous specimens in the laboratory, so that we need a method for delineating the crack perimeter within the sample. Ultrasonic scanning tools were already available but we have also explored the potential of various electromechanical, thermal and deformation field mapping techniques. Meanwhile, we have developed some effective but lab-specific marker techniques for delineating successive perimeters from induced surface markings, corresponding to timed pressure pulses.

A last group of problems--which seem to have received little attention (with few exceptions, Daneshy, 1978; Warpinski et al, 1979) but which are central to our objectives at MIT, and have foremost importance for a great variety of resource extraction endeavors -- are those of sustaining multiple crack propagation and achievement of linkage, in order to generate underground fracture networks. Our early experiments (Crichton, 1980) have shown us that it is quite feasible to make fractures from adjacent boreholes link up, by picking the sequence of borehole pumping in a reasonably careful fashion -- so that the interacting stress fields always favour curving toward each other, rather than away. The broader fundamental relevance of experimentation on this phenomenon, from the viewpoint of three-dimensional crack evolution in solid materials at all size-scales, should also be emphasised. On the other hand, the generation and sustenance of multiple cracks from a single borehole, especially in a biased tectonic field, seems also to be in the realm of practical achievability and we are studying a number of possible artifacts to do this, even quasi-statically: the most promising approach may be a rapid (or laboratory jacketed) pressurisation technique for initiation, followed by successive selective pumping and propping of each fracture thereby created. Multiple crack initiation can certainly be achieved by suitable explosive pressure rise times (e.g. Young, 1978, Schmidt et al., 1979) but, even if the obvious drawbacks of explosives are overcome, this must be followed by conventional hydrafrac; we will refer to this combination as the "hybrid fracturing technique", and note that it may alone (among dynamic techniques) have potential for reservoir stimulation. Assistance for such multiple cracking can also be had from the induction of thermal stresses e.g. created by steam injection -- or by cooling near the borehole on the main fracture surface, phenomena which we have recently examined in some depth (Barr, 1980).

These and many other mechanisms are being studied further (in MIT UFRAC) toward confident application in the field. However, our presentation here will concentrate on just one major technique, hydraulic fracturing, which seems to be central to most reasonable plans for (enhanced) resource extraction. As well, we will limit ourselves to some basic models, which adequately illustrate the principal features of the process and of our other more complex simulations; those allowing clear verification in the laboratory and subsequent confident application to design of field operations.

2. FIRST-ORDER MODELS AND DESIGN OF HYDRAULIC FRACTURES

2A. Simple lumped models of hydraulic fracturing

Our experience with more complicated and general models has allowed/motivated us to reduce the models as simple first-order differential equations for length and height growth in time; here a characteristic time appears naturally to set the scale for non-dimensionalisation of real time, and it shows very readily the primary roles of excess pressure (injection pressure minus confining stress), moduli of the strata in the reservoir, and rheology (viz., viscosity) of the frac fluid, appearing with powers 2n+2-m, 2n+1-m and 1, respectively.

These equations can actually be integrated analytically, under certain restrictive simplifying assumptions, and this leads to useful insight on the phenomenological behavior of hydrafracs. More generally, they can be integrated numerically for any well-bore and fluid loss conditions: this leads to an interesting pressure variation in time for specified fluid injection rate (e.g., depending strongly on height/length growth rates) and vice-versa. The resulting pocket-calculator programs are very simple and effective, having more potential than many existing industrial models: it can quite reasonably be claimed that they solve the fracture height growth (containment) problem, which many researchers have felt to require a cumbersome fully 3-D analysis. However, the rational use of these simple lumped models does require experience with more complex schemes (Sec. 2B)

2A.1. Equations governing lumped P3DH-type models

With reference to fig. 2A.1, we assume that growth in one direction (say height H) is governed by a CGD-type model, while the other (say length L) is described by a PKN-type model; this is a reasonable description if, as we wish, the fracture is well contained but these interpretations, assumed henceforth, can be reversed (i.e. L interchanged with H) if the fracture is very poorly contained (H > L). The model seems to work quite well also in the middle ground of relatively equiaxed geometries ($L \simeq H$) but a better description (taking account of limited flow through perforations) may be circular growth (Cleary and Wong, 1982), which captures the higher pressures needed for flow.

Based on these fairly unrestrictive geometric assumptions, we have been able to derive some very simple formulae for the growth of H and L in time. Without any loss in generality for the moment, the height-growth expression can be written as

$$\dot{H} = \overline{\gamma}_2 H/\tau_c, \tag{2A.1}$$

which may be applied to either wing (upper H_U or lower H_L, fig. 2A.1). Here the characteristic time τ_c has been obtained by a combination and non-dimensionalisation of the governing equations (3A-3C) for a hydraulic fracture, (e.g. Cleary, 1980), leading to

$$\tau_c^m \equiv \overline{\eta}/(\overline{E}\hat{\sigma}^{2n+2-m}\ H^{2n-2m}) \tag{2A.2a}$$

in which the elastic crack opening relation has appeared through

$$\Delta_E/H = \gamma_1\hat{\sigma} \equiv \gamma_1\sigma/\overline{E},\ \ \sigma \equiv p_f - \sigma_c \tag{2A.2b}$$

(Obviously, any power-law relation can also be used for crack-opening, still leading to formulae of the same type; we specialize for simplicity). Here the coefficient γ_1 is obtained by solving for crack opening (e.g. Cleary et al., 1981) with the relevant distribution of excess pressure $\sigma = p_f - \sigma_c$ in the fracture: clearly, it will vary somewhat in time (e.g. from $2/\pi$ to 1 in going from L=H to L>>H) but so will most other coefficients, especially γ_2 (discussed later, see figure 2A.1b).

Actually, the form in eqn. (2A.1) can be written to describe the growth of any fracture dimension, including the length. However, it is more appropriate to conduct a specific analysis for lateral flow, assuming only that width depends dominantly on height H (as in eqn. (2A.2b)), as detailed in Sec. 2B; the result is an important variation on eqn.(2A.1):

$$\dot{L}^m = \overline{\Gamma}_2^m\ H^{m+1}/L\tau_c^m \tag{2A.3}$$

in which H/L clearly modifies the propagation rate. Eqns. (2A.1,3) are the most general forms of the equations governing CGD-type and PKN-type models, respectively: all other forms and solutions can be derived from them, and they are also remarkably simple (even by comparison to schemes presently used in the industry). Particular forms are given in SPE 9260 (Cleary, 1980).

Implementation of eqn. (2A.1,3) requires a knowledge of the coefficients $\bar{\gamma}_2$ and $\bar{\Gamma}_2$. These may be obtained numerically, e.g. as described in Sec. 2B for $\bar{\Gamma}_2$ and by quite accurate self-similar approximations for $\bar{\gamma}_2$ (Settari and Cleary, 1982). However, it is worth having expressions which relate those to another pair of physically transparent parameters, namely the slopes of the pressure distribution at the well-bore (x=0) and center of the fracture (z=0), Γ_2 and γ_2 respectively (fig. 2A.1b). Straightforward implementation of mass conservation for the whole fracture (e.g. using the equations in Section 2B) leads to an identity, which actually follows directly also by combining eqns:

$$\bar{\Gamma}_2^m = \frac{\Gamma_2\Gamma_4}{\Gamma_1\gamma_4}\left[\frac{1-\dot{W}_L/\dot{W}}{\Gamma_3+\Gamma_3\Lambda_L}\right]^m, \tag{2A.4a}$$

The change of cross-sectional mass-content is parameterised by

$$\Lambda_L = \frac{L\ d(\Gamma_3\rho\hat{\sigma}H^2)/dt}{(dL/dt)\Gamma_3\rho\hat{\sigma}H^2} \tag{2A.4b}$$

and this must be determined (using eqn.(2A.1)) at each instant in marching out eqn. (2A.3). Here Γ_3 is the (volume x density) factor for the overall fracture (fig. 2A.1), namely

$$W-W_L = 2\Gamma_3\rho\hat{\sigma}H^2L\Big|_0^t \tag{2A.4c}$$

and the mass injection rate $\dot{W}$ is determined by the flow law

$$(\pm\dot{W}/2H\rho)^m = (\Gamma_4/\gamma_4\bar{\eta})\Delta^{2n+1}(\pm\Gamma_2\sigma/L),\ \Delta = \Delta_E + \Delta_A \tag{2A.4d}$$

in which Δ_A is the anelastic component of displacement. The ratio Γ_4/γ_4 accounts for the difference in channel-flow factor for the crack profile in question (e.g. elliptical channel/parallel plates gives $12\pi/64$). The mass loss-rate $\dot{W}_L$ is to be obtained from a separate computation. Although the new coefficients, Γ_2 and Γ_3, are not constants, they are much more readily determinate than $\bar{\Gamma}_2$, as in Section 2B, being relatively independent of fluid loss and pumping conditions.

By an entirely analogous argument for vertical flow along the cross-section defined by H, we may obtain the relation

$$\overline{\gamma}_2^m = \frac{\gamma_2}{\gamma_1}\left[\frac{1-\dot{w}_L/\dot{w}}{2\gamma_3+\gamma_3\Lambda_v}\right]^m \tag{2A.5a}$$

with pressure alteration at the cross-section is parameterized by

$$\Lambda_v \equiv \frac{H\ d(\gamma_3\rho\hat{\sigma})/dt}{(dH/dt)\ \gamma_3\rho\hat{\sigma}} \tag{2A.5b}$$

This equivalency follows from mass-conservation relation for the cross-section and the vertical flow law, respectively given by

$$w - w_L = \gamma_3\rho\hat{\sigma}H^2 \Big|_0^t \tag{2A.5c}$$

$$(\pm\dot{w}/\rho)^m = \Delta^{2n+1}(\pm\gamma_2\sigma/H)/\overline{\eta} \tag{2A.5d}$$

in which Δ_F is given by eqn. (2A.2b) with $\Gamma_1 \equiv \gamma_1$ (having distinct symbols only to clearly identify their context).

The slope coefficient γ_2 depends primarily on contrasts in confining stress and moduli between strata in the reservoir and it may best be phrased for applications as follows:

$$\gamma_2 = \overset{\circ}{\gamma}_2 S(h);\ S(0)=1,\ S(\infty) = \overset{\infty}{\gamma}_2/\overset{\circ}{\gamma}_2;\ S=S_D S_S\ ;\ h=H/H_R \tag{2A.5e}$$

in which the shape of the slope function S depends on the details of the barriers impeding vertical growth: specifically, it has a different form for moduli vs. stress contrast, which requires the minimum level of rationalisation just provided -- i.e. different functions S_D and S_S for "deformability" and confining stress contrasts (e.g. as in fig. 2A.1a). The values γ_2° and γ_2^{∞} are those for the limits of very small and very great heights respectively; these would typically be the values in a homogeneous region (γ_2°, fig. 2A.2b) and that in a stratified region (γ_2^{∞}). For stress and moduli contrast only, such limits may be related by an expression as simple as

$$(\overset{\infty}{\gamma}_2/\overset{\circ}{\gamma}_2)^m = (\sigma_A/\sigma_R)^{2n+2-m}\ (\overline{E}_R/\overline{E}_A)^{2n+1-m} \tag{2A.5f}$$

which derives directly from the definition of characteristic time in eqn. (2A.2a). When the effective moduli of the adjacent strata are not the same, $E_L \neq E_U$, then the operative adjacent modulus E_A must be a suitable combination of E_L and E_U (which interchange in going from $E_A{}^U$ to $E_A{}^L$); this and other exact details of eqn. (2A.5e) can be worked out by using a self-similar model, for a cross-section like fig. 2A.2 (Cleary et al., 1981).

Reformulation for Specified Flow-Rate. To allow for the more typical field conditions of specified flow rate, we must transform from pressure to flow in eqns. (2A.1,3), which is achieved by use of eqn. (2A.4c), leading to

$$\frac{\dot{L}^m LH}{\bar{\Gamma}_2^m} = \frac{\bar{E}}{\bar{n}} \left[\frac{W-W_L}{2\Gamma_3 \rho HL}\right]^{2n+2-m} = \frac{\dot{H}^m H^2}{\bar{\gamma}_2^m} \qquad (2A.6)$$

Again, for any specified injected mass W, the evolution of L and H can be computed, simultaneously with loss W_L, by algebraic or numerical means, as follows.

2A.2. Algebraic solutions of lumped model equations.

Some broad features and specific algebraic formulae can be extracted, with which results of more general numerical solution schemes can be compared and verified. The most general result is obtained by direct comparison of eqns. (2A.1,3), namely

$$\bar{\gamma}_2 \frac{d}{dt} L^{1+1/m} = \bar{\Gamma}_2 \frac{d}{dt} H^{1+1/m} \qquad (2A.7a)$$

If the ratio of coefficients is time-independent, then we obtain

$$H^{1+1/m} - H_o^{1+1/m} = \bar{\gamma}_2 \left(L^{1+1/m} - L_o^{1+1/m}\right)/\bar{\Gamma}_2 \qquad (2A.7b)$$

Various special solutions have been worked out, based on eqn. (2A.7), by Cleary et al., 1981 and by Cleary, SPE 9259, 1980.

2A.3 Numerical Solutions of Lumped Model Equations

The equations (2A.1,3,6) can more generally, of course, be solved numerically for arbitrarily complicated initial conditions and time-dependence of the coefficients in $\bar{\gamma}_2$, $\bar{\Gamma}_2$ -- especially due to fluid loss W_L, which will not typically follow the same behavior as W and therefore will not allow homogeneous (e.g. power-law) solutions of the kind we often use (Cleary, 1980) for demonstration. Sample results are shown in fig. 2A.3; dominant effects of moduli and stress contrasts are clearly demonstrated.

2B. Summary of P3DH model equations and results.

For many practical reservoir conditions of fairly good containment (i.e., restricted height growth), the much-quoted outstanding problem of fracture height determination is solved correctly (for the first time, we believe) by the pseudo-three dimensional hydrafrac (P3DH) modelling schemes developed by Cleary (SPE 9259, 1980); as well, these are sufficiently simple, yet general, that realistic (quite complex) reservoir properties can be incorporated--a feature not soon promised in any of the more complex 3-D model developments that we know of (even including our own efficient version, Cleary et al. 1981). The basic assumption is that one-dimensional lateral flow can be coupled to a suitable two-dimensional model for crack-opening and vertical fluid flow at each cross- section, but that a fully 3-D grid is unnecessary, thus saving greatly on computation.

A suitable mesh spacing distribution is used, one which moves with the fracture (in contrast to Settari and Cleary, 1982); we also use a convenient expression for height growth at each cross-section (based on the experience with models for eqn. (2A.1)) and thus march out the shape of the fracture after each pressure solution is obtained. In this way, the evolution of fracture shape can be traced and the (downhole) pressure history can be obtained for any specified injection history--and vice-versa; an interesting outcome is the potential for type-matching the pressure curve to determine how much vertical fracture growth is occurring during a stimulation operation. Such applications and many others have been described by Settari and Cleary (1982) and by Cleary et al (1981).

The basic assumptions and derivations have already been outlined in SPE 9259, so we merely summarize the governing equations, with reference to fig. 2B.1. Mass conservation is expressed as follows, for the net lateral fluid flow Q.

$$\frac{\partial}{\partial x}(\rho Q) + 2\rho H\overline{q}_L = -\frac{\partial}{\partial t}(2\hat{\gamma}_3 \rho H \Delta) \tag{2B.1}$$

where $\overline{q}_L$ is the average fluid loss rate and $\hat{\gamma}_3$ is the volume (shape) factor for the cross-section at x. The central crack opening Δ is still given by eqn. (2A.2b), where H and σ are now the height and excess pressure at any cross-section (at x). The system description is completed by the rheological equation for the frac-fluid in channel-flow:

$$\left(\pm\frac{Q}{2H}\right)^m = \mp \frac{\Gamma_4}{\gamma_4} \frac{\Delta^{2n+1}}{\overline{n}} \frac{\partial p_f}{\partial x} \tag{2B.2}$$

Combination of eqns. (2B.1,2; 2A.2b) gives a single equation in the pressure distribution, describing non-linear diffusion on the region (0,L), for any wing of length L:

$$\frac{\partial}{\partial x}\left[\pm \overline{\Gamma}_5 \left(\mp \frac{\partial \hat{f}}{\partial x}\right)^{\frac{1}{m}}\right] + \rho H\overline{q}_L = -\frac{\partial}{\partial t}\left(\gamma_3 \rho H^2 \hat{\sigma}\right) \tag{2B.3a}$$

in which the flow is dominated by the new pressure function and transmissivity, defined by

$$\hat{f} = \hat{\sigma}^{\,2n+2} \quad \text{and} \quad \overline{\Gamma}_5^m = \frac{\Gamma_4 \overline{E}}{\gamma_4 \overline{n}} \frac{H^{2n+1+m}}{(2n+2)\Gamma_1} \left[1 + \frac{\Delta_A}{\Delta_E}\right]^{2n+1} \tag{2B.3b}$$

However, completion of the description requires the determination of the boundaries $H(x,t)$ and $L(t)$, and this is where options begin to appear (Cleary et al., 1981).

The length growth requires a criterion, for what we have called the leading-edge (Fig. 2B.1), which is just the segment of the fracture near the front that cannot be completely described by eqns. (2B.1, 2, 3). We denote by σ_F the excess pressure required (at the front $x = L$) to drive this leading-edge at speeds compatible with the growth rate $\dot{L}$ deduced from solving (2B.1 2, 3) in the main body of the fracture; this rate can be derived from overall mass conservation for the fracture which may be manipulated into an equivalent expression

$$dL/dt = Q_F/2\hat{\gamma}_3 H_F \Delta_F \;; \quad \Delta_F = \hat{\sigma}_F H_F + \Delta_A^F \qquad (2B.4a)$$

provided that eqn. (2B.3a) is satisfied at all points x. The leading-edge criterion itself has a similar character to the models which produced eqn. (2A.1), because it describes a segment of fracture (e.g. semi-circular) at the front, having width dominated by effective length (e.g. radius); it may also look like eqn. (2A.3), if we wish to lump part of the channel (described by eqns. (2B.1, 2, 3)) into it. The best way to write a propagation equation for it is as follows

$$dL/dt = \bar{\gamma}_2^F \, H_F/\tau_c^F \qquad (2B.4b)$$

where τ_c^F has the definition in eqn. (2A.2a), with $\sigma = \sigma_F$. Combination of eqns. (2B.4 a, b) leads to the boundary condition

$$\left(\frac{Q_F/H_F^3}{2\hat{\gamma}_3\bar{\gamma}_2^F}\right)^m = \frac{\bar{E}}{\bar{\eta}} \frac{\hat{\sigma}_F^{2n+2}}{H_F^{2m-2n}} \qquad (2B.4c)$$

Specification of Q_W or σ_W, or a combination, completes the model.

Now, it is possible to satisfy eqn. (2B.3a) exactly by using established numerical solution techniques for (nonlinear parabolic) equations which are now common practice in the business of reservoir engineering. However, these techniques are quite demanding computationally and have been developed only with fixed meshes, through which the fracture has to move; this means that detailed solution near the fracture front is not feasible, since a prohibitively fine mesh would be needed over the whole trajectory of the fracture (Settari and Cleary, 1982). A mesh which moves with the fracture avoids this need, except near the front; some sample solutions are given by Cleary et al., 1981. The resulting shapes look like those on the "self-determining height" side of fig. 2B.1; when these are unrealistic (e.g. for $L \sim H$), the shapes may be specified, as shown, in our programs.

3. DEVELOPMENT OF MORE DETAILED HYDRAFRAC MODELS

There were a number of special tasks which required out attention, both in employing the approximate models mentioned earlier in this section and in adapting/extending them to cover more complicated circumstances: (A) The need for baseline precision solutions (Petersen 1980,1982) of the equations governing any generic cross-section of a fracture: without these, we cannot make any estimates of how complex/expensive would be a full 3-D simulation or indeed of how good are the various (e.g., self-similar) approximations which have been used to model such cross-sections. Simpler models can then be employed judiciously as components of a more realistic overall description (e.g., P3DH in Sec. 2B). (B) We need the exact solution for one geometry which can be precisely reproduced in the laboratory and allows very rigorous experimental testing of model predictions. A circular crack (Wong, 1981) fulfills this model role, it has practical applications in the field, it allows further testing of self-similar approximations and it serves as a first check for more general 3-D simulators (Cleary et al., 1981). (C) The extension of our coupled fluid-flow and crack opening solutions to multiple cracks, emanating from one or more (adjacent) wellbores and interacting with various reservoir features. This problem has profound importance for our long-standing ideas of mechanical guidance and creation of fracture networks underground. The results to date (Narendran and Cleary, 1982) compare favorably with experiments. (D) Analysis of mechanisms (e.g., blunting, turning) which could control the growth of these various fracture geometries in non-homogeneous regions, especially the role of stratification, natural fractures/faults and lenses (Lam and Cleary, 1982).

4. SUMMARY OF LABORATORY SIMULATION AND CONCLUSIONS

We have long maintained that much could be done in the laboratory to verify developing models and gain new insight on phenomenology, before field testing of the purely research kind (e.g., not tied to a well which would be drilled for production anyway) should be pursued at great expense. Our experimental work to date strongly supports that contention, both in the progress we have made and the failures we have had. We believe we now have the potential to simulate field conditions up to a high level of complexity, with precise control of all the parameters, but we have also learned how many things can go wrong in a nominally perfect arrangement of wellbore, perforated casing and prefractured walls--most of which we have eventually remedied, owing to the accessibility of the "test location." Progress has been made on the following primary projects (Cleary et al., 1981):

A. Construction of general-purpose simulator for highly repeatable observation of hydrafrac growth at low expense of time and materials; we achieve this by interface separation, a novel concept in the context, solidly based on scaling of calculations for field conditions--which show, for instance, that fracture toughness can usually be neglected! We have taken repeatable predictable data on a wide variety of configurations (Papadopoulos et al., 1980-81).

B. Examination of crack growth and interaction in a biased stress field, where details of material decohesion do play an important role. We should note that our results (Cleary et al., 1981) seem to be substantially different from those of previous researchers who have conducted tests without suitable scaling of toughness, pressures and confining stress: for instance, smooth turning and strong interaction are observed.

C. Development of the capabilities for facile data acquisition and control in each of the relevant experiments, mainly employing new effective software and combining a number of purchased or built hardware components (Cleary et al., 1981).

D. Testing of various schemes for monitoring the growth of cracks, using acoustic or deformation or electrical signals.

E. Investigation of the rheological behavior of various fluids to be used both in our lab simulators and in the field; one final goal is to design new fluids with suitable mechanical properties. We have built some innovative rheometers to measure the properties (e.g., using reciprocating pipe-flow).

F. Studies of pore-pressure-induced-cracking under triaxial conditions in a pressure vessel and comparison to theoretical predictions for such processes: samples are being tested under conditions of homogeneous pore-pressure, axial flow and radial flow so that any induced cracking can be detected through changes in permeability (and other properties, e.g., ultrasonic wave speeds).

G. High temperature triaxial tests to examine cracking induced by thermal stresses and phase changes.

An overall conclusion of ours is that, despite all the laboratory work (and computer modelling) in this area, the dominant features of underground fracturing are just now beginning to be understood, based on an evaluation of the more basic components.

Acknowledgements

The work reported here is part of the M.I.T. UFRAC project which is supported by a number of production and service companies in the energy industry, and by grants from the NSF.

REFERENCES

Abou-Sayed, A.S., "Fracture Toughness K_{IC} of Triaxially-Loaded Limestones," pp. 2A3, 1-8 in Proc. 18th Symp. on Rock Mech., June, 1977.

Abou-Sayed, A.S., C.E. Brechtel and R.J. Clifton, "In-Situ Stress Determination on Hydrofracturing--a Fracture Mechanics Approach," Jour. Geophys. Res., 83, B6, 2851-2862, 1978.

Barr, D.T., "Thermal Cracking in Nonporous Geothermal Reservoirs," S.M. Thesis at M.I.T., May 1980.

Cleary, M.P., "Primary Factors Governing Hydraulic Fractures in Heterogeneous Stratified Porous Formations," Paper No. 78-Pet-47 presented at Energy and Technology Conference of the Petroleum Division, ASME, Houston, TX, November 1978.

Cleary, M.P., "Rate and Structure Sensitivity in Hydraulic Fracturing of Fluid-Saturated Porous Formations," pps. 127-142 in 20th U.S. Symp. on Rock Mech., June 1979.

Cleary, M.P., "Comprehensive Design Formulae for Hydraulic Fracturing," SPE 9259, and "Mechanisms & Procedures for Producing Favourable Shapes of Hydraulic Fractures," SPE 9260, presented at Fall Annual Meeting of SPE-AIME, Sept. 1980.

Cleary, M.P., et al., "Theoretical and Laboratory Simulation of Underground Fracturing Operations," First Annual Report of the M.I.T. UFRAC project, Res. Extr. Lab., August 1981.

Cleary, M.P. and S.K. Wong, "Self-similar and spatially-lumped models for tracing the evolution of circular hydraulic fractures", sub. to Jour. App. Mech., 1982.

Crichton, A., "Crack Interaction in Hydraulic Fracturing of Cement Blocks," B.S. Thesis, M.I.T., 1980.

Daneshy, A.A., "On the Design of Vertical Hydraulic Fractures," Jour. Petroleum Technology, pps. 83-93, 1973.

Daneshy, A.A., "Hydraulic Fracture Propagation in Layered Formations," Soc. Pet. Eng. Jour., 18, 1, 33-41, 1978.

Geertsma, J. and F. de Klerk, "A Rapid Method of Predicting Width and Extent of Hydraulically-Induced Fractures," Jour. Petroleum Technology, p. 157, December 1969.

Greenfield, R.J., L.Z. Schuck and T.W. Keech, "Hydraulic Fracture Mapping Using Electrical Potential Measurements," In Situ, 1 (2), pps. 146-149, 1977.

Haimson, B.C. and C. Fairhurst, "In-Situ Stress Determination at Great Depth by Means of Hydraulic Fracturing," in Rock Mechanics--Theory and Practice (ed. W.H. Somerton), pub. by Soc. Mining Engineers (AIMMPE Inc.), pps. 559-584, 1970.

Haimson, B.C., "The State of Stress in the Earth's Crust," Rev. Geophys. Space Physics 13 (3), pps. 350-352, 1975.

Keck, L.J., and C.L. Schuster, "Shallow Formation Hydrofracture Mapping Experiment," Trans. ASME Jour. Pres. Ves. Tech., February 1978.

Komar, C.A., "Strategy for Stimulation Technology in the Devonian Shale," pps. G-1/1-15 in Proc. DOE Symposium on Enhanced Oil Recovery, 1978.

Lam, K.Y. and M.P. Cleary, "Numerical Analysis of Crack-Branching and Slippage at a Frictional Interface," Dept. of Mech. Eng., M.I.T., REL-82-1, January 1982.

Medlin, W.L. and L. Masse, "Laboratory Investigation of Fracture Initiation Pressure and Orientation," Paper No. SPE 6087 of AIMMPE, presented at 51st Annual Fall Conference, October 1976.

Morita, N., K.E. Gray and C.M. Kim, "Stress-state, Porosity, Permeability and Breakdown Pressure around a Borehole during Fluid Injection," pps. 192-197 in Proc. 22nd U.S. Symposium on Rock Mechanics, M.I.T., June 1981.

Narendran, V.M. and M.P. Cleary, "Elastostatic Interaction of Multiple Arbitrarily Shaped Cracks in Plane Inhomogeneous Regions," Dept. of Mech. Eng., M.I.T., REL-82-6, June 1982.

Nordgren, R.P., "Propagation of a Vertical Hydraulic Fracture," Jour. Soc. of Petroleum Engineers, p. 306, 1972.

Papadopoulos, J.M., B.A. McDonough, D. Summa and M.P. Cleary, "Laboratory Simulation of Hydraulic Fracturing," Reports to Lawrence Livermore Laboratories, and Reports of Resource Extraction Laboratory, MIT, March 1980, 1981, 1982.

Perkins, T.K. and L.R. Kern, "Widths of Hydraulic Fractures," Journal of Petroleum Technology, p. 937, 1961.

Petersen, D.R., "Numerical Analysis of Hydraulic Fracturing and Related Crack Problems," S.M. and M.E. Theses at MIT, 1980, 1982

Pierce, A.E., S. Vela and K.T. Koonce, "Determination of the Compass Orientation and Length of Hydraulic Fractures by Pulse Testing," SPE 5132, SPE Annual Fall Meeting, 1974.

Power, D.V., C.L. Schuster, R. Hay and J. Twombley, "Detection of Hydraulic Fracture Orientation and Dimensions in Cased Wells," SPE Annual Fall Meeting, 1975.

Pugh, T.D., B.W. McDaniel and R.L. Seglem, "A New Fracturing Technique for Dean Sand," Jour. Pet. Tech., pps. 167-172, Feb. 1978.

Roegiers, J.C. and D.W. Brown, "Geothermal Energy: A New Application of Rock Mechanics?", Proc. 3rd Congr. Int'l Soc. Rock Mech., Denver, CO., pub. by Nat'l Acad. Sci., vol. 1, 1974.

Schmidt, R.A., R.R. Boade, and R.C. Bass, "A New Perspective on Well-Shooting--the Behavior of Contained Explosions and Deflagrations," Paper No. SPE 8346, Fall Annual Meeting, 1979.

Settari, A. and M.P. Cleary, "Three-Dimensional Simulation of Hydraulic Fracturing," Paper No. SPE 10504, and "Development and Testing of a Pseudo-Three-Dimensional model of Hydraulic Fracturing," Paper No. SPE 10505, 1982.

Simonson, E.R., A.S. Abou-Sayed and R.J. Clifton, "Containment of Massive Hydraulic Fractures," Soc. Pet. Engr. Jour., 18, 1, pps. 27-32, 1978.

Smith, M.B., G.B. Holman, C.R. Fast, R.J. Covlin, "The Azimuth of Deep, Penetrating Fractures in the Wattenberg Field," JPT, 1978.

Swolfs, H.S. and C.E. Brechtel, "The Direct Measurement of Long-Term Stress Variations in Rock," Proc. 18th Symposium on Rock Mech., pub. by Colorado School of Mines Press, Golden, CO, 80401, pp. 4C5-1 to 4C5-3, June 1977.

Voegele, M.D. and A.H. Jones, "A Wireline Hydraulic Fracturing Tool for the Determination of In-Situ Stress Contrasts," Paper No. SPE-DOE 8937, presented at Unconventional Gas Recovery Symp., Pittsburgh, PA, May 1980.

Warpinski, N.R., R.A. Schmidt and D.A. Northrop, "In-Situ Stresses: The Predominant Influence on Hydraulic Fracture Containment," Paper No. SPE/DOE, 8392, 1980. (JPT, pps. 653-664, March 1982).

White, J.L. and E.F. Daniel, "Key Factors in MHF Design," Jour. Pet. Tech., 33, 8, pps. 1501-1512, August 1981.

Wood, M.D., D.D. Pollard and C.B. Raleigh, "Determination of In-Situ Geometry of Hydraulically Generated Fractures Using Tilt-Meters," SPE 6091, Annual Fall Meeting, October 1976.

Wong, S.K., "Numerical Analysis of Axisymmetric & Other Crack Problems Related to Hydraulic Fracturing," M.S. Thesis, M.I.T., 1981.

Young, C., "Evaluation of Stimulation Technologies in the Eastern Gas Shales Project," pps. G5/1-16, Proc. DOE Symposium on Enhanced Oil Recovery, vol. 2, 1978.

Zoback, M.D., F. Rummell, R. Jung and C.B. Raleigh, "Laboratory Hydraulic Fracturing Experiments in Intact and Pre-Fractured Rock," Int. Jour. Rock Mech., 14, pps. 49-58, 1978.

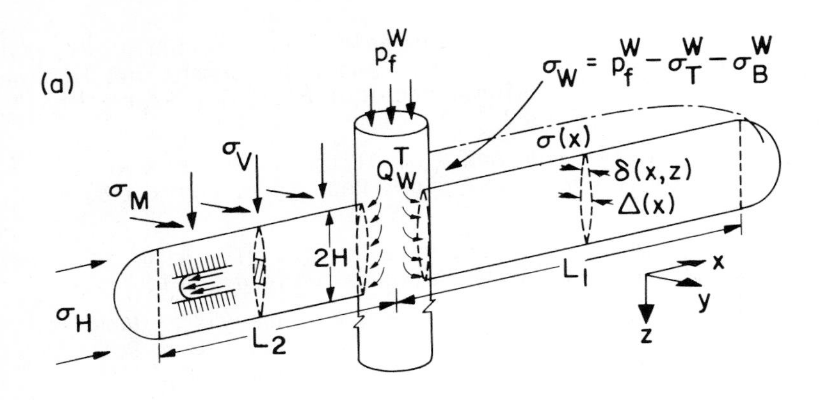

Fig. I.1a. Schematic of conventional PKN geometry used in making sample calculations for field applications.

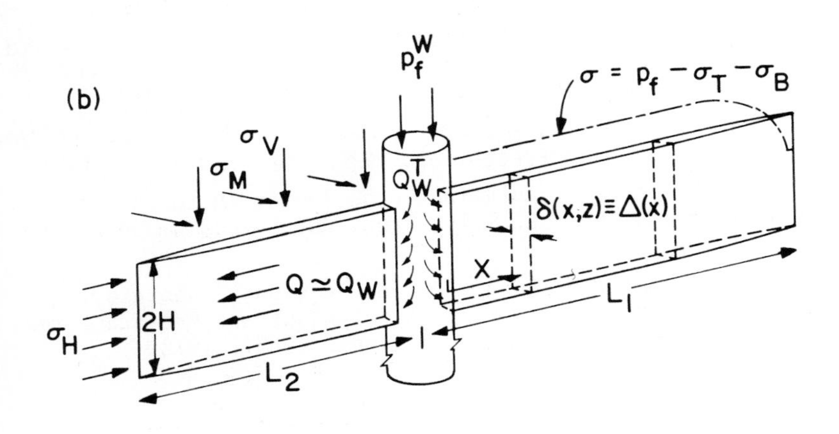

Fig. I.1b. Demonstrates conventional implementation of CGDD model for computation of lateral extent in field operations.

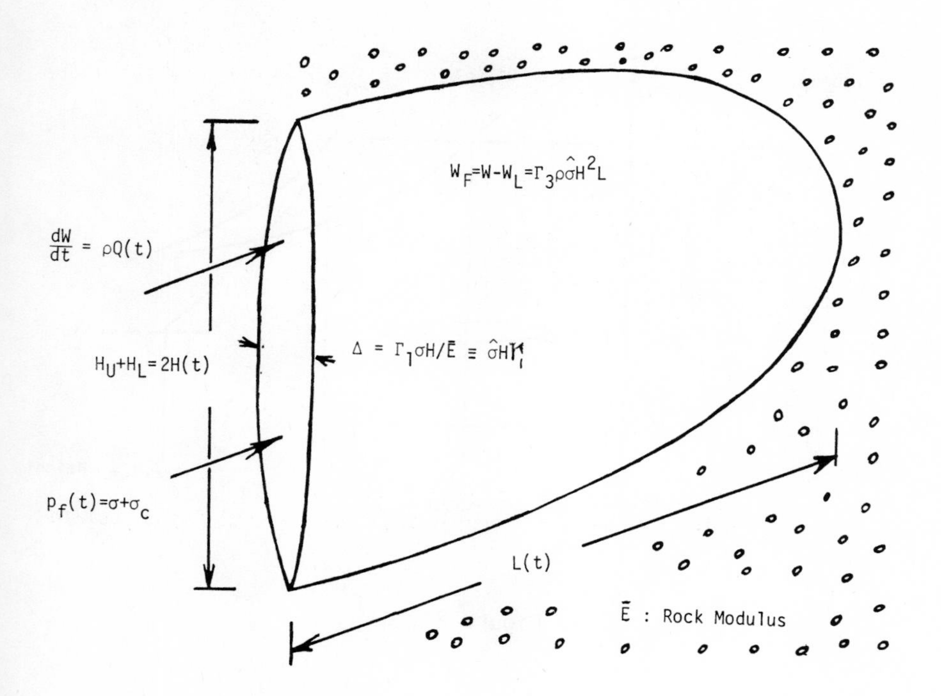

Figure 2A.1a

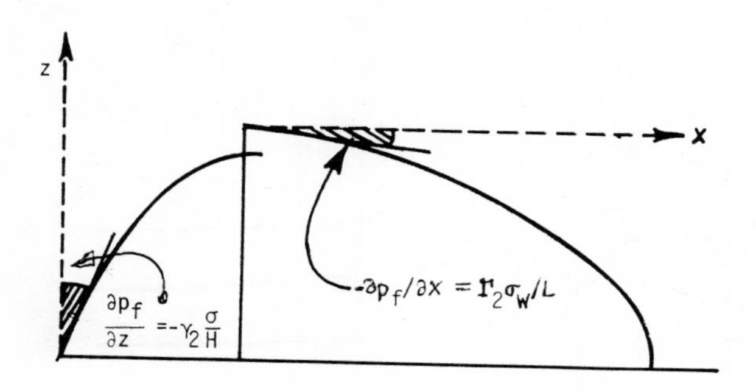

Figure 2A.1b

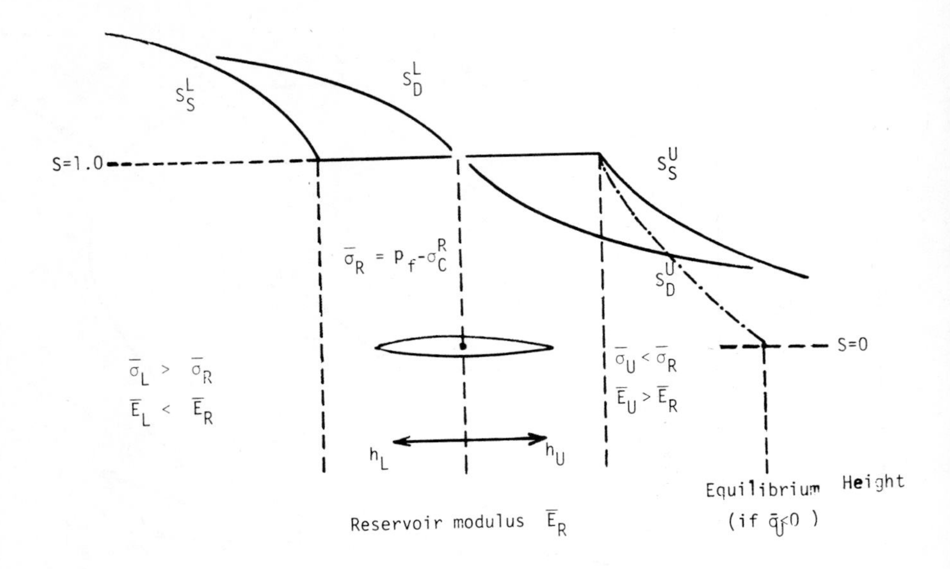

Figure 2A.2a

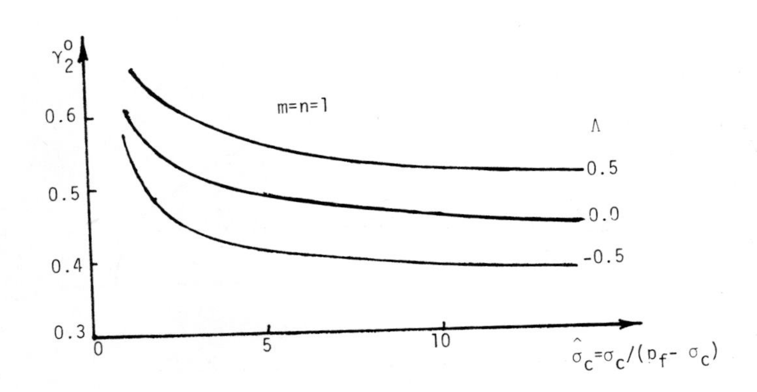

Figure 2A.2b

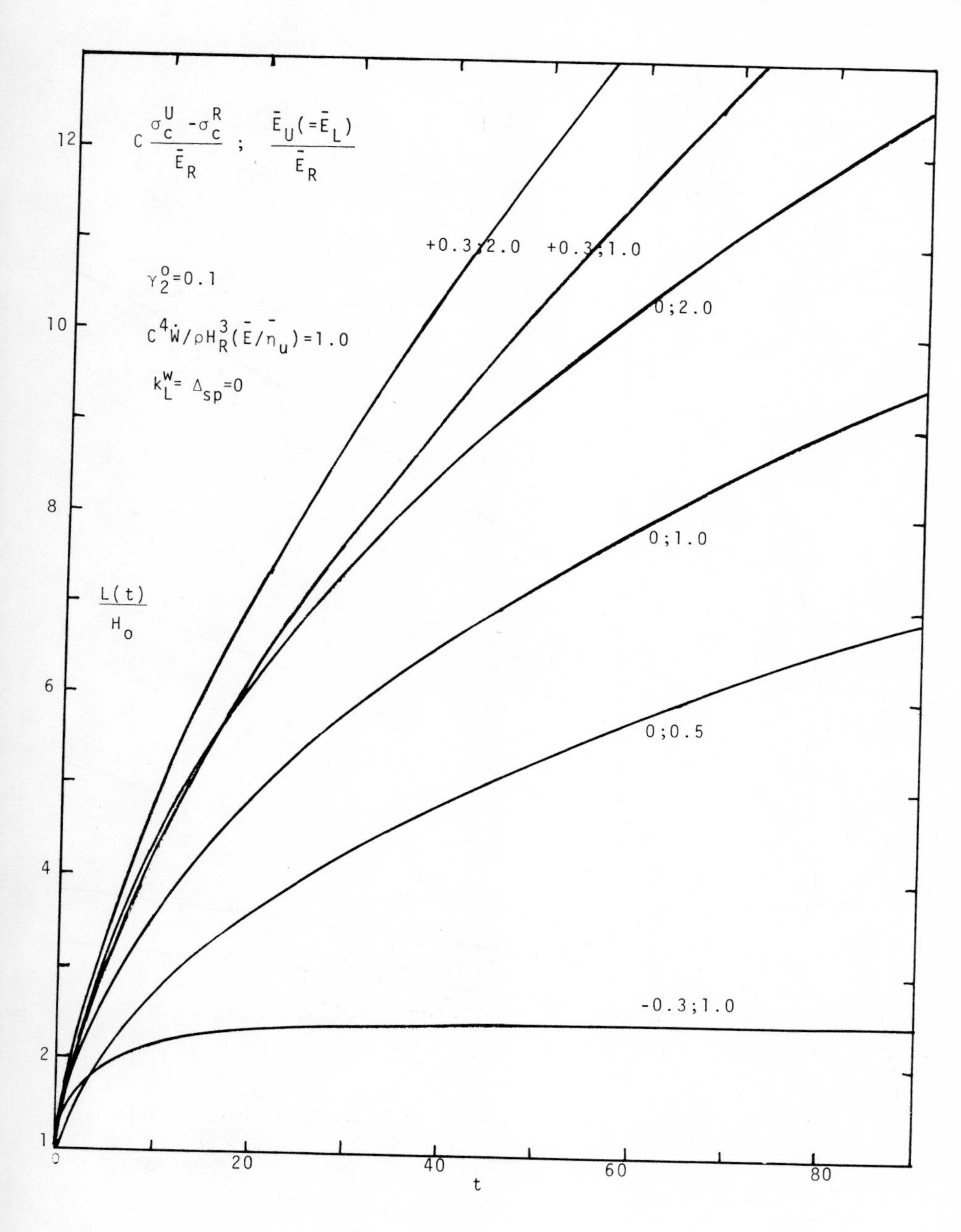

$C\frac{\sigma_c^U - \sigma_c^R}{\bar{E}_R}$; $\frac{\bar{E}_U(=\bar{E}_L)}{\bar{E}_R}$
$\gamma_2^o=0.1$
$C^4\dot{W}/\rho H_R^3(\bar{E}/\bar{\eta}_u)=1.0$
$k_L^w=\Delta_{sp}=0$
$\frac{L(t)}{H_o}$
+0.3;2.0
+0.3;1.0
0;2.0
0;1.0
0;0.5
-0.3;1.0
t

Figure 2A.3a

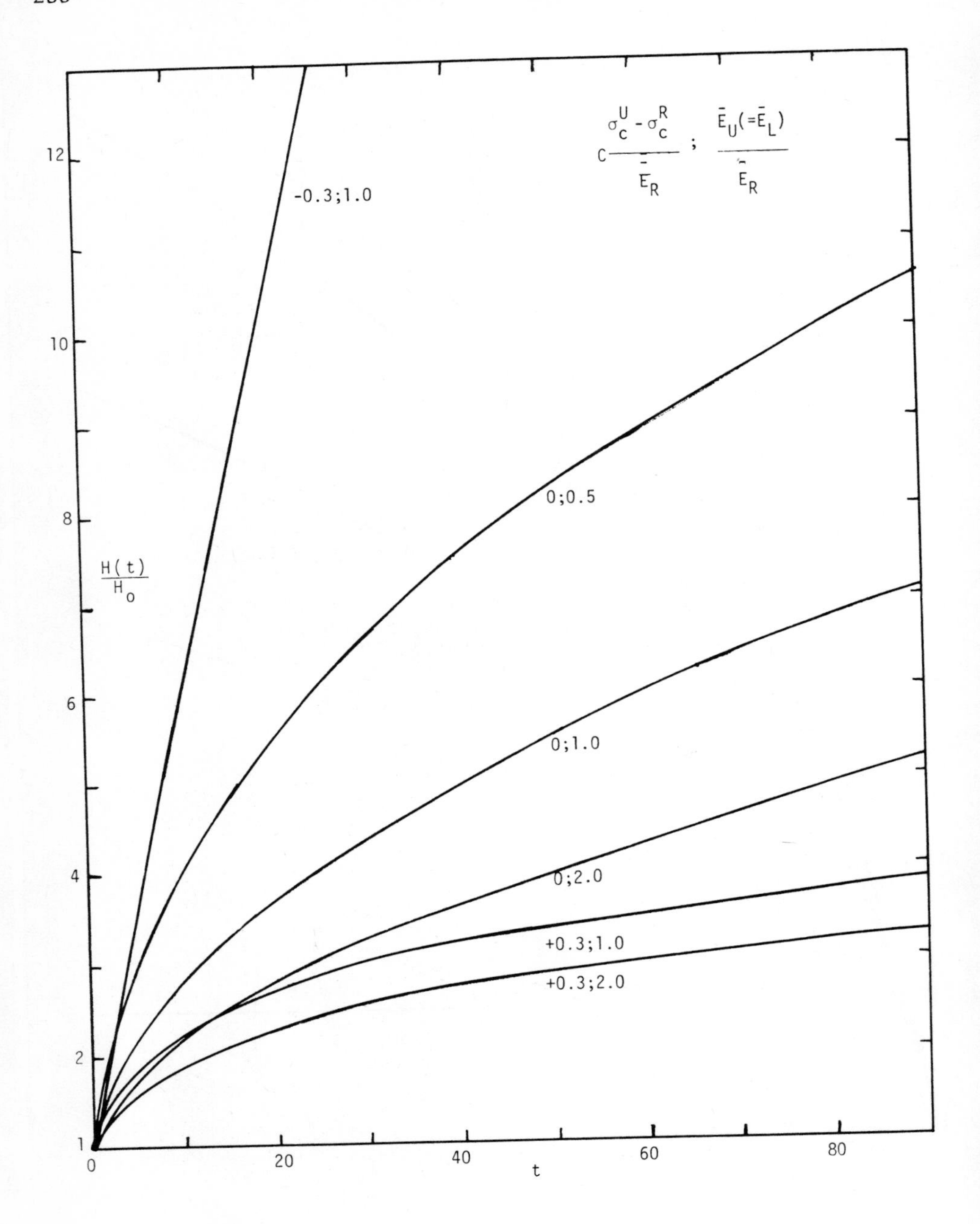

$C\frac{\sigma_c^U - \sigma_c^R}{\bar{E}_R}$; $\frac{\bar{E}_U(=\bar{E}_L)}{\bar{E}_R}$
-0.3;1.0
0;0.5
0;1.0
0;2.0
+0.3;1.0
+0.3;2.0
$\frac{H(t)}{H_o}$
t
12
10
8
6
4
2
1
0
20
40
60
80

Figure 2A.3b

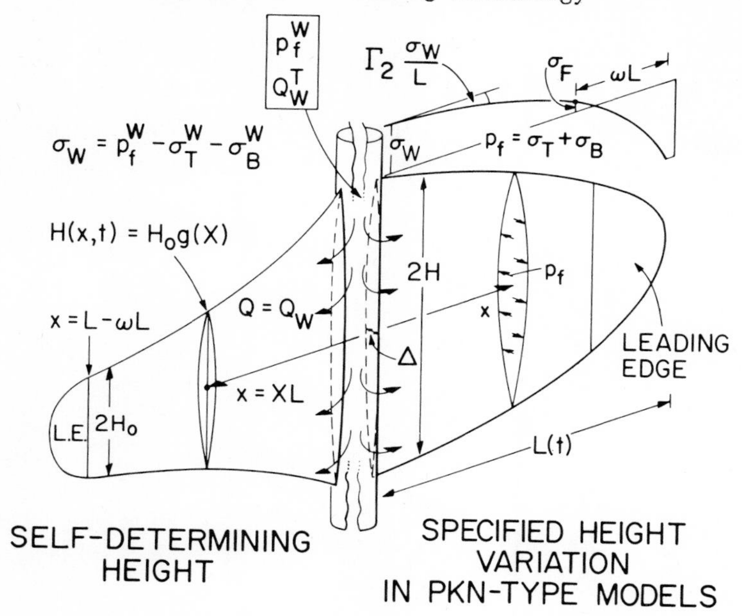

Fig. 2B.1a. Schematic of fracture shapes which can be described by PKN-type models for lateral extension, showing two distinct kinds of vertical height profiles.

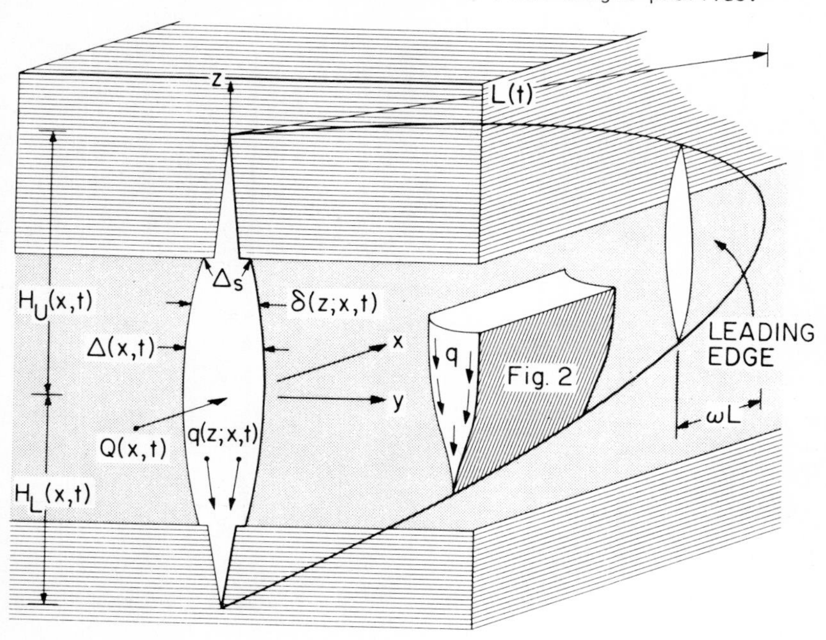

Fig. 4. Illustration of concepts in pseudo-three-dimensional hydrofrac (P3DH) models.

FRACTURE TOUGHNESS EVALUATION OF ROCKS IN THE PRESENCE OF PRESSURIZED WATER AT ELEVATED TEMPERATURE

by Hideaki Takahashi

Research Institute of Strength and Fracture of Materials

Faculty of Engineering, Tohoku University
Sendai, Japan.

ABSTRACT

The presence of pressurized water or steam in the fractures (crack) of geothermal reservoirs can facilitate crack propagation by stress corrosion phenomena. Thus it is suggested that such a subcritical crack growth in rocks plays an important role in fracture extension during both stimulative hydraulic fracturing and service performance in the geothermal reservoir. This paper describes the development of a fracture mechanics test method to evaluate fracture properties of rock samples under realistic conditions which simulate water chemistry, temperature, and pressure in a geothermal reservoir.

INTRODUCTION

In recent years, in order to understand the fracture behavior of rocks, a number of experimental studies on fracture mechanics have been carried out, involving two main topics: fast brittle fracture and subcritical crack propagation (stable crack growth). In relation to the onset of fast fracture in rock materials, specimen size effect on fracture behavior has been extensively discussed [1][2][3]. Although the rock consists of various brittle minerals, the deformation and fracture behavior show a significant non-linearity. Thus, until now it is still an open question whether or not fracture mechanics parameters can govern the fracture extension at the crack tip in rock. Therefore the development of reasonable tests and evaluation methods which are directly related to the fracture behavior of rock is highly desirable. Furthermore, the presence of liquid water, water vapor, or some other reactive species in the crack tip environment can facilitate crack propagation by promoting a weakening reaction. For the quartz/water system reaction of the form

$$(-Si-O-Si-) + H_2O \rightarrow (-Si-OH \cdot OH-Si-)$$

may occur. This phenomenon is known as stress corrosion cracking [4].

An idea of the potential importance of stress corrosion cracking can be gaged from the fact that it has been suggested as a key factor in both some time-dependent earthquake phenomena, and in limiting the potential heat extraction of a geothermal reservoir even in a natural hydrothermal system as well as in a hot dry rock one. This paper describes a basic development of fracture mechanics methodology for evaluating fracture behavior in a geothermal reservoir.

BACKGROUND OF FRACTURE MECHANICS APPROACH

Linear elastic fracture mechanics has been shown to be a powerful tool for the study of cracking behavior by material scientists, who have analysed the brittle failure of structural engineering components. Now the fracture mechanics approach is being employed to provide new insight into the fracture of rocks, especially in designing fracture in a geothermal reservoir, where evaluation of the cracking behavior of rocks in the earth's crustal condition is urgently needed.

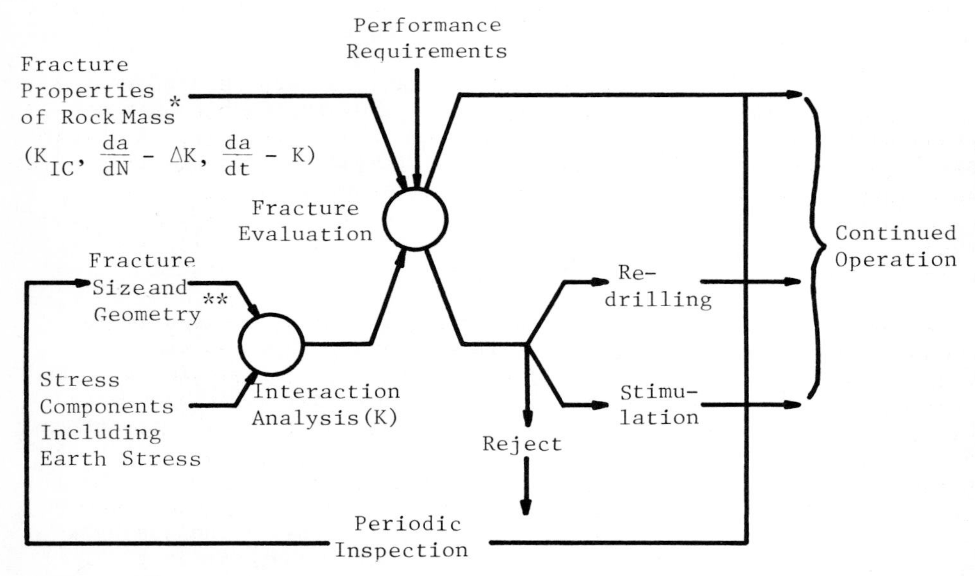

* Includes effect of operating environment, etc.
** Includes expected crack growth due to corrosion, fatigue, etc.

Fig. 1 Schematic Representation of an Idealized Fracture Control for a Geothermal Reservoir

Figure 1 shows a fracture evaluation procedure based on fracture mechanics, where a full understanding of three basic quantities, the fracture properties of a rock mass under the crustal condition, the location and size (geometry) of the crack, and the stress components including earth stress, is prerequisite. Fracture extension can be analysed in terms of the stress intensity factor. If the stress intensity factor is raised above a critical value, K_c, which is a material constant, then the crack will propagate at large velocities. The value K_c is influenced by the aggressive environment, when a crack tip strain rate is low enough for a chemical reaction. In addition, for many rocks, crack extension can occur at much lower values of K than K_c. Therefore, a variety of environmentally dependent mechanisms, notably stress corrosion cracking, can facilitate this stable, quasistatic subcritical crack propagation. Among these three basic quantities, the fracture properties of the rock mass under the realistic crustal condition should be determined experimentally from the fracture mechanics testing, where the environmental conditions such as temperature and confining pressure must simulate the conditions of a geothermal reservoir. An extensive development of a simple predictive test using a small specimen is highly desirable. The predictive testing is an experiment which results in knowledge concerning the condition of the fracture initiation and propagation under a realistic geothermal environment. This test method should be cheap, fast, reliable, and widely applicable. Thus, a predictive test is in contrast to service experience with the actual geothermal reservoir, because such service experience is neither cheap nor fast.

Figure 2 [5] lists various predictive test techniques used usually in corrosion engineering, where the degree of different complexity is arranged in the form of a ladder. The two bottom rungs indicate tests with simple specimen, either standard or specially conceived and oriented toward a specific problem. The two middle rungs indicate tests with built-up components, the upper including simulated service conditions. The two top rungs indicate the most complex testing device, namely simulation of the actual performance.

Figure 3 gives examples of predictive tests of differing complexity, as applied to rock mechanics fracture testing for a geothermal reservoir. One of the essential questions here is how we can predict the service behavior (top rung of the ladders in Figs. 2 and 3), based on simple one-parameter tests (bottom rung of the ladders in Figs. 2 and 3). Thus we have to focus our attention on the simple tests (one-parameter tests), their predictive quality, and their limitations. If these problems are resolved, we can predict the fracture behavior in a field during hydraulic fracturing and the future service performance with a resonable knowledge of the safety margin.

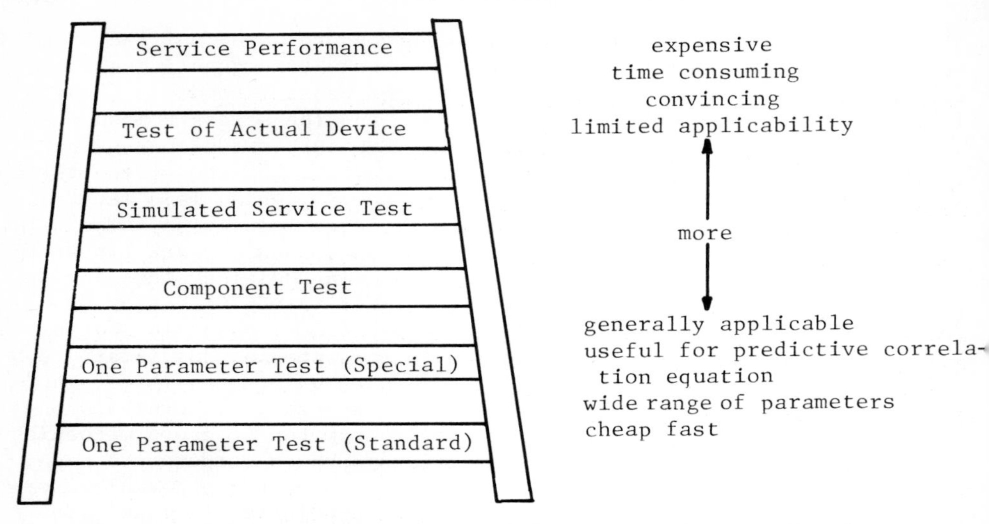

Fig. 2 Illustrative Drawing of Various Predictive Testing Showing Degrees of Complexity

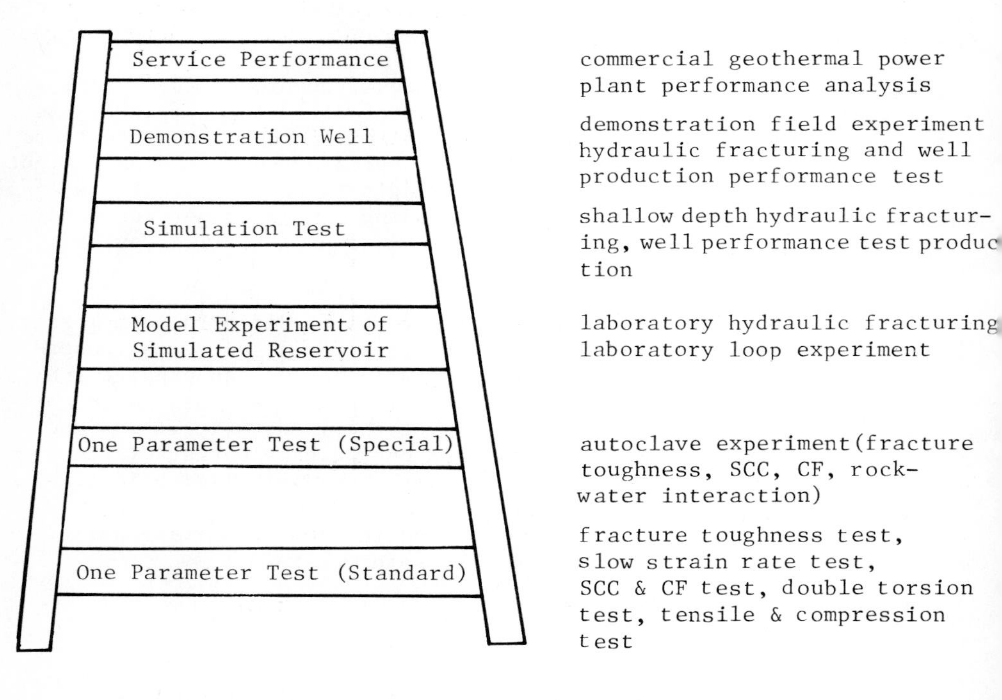

Fig. 3 Predictive Testing for Geothermal Reservoir Engineering

ACOUSTIC EMISSION STUDY OF FRACTURE BEHAVIOR OF GRANITE

Acoustic Emission and Microfracture Processes in Granite [6]

In order to perform a one-parameter FM testing, we have to use usually a small specimen. Therefore, specimen size effect on fracture behavior is first of all taken into consideration, where the main focus is placed on a quantitative evaluation of the onset of crack extension and subsequent stable growth. The AE technique is very useful for detecting microfracture in rocks as well as in metallic materials. This section describes the feasibility of non-linear fracture mechanics to monitor a crack initiation and extension behavior in granitic rock by use of the AE technique. A combination of continuum fracture mechanics and acoustic emission provides a suitable procedure for determining the fracture toughness of rocks.

The rock material used in this study was a coarse-grained granitic rock, composed of quartz, plogioclase, orthoclase and biotite, obtained from a quarry in Fukushima. Notched three-point-bend specimens of granite were machined in accordance with ASTM standards (E813-81). Dimensions of all specimens are given in Fig. 4, where three different specimen sizes, small, medium and large, were prepared with an artificial notch of 0.15mm root radius. A crack gage was placed on the side surface of these specimens to monitor crack growth. Fracture toughness tests were performed in a screw-driven testing machine at room temperature. A recent standard for an elastic-plastic fracture toughness test

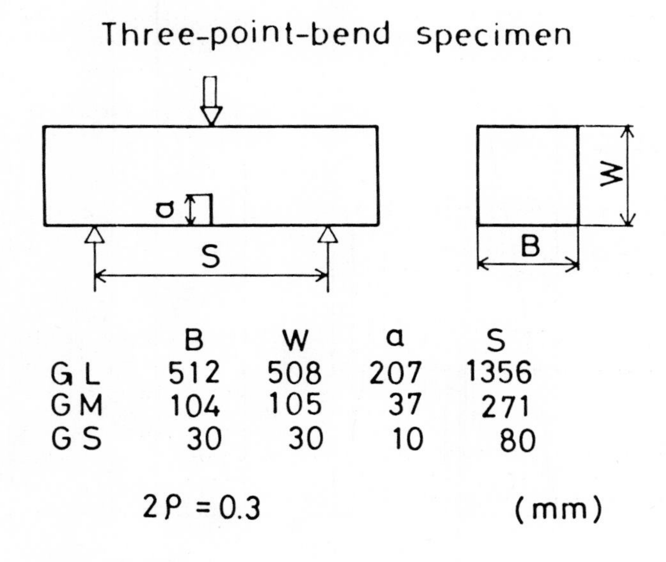

Fig. 4 Geometry and Dimensions of Three Point Bend Fracture Toughness Specimen

(ASTM E813-81) was used to evaluate the significant non-linearity of the load-displacement relation. A broad band AE transducer was mounted to the side surface of a test specimen. AE signals were amplified at 40-60dB by a wide band charge sensitive amplifier with band pass filter from 2kHz to 1MHz.

The load clip-gage displacement (P-Vg) record for a small size specimen is shown in Fig. 5. The load-displacement record indicates significant non-linear deformation behavior. After the maximum load, the load-displacement record was characterized by a decrease in the load with an increase in displacement due to crack extension. As understood in Fig. 5, the measurement of the crack extension by the crack gage indicates that an onset of crack extension occurs near the maximum load, the crack extension rate increases gradually, and the specimen is broken in an unstable fashion. This evidence indicates that the fracture process in rock is characterized by stable crack growth. An acoustic emission record for the granite is also given in Fig. 5. Considerable acoustic activity which seems to be attributed to the cracking, was detected at loads well below the maximum load. A few acoustic emission were detected before the extension of the main macro-crack was observed by the crack gage, whereas the abrupt increase of AE with high ampli-

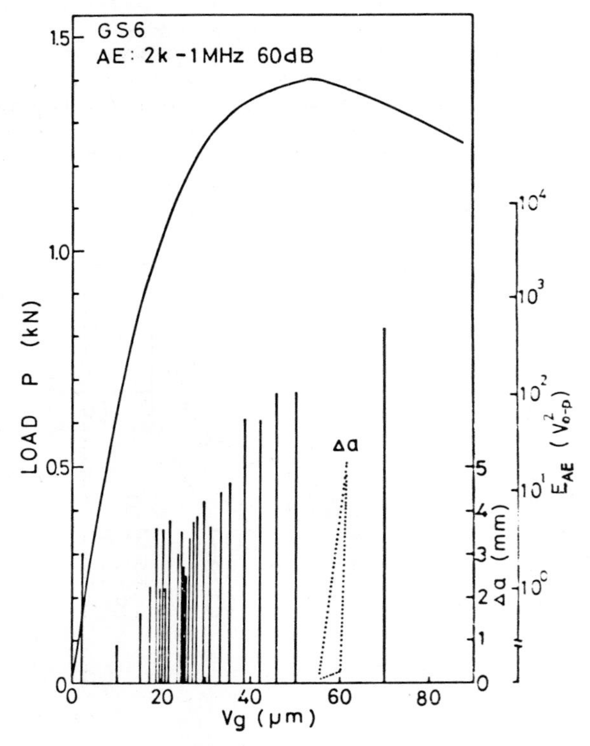

Fig. 5 A Typical Example of Load-Crack Mouth Displacement Record (P-Vg) and Acoustic Emission Data

tude corresponds to the onset of macro-crack growth by the crack gage on both sides of the specimen surface.

Note that the first AE activity above background noise occurred at the load level at which the load-displacement record becomes non-linear. This evidence suggests the development of numerous micro-cracks in the volume surrounding the notch tip. The medium and large-size notch samples were also tested by the same procedure. The same AE activity and fracture process were observed. In the case of large samples there was also detected the significant non-linearity of the P-Vg curves and stable crack growth, before the onset of the final fracture.

Determination of Fracture Toughness of Granite [6]

Based upon the data on metallic materials [7][8], it is concluded for the case of rock samples that an abrupt increase of AE activity corresponds to the onset of macro-crack extension at the notch or crack tip. Thus the onset of macroscopic crack extension was defined as a critical point of loading level, where the J-integral procedure was used because of the non-linear load-displacement relationship. Figure 6 shows the relationship between the accumulated summation of AE energy ΣE_{AE} and the J-integral, where a change of the slope corresponds to the onset or crack growth, denoted by J_{iAE}. In Table 1 there are summarized the fracture toughness data. As shown clearly, the values of J_{iAE} determined on the different size samples are independent of the specimen size, whereas J_{in}-values determined by the crack gage are strongly dependent on the specimen size. Thus J_{iAE} can be defined as

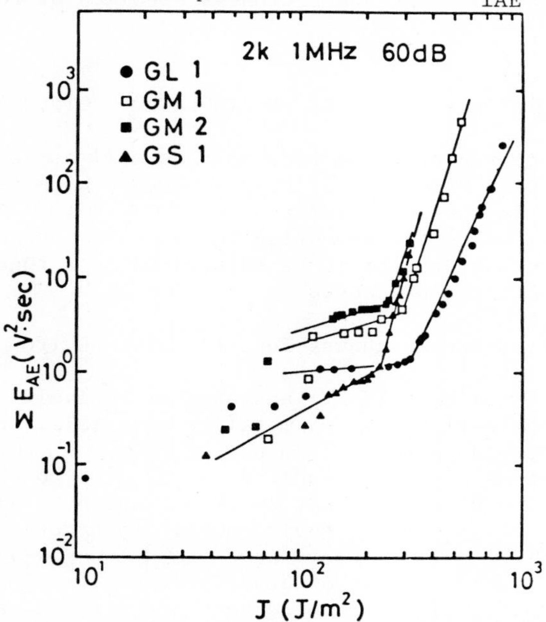

Fig. 6 A Relationship between J-Integral and Accumulative AE Energy (J-ΣE_{AE})

Table 1 Summary of Fracture Toughness Data

	$J_{iAE}(J/m^2)$	$J_{in}(J/m^2)$
GS 1	230	260
2	250	460
3	300	340
4 (Fatigue Precrack)	255	310
6	280	340
11	210	390
GM 1	300	430
2	250	260
GL 1	320	520

a material proper fracture toughness parameter. It is also noted that the J_{iAE} value of the pre-cracked sample coincides with that of the notched sample, and there is no notch root radius effect on the fracture toughness. This evidence suggests that the notched sample could be successfully used as the fracture toughness specimen of the rock material.

FRACTURE BEHAVIOR OF GRANITIC ROCK IN SIMULATED GEOTHERMAL ENVIRONMENT

In the hot dry rock geothermal reservoir, as well as in the natural hydrothermal geothermal reservoir, the presence of high temperature and pressure water in the crack surface of the rock mass facilitates stress corrosion cracking, where the reduction in fracture toughness and the time dependent crack growth rate should be determined as a function of temperature and confining pressure.

Slow Strain Rate Fracture Toughness Test in Water Environment [9]

The slow strain rate test (SSRT) has been extensively used as a simple predictive corrosion test of metallic materials. Recently Shoji et al. [10] have proposed the significance of the slow strain rate fracture toughness test as a simple prediction procedure to evaluate the environment-sensitive time dependent crack extension of alloy steels in a light water nuclear reactor environment. In order to verify the feasibility of the SSRT procedure for the rock mechanics experiment, the standard fracture toughness test of granite in a water environment at room temperature was carried out, where both acoustic emission and crack gage technique were also applied, using 3pt. bending small specimen as the one given in Fig. 4. Figure 7 shows some examples of

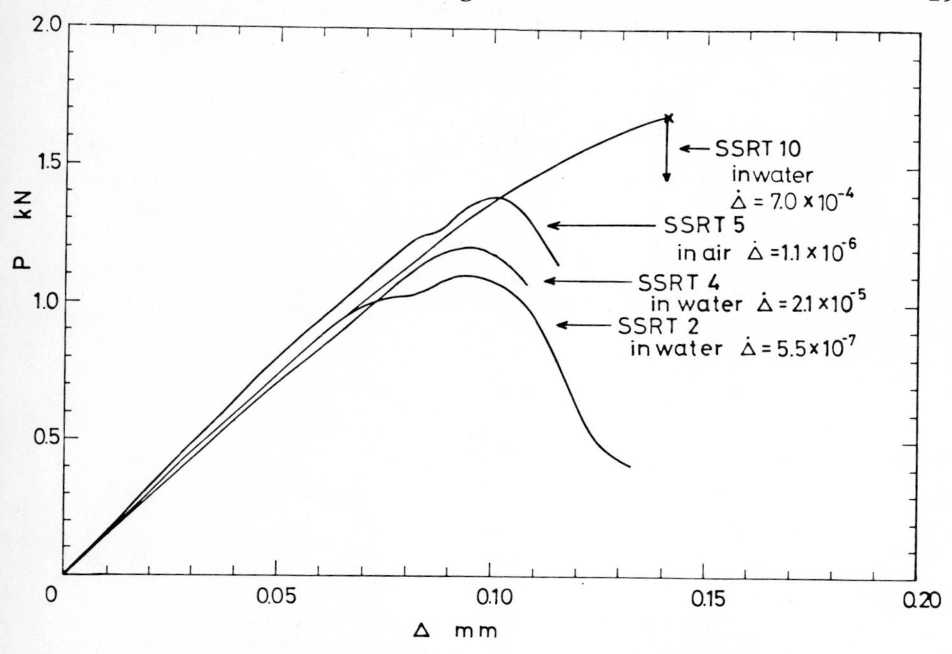

Fig. 7 Some Examples of Load-Load Line Displacement Records (P-Δ) in Slow Strain Rate Fracture Toughness Tests

load-load line displacement records (P-Δ) obtained from the SSRT experiment, where a significant reduction of the maximum load in the water environment is observed with decreasing loading rate, in comparison with the air environment.

Figure 8 demonstrates a relationship between the J-integral and a summation of AE energy. It is also noted that an abrupt increase of ΣE_{AE} corresponds to the extension of the macrocrack in both the air and the water environment, as shown by arrows in Fig. 8, and a remarkable reduction in the AE activity was observed in the water environment. This evidence suggests that the water environment can reduce the friction resistance between gains and fracture surfaces. The critical J values, J_{iAE}, determined by the AE technique agree fairly well with the value of J_{in} determined by the crack gage technique.

Figure 9 shows that the J_{in}-value decrease significantly with decreasing deflection rate ($\dot{\Delta}$) in the water environment, and a slight reduction of the fracture toughness is observed presumably due to the presence of water at the crack tip even in the air environment. Since the deflection rate is proportional to the crack tip deformation rate, here $\dot{\Delta}$ is used as a measure of the crack tip strain rate, which plays the most important role in the environment-sensitive cracking. Although there is large scatter of the J_{in}-values in the air environment, as shown in Fig. 9, it is suggested that the J_{in}-data in water increase with increasing strain rate, and at the strain rate of $\dot{\Delta}=10^{-3}-10^{-4}$mm/sec,

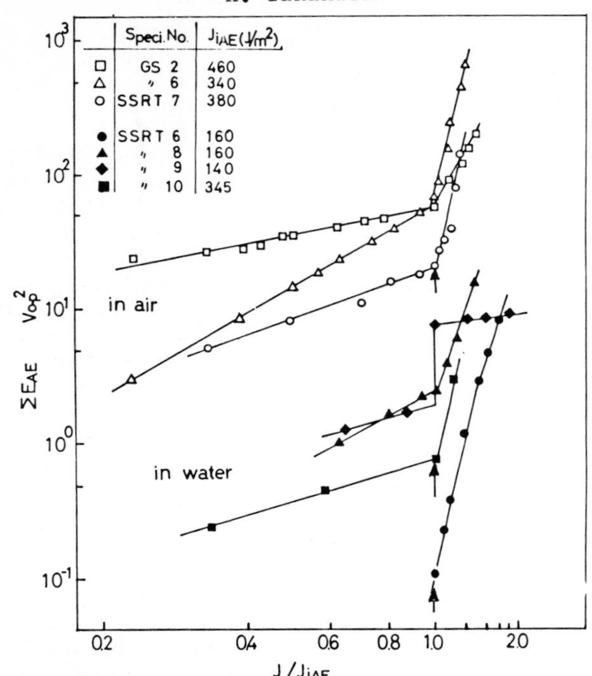

Fig. 8 J-ΣE_{AE} Relationship in Slow Strain Rate Fracture Toughness Tests

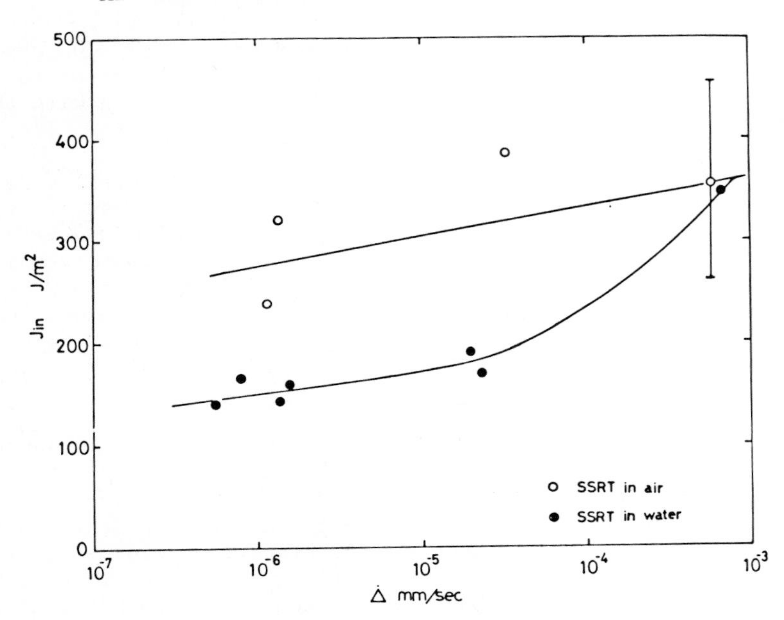

Fig. 9 Reduction of Fracture Toughness in SSRT Experiments as a Function of Deformation Rate ($\dot{\Delta}$)

J_{in} in the water is almost equal to that in the air. This evidence is very similar to the experimental data of SCC and SSRT for a combination of nuclear pressure vessel and piping alloy steels, and high temperature water [6].

Effect of Confining Pressure, and Hot Water Environment on Fracture Toughness of Granitic Rock

Effect of Confining Pressure [11-13]: Schmidt [11] has shown that the fracture toughness of rock material at room temperature increases with increasing confining pressure, where the pressurized fluid was not immersed into the crack tip by jacket. Figure 10 shows the summarized data of the effect of confining pressure on the fracture toughness of various rock materials, depicted from published literature [12][13]. These experimental results indicate that only jacketed samples show a strong effect of pressure on the fracture toughness, whereas unjacketed specimens have no confining pressure effect.

Effect of High Temperature and Pressurized Water [14]: Figure 11 demonstrates the test equipment and details of an autoclave for simulating the condition of a geothermal reservoir (max. confining pressure ~30MPa, max. temperature ~350°C), where a cylindrical specimen (47mm outer diameter, 50mm length) having a bore hole of 15mm length was used. Internal pressure was supplied by the pressurized water, and the pressure vs. time was recorded in an X-Y recorder. The fracture tests were conducted at room temperature and at 200°C in the water environment, where the water environment simulates that of the geothermal reservoir. Here AE

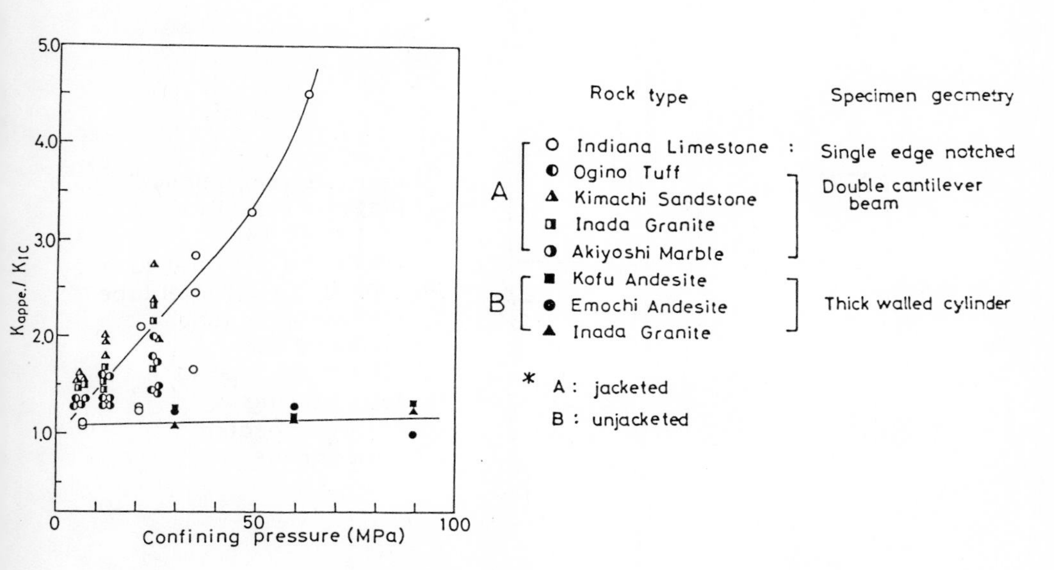

Fig. 10 Effect of Confining Pressure on Fracture Toughness (K_{IC}) of Various Rocks at Room Temperature

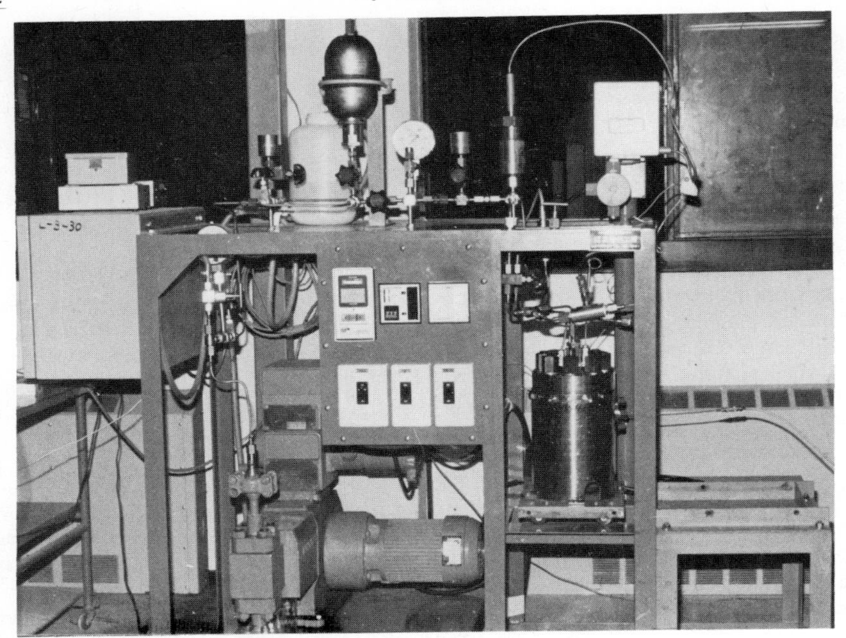

Fig. 11 (a) Experimental Set-up for Autoclave Fracture Tests under the Simulated Geothermal Reservoir Condition

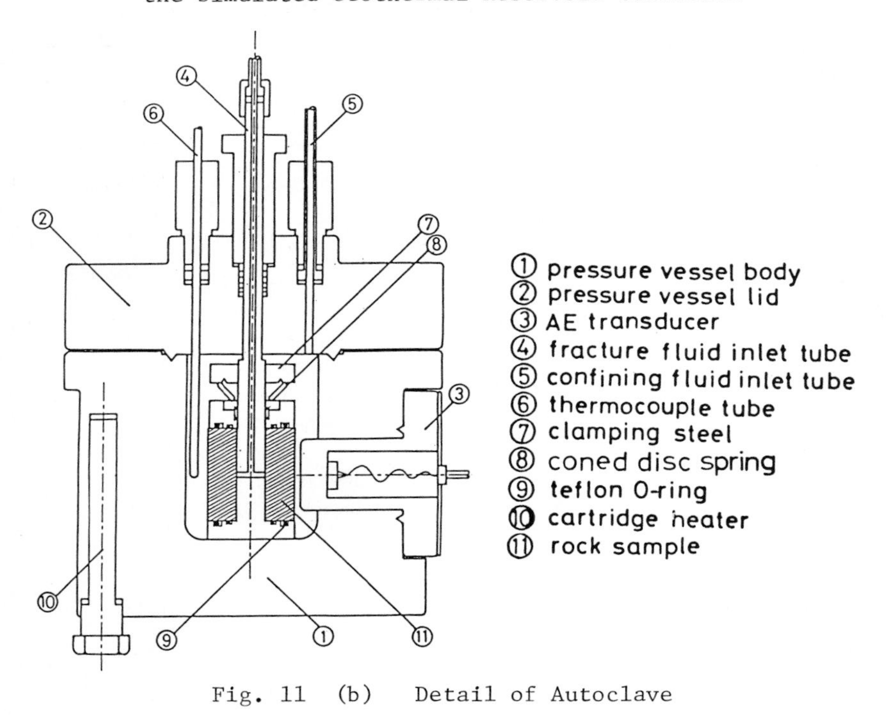

Fig. 11 (b) Detail of Autoclave

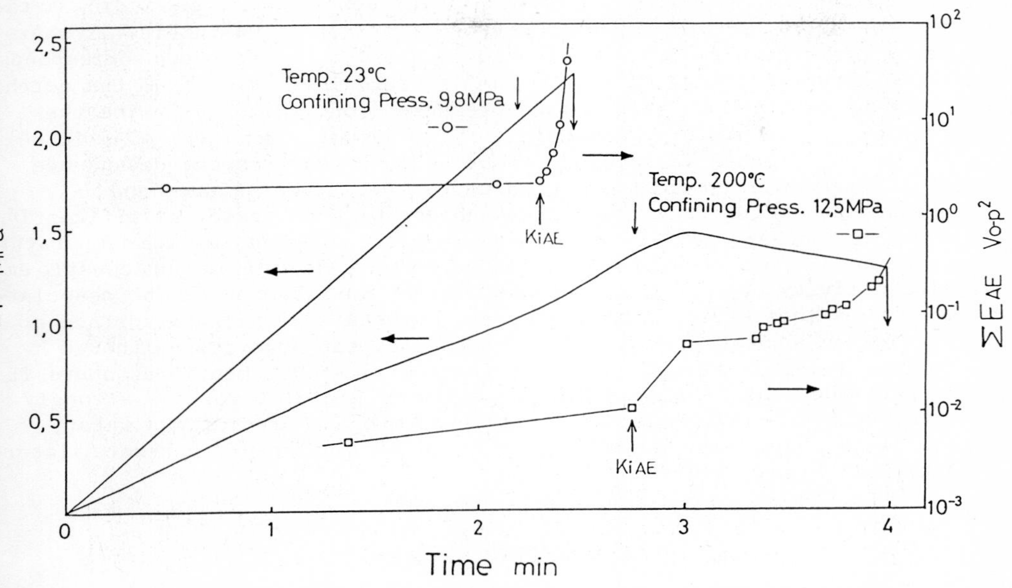

Fig. 12 Typical Examples of Pressure (Pressure Difference ΔP) versus Time Record and AE Data (Granitic Rock)

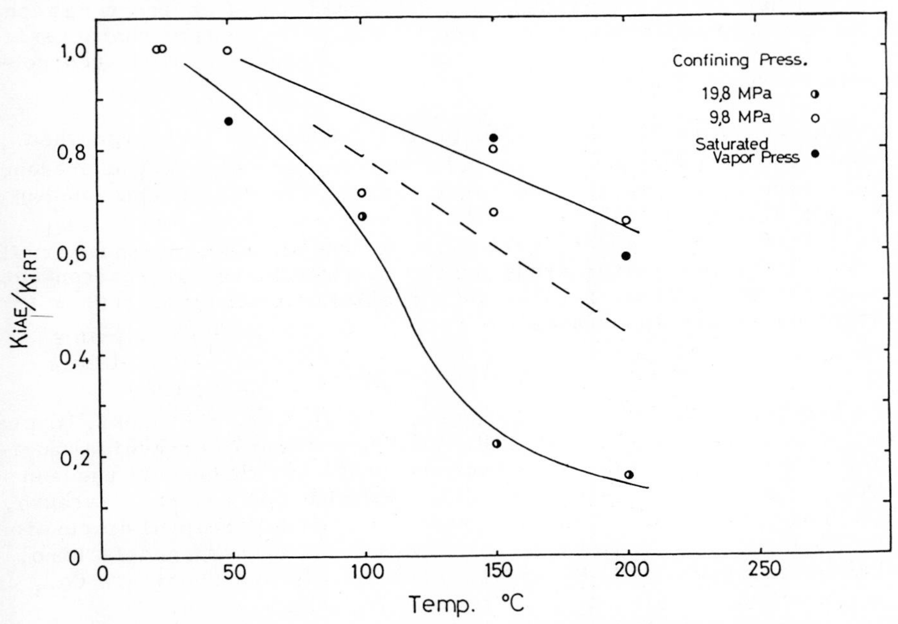

Fig. 13 Effect of Temperature, Confining Pressure and Water Environment on Fracture Toughness of Granitic Rock

crack monitoring was also done using a high temperature AE transducer, where the AE signal could be transmitted through the confining medium to the AE transducer. A typical pressure-time record is given in Fig. 12, where clear break-down phenomena were observed, and each breakdown corresponds to an abrupt increase of ΣE_{AE}. The fracture initiation from the notch could be detected by both AE and pressure drop, which determine the fracture toughness of the samples, K_{iAE}, assuming a linear elastic deformation. Figure 13 shows the temperature and environment dependence of fracture toughness of the granite. At high temperatures and pressurized water, the fracture toughness K_{iAE} decreases significantly, due to the mechanochemical reaction between the rock and the high temperature water. It is also considered that a significant reduction in the acoustic emission activity under a high temperature water environment may be associated with a loss of friction resistance between the fracture surface of granite and a chemical weakening of the crack tip bonding. Although further study is needed on the environmental effect mentioned above, it is concluded that the fracture toughness of granitic rock is strongly affected by the presence of high temperature and pressurized water, and it is suggested that an extensive investigation of rock-water interaction is highly desirable.

CONCLUDING REMARKS

With special reference to the microfracture mechanism of granitic rock, a suitable predictive test procedure for evaluating the fracture of a geothermal reservoir has been developed based on fracture mechanics, where the main focus was placed on a fracture mechanics characterization of fracture behavior of the granitic rock under a simulated geothermal reservoir environment.

Further analysis of the results of the predictive test described here may lead not only to a detailed understanding of how the presence of high temperature pressurized water reduces the fracture toughness of rocks, but also to a full evaluation of fracture extension during hydraulic fracturing and/or service operation in a geothermal reservoir. More detailed experimental study on the micromechanism of environmentally sensitive cracking of rocks in a realistic geothermal reservoir condition is highly encouraged.

ACKNOWLEDGEMENTS

The author wishes to express his gratitude to Prof. M. Suzuki, Toyota Technological Institute and Prof. H. Abé, Department of Mechanical Engineering, Tohoku University for their encouragement throughout the course of this investigation. Thanks are also extended to Drs. K. Nakatsuka, H. Niitsuma, and T. Shoji, Tohoku University, for their helpful discussions.

The assistance of Messers T. Wakabayashi, T. Hashida, and I. Hino, Tohoku University, and S. Miyazaki, Japan Metals and Chemicals Co., in the preparation of this paper is gratefully acknowledged.

Financial support for much of the work reported here was provided by the Ministry of Education, Science and Culture under grant-in-aid 455031, 555040, 505514, 56045015 and 57045014.

REFERENCES

1. Schmidt, R. A., "Fracture Toughness of Limestone," Experimental Mechanics, 16-5 (1976), pp. 161-167.
2. Schmidt, R. A. and Lutz, T. L., "K_{IC} and J_{IC} of Westerly Granite-Effects of Thickness and In-Plane Dimensions," ASTM STP 678 (1979), pp. 166-182.
3. Weisinger, R., Lostin, L. S. and Lutz, T. L., "K_{IC} and J Resistance Curve Measurements on Nevada Tuffs," Experimental Mechanics, 20-2 (1980), pp. 68-72.
4. Atkinson, B. K., "Subcritical Crack Propagation in Rocks: Theory, Experimental Results and Applications," J. Structural Geology, 4-1 (1982) pp. 41-56.
5. Speidel, M. O., "Overview of Methods for Corrosion Testing as Related to PWR Steam Generator and BWR Piping Problems," in "Predictive Methods for Assessing Corrosion Damage to BWR Piping and PWR Steam Generators" (Eds. H. Okada and R. Staehle), NACE,1982,pp.31-44
6. Hashida, T. et al., "Determination of Fracture Toughness of Granitic Rock by Means of AE Technique," Proc. 6th International AE Symposium, NDI Japan, 1982, in press.
7. Takahashi, H., et al., "Acoustic Emission Crack Monitoring in Fracture Toughness Tests for AISI 4340 and SA533B Steels," Experimental Mechanics, 21-3 (1981) pp. 89-99.
8. Khan, M. A., Shoji, T. and Takahashi, H., "Acoustic Emission from Cleavage Microcracking in Alloy Steels," Metal Science, 16-2 (1982) pp. 118-126.
9. Hashida, T., "Stress Corrosion Cracking in a Granitic Rock," Master Thesis, Tohoku University, 1982.
10. Shoji, T. et al., "Role of Loading Variables in Environment Enhanced Crack Growth for Water Cooled Nuclear Reactor Pressure Vessels," Proc. IAEA Specialists' Meeting on Cyclic Crack Growth in LWR Pressure Components, Freiburg, (1981),pp. 147-159.
11. Schmidt, R. A. and Huddle, C. W., "Effect of Confining Pressure on Fracture Toughness of Indiana Limestone," Int. J. Rock Mech. Min. Sci. and Geomech. Abstr., 14 (1977) pp. 289-293.
12. Kobayashi, R and Otsuka, N., "Fracture Toughness of Rocks at Confining Pressure," J. of the Mining and Metallurgical Institute of Japan, 96-12 (1980), pp. 879-884.
13. Kosugi, M. and Hayamizu, H., "Laboratory Simulation of Hydraulic Fracturing," Preprint of Annual Meeting of Mining and Metallurgical Institute of Japan, (1980), pp. 15-16.
14. Wakabayashi, T. et al., "Fracture Toughness Test of Granite in a Simulated Geothermal Reservoir Environment," Preprint of Annual Meeting of the Geothermal Research Society of Japan, 1982, p. 41.

APPLICATION OF SEISMIC INVERSE METHODS TO MAPPING OF SUBSURFACE FRACTURES

Norman Bleistein and Jack K. Cohen

Department of Mathematics and Computer Science
University of Denver
Denver, Colo. 80208 USA

INTRODUCTION

The purpose of seismic inverse methods is to map the interior of the earth from observed responses to signals transmitted into the earth. The terms, "map" and "signal," must be taken in the context of the model of signal propagation being used. The methods to be discussed, here, use elasto-dynamic sources -- electro-mechanical transducers, airguns or explosives -- as signal generators. Observations are usually made relatively near the source so that mode conversion can be neglected and the problem can be modelled as one of acoustic wave propagation. A further simplifying assumption is that the density is constant. Consequently, "mapping" is reduced to determining the soundspeed from the near "backscatter" response of an acoustic source.

In a "direct scattering problem" one is given a source distribution and information about the medium. The objective is to determine the subsequent propagation of the signal and the "scattering" due to the inhomogeneities of the medium. In an "inverse scattering problem," one is given information about the source and observations of the scattered signal. The objective, then, is to determine one or more of the parameters in the equation(s) modelling the propagation into the unknown medium. As noted above, the seismic inverse problem is most often reduced to the problem of determining one parameter, soundspeed from near backscattered signals.

Even for this simplest problem, the nature of the host medium, the earth, precludes determination of "complete" information, even for this simplified model. It is difficult to provide adequate energy to allow extremely low frequency signals to propagate. There is also an attenuation factor in signal propagation in the earth. This effect increases monotonically with frequency. Thus, there is a loss of high frequency information, progressively increasing with depth. Of course, these effects only compound the effect of the practical bandwidth limitations of the signal generator.

Thus, at best, one can hope to obtain information about the unknown parameter(s) in the earth only in some _limited bandwidth_ of its Fourier representation. For many of the length scales of interest in the seismic inverse problem, the limited bandwidth of the data may be thought of as being in the _high frequency_ regime. To understand this, we introduce the _dimensionless_ frequency parameter,

$$\Lambda = \frac{4\pi fL}{c}. \tag{1}$$

Here, f is a typical frequency, c is a typical soundspeed and L is a typical wavelength. There is an extra factor of 2 in this equation as compared to the same dimensionless parameter for the direct scattering problem. The reason is that in the inverse problem, it is the _two way travel time_ -- down into the earth and back up again -- which is of interest. This manifests itself in the analysis in a halved propagation speed.

Let us consider a frequency of 6 hz and a propagation speed of 1500m/s. For length scales of interest, we consider the _range_ down to the inhomogeneities and the _Gaussian curvature_ of a reflector of interest. For a value of L no larger than 400m, Λ is larger than 4, which suffices to characterize the problem as one amenable to analysis by high-frequency-approximate techniques. For larger values of f or L encountered in practice, high frequency analysis is all the more justified.

An elementary fact about bandlimited high frequency Fourier data is that it does not contain information about _trends_ of the original function -- which information resides in low frequencies -- but it does contain information about discontinuities. Thus, _these techniques are ideally suited to exploration for fractures in the earth_.

In summary, then, the objective of seismic inverse methods is to extract information about the interior of the earth consistent with the model of the earth and in the context of the practical constraints of the data gathering process. In developing methods of inversion, it is reasonable to exploit high frequency methods of analysis to gain insight, to interpret, to derive techniques and to simplify algorithms. This is regularly done.

In the broadest sense, then, any technique for extracting information about the interior of the earth from the response to signals transmitted into the earth is an inverse method. The narrower usage presently extant in the seismic exploration literature reserves that term for methods based directly on the wave equation, as opposed to methods based indirectly on it through physical interpretations of the wave equation, such as Huygens principle or ray optics.

Our own work [1-5] and the work of our associates [6-10] falls in the category of the narrower definition of inverse methods. However, some attempt will be made here to cite contributions to inversion in the larger sense, with no offense meant to those authors who prefer to think of their contribution to the extensive literature of <u>wave equation migration</u> [11-24] *. and other earlier methods as being distinct from inversion.

The literature on seismic inverse methods is significantly younger, but burgeoning. This is an area of intense ongoing research at the present time; see [24-32]. In particular, there are some results on inversion for two or more parameters; see [24,26,27,31].

THE SEISMIC EXPERIMENT

The seismic experiment of interest here is as follows. A source is set off at some point on the surface of the earth. The upward scattered response of the earth is recorded at an array of receivers, typically lying along a straight line, either centered at the source or to one side of the source. The usual number of receivers is twelve, twenty four, or forty eight. If the response were recorded over a wide enough area on the surface, then the data of such an experiment could be used to solve <u>an inverse problem for the wave equation</u>. Indeed, Claerbout demonstrated with a synthetic example [11,12] that such a method of inversion was valid. In his example, the upward scattered response was generated for a single reflector in a "host medium" of known soundspeed. The source and receiver signals were then "downward continued" using the parabolic approximation to the reduced wave equation. The reflector was reproduced by the interference patterns of the solutions at five different frequencies, superimposed on one another.

---

* No attempt is made here to provide a complete bibliography. However, it is hoped that the cited references provide the reader with a point of departure into the literature. This remark applies equally to the citations on migration and the citations on inversion.

As noted above, the data is not gathered over a sufficiently broad array on the upper surface to use this technique on a single experiment in practice. Furthermore, the data from a _single_ experiment is typically to noisy to use reliably in this manner Instead, an _ensemble_ of such experiments is performed, most often over a line on the surface, but, increasingly more often, over an areal array on the upper surface. The data from a set of experiments for which the source and receiver have a common midpoint are then averaged to produce an _equivalent backscatter_ response at that point. In producing this average, it is necessary to use some _a priori_ estimate of velocity in order to adjust temporal delay to spatial offset. Here, again, the offset experiments for a single source can be used [34]. For horizontal reflectors, the delay from trace (time record) to trace can be predicted as a function of "average velocity." A "best fit" to the data for reflectors identified as being _horizontal_ is then used to estimate velocity for the averaging -- _stacking_ -- of the _common midpoint traces_. (There are a host of other processes performed on the data which we shall not even mention, here.)

The result of the pre-processing is to produce an "equivalent backscatter impulse (hopefully) response," either along a line or on an areal grid on the upper surface.

Given an areal array of backscattered responses, one can proceed to implement algorithms to determine a three dimensional map of the interior of the earth. Given a linear array of responses, this is not possible. One must assume, in this case, that the earth parameters depend only on depth and the one transverse variable along the direction of the array. Of course, there will be situations in which this assumption is so completely invalid as to lead the user seriously astray. On the other hand, this assumption has proven to be adequate in many situations of practical interest.

For this latter case, it is still possible to model the wave propagation as three dimensional, even though the earth model is essentially two dimensional. This model is often designated by the term _two and a half dimensional_, to distinguish the model from the case in which the wave propagation is also taken to be two dimensional.

> Wave equation migration is based on the premise that the sparsity of data in a single experiment (backscatter response) may be compensated for by the redundancy of many experiments. Therefore, it is assumed that the ensemble of backscatter responses is itself a wave which can be continued downward in space and backward in time by the wave equation to reproduce the reflectors.

In this method, the objective is to locate the reflectors in the earth in the context of an _a priori_ velocity distribution. While all of the assumptions of the method might make the reader wary, the overwhelming success of the method speaks for itself. It is more a matter of understanding why the method works, when it does, and why it does not, when it does not.

VELOCITY INVERSION

In a series of papers [1-3] we have developed a perturbation method for determining the soundspeed in the earth from the ensemble of backscattered responses. In this method, it is assumed that the soundspeed in the earth can be represented as a "small" perturbation from some known reference speed. Thus, we write for the soundspeed v,

$$\frac{1}{v^2} = \frac{1}{c^2}[1+\alpha]. \tag{2}$$

It is assumed that c is "known" and the objective is to determine α. An incident wave is assumed, which is the response to the source in the context of the known reference velocity. With U_I denoting this incident wave, U_S denoting the scattered wave and U the total wave field,

$$U = U_I + U_S, \tag{3}$$

the two components of the total wave field must satisfy the wave equation with soundspeed c:

$$\nabla^2 U_I - \frac{\partial}{\partial t^2} U_I = -\delta(\underline{x}-\underline{\xi}), \tag{4}$$

$$\nabla^2 U_I - \frac{\partial}{\partial t^2} U_S = \frac{\alpha}{c^2}\left[U_I + U_S\right]. \tag{5}$$

In the first equation, δ is the Dirac delta function and ξ is the source location. The range of $\underline{\xi} = (\xi,\eta,0)$ characterizes the ensemble of source locations. In the second equation, it can be seen that the scattered field is proportional to α. On the right side, the first term in the source is proportional to α, while the second term is quadratic in α. Thus, assuming that α is small, we neglect the second term. The Green's function representation of the solution to this problem then provides an integral equation for α.

$$U_S(\xi,\eta,0,\tau) = \int dx\,dy\,dz\,dt \frac{\alpha}{c^2} U_I(x,y,z,t;\xi,\eta,0,0)\, U_I(x,y,z,\tau;\xi,\eta,0,t). \tag{6}$$

In this equation, we have used the fact that the incident wave, itself, is the Green's function. Also, we have included the second set of arguments which characterized the source/receiver point. When observations are taken over an areal array, ξ, η for all time τ, we seek a solution, $\alpha(x,y,z)$. When only a linear array of data ξ is given for all time τ, we seek a solution, $\alpha(x,z)$.

When the reference speed is taken to be a constant, this integral equation can be inverted in closed form by integral transform techniques [2]. For the fully three dimensional inverse problem, the result is

$$\alpha(x,y,z) = \frac{8ic^3}{\pi}\int d\xi d\eta \int_{-\infty}^{\infty} dk_1 dk_2 dk_3 \int_0^{\infty} d\tau \int_0^{\tau} dt \Big[k_3(\tau^2 - t\tau)\, U_S(\xi,\eta,0,\tau)$$

$$\cdot \exp\big[2i[k_1(x-\xi) + k_2(y-\eta) - k_3 z] - i\omega\tau\big]\Big], \tag{7}$$

$$\omega = c(\mathrm{sign}k_3)\left[k_1^2 + k_2^2 + k_3^2\right]^{1/2}.$$

In the two and one half dimensional case,

$$\alpha(x,z) = \frac{8ic^3}{\pi}\int d\xi \int_{-\infty}^{\infty} dk_1 dk_3 \int_0^{\infty} d\tau \int_0^{\infty} dt \Big[k_3(\tau^2 - t\tau)\, U_S(\xi,0,0,\tau)$$

$$\cdot \exp\big[2i[k_1(x-\xi) - k_3 z\big] - i\omega\tau\big]\Big], \tag{8}$$

$$\omega = c(\mathrm{sign}k_3)\left[k_1^2 + k_3^2\right]^{1/2}.$$

These integral formulas are quite similar to those of the Fourier migration technique of Stolt [14]. In particular, the Fourier superpositions are the same. However, the amplitudes differ. In the migration technique, the amplitude is deduced from a physical premise equivalent to the one stated above. In the inversion method stated here, the amplitude is deduced from an integral representation of the solution to the governing equation of propagation of the model.

In using these formulas, we do not process for α itself. This requires full bandwidth data which is not available. In a series of papers, [35-38], a method is developed for processing the data to produce a series of bandlimited Dirac delta functions which peak on the reflecting surfaces and have reflection strength in known proportion to the reflection strength. This technique also allows us to process the data in different domains with different reference velocities, appropriate to each domain.

Thus, the output of the velocity inversion method is a reflector map and also an estimation of the reflection strength from which velocity increments across the reflector can be estimated. The former is similar to the output of wave equation migration, while the latter constitutes an improvement over that method. When true amplitude information has not been preserved in the recording and preprocessing, velocity inversion produces a reflector map equivalent to the output of migration. When relative true amplitude data is provided, the output provides relative reflection strength.

REFERENCES

[1] Cohen, J. K., and Bleistein, N., An inverse method for determining small variations in propagation speed: Soc. Ind. Appl. Math. J. on Appl. Math., 32, 4, p. 784-799, 1977.

[2] Cohen, J. K., and Bleistein, N., Velocity inversion procedure for acoustic waves: Geophysics, 44, 6, p. 1077-1085, 1979.

[3] Bleistein, N., and Cohen, J. K., The velocity inversion problem: present status, new directions: Geophysics, to appear, Nov., 1982.

[4] Lahlou, M., Bleistein, N., and Cohen, J. K., Highly accurate inverse methods for three dimensional stratified media, preprint, 1982.

[5] Gray, S. H., Bleistein, N., and Cohen, J. K., A comparison of first and second order velocity inversion methods: preprint, 1982.

[6] Gray, S. H., A second order procedure for one dimensional velocity inversion: SIAM J. Appl. Math., 39, 3, p. 456-462, 1980.

[7] Gray, S. H., Velocity inversion in a stratified medium with separated source and receiver, II: J. Acoustical Soc. Amer., 69, 3, p. 661-661, 1981.

[8] Gray, S. H., and Hagin, F. G., Toward precise solution of one-dimensional inverse problems: SIAM J. Appl. Math., to appear.

[9] Hagin, F. G., Some numerical approaches to solving one-dimensional inverse problems: J. Comp. Phys., 43, 1, 1981.

[10] Hagin, F. G., and Gray, S. H., Travel-time-like variables and the solution of velocity inverse problems: preprint, 1981.

[11] Claerbout, J. F., Toward a unified theory of reflector mapping: Geophysics, 36, 3, p. 467-481, 1971.

[12] Claerbout, J. F., Fundamentals of Geophysical Data Processing: McGraw-Hill, New York, 1976.

[13] Claerbout, J. F., and Doherty, S. M., Downward continuation of moveout corrected seismograms: Geophysics, 37, 5, p. 741-768, 1972.

[14] Stolt, R. H., Migration by Fourier transform: Geophysics 43, 1, p. 23-48, 1978.

[15] Schneider, W. A., Integral formulations for migration in two and three dimensions: Geophysics, 43, 1, p. 49-76, 1

[16] Larner, K. L., Hatton, L., Gibson, B. S., and Hsu, I-C., Depth migration of imaged time sections: Geophysics, 46, 5, p. 734-750, 1982.

[17] Hatton, L., Larner, K. L., and Gibson, B. S., Migration of seismic data from inhomogeneous media: Geophysics, 46, 5, 751-767, 1981.

[18] French, W. S., Computer migration of oblique seismic reflection profiles: Geophysics, 40, p. 908-961, 1975.

[19] Lowenthal, R., Roverson, L., Lu, R., and Sherwood, J., The wave equation applied to migration: Geophysical Prospecting, 24, p. 380-399, 1976.

[20] Judson, D. R., Lin, J., Schultz, P. S., and Sherwood, J. W. C., Depth migration after stack: Geophysics, 45, 3, p. 361-375, 1980.

[21] Bleistein, N., and Cohen, J. K., Direct inversion procedure for Claerbout's equations: Geophysics, 44, 6, p. 1034-1040, 1979.

[22] Berkhout, A. J., Seismic Migration: Elsevier, New York, 1980.

[23] Berkhout, A. J., Wave field extrapolation techniques in seismic migration, a tutorial: Geophysics, 46, 12, p. 1638-1656, 1981. J. Geophys. Res., 85, p. 5364-5366, 1980.

[24] Clayton, R. W., and Stolt, R. H., A Born-WKBJ inversion method for acoustic reflection data, Geophysics, 46, 11, p. 1559-1567, 1981.

[25] Coen, S., The inverse problem for a three-dimensional acoustic medium: SEG 50th Annual International SEG Meeting, Houston, Nov. 17, 1980.

[26] Coen, S., On the elastic profiles of a layered medium from reflection data: J. Acoust. Soc. Amer., 70, 1, p. 172-175.

[27] Coen, S., Inverse scattering of the permittivity and permeability profiles of a plane stratified medium, J. Math. Phys., 22, 5, p. 1127-1129, 1981.

[28] Coen, S., and Mei, K. K., 1981, Inverse scattering technique applied to remote sensing of layered media, IEEE Trans. Ant. and Prop., AP-29, 2, p. 298-306, 1981.

[29] Coen, S., and Yu, M. W. H., The inverse problem of the direct current conductivity profile of a layered earth, Geophysics, 46, 12, p. 1702-1713, 1981.

[30] Raz, S., Sloping layers: a direct reconstruction, preprint.

[31] Raz, S., Three dimensional velocity profile inversion from finite offset scattering data, Geophysics, 46, 6, p. 837-842, 1981.

[32] Raz, S., Direct reconstruction of velocity and density profiles from scattered field data, Geophysics, 46, 6, 832-836, 1981.

[33] Raz, S., An explicit profile inversion: beyond the Born model, Radio Science, to appear.

[34] Waters, K. H., Reflection Seismology, A Tool for Energy Resource Exploration: Wiley, New York, 1978.

[35] Bojarski, N. N., A survey of electromagnetic inverse scattering: Univ. Res. Corp., Special Projects Lab. Rep., DDC AD-813-581, 1966.

[36] Mager, R. D., and Bleistein, N., An examination of the limited aperture problem of physical optics inverse scattering: IEEE Trans. Ant. Prop., AP-25, 5, p. 695-699, 1978.

[37] Armstrong, J. A., and Bleistein, N., An analysis of the aperture limited Fourier inversion of characteristic functions: University of Denver Research Report MS-R-7812, 1978.

[38] Cohen, J. K., and Bleistein, N., the singular function of a surface and physical optics inverse scattering: Wave Motion, 1, 1, p. 153-161, 1979.

DETECTION OF THE NEWLY FORMED FRACTURE ZONE BY THE SEISMOLOGICAL METHOD

by K. Yuhara*, S. Ehara* and H. Kaieda**

* Faculty of Engineering, Kyushu University
Fukuoka, Japan

** Central Research Institute of
Electric Power Industry
Abiko, Chiba Prefecture, Japan

ABSTRACT

A seismological method to detect water-saturated fracture zones was investigated. Observations of natural earthquakes above geothermal systems exhibit anomalously high seismic wave attenuation, which is extremely useful in estimating water-saturated fracture zones. Two tripartite arrays were installed at geothermal areas in central Kyushu. Seismograms were analyzed using the reduced spectral ratio technique to determine the seismic wave attenuation factor Q. As a result, lower Q values in the upper crust of geothermal areas were found. This fact may correspond to the existence of more water-filled fractures in the upper crust of geothermal areas in central Kyushu. The same method was applied to as small an area as a hydraulic fracturing test site. As a result, it is clarified that seismic waves passing through the newly formed fracture zone attenuate much more rapidly than those passing through the other path. This fact shows that it is possible to detect the newly formed fracture zone by the observation of seismic wave attenuation. Q values of the rocks beneath the hydraulic fracturing test site were measured in the laboratory by the pulse transmission method. By using such laboratory experimental data, we proposed a method to estimate quantitatively the increasing rate of porosity from the change in the observed Q.

INTRODUCTION

Water-saturated fracture zones are newly formed in hot dry rock through the hydraulic fracturing process. Changes of various physical properties in the rock occur with such a change of structure. We consider that one of the physical properties which shows the most remarkable change is the seismic wave attenuation factor Q. Laboratory studies show that the attenuation of elastic waves in rocks strongly depends on the physical state and saturation conditions.[1] Generally, attenuation varies much more than the elastic wave velocities as a result of a change in the physical state of materials. A similar situation is also observed in field experiments.[2] A seismological study for the earthquake swarms which occurred at Matsushiro, central Japan in the period of 1965 to 1967, clarified that the seismic wave attenuation factor Q beneath where many earthquakes occurred was much lower (1/20) than that beneath where no earthquakes occurred. However, the decrease of the seismic P wave velocity is only 6%.[3] In this case, the relation between the occurrence of the crack and the permeation of the underground water was discussed.[4] Therefore, we consider that the observation of Q is extremely useful in the detection of newly formed water-saturated fracture zones.

Generally speaking, the most important characteristic structure of the geothermal reservoir is the existence of fracture zones filled with high temperature water. Accordingly, it is expected that Q values in geothermal areas are lower than those of their surroundings. In this paper we show, first, the difference in Q values between geothermal areas and their surroundings in central Kyushu. Second, we point out the possibility of detecting an anomalous Q structure beneath even such a small area as a hydraulic fracturing test site. Finally, we discuss the relation between porosity and Q value, using our laboratory experimental data.

SEISMIC WAVE ATTENUATION BENEATH GEOTHERMAL AREAS IN CENTRAL KYUSHU

General tectonic setting in central Kyushu

There are many active and other Quaternary volcanoes in central Kyushu. Many fumaroles and hot springs also exist in connection with the volcanoes. A high heat flow zone higher than 2 HFU ($84mW/m^2$) surrounds the above geothermal manifestations as shown in Fig.1, where shallow seismic activity is high[5] and gravity is low.[6] Three geothermal power plants are operated at present there. As mentioned above, central Kyushu is tectonically active. Therefore, there may be great differences between underground structures beneath central Kyushu and their surroundings.

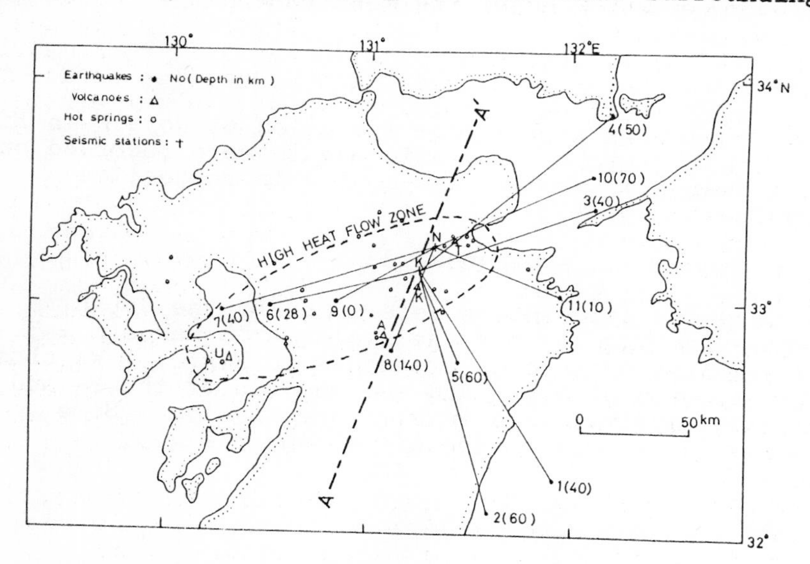

Fig. 1 Tectonic map in central Kyushu and seismic arrays. solid circle: earthquakes, No(Depth in Km), open triangle: active volcanoes, open circle: hot springs, cross: tripartite arrays(K: Kuju and N: Noya).

Determination of Q value

The spectral ratio technique was used to infer Q structure.[7] The amplitude spectra of a seismic wave observed at a station can be written as

$$A(f) = O(f) \cdot C(f) \cdot S(f) \cdot G \cdot e^{-\pi \frac{T}{Q} f} \quad \cdots\cdots (1)$$

where O(f) is the source spectrum, C(f) the crustal transfer function, S(f) the instrumental transfer function, G the geometrical spreading, T the travel time, Q the quality factor, and f the frequency. The spectral ratio of S wave to P wave is expressed as

$$R(f) = (O_S(f) \;/\; O_P(f)) \; e^{-\pi f\left(\frac{T_S}{Q_S} - \frac{T_P}{Q_P}\right)} \quad \cdots\cdots\cdots\cdots (2)$$

Therefore, we can obtain Q_S(Q value for S wave) from the gradient of the straight line $d(\ln R(f))/df$, assuming that $O_S(f)/O_P(f)$ = const. and $Q_S/Q_P = 4/9$.[8]

Observation

Two tripartite arrays were installed at Noya(N in Fig.1) and Kuju(K in Fig.1). Seismic events were recorded on magnetic tape for about three months in respective observations.

Results from small earthquake data

We used the hypocenters determined by the Japan Meteorological Agency for the small earthquakes($m = 3 \sim 5$). A few examples of records are shown in Fig.2. We obtained Fourier spectra of 0.5-s and 2-s records of the P- and S-waves first arrival of events, respectively. Some examples of spectral ratios are shown in Fig.3.

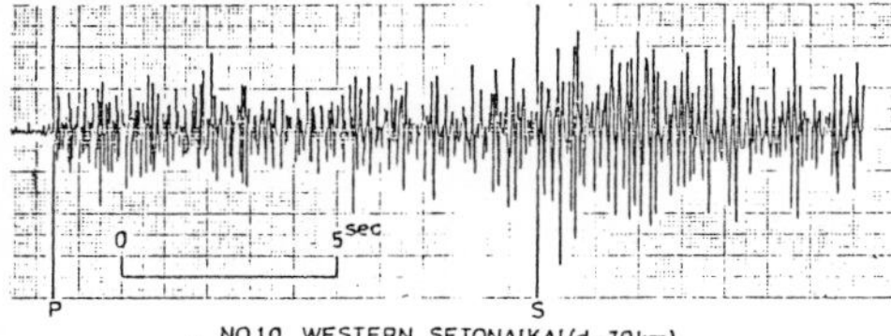

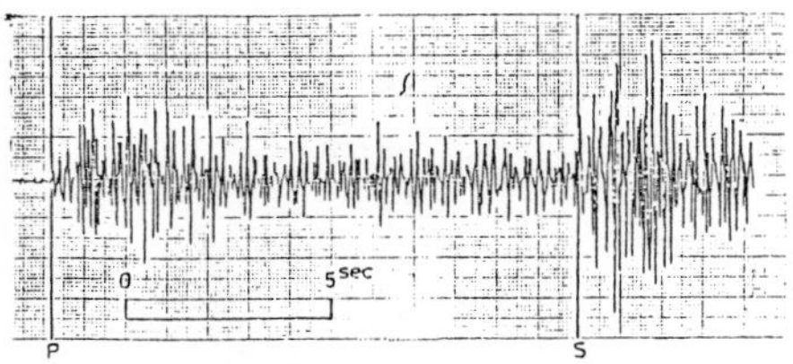

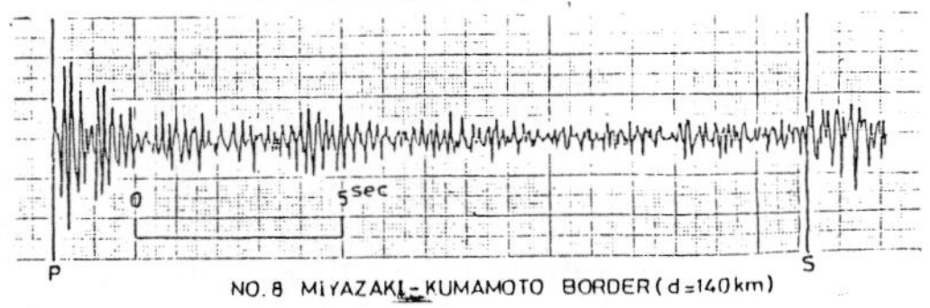

Fig. 2 Sample events recorded by Kuju or Noya tripartite arrays.

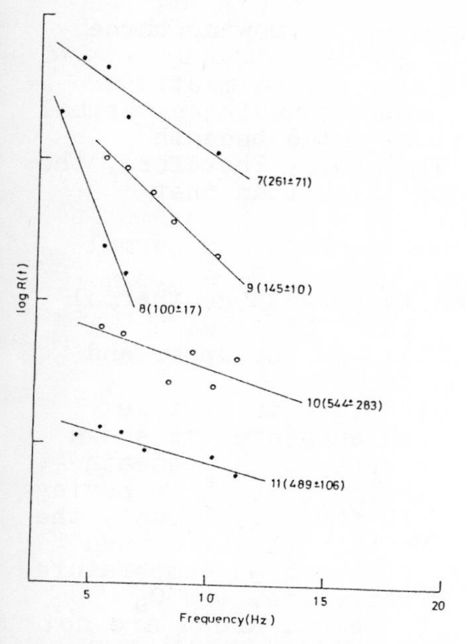

Fig. 3 Examples showing spectral ratio.

From the gradient of the straight line, we can get the Q_s value, using travel times of P and S-waves. Q_s values along the respective wave paths are shown in Fig. 4. "Low Q" and "High Q" show the large scale Q structure inferred from other seismological studies.[9]

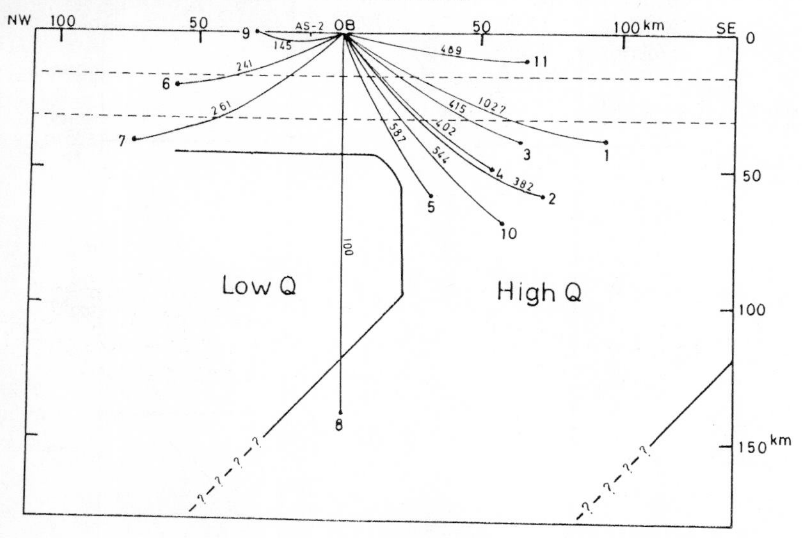

Fig. 4 Q_s value along each ray path. This vertical section is perpendicular to A-A' in Fig. 1.

From the figure, we can point out the following three features:

1) Q_s values in the crust and the uppermost mantle are in the several hundreds. The Q_s values beneath geothermal areas are about 250. However, the Q_s value beneath nongeothermal areas is 470 on the average. Therefore, the Q_s value beneath geothermal areas is lower than that beneath non-geothermal areas.

2) The Q_s value in the upper crust beneath geothermal areas is about 150.

3) The Q_s value in the upper mantle beneath geothermal areas is extremely low, i.e., 70. This value was calculated, assuming that the Q_s values of the upper and lower crust are 150 and 500, respectively.

As a result, a model of the Q_s structure in profile across Kyushu island from SE to NW was suggested as shown in Fig. 5. The low Q_s value in the upper crust beneath geothermal areas in central Kyushu is interpreted as having more water-filled fracture zones. In the lower crust, the cracks in the rocks are closed by the high pressure,[10] and liquid water does not exist because of the high temperature inferred from the heat flow data.[11] Therefore, the Q_s values in the lower crust beneath geothermal areas are not so low, and they are not so different between geothermal and non-geothermal areas. The low Q_s value in the uppermost mantle beneath geothermal areas may originate from partial melting which is inferred from the heat flow data.

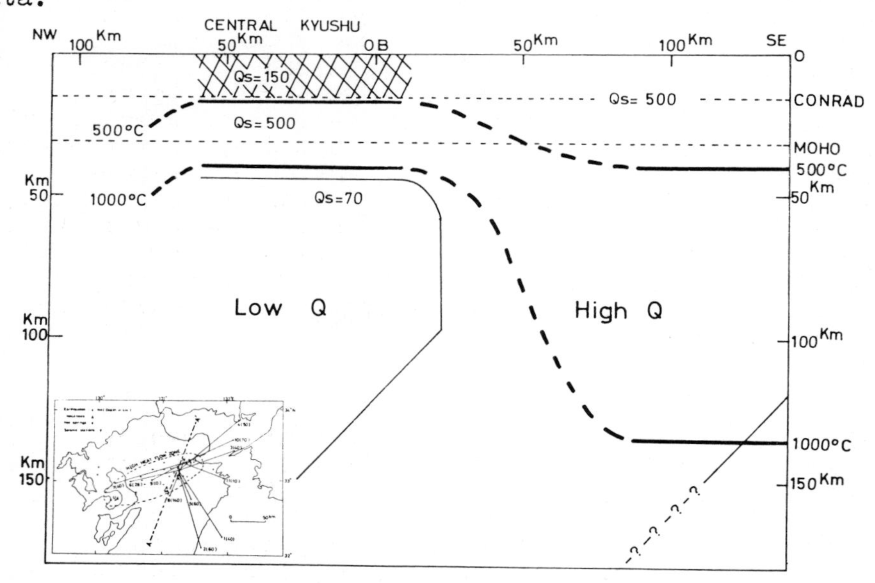

Fig. 5 Temperature and Q_s model beneath Kyushu along the axis perpendicular to A-A'.

Results from microearthquake data

Some of the hypocenters of microearthquakes were determined by the seismic array of the Kyushu Electric Power Company and by our tripartite nets. The depths of the hypocenters are shallower than 15 Km. The average Q_s values around Noya and Kuju are 90 ± 64 and 105 ± 81, respectively. These values are a little smaller than those of the upper crust in central Kyushu deduced from small earthquake data. As Noya and Kuju are more active geothermal areas in central Kyushu, we deduce that there are more water-filled cracks beneath Noya and Kuju than other areas.

SEISMIC WAVE ATTENUATION BENEATH THE HYDRAULIC FRACTURING TEST SITE NEAR YAKEDAKE VOLCANO, CENTRAL JAPAN

Hydraulic fracturing experiment

In Japan, hydraulic fracturing experiments of hot dry rock have been conducted at a site near the Yakedake volcano(active volcano),central Japan, as one of the Sunshine Project-New Energy Projects by the Japanese Government since 1979(Fig. 6). In 1980, an injection well was drilled to 300 m deep for the hydraulic fracturing experiment. The pressurized part is changed freely, using a dual packer set up about 200 to 300 m deep. The maximum pumping pressure of 19 MPa and the maximum flow rate exceeding 0.3 m^3/min were attained. Around the injection well(HY hole), four holes about 300 m deep were drilled for geophysical measurements. Hydraulic fracturing was accomplished repeatedly every year. Breakdown of pressure and many acoustic emissions were observed to follow the fracture of the rock.

Our observation

Three highly sensitive seismometers are installed at 5.2 m(Pt.1), 59.3 m(Pt.2), and 122.4 m(Pt.3) distant from the injection well(Fig.7). The observation was conducted continuously for 20 days each year(1980 and 1981) including the periods of the hydraulic fracturing experiment. Thirty-three near-microearthquakes(S-P time < 2.0 sec) were observed in two years. Among these events the Q_s values at Pt.1 and Pt.3 were determined for the six

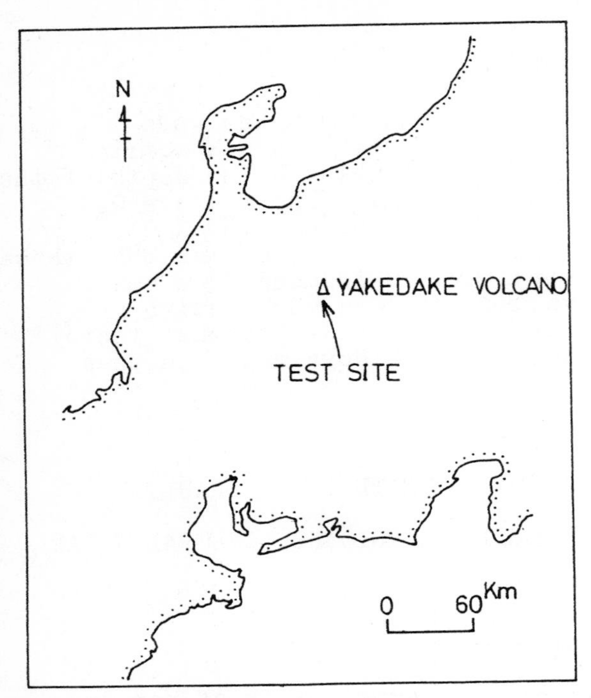

Fig. 6 Location map of Yakedake volcano and test site.

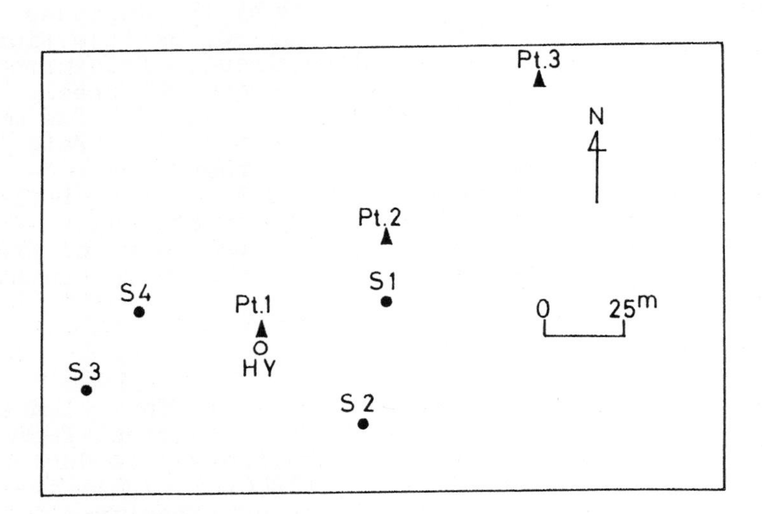

Fig. 7 Arrangement of wells and seismometers. HY: injecting well, S1~S4: observation wells, Pt. 1-3: Seismometers.

microearthquakes using the spectral ratio technique. Pt. 1 is very near the injection well, and water-filled fracture zones may be newly formed by the hydraulic fracturing beneath it; therefore there is a possibility of detecting the high seismic wave attenuation beneath Pt.1 from comparing the earthquakes observed at Pt.1 with those observed at Pt.3. The relation between QPt.1 and QPt.3 is shown in Fig. 8. The Q_s values observed at Pt.1 are smaller than those at Pt.3, although standard deviations are so large.

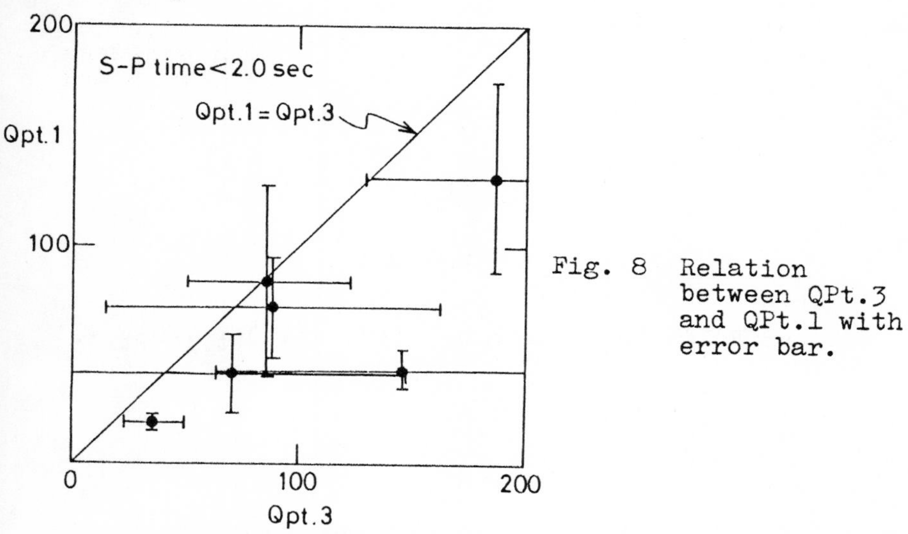

Fig. 8 Relation between QPt.3 and QPt.1 with error bar.

Fig. 9 shows the relation between S-P time and the ratio QPt.1/QPt.3. The shorter the epicentral distances become, the smaller the ratios QPt.1/QPt.3 are. The distance between Pt.1 and Pt.3 is about 120 m and it is much shorter than the epicentral distance, therefore in short, the waves propagating from a source to each station are considered to be identical at the base of the hydraulic fracturing test site. Then we may suggest that the difference of the observed Q_s values at Pt.1 and Pt.3 depends on the difference of Q values in the near surface rocks beneath the test site. Accordingly, the lower Q_s value observed at Pt.1 may correspond to the existence of the water-filled fracture zones newly formed by the hydraulic fracturing. Such a situation is illustrated in Fig. 10. If the ratio of Q_s between newly formed water-saturated rocks and surrounding dry rocks is 1/20, the thickness of the fracture zone beneath Pt.1 is estimated to be about 120 m. Though there are many uncertainties in the above estimates, it is very interesting to detect the fracture zone from the observation of seismic wave attenuation.

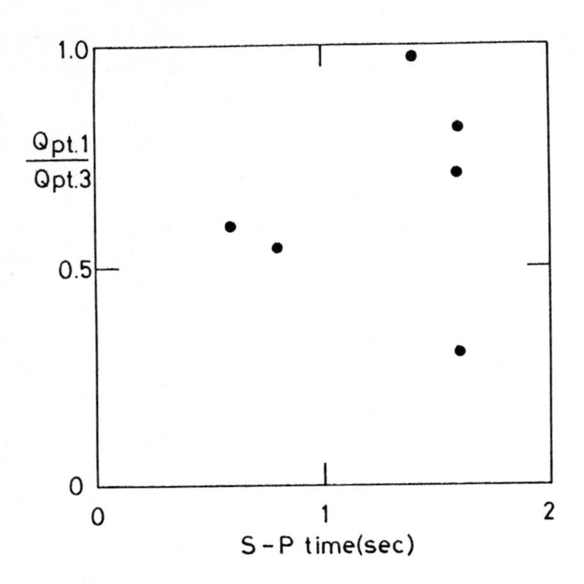

Fig. 9 Relation between S-P time and $Q_{Pt.1}/Q_{Pt.3}$.

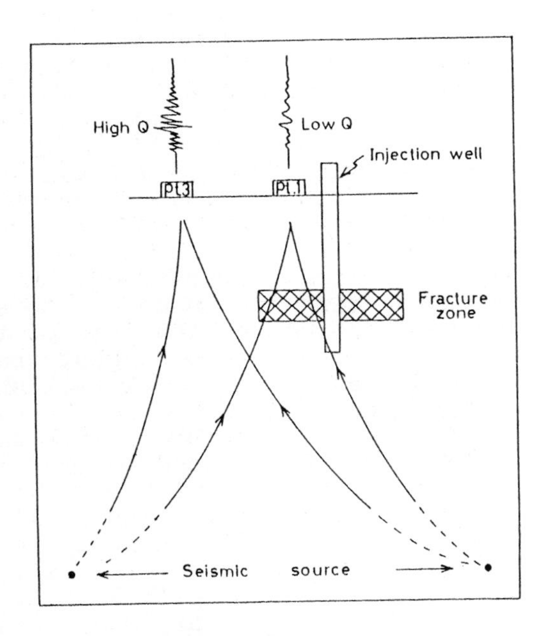

Fig. 10 A scheme for seismic wave attenuation at the hydraulic fracturing test site.

ELASTIC WAVE ATTENUATION IN ROCK SAMPLES

Experiment

Q_P values of the rocks beneath the hydraulic fracturing test site were measured relative to a reference sample with very low attenuation, by the pulse transmission technique. The rocks were taken from near surface to 300 m deep and were Paleozoic cherts, slates, sand stones and Quaternary andesites. The density is 2.60 to 2.75 × 10^3 kg/m^3, the P wave velocity 4.2 to 5.0 km/sec and the porosity 0.3 to 1.5 %. We measured Q_P values of rocks in three different states: One is dry, another is water-saturated and the third is water-saturated after fracturing, which is achieved by rapid cooling in water after heating in the air.

Results

Q_P values in water-saturated rocks are smaller than those in dry rocks, and after fracturing they become smaller than those before fracturing. The relation between Q_2/Q_1 (Q_2 and Q_1 are values of Q in water-saturated rocks after fracturing and that in dry rocks, respectively) and $(P_2 - P_1)/P_1$(P_2 is porosity of rocks after fracturing and P_1 the initial porosity) is shown in Fig. 11. The larger the rate of increase of porosity is, the smaller the Q_2/Q_1 value becomes. In such a relation, if $Q_2/Q_1 = 1/20$, $P_2/P_1 = 3$. This means that when a dry rock with initial porosity 0.5% is fractured, the porosity becomes 1.5%, and when it is furthermore water-saturated, the Q value decreases to 1/20 of the initial value. Conversely, if the decreasing rate of the Q value is known by field observation, we can estimate the rate of increase of porosity, although laboratory data are not always extrapolated to seismic frequencies. As mentioned above, even if the rate of increase of porosity is small, the Q value decreases very much, when the rock is water-saturated. Therefore, it is very useful to detect the fracture zone by the change of Q value.

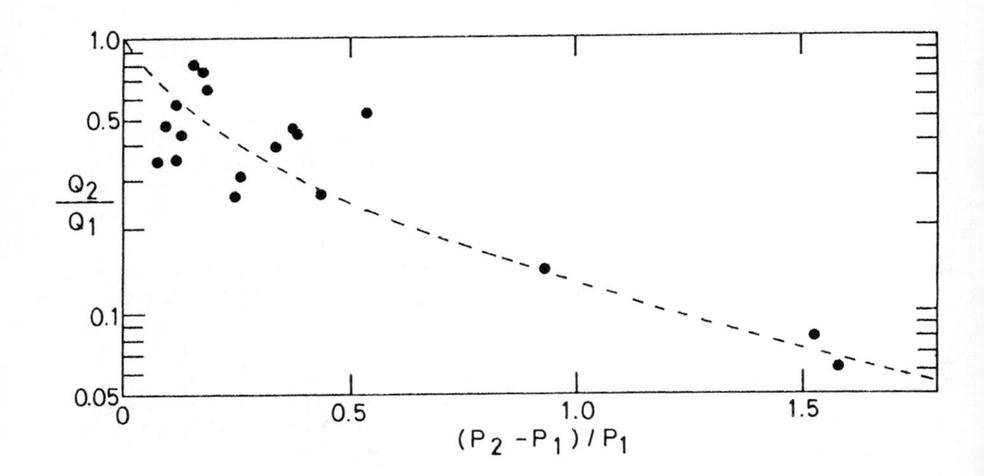

Fig. 11 Relation between $(P_2 - P_1)/P_1$ and Q_2/Q_1. P_1: initial porosity of rocks, P_2: porosity of rocks after fracturing, Q_1: Q value in dry rocks, Q_2: Q value in water-saturated rocks after fracturing.

CONCLUSION

We discussed the possibility of detecting the water-saturated fracture zones by using the seismic wave attenuation factor Q. At first, we showed the lower Q value of the upper crust beneath geothermal areas in central Kyushu. This is interpreted by the existence of more water-saturated fracture zones. We applied the same method to as small an area as a hydraulic fracturing test site. As a result, it is clarified that the seismic wave passing through the newly formed fracture zone attenuates much more rapidly than that passing through the other path. This fact shows that it is possible to detect the newly formed fracture zone by the observation of seismic wave attenuation. We measured the rate of decrease of the Q value of water-saturated rock samples accompanying the new crack formation. By using such laboratory experimental data, we proposed a method to estimate quantitatively the increasing rate of porosity from the change in observed Q.

REFERENCES

1. Toksöz, M.N., Johnston D.H. and Timur A., "Attenuation of Seismic Waves in Dry and Saturated Rocks: I. Laboratory Measurements", Geophysics, Vol. 44, No. 4, 1979, PP. 681-690.

2. Suzuki, S., "Anomalous Attenuation of P Waves in the Matsushiro Earthquake Swarm Area", J. Phys. Earth, Vol. 20, 1971, PP. 1-21.

3. Okada, H., Suzuki, S. and Asano,S., "Anomalous Underground Structure in the Matsushiro Earthquake Swarm Area as Derived from Fan Shooting Technique", Bull. Earthq. Res. Inst., Vol. 45, 1970, PP. 417-471.

4. Ohtake, M., "Seismic Activity Induced by Water Injection at Matsushiro, Japan", J. Phys. Earth, Vol. 22, 1974, PP. 163-176.

5. Mitsunami, T., Kubotera, A., Omote, S., and Kinoshita, Y., "Seismicity of the Hohi Geothermal Area, Kyushu, Japan, J. Geotherm. Res. Soc. Japan, Vol. 3, No. 1, 1981, PP. 43-53. (in Japanese with English abstract)

6. Kubotera, A., Tajima, H., Sumitomo, H., Doi, H. and Izutsuya, S., "Gravity Surveys on Aso and Kuju Volcanic Region, Kyushu District, Japan, Bull. Earthq. Res. Inst., Vol. 47, 1969, PP. 215-255.

7. Moriya, T., "Folded Structure of Intermediate Depth Seismic Zone and Attenuation of Seismic Waves beneath the Arc-Junction at Southwestern Hokkaido", Proceedings, Symposium on Subterranean Structure in and around Hokkaido and Its Tectonic Implication, 1976, PP. 13-27. (in Japanese with English abstract)

8. Anderson, Don, L., Ben-Menahem, Ari, and Archambeau, C.B., "Attenuation of Seismic Energy in the Upper Mantle", J. Geophys. Res., Vol. 70, No. 6, 1965, PP. 1441-1448.

9. Utsu, T., "Anomalous Seismic Intensity Distributions in Western Japan", Geophys. Bull. Hokkaido Univ., Vol. 21, 1969, PP. 45-52. (in Japanese with English abstract)

10. Nur, A., "Viscous Phase in Rocks and the Low-Velocity Zone", Jour. Geophys, Res., Vol. 76, No. 5, 1971,

PP. 1270-1277.

11. Yuhara, K. and Ehara, S., "Geothermal Fields of Arc Volcanism in Japan", Bull. Volcanol. Soc. Japan, Vol. 26, No. 3, 1981, PP. 185-203. (in Japanese with English abstract)

STRIKE-DIP DETERMINATION OF FRACTURES IN DRILLCORES FROM THE OTAKE-HATCHOBARU GEOTHERMAL FIELD

by Masao Hayashi and Tatsuo Yamasaki

Research Institute of Industrial Science, Kyushu University 86

Kasuga City, Fukuoka, 816 Japan

ABSTRACT

The strike and dip of fractures in drillcores from some wells in the Otake-Hatchobaru geothermal field, Japan, have been determined after measuring the thermo-remnant magnetism of the cores with an astatic magnetometer. This experiment has confirmed that such important fractures as slickensides and hydrothermal veins dip mostly southward, in one plane of the conjugate set of faults. Moreover, the analysis of these fractures, including joints, has shown the tectonic history of the field to be as follows: at an early stage, the field was subjected to the stress that formed the NE trending "normal" faults. Then, it was followed by the formation of the NW trending "normal" faults which correspond to the breeding ones for geothermal fluids. At present, a "lateral" stress seems to control the geothermal field, probably acting on the NW fractures to increase their permeability.

INTRODUCTION

The Otake-Hatchobaru geothermal field is in the Beppu-Shimabara Graben (Matsumoto, 1979) , which traverses Kyushu island from ENE to WSW (Fig. 1). The geothermal field is located very near the point at which the graben changes direction from ENE-WSW to NE-SW. The geology of the geothermal field has been described by Yamasaki et al. (1970) and others. It consists of the Neogene Usa Group, the early Pleistocene Hohi Volcanic Rocks, and the middle to late Pleistocene Kujyu Volcanic Rocks in ascending order. However, the geologic structure of the field has not been entirely clarified yet in detail .

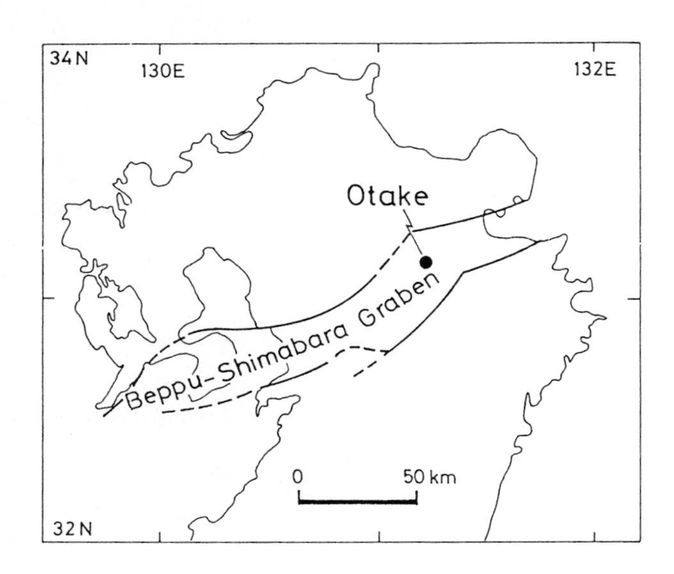

Fig. 1. Location of the Otake-Hatchobaru geothermal field.

The purpose of this paper is to determine the strike and dip of fractures in drillcores obtained from the Otake - Hatchobaru geothermal field, and then to analyse the stress acting on rocks that form geothermal reservoirs. Since original rocks of the geothermal field are made up mostly of younger volcanic rocks, their thermo-remnant magnetism can be assumed to coincide with the present geomagnetic field. Therefore, the measurement of the declination and inclination of the drillcores will give the strike and dip of fractures in it. Then, the tectonic history of the geothermal field can be traced using the obtained data together with other available geological data.

GEOLOGIC SETTING OF SAMPLES

Fig. 2 shows the location of wells drilled in the Otake - Hatchobaru geothermal field. Solid circles indicate the wells whose cores have been examined. The number of samples are 73 from Well HT-4, 14 from Well HT-1, and 3 from Well O-17. The lithology and hydrothermal alteration of the former two wells are illustrated in Figs. 3 and 4, respectively.

In Figure 2, faults estimated geologically and geophysically (electric) are also shown. Of these, the Sujiyu-Takenoyu fault which strikes NW to SE is the largest in the field and is considered to be

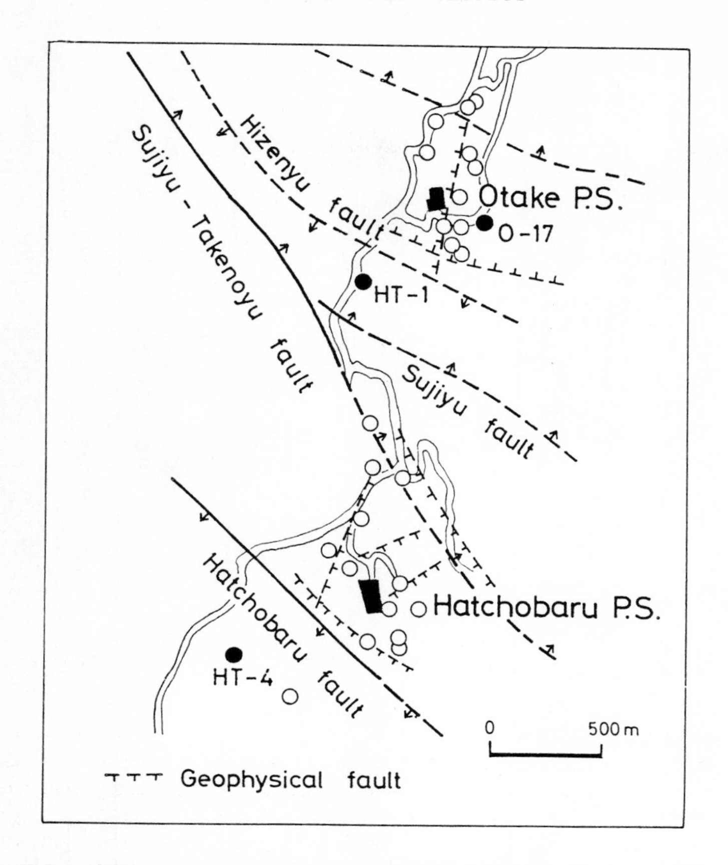

Fig. 2. Location map of wells drilled in the Otake-Hatchobaru geothermal field and faults estimated geologically and geophysically. The solid circles indicate the wells whose cores were examined.

the main breeding fault. This fault diverges at a place near Well HT-1 to form the subordinate Sujiyu fault. The Hatchobaru fault has strikes similar to those of the main fault, and divides the field into a very active Type A part (Hayashi et al., 1981) on the northeast and a less active Type B part on the southwest. The Hizenyu fault also strikes NW to SE and probably dips southwest. On the other hand, the geophysical survey suggested the existence of NE to SW or ENE to SWS trending faults that have been confirmed by this experiment.

EXPERIMENTAL

Fig. 5 illustrates an astatic magnetometer, Type NY-2, made by Kyoritsusha Ltd, Japan. The device consists of two small bar magnets

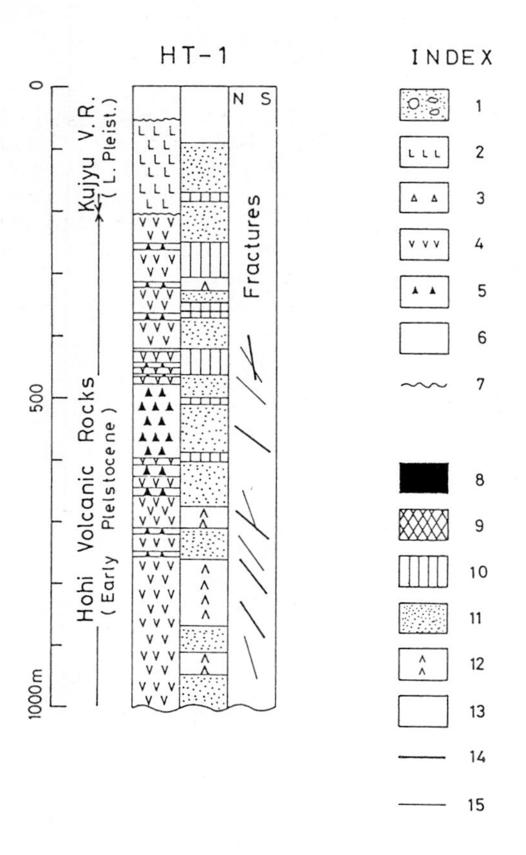

Fig. 3. Geology and hydrothermal alteration of Well HT-1. 1. Tallus deposits, 2. Hornblende andesite lava, 3. Hornblende andesite tuff breccia, 4. Pyroxene andesite lava, 5. Pyroxene andesite tuff breccia, 6. Unknown, 7. Unconformity, 8. Silicified zone, 9. Alunite zone, 10. Kaolin-pyrophyllite zone, 11. Aluminosilicate clay zone, 12. Partially altered zone, 13. Unaltered zone, 14. Slickenside, 15. Hydrothermal vein.

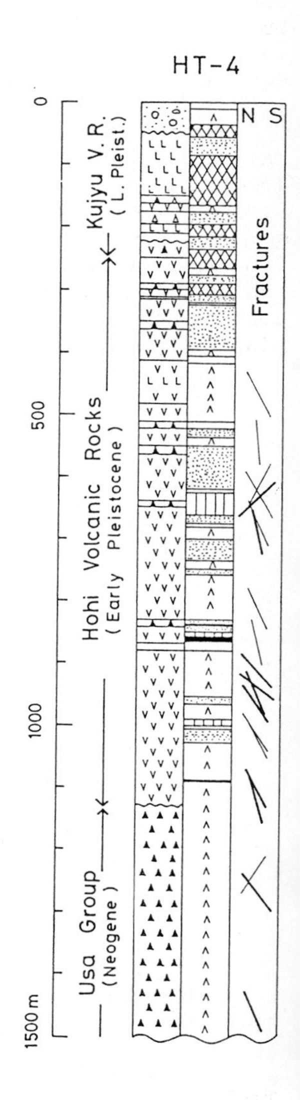

Fig. 4. Geology and hy
thermal alteration of
HT-4.

with nearly the same magnetic monment but opposite polarity, suspended by a thin phosphor-bronze wire. When these two magnets are so placed that the sum of their magnetic moment vanishes, the magnetic system does not move when the magnetic field changes uniformity.

When a rock specimen is brought close to the lower magnet, the magnetization of the rock specimen acts practically only upon the lower magnet. As a result of this action, the magnet-system turns around the vertical axis until it is balanced by the torsion of the suspension wire. The deflection angle can be read by the attached lamp-and-scale. Thus, the direction of magnetization of the specimen can be obtained by measuring the deflection angle, with the remote control of the attached turning table, on which the rock specimen is placed (Kyoritsusha, 1980).

ASTATIC-MAGNETOMETER

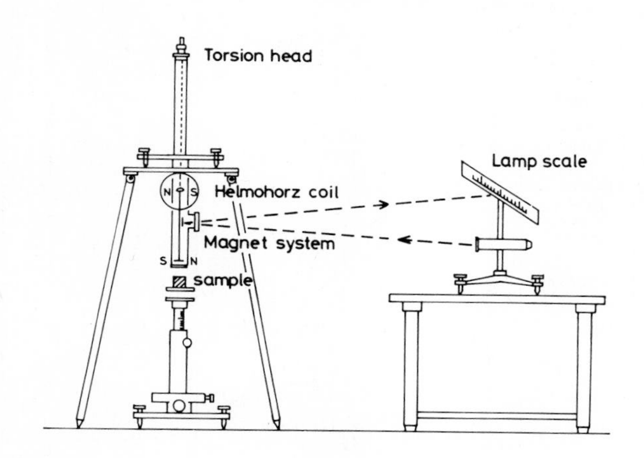

Fig. 5. An astatic-magnetometer used in this study.

RESULTS

The results of strike-dip determination of fractures are shown in Tables 1 to 3 and Figs. 6 and 7, and the frequency distribution of dip-angles is represented in Fig. 8. The strikes of slickensides are classified into three groups; NE-SW, NW-SE and E-W, in which the second is sometimes accompanied by nearly horizontal streaks and the others by almost vertical or oblique ones.

As is obvious from Figs. 6 and 7, such important fractures as slickensides and hydrothermal veins dip mostly southward. The dip-angles of all fractures average 57.7 ± 13.5 (standard deviation), in which those of the slickensides (55.7 ± 12.9) are a little smaller than those of the veins (59.2 ± 13.9), and those of the joints are in-between the two (57.2 ± 14.6).

Table 1. Outlines of fractures found in drillcores from Well HT-1. The degrees in the parentheses of a slickenside represent the angle from the top of a fracture which cuts a drillcore to the direction of movement of its hanging wall on a right-handed rotation.

Depth(m)	Rock name	Magnetism	Fracture	Strike-Dip	Remarks
436.0	andesite lava	N(medium)	slickenside (290°, weak)	N71°E, 81°S	later slickenside
446.0	weakly altered andesite lava	N(medium)	white vein (1 mm)	N67°E, 60°S	
475.0	andesite lava	N(medium)	white vein (1 mm)	N58°E, 44°S	
564.0	altered andesite lava	N(weak)	S(140°, strong) + vein (10 cm)	N44°E, 37°S	
567.0	altered andesite lava	N(weak)	S(110°, weak) + vein (1 cm)	N60°E, 46°S	later slickenside
569.0	weakly altered andesite lava	N(medium)	slickenside (140°, strong)	N58°E, 48°S	undulatory surface
673.0	andesite lava	N(medium)	white vein (2 mm)	N87°W, 70°S	
704.0	andesite lava	N(medium)	slickenside (260°, weak)	N24°E, 42°S	
786.0	andesite lava	N(strong)	slickenside (60°?, weak)	N68°W, 54°S	
802.0	altered andesite lava	N(weak)	vein (1 mm)	N84°E, 55°S	conjugate set ?
			joint	N33°E, 85°S	
803.0	andesite lava	N(weak)	joint	N5°W, 59°E	conjugate set ?
			vein (1 mm)	N5°W, 75°W	
813.0	andesite lava	N(strong)	vein (5 mm)	N4°E, 59°E	
			joint	N44°E, 23°N	
855.0	andesite lava	N(strong)	S(70°?, weak)	N61°W, 58°S	conjugate set
			joint	N29°E, 68°S	

Table 2. Outlines of fractures found in drillcores from Well HT-4. The degrees in the parentheses of a slickenside represent the angle from the top of a fracture which cuts a drillcore to the direction of movement of its hanging wall on a right-handed rotation.

Depth(m)	Rock name	Magnetism	Fracture	Strike-Dip	Remarks
633.6	altered andesite	N(weak)	slickenside (45°, weak)	N61°E, 43°N	
686.2	andesite lava	N(medium)	S(45°, strong) + vein (1 cm)	N40°W, 78°S	
916.8	andesite lava	N(strong)	S(135°?, weak) + vein (1 mm)	N24°W, 48°S	
923.0	andesite lava	N(strong)	S(165°, weak) + vein (1 mm)	N76°W, 48°S	
942.3	andesite lava	N(strong)	S(165°, weak) + vein (1 mm)	N85°E, 40°S	undulatory surface
951.2	andesite lava	N(strong)	S(180°?, weak) +	N73°W, 56°S	
			joint	N67°E, 67°S	
964.4	andesite lava	N(medium)	S(170°, weak) +	N86°E, 58°S	
			joint	N9°W, 59°E	
1098.0	andesite lava	N(strong)	slickenside (190°, weak)	N24°W, 64°S	
1121.0	andesite lava	N(strong)	slickenside (220°, weak)	N55°E, 73°S	
1265.3	tuff breccia	R(weak)	S(180°, strong) + vein (5 mm)	N50°E, 54°S	undulatory surface
1456.2	weakly altered andesite lava	R(weak)	S(70°, weak) + vein (1 mm)	N51°W, 67°S	later slickenside

Table 3. Outlines of fractures found in drillcores from Well O-17. The degrees in the parentheses of a slickenside represent the angle from the top of a fracture which cuts a drillcore to the direction of movement of its hanging wall on a right-handed rotation.

Depth(m)	Rock name	Magnetism	Fracture	Strike-Dip	Remarks
333.0	weakly altered andesite lava	N(medium)	slickenside (285°, strong)	N4°E, 51°W	
473.0	weakly altered andesite lava	N(medium)	white vein (2 mm)	N84°E, 65°N	
547.0	andesite lava	N(strong)	slickenside ? + vein (1 mm)	N10°E, 49°W	

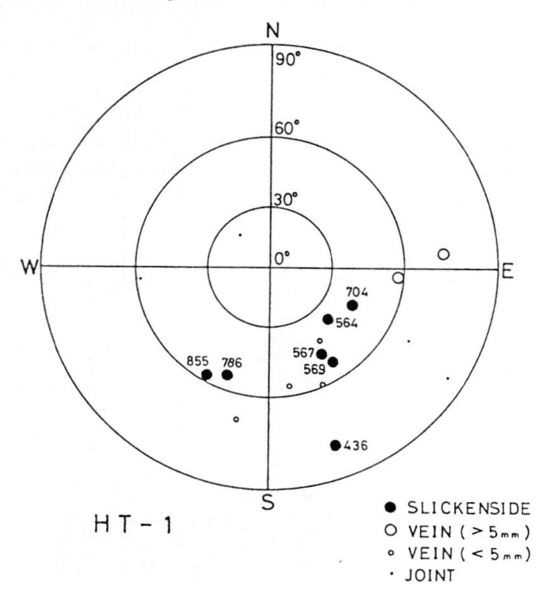

Fig. 6. Wulff's net projection of fractures in drillcores of Well HT-1. The upper hemisphere is used.

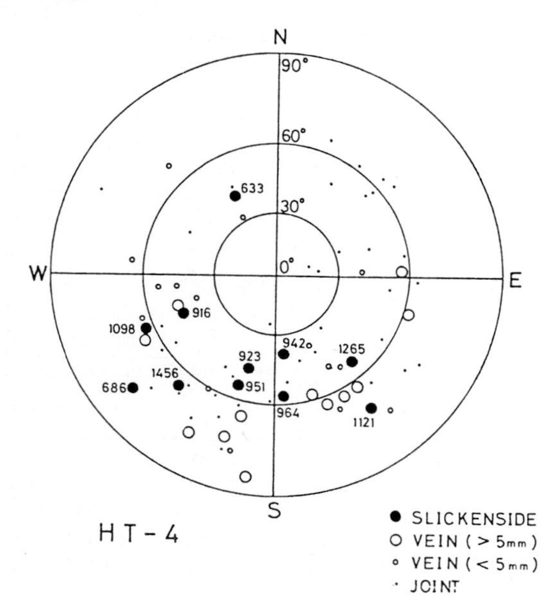

Fig. 7. Wulff's net projection of fractures in drillcores of Well HT-4. The upper hemisphere is used.

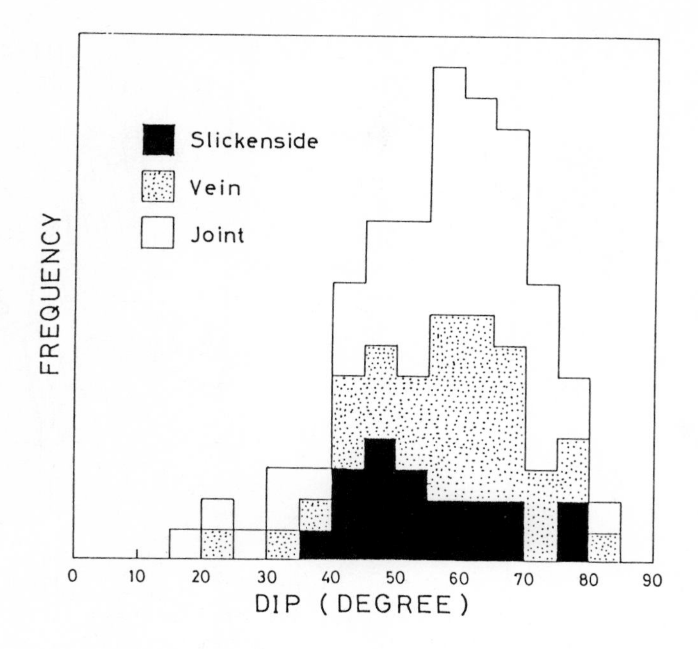

Fig. 8. Frequency distribution of the dip-angle of fractures in drillcores from Wells HT-1, HT-4 and O-17.

DISCUSSION

It is a little surprising that most slickensides and hydrothermal veins dip southward with only a few exceptions. This may lead to the important conclusion that one side of the conjugate set of a fault system develops many more fractures than the other under natural stress fields, and that only these fractures generally allow the up-flow of geothermal fluids.

The stress fields acting on rocks can easily be estimated when the strike and dip of a slickenside are known and the direction of movement is available from the observation of the slickenside. That is, the direction of the intermediate principal stress should be at a right angle to the streaks on the slickenside. Furthermore, since the angle of shear is usually in the range of 40 to 60 degrees, the directions of the other two principal stresses can be determined using Wulff's net. If the other plane is similar in strike and dip to one of the previously measured fractures, it confirms the existence of the presumed stress fields.

The stress fields of slickensides are classified into one of four representative types shown in Fig. 8, and as follows:

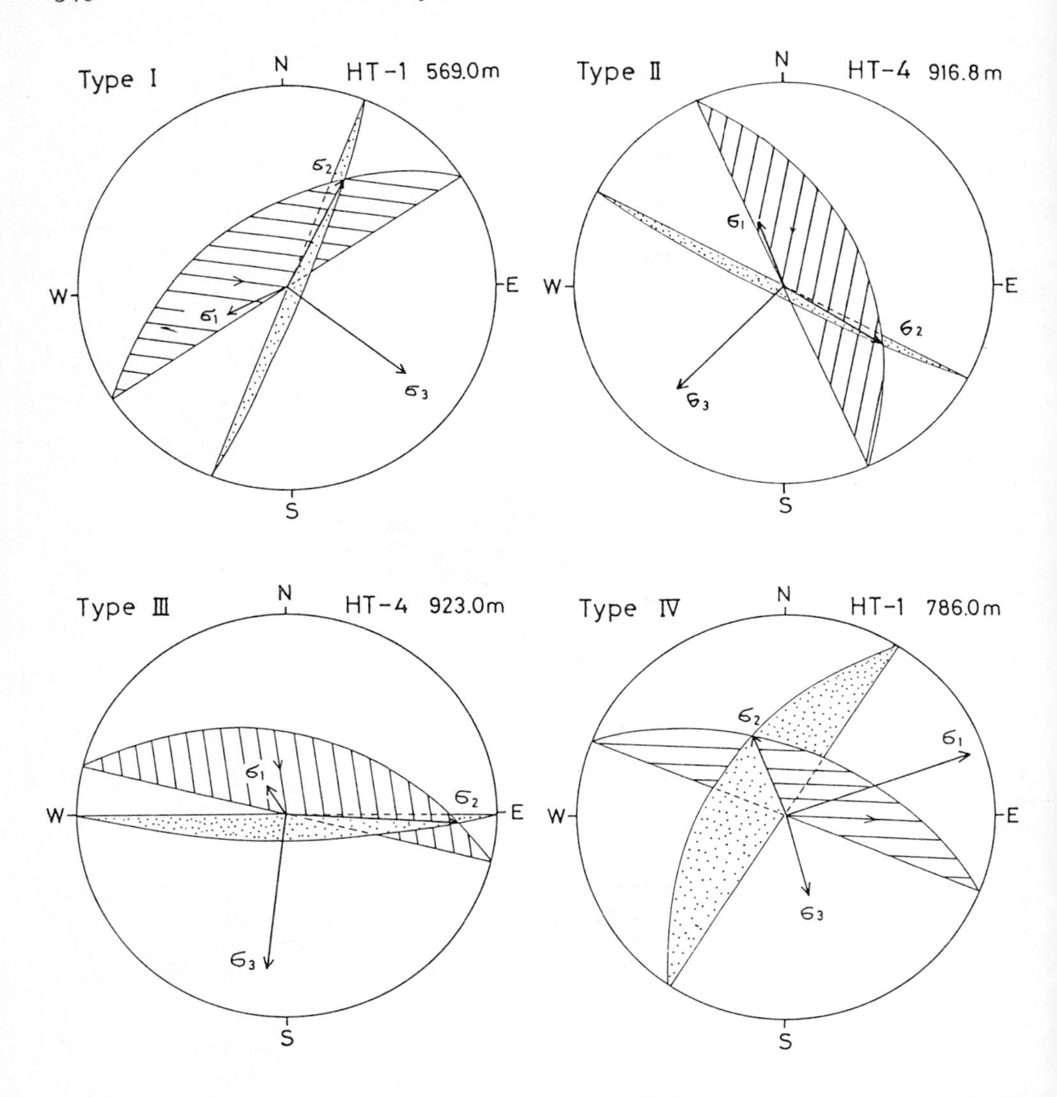

Fig. 9. Four representative kinds of stress fields estimated by the analysis of slickensides. The great-circle slanted with lines indicates a slickenside and the arrow the direction of movement of the hanging wall. The dotted great-circle is the other plane of the conjugate set of a fault system.

I : Normal stress fields whose minimum principal stress strikes roughly NW to SE (Well HT-1 564.0 m, 567.0 m, 569.0 m, 704.0 m; Well HT-4 633.3 m, 1265.0 m, 1121.0 m).

II : Normal stress fields whose minimum principal stress strikes roughly NE to SW (Well HT-4, 686.2 m, 916.8 m, 1098 m).

III: Normal stress fields whose minimum principal stress strikes roughly N to S (Well HT-4 923.0 m, 942.0 m, 951.2 m, 964.4 m).

IV : Lateral stress fields whose maximum principal stress strikes roughly E to W (Well HT-1 436.0 m, 786.0 m, 855.0 m; Well HT-4 1456.0 m).

Of these stress fields, Type I may be the oldest and probably corresponds to those which caused the formation of the Beppu-Shimabara Graben in Fig. 1. This type formed the NE-SW trending faults which could not be revealed by a geological survey , but were found by a geophysical survey. Its minimum principal stress, trending NW to SE, later intensified, and Type II stress fields whose intermediate stress strikes NW to SE resulted. Type II stress fields produced the NW-SE trending faults that presently allow the passage of geothermal fluids. On the other hand, the NE-SW fractures might have been so compressed by the Type II stress fields that only a small amount of geothermal fluids was able to flow up along them at that time. Type III stress fields would have occurred during the time of Types I and II on a small scale.

At present, Type IV stress fields seem to control the Otake - Hatchobaru geothermal field. They may act so as to increase the up-flow of geothermal fluids along the NW-SE trending faults, but not along the NE-SW faults, because slickensides dominate on the former fault planes.

From the above discussion, it may be concluded that where two fault systems intersect with high angles only one of the fault systems will be a breeding fault of geothermal fluids under certain stress fields, so that only one of them is of interest for geothermal exploration.

REFERENCES

1. Hayashi, M., Taguchi, S. and Yamasaki, "Activity index and thermal history of geothermal systems", Geothermal Resources Council Transactions, vol. 5, 177-180, 1981.
2. Kyoritsusha, Ltd. "An astatic - magnetmeter, Type NY-2 (An instruction book), 1980.
3. Matsumoto, Y. "Some problems on volcanic activities and depression structures in Kyushu, Japan", Mem. Geol. Soc. Japan, vol. 16, 127-139, 1979.
4. Yamasaki, T., Matsumoto, Y. and Hayashi, M. "The geology and hydrothermal alteration of Otake geothermal area, Kujyu volcano group, Kyushu, Japan", Geothermics, Special Issue 2, 197-207, 1970.

HYDRAULIC FRACTURING AND FRACTURE MAPPING AT YAKEDAKE FIELD

Shogo Suga and Toshinobu Itoh

Mitsui Mining & Smelting Company Ltd.
Tokyo, Japan

Japan Petroleum Exploration Company
Tokyo, Japan

ABSTRACT

The feasibility study of the HDR system in Japan has been made under the Sunshine Project. The experiment field lies in the Yakedake area, where the mapping of artificial fractures due to hydraulic fracturing has been mainly studied with measurement of AE signals. This paper presents the fracture mapping results based on the AE measurement associated with field experiments in 1981.

INTRODUCTION

The concept of the economical extraction of the energy contained in HDR (Hot Dry Rock) systems at accessible depths within the earth's crust was established in 1972 at the Los Alamos Scientific Laboratory in the United States[1]. The feasibility study of the HDR system in Japan started in 1974 under the Sunshine Project. Field tests have been made at the Yakedake area in middle Japan since 1977.

For effective energy extraction through the HDR closed-loop, injection and production wells must be well connected with the reservoir fracture. It is thus of great significance to grasp the precise fracture geometry created. Hydraulic fracturing which has already been utilized in petroleum fields, may be the best in creating the fracture from not only the technical point of view but also the economical standpoint.

In the application of the hydraulic fracturing to the HDR system, it is thought that the so-called Griffith cracks propagate to a large fracture[2,3]. Due to the rock fracture around the crack, infinitesimal earthquakes called AE (Acoustic Emission) occur. The AE wave analysis may then enable us to estimate the fracture geometry[4]. In this paper, we report the fracture mapping results associated with measured AE signals through the hydraulic fracturing at the Yakedake area.

YAKEDAKE TEST FIELD

Our test field lies 4 km west-southwest from Mt. Yakedake, an active volcano in the middle Japan mountainous region (Fig. 1). Host rock to be hydraulically fractured belongs to the Hirayu formation in the Permian Mino group, and is mainly composed of the alternation of sandstone and clay slate.

Fig. 2 shows the placement of test equipment in the Yakedake field. We practice hydraulic fracturing in the HY-well in the figure. The AE data are measured at four wells from S-1 to S-4 by using downhole sondes with AE sensors. To estimate the AE sources by assuming the isotropic AE wave velocity appropriately, at least four detecting points are needed. In Fig. 3, we show the schematic diagram of the fracture mapping based on the AE measurement.

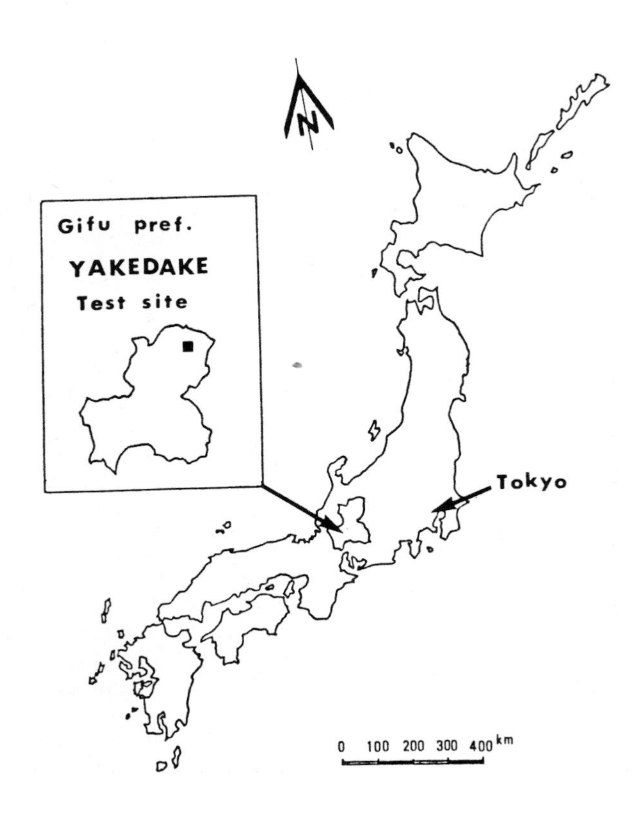

Fig. 1. Location of Yakedake hydraulic fracturing test site in northern Gifu prefecture.

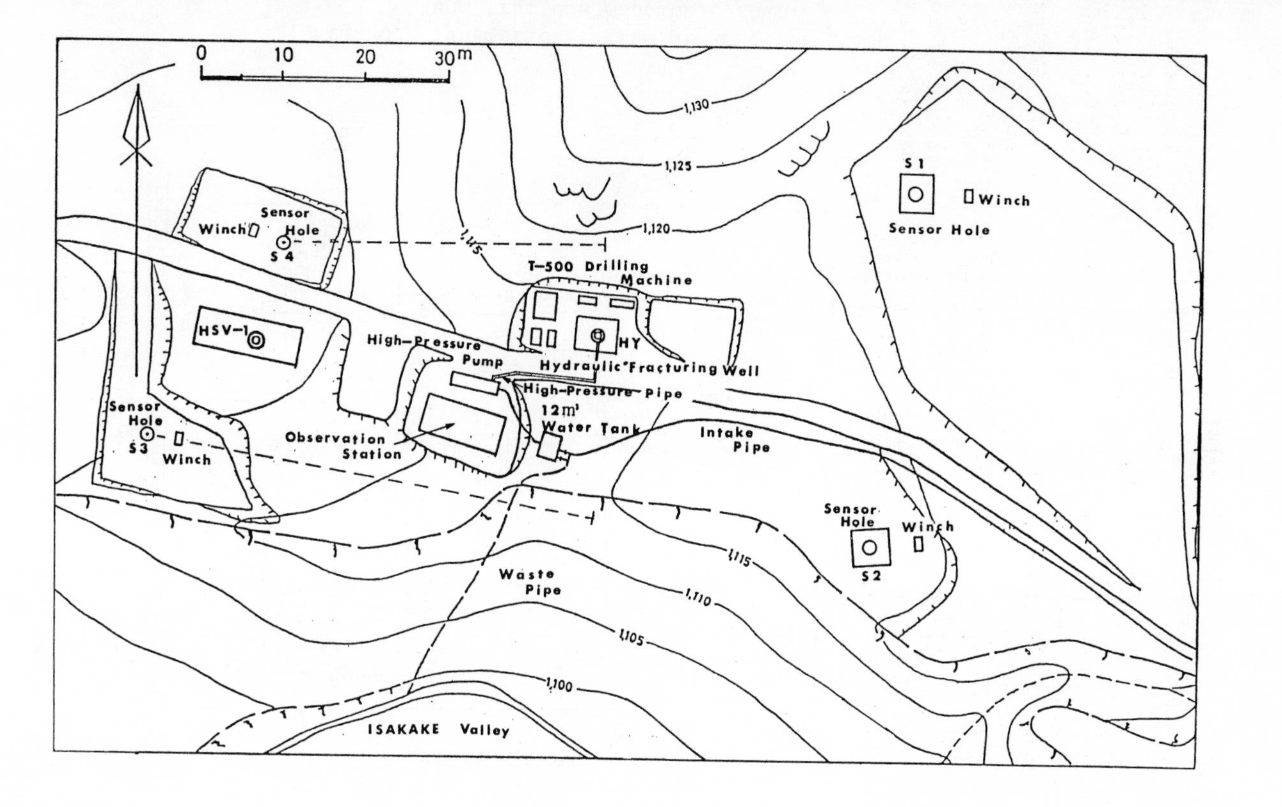

Fig. 2. Placement of test equipment in Yakedake field.

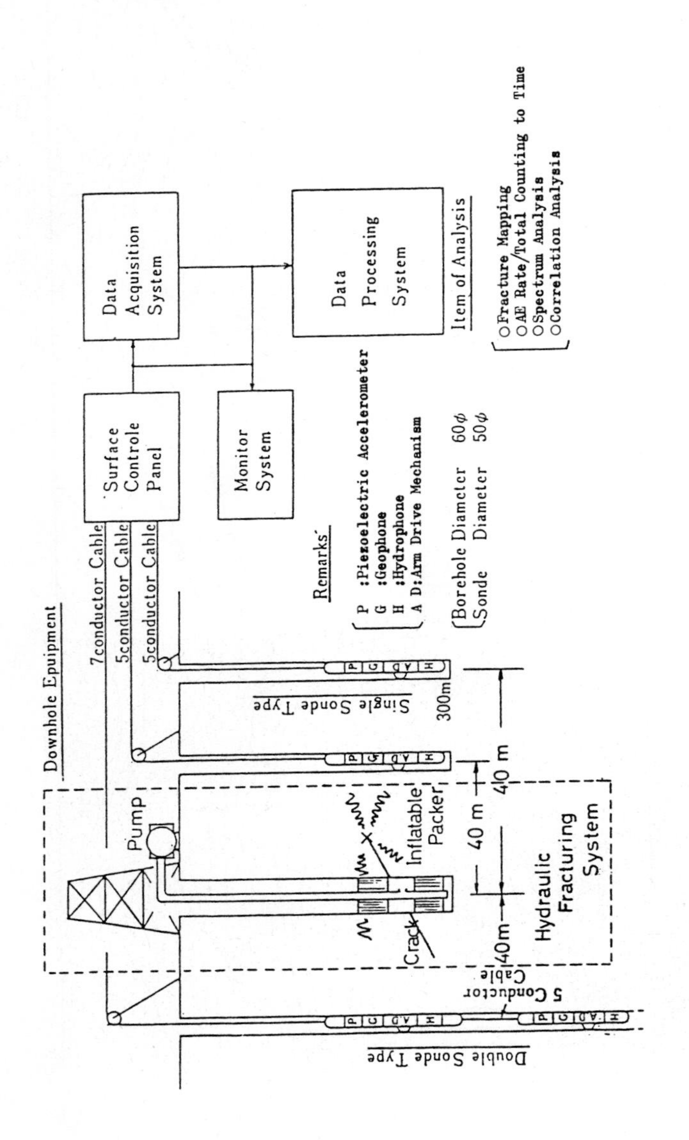

Fig. 3. Schematic diagram of AE observation and fracture mapping system.

HYDRAULIC FRACTURING

In hydraulic fracturing with a high pressure pump, we press the rock through a packer and pour water into the earth in accordance with fracture propagation. The pressure and flow rate are measured and recorded during the fracturing at the ground station.

Fig. 4 shows the Halliburton HT-4000 pump with maximum pressure of 980 kg/cm^2 and with a capacity of 1.36 m^3/min. We use the Lynes dual packer of Fig. 5, the working principle of which is schematically drawn in Fig. 6.

We remark that the pressure history is very important in the stress analysis. Fig. 7 shows a typical p-t diagram. The measurement of reopening pressure is further of significance. By using the breakdown pressure P_b and shut-in pressure P_s during the hydraulic fracturing (see Fig. 7), as well as the reopening pressure, say P_b', we can compute not only the principal stress within the earth but also the tensile strength of the rock.

Fig. 4. Halliburton HT-4000 pump.

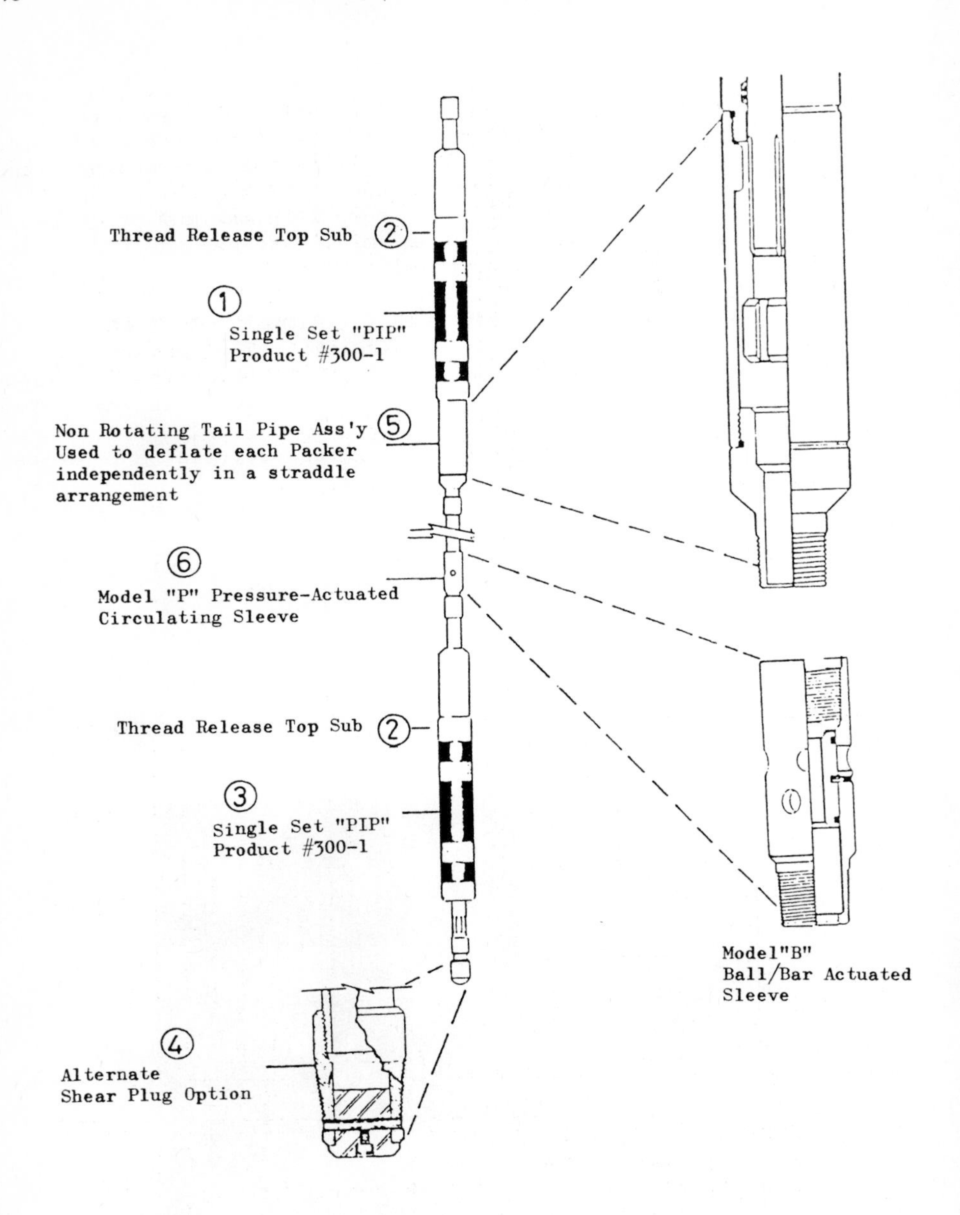

Fig. 5. Assembly diagram of Lynes packer.

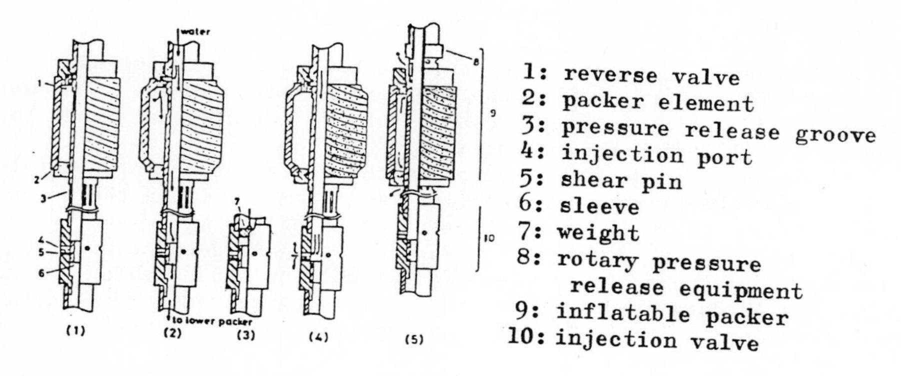

Fig. 6. Schematic diagram for injection due to inflation and deflation of packer.

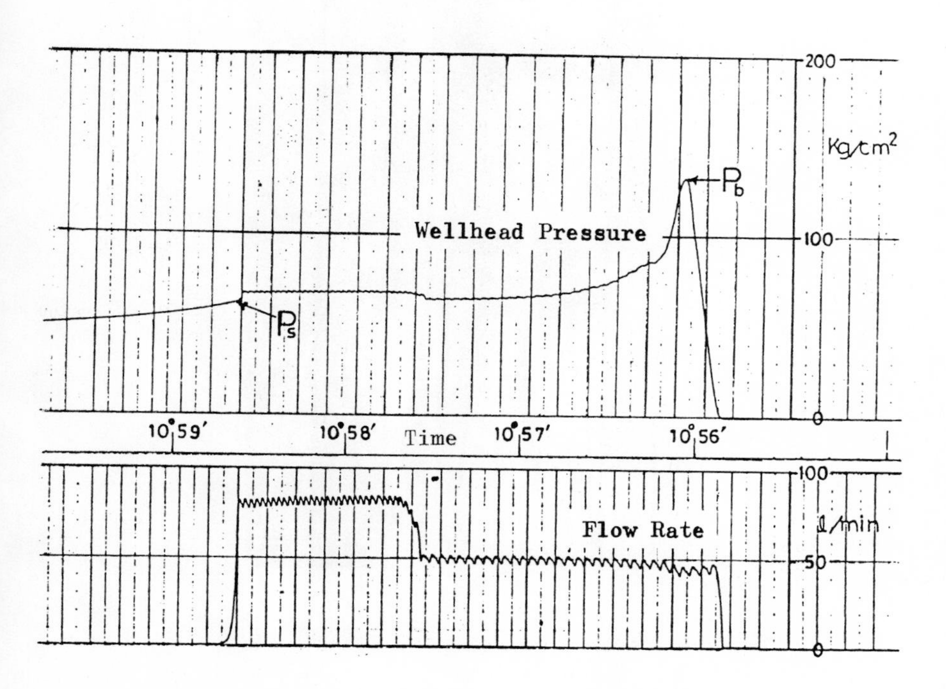

Fig. 7. Pressure profile through hydraulic fracturing.

AE MEASUREMENT

In measuring AE signals, we use the downhole sondes of Fig. 8 with installed AE sensors such as the hydrophone and geophone. The water-proof sonde with resisting pressure of 100 kg/cm^2 is attached to the borehole wall during the measurement.

The downhole sonde is connected with the armored cable containing signal and power lines. In the field measurement within the borehole, the winch of Fig. 9 is used with a capacity of 130 kg.

We connect the armored cable with the AE data acquisition and processing systems by the use of a doubly shielded cable. Fig. 10 shows the AE data acquisition and processing systems, with detailed configurations of Figs. 11 and 12, respectively.

Fig. 8. Downhole sondes after assembly.

Fig. 9. Winch system for downhole sonde.

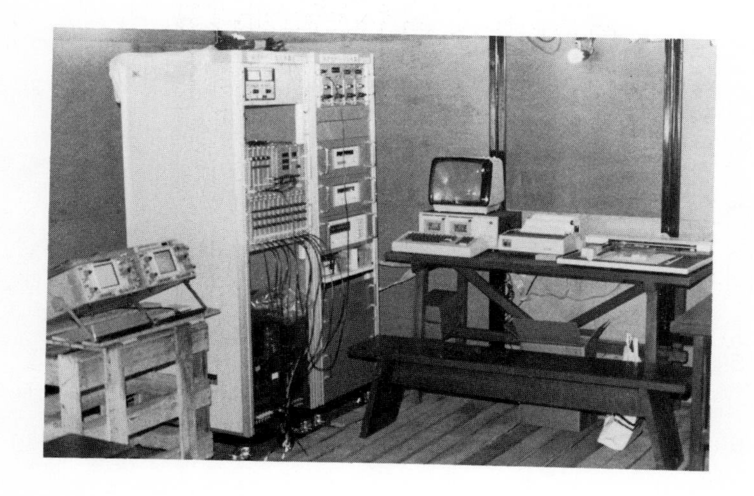

Fig. 10. AE data acquisition and processing systems.

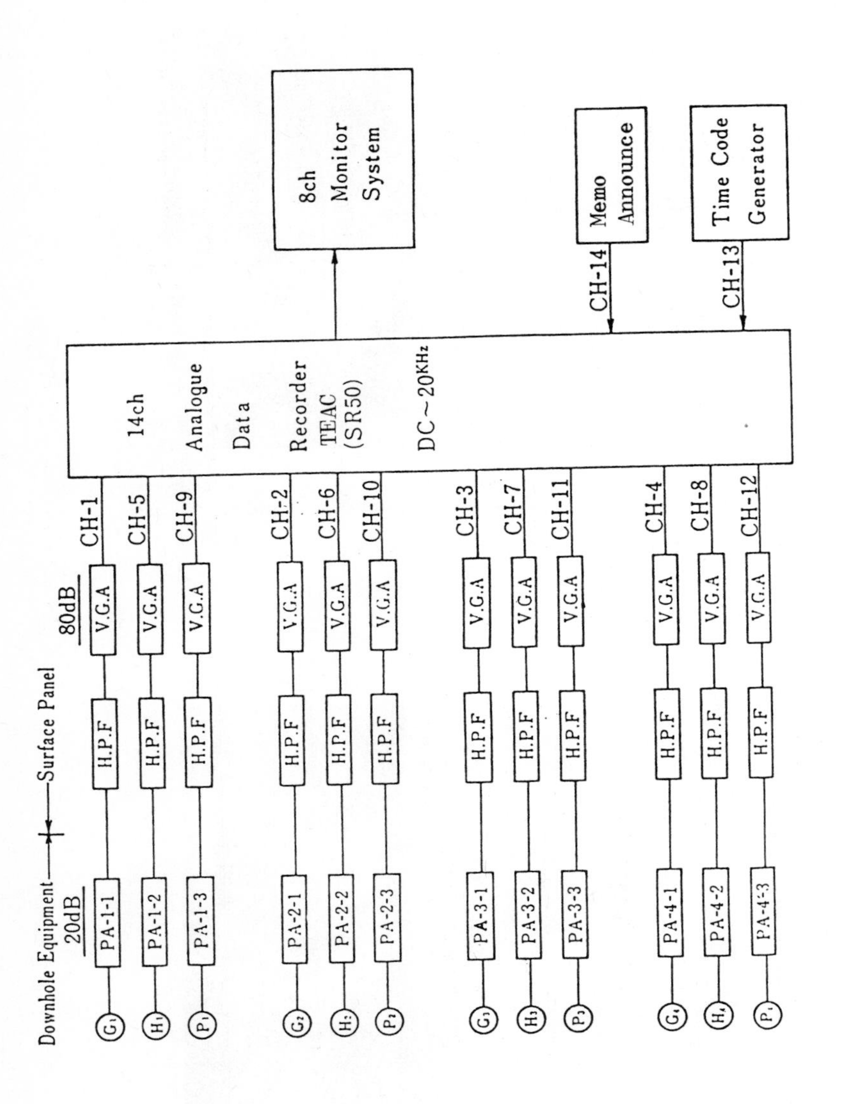

Fig. 11. AE data acquisition system.

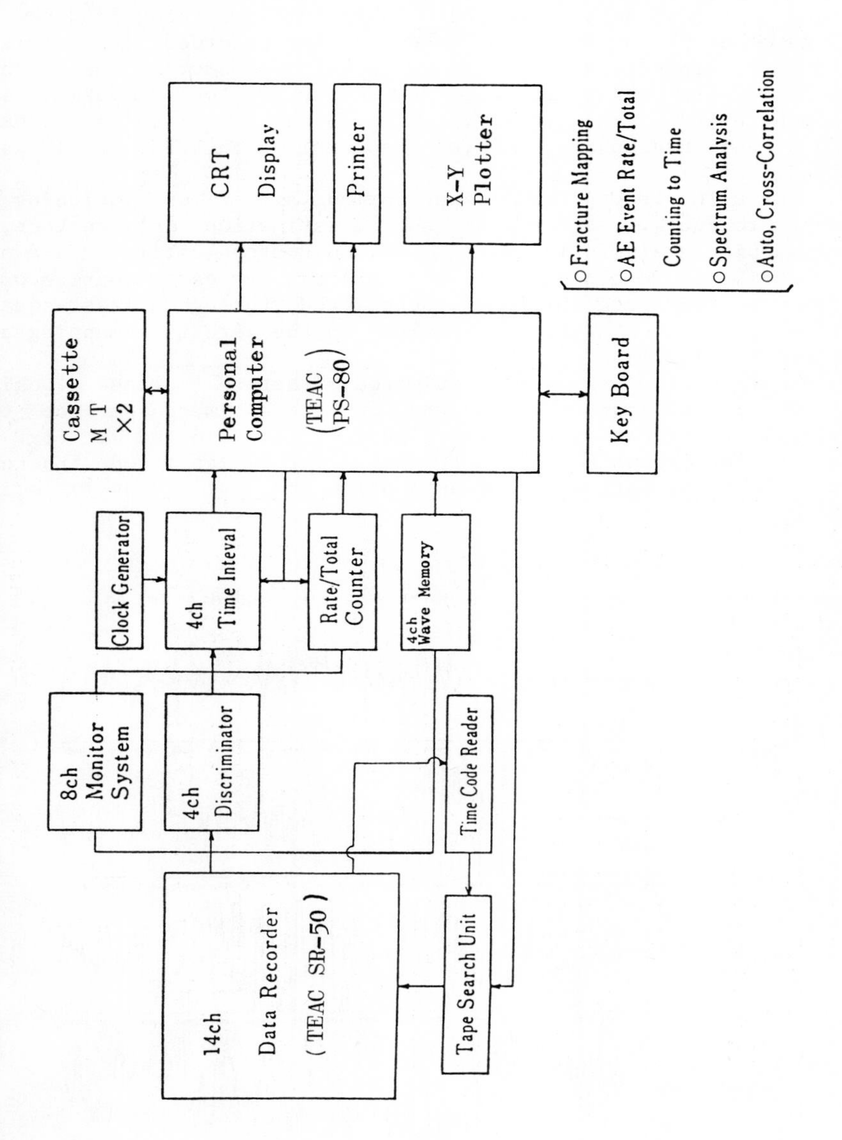

Fig. 12. AE data processing system.

FRACTURE MAPPING

Among measured AE data, the 985 events that are recognized not as noise but as signal over three channels are recorded in a digital cassette tape. In estimating the corresponding AE sources, we further pick up 127 events with clear first break for all four channels, and plot them by using an X-Y plotter. Fig. 13 shows typical AE signals measured by the hydrophone. Here the sampling interval of 10 μsec is adopted.

The hydraulic fracturing is performed six times including a reopening procedure. For the AE source estimation in each test, a fracture mapping analysis is made. The AE sources estimated are not always distributed on a plane. In our analysis, we can recognize only three fracture surfaces which are determined through a least-square regression analysis. Typical results on the fracture mapping are illustrated in Fig. 14.

The cross-section shows an estimated fracture surface looking N3°E. We also contour the fracture on the estimated surface for clarity, see the front view in the figure. It is noted in Fig. 14 that the packer point P does not fall on the estimated fracture surface. This may be due to our assumption that the AE wave velocity is isotropic.

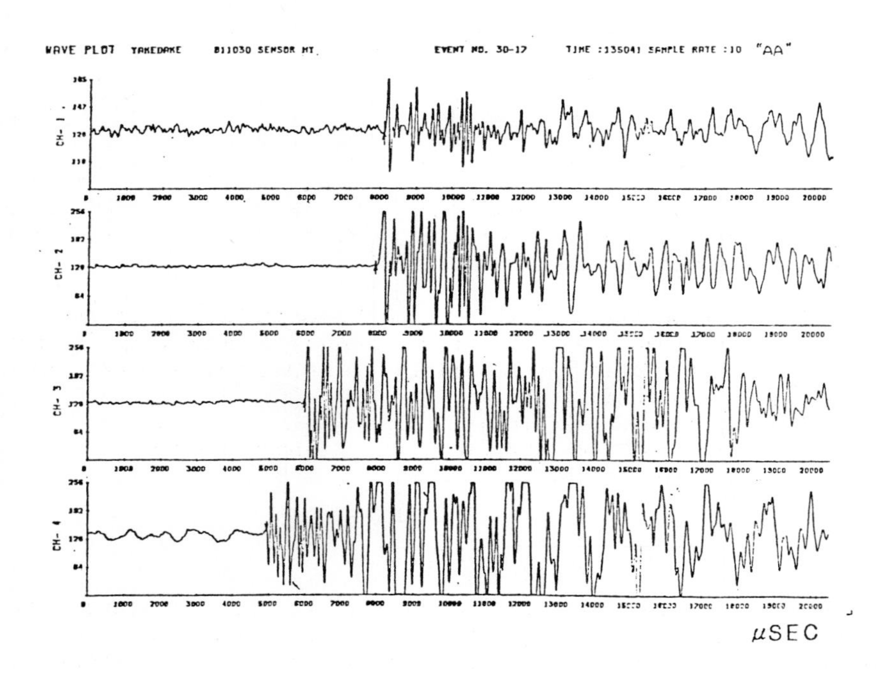

Fig. 13. Example of AE signals measured by hydrophone.

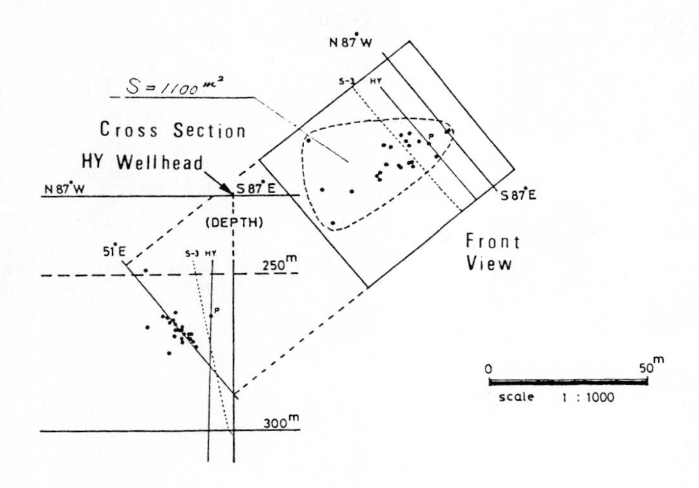

Fig. 14. Fracture mapping example.

CONCLUDING REMARKS

We performed hydraulic fracturing at the Yakedake field. It is proved that fracture mapping can successfully be made by the AE wave observation. However, we have many problems to solve:

1) In our experiments, at least four detecting points are needed for the fracture mapping. A three-axis geophone sonde, like the one of LASL, will enable us to perform the fracture mapping analysis with only one detecting point.
2) Effects of natural fissures should be clarified.
3) Further remedies for the high temperature in drilling and packing must be found.

ACKNOWLEDGMENT

The authors are grateful to members of the Sunshine Project Promotion Headquarters, Agency of Industrial Science and Technology, Ministry of International Trade and Industry, for permission to publish this paper.

REFERENCES

1. Brown, D.W., Smith, M.C. and Potter, R.M., "A New Method for Extracting Energy from Dry Geothermal Reservoirs", LADC-72-1157, 1972, Los Alamos Scientific Laboratory, NM.

2. Sneddon, I.N., "The Distribution of Stress in the Neighbourhood of a Crack in an Elastic Solid", Proc. Roy. Soc., A187, 1946, pp. 229-260.

3. Hubbert, M.K. and Willis, D.G., "Mechanics of Hydraulic Fracturing", Transactions AIME, vol. 210, 1957, pp. 153-166.

4. LASL HDR staff, "Hot Dry Rock Geothermal Energy Development Program, Annual Report, Fiscal Year 1978", LA-7807-HDR, 1978, Los Alamos Scientific Laboratory, NM.

IN-SITU AE MEASUREMENT OF HYDRAULIC FRACTURING AT GEOTHERMAL FIELDS

by Hiroaki NIITSUMA, Katsuto NAKATSUKA, Hideaki TAKAHASHI
Noriyoshi CHUBACHI, Mamoru ABE, Hidekichi YOKOYAMA
and Risaburo SATO

Faculty of Engineering, Tohoku University
Sendai, 980 Japan

ABSTRACT

A triaxial acceleration-sense down-hole AE measurement system was developed and successfully applied to the monitoring of both man-made and natural hydraulic fracture in a geothermal field. The acceleration-sense instrumentation allows long-distance measurement of subsurface crack extension. AE events during hydraulic fracturing in Nigorikawa geothermal field could be detected at a distance of 1600m from the AE source. AE events during a closure operation of a well-head valve in the Kakkonda geothermal field could also be detected by the system. In this paper, the feasibility of the simple measurement system for *in-situ* monitoring of subsurface crack extension in a geothermal field is demonstrated with reference to the fracture mechanics model of a geothermal reservoir.

INTRODUCTION

Hydraulic fracturing has been effectively applied to a well-stimulation technique for both improvement and recovery of a natural hydro-thermal reservoir, as well as fracture formation in a HDR development [1]-[3]. It is important for fracture mechanics design and control of the hydraulically induced reservoir to monitor the fracture process of the rock mass and to determine the configuration of the fracture created.

On the other hand, it has become evident recently for the natural hydro-thermal energy extraction system, that the productivity of geo-thermal wells is highly dependent upon the existence of a fracture or a fault in deeper resevoir rocks[4]. A fracture mechanics study on fracturing behavior of a rock mass[5][6] has shown a decrease of fracture toughness and subcritical crack growth in a rock mass in elevated temperature hot-water environment. These facts indicate that the fracture or fault in a reservoir is always active and there exists a possibility of harmful failure of the geothermal system under inappropriate operations, e.g. failure of the production well, connection between production and injection well, etc.. It is also important for normal operations of the system to estimate the subsurface crack extension in service.

For the last four years, the research group in Tohoku University has investigated the potential for monitoring the subsurface crack extension by means of an acceleration-sense triaxial AE measurement, where a long-distance single sonde measurement in a shallow well has been employed from the viewpoint of convenience and cost.

The measurement system was applied to AE detection in geothermal fields. AE during a massive hydraulic fracturing which was carried out for well stimulation in Nigorikawa, Hokkaido, in May 1, 1980, was measured. Clear AE events were detected at a distance of 1600m from the AE source, and this fact shows the feasibility of a simple measurement system for *in-situ* monitoring of man-made subsurface crack extensions. Another application of the system was the detection of subsurface crack extension during a closure operation of a production well-head valve (named "build-up test") in the Kakkonda geothermal power plant, in July 15 - August 8, 1982. Many clear AE events which suggest the activity of the natural subsurface crack were detected, corresponding to the valve operation.

This paper describes the acceleration-sense triaxial AE measurement system and its applications to monitoring of both man-made and natural hydraulic fracturing in geothermal fields, with reference to the fracture mechanics model of geothermal reservoirs.

FRACTURE MECHANICS MODEL OF GEOTHERMAL RESERVOIR AND FRACTURE MAPPING

Fracture extension behavior of a geothermal reservoir is schematically represented in Fig. 1, on the basis of the fracture mechanics concept. Figure 1 (a) and (b) represent crack extension during hydraulic fracturing, and (c) and (d) show it during in-service operation and closure operation of a well-head valve called a "build-up test". Crack extension of both Mode I and II are shown in the figure. In hydraulic fracturing of Mode I, the "penny shaped" reservoir is created from the pre-existing crack a_0. Crack extensions of Mode II are also expected[7], where a decrease of fracture toughness of the rock mass, ΔK_{IICT}, due to fluid permeation during the fracturing operation, and an increase of K_{II} due to an increase of downhole pressure cause a shear mode crack extension by earth stress. During in-service operation the reservoir is extended by subcritical crack growth $\Delta a(t)$, because the fracture toughness of the rock mass decreases by a chemical attack, which is described as $-\Delta K_{IC}(t)$ and $-\Delta K_{IIC}(t)$ in the figure. Therefore, an increase of downhole pressure by the closure of the well-head valve causes further crack extension of Mode I. Neverthless, the downhole pressure P_T is smaller than the fracturing pressure P_f in Fig. 1 (b). Similarly, Mode II crack extensions accompanied by well-head valve closure are also expected. In this case, a change of hot

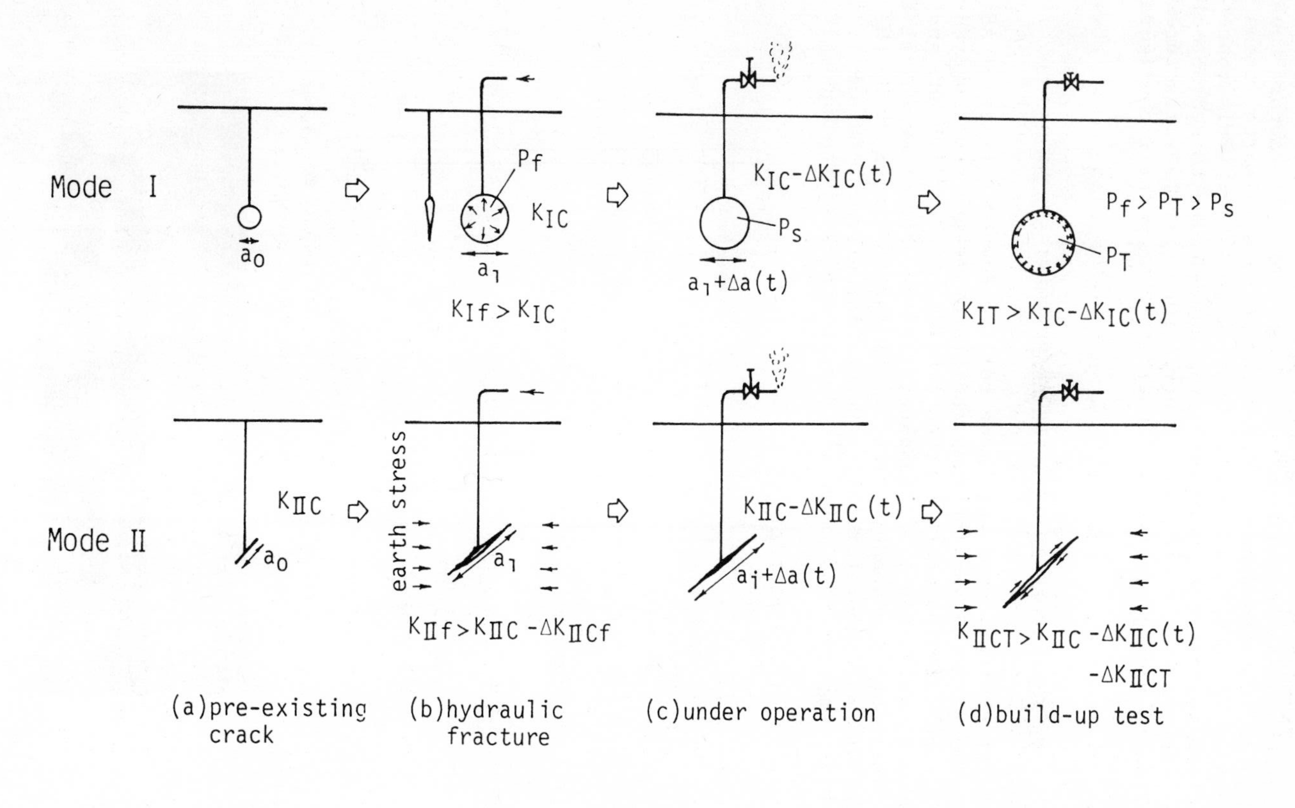

Mode I
Mode II
a_0
K_{IIC}
P_f
K_{IC}
a_1
$K_{If} > K_{IC}$
earth stress
$K_{IIf} > K_{IIC} - \Delta K_{IICf}$
$K_{IC} - \Delta K_{IC}(t)$
P_s
$a_1 + \Delta a(t)$
$K_{IIC} - \Delta K_{IIC}(t)$
$P_f > P_T > P_s$
P_T
$K_{IT} > K_{IC} - \Delta K_{IC}(t)$
$K_{IICT} > K_{IIC} - \Delta K_{IIC}(t) - \Delta K_{IICT}$
(a)pre-existing crack
(b)hydraulic fracture
(c)under operation
(d)build-up test

Fig. 1.

water flow near the crack causes a decrease of the fracture toughness $-\Delta K_{IICT}$, so that the shear mode fracture occurs by earth stress and the increase of K_{II} due to increased downhole pressure. This concept can be applied to both HDR and steam- or liquid-dominated geothermal reservoirs. In order to maintain or control the subsurface fractures in geothermal systems, it is very important to know their configuration and behavior.

DOWNHOLE TRIAXIAL AE MEASUREMENT SYSTEM

Two types of AE sonde were made for the downhole triaxial AE measurement. Figure 2 (a) and (b) show the prototype and the improved-type AE sonde, respectively. Both of them consist of a cable head, and electronic circuit package, a transducer housing, and a mechanical fixing arm. The prototype sonde, which is similar to that made by the Los Alamos National Laboratory[8], has one fixing arm. The arm is driven by a high temperature DC motor so that the back of the sonde is pressed

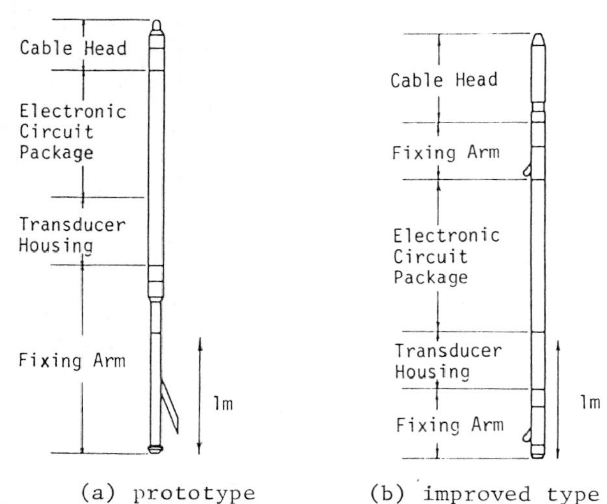

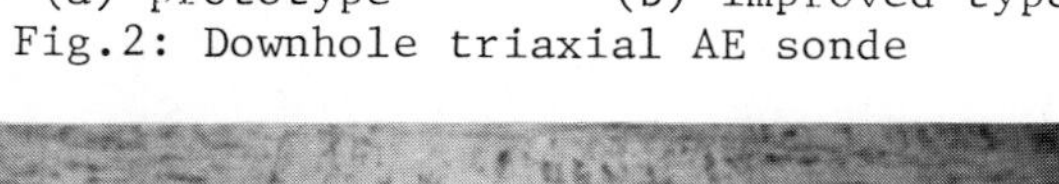

(a) prototype (b) improved type
Fig.2: Downhole triaxial AE sonde

Fig. 3: Fixer of the improved-type sonde

against the borehole wall. The weight of the sonde also increases the holding pressure. Although this fixer is simple and applicable to bore holes of various sizes, it has the following difficulties: (a) holding characteristic is sometimes insufficient against the motion perpendicular to the arm, and (b) bending or precessional motion occurs, if the borehole is not straight and smooth. In order to avoid these difficulties mentioned above, four fixing arms were set in the improved AE sonde. Two fixers, each of which has two arms mounted at an angle of 120° with each other, are assembled at both the upper and the lower part of the sonde. Figure 3 shows the two arm fixer. The performance of the fixing system was satisfactorily confirmed by the model experiment described later. Type 316 stainless steel pipe is used as an outer case on both types of sonde which were designed for the operation at maximum 150°C and 20MPa.

Piezoelectric accelerometers were used as AE transducers here. The acceleration-sense instrumentation allows the highly sensitive measurement for higher frequency components of AE, which is useful for the detection and the location of the fracture. On the other hand, lower frequency background noises and micro seismic waves from other distant places can be suppressed in the acceleration-sense instrumentation. Twelve piezoelectric accelerometers (Node A-26 type, charge sensitivity : $30pCs^2/m$, resonant frequency: higher than 10kHz, maximum operating temperature: 150°C) were mounted to the transducer housing. For triaxial measurement, three groups of transducers were directed at right angles from each other. In each group, four transducers were connected electrically in parallel. Three charge sensitive preamplifiers were mounted in the electronic circuit package. The gain of the preamplifier can be changed by a command signal from ground surface. Overall sensitivity and frequency range of the sonde are 270 or $27vs^2/m$ and 7Hz-7kHz, respectively. The AE signal is amplified by 10-50dB in the ground surface station where the data recording, waveform monitoring, audible sound monitoring, event-ringdown counting, and data printing out are carried out. The recorded signals are processed on a personal computor.

Three types of instrument cables were prepared. The first is commonplace cabtyre cable of 30m containing three coaxial cables and twenty-four lines, and used for measurements in low-temperature and shallow boreholes. The second and third are multipurpose armored cables which have been prepared recently. The maximum operating temperature of the cables is about 260 °C. The former is 150m in length and contains three coaxial cables and ten lines. The latter is 1050m in length and contains three coaxial cables and seven lines. The cross section of the last type of cable is shown in Fig. 4.

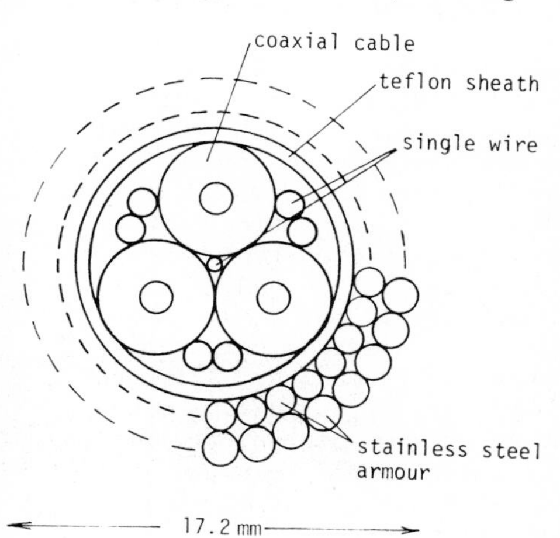

Fig. 4: Cross section of the instrument cable

Fig. 5: Instrument car carried a duplex winch

An instrument car which carries a duplex winch for the armored cables has been developed quite recently and is shown in Fig. 5.

In order to investigate the fixing performance of the sonde, a model experiment was carried out in a quarry. A blasting at a distance of 40-80m was used as an AE source. The AE sonde was set at a depth of 4m by the fixing arm. A small triaxial sonde was made and fixed for calibration to an adjacent borehole by cementing. Results for the improved-type sonde are shown in Figs. 6 and 7. The resonance in the Z axis shown in Fig. 6 is considered to be borehole resonance. Although there exists a spurious movement to some degree, the motion of the AE sonde is approximately similar to that of the calibration sonde.

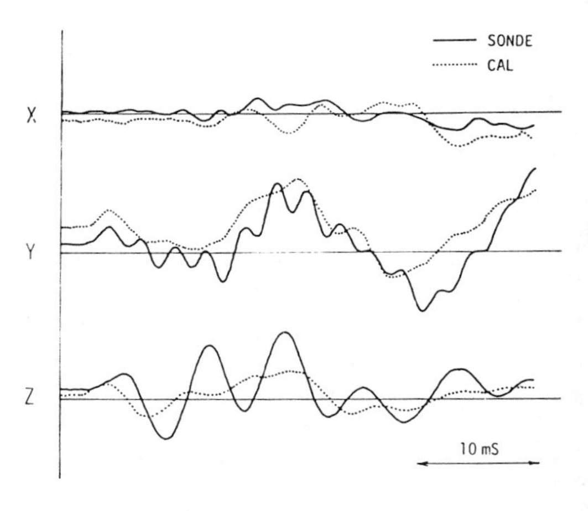

Fig. 6: Comparison of detected waveform

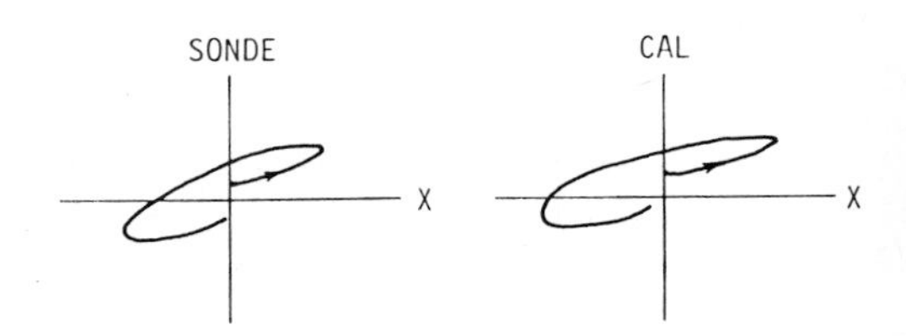

Fig. 7: Comparison of lissajou pattern

DETECTION OF AE DURING HYDRAULIC FRACTURING IN THE NIGORIKAWA GEOTHERMAL FIELD

Hydraulic fracture was applied to well-stimulation of a natural hydro-thermal reservoir (liquid- and steam-dominated reservoir) in the Nigorikawa geothermal developing field, Hokkaido, in May, 1980 [1][9][10]. Long distance AE measurement during the hydraulic fracture was carried out. Figure 8 shows a schematic view of the fracturing well, the subsurface structure, and the expected fractured location with respect to the AE sonde emplacement near the ground surface. The maximum lost circulation during well drilling operation, which indicates an existence of a large pre-existing crack, was observed at the depth of about 1400m. The crack extension from this location during the hydraulic fracturing was expected. The prototype 1-arm AE sonde was locked into the surface rock at the bottom of a 30m well. The distance between sonde and the expected fracture location is about 1600m. AE monitoring had been continuing for 3.5hr during the hydraulic fracturing operation, where the duration of hydraulic fracturing operation was about 1.5hr, and before and after the operation the additional AE monitoring was also done to check the background noise level. Both well head pressure and flow rate were also monitored, and the position of breakdown was determined.

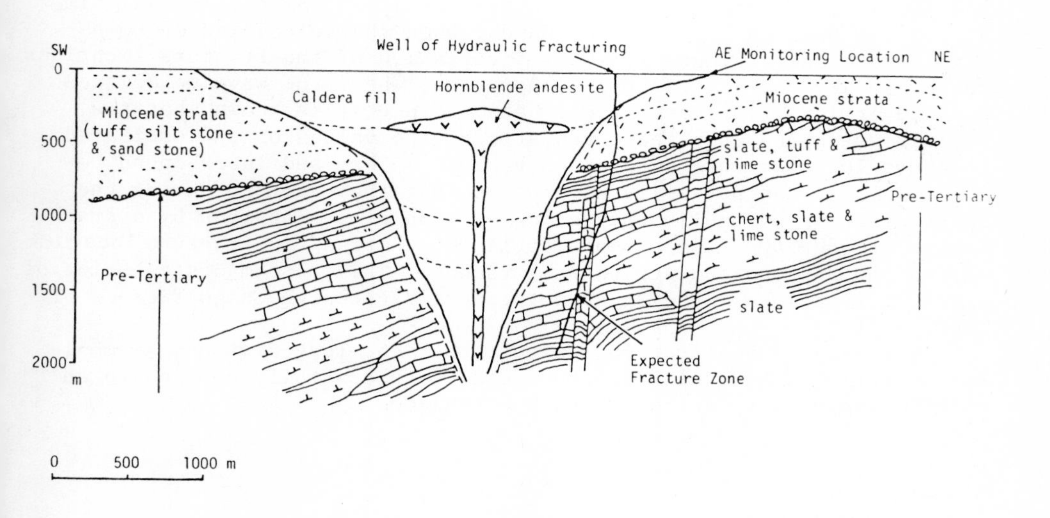

Fig. 8: Schematic view of the fracturing well, the subsurface structure, and the expected fracture zone in the Nigorikawa geothermal field.

Twenty-one AE events could be detected during the hydraulic fracturing by the waveform monitoring. The maximum value of their amplitude is 0.86gal. A typical AE which can be clearly distinguished from noise is shown in Fig. 9. The waveform is a damped wave with fundamental frequency of 37.5Hz. In order to detect the signals covered with background noise, tone color was monitored by ear where the reproducing speed of data recorder was set at four times faster than the recording speed. Twenty-two additional small signals can be detected by this method. Since no AE event was observed during one hour before and after fracturing, the observed AE signals were proved to be generated during the hydraulic fracturing process. Figure 10 shows the pressure- and flow rate- time curves together with AE events and their ringdown counts. Normally a hydraulically induced fracture formation is estimated by a pressure drop, denoted as a breakdown, when the pressure drops abruptly with increasing flow rate. As understood in this figure, however, one cannot determine the breakdown point clearly only from the pressure change.

An AE event would be generated potentially by tension elastic rebound at the fracture boundary and rock bursts near the fracture zone. Therefore, the AE technique is effectively used for the monitoring of the abrupt breakdown during hydraulic fracturing.

Figs.11 (a) and (b) show the lissajou patterns between the X and Y waves, where (a) is the pattern of the first 9-10 cycles and (b) is that of the next 10 cycles. The arrival of a shear wave could clearly be detected by the hodogram. The distance between the fracture zone and the AE monitoring location can be estimated from the arrival time difference between the P- and S-waves, when the velocities of both waves are known. Figure 12 shows a variation of the fracture locations during the hydraulic fracturing operation, where the wave velocities were determined from core sample data. Since it is known that the velocities depend upon the earth stress, the values of the estimated fracture location were determined by both upper- and lower-bound velocities. It can be easily understood that the fracture during the first stage of hydraulic fracturing extends always upwards by a crack length of 480-560m from a depth of 1400m, whereas the fracture locations during the 2nd stage distribute at random. This random distribution of fracture locations in the 2nd fracturing indicates the significance of blocking[1].

These experimental results suggest that the long-distance AE monitoring during a hydraulic fracture can be successfully applied to an *in-situ* evaluation of the fracture extension.

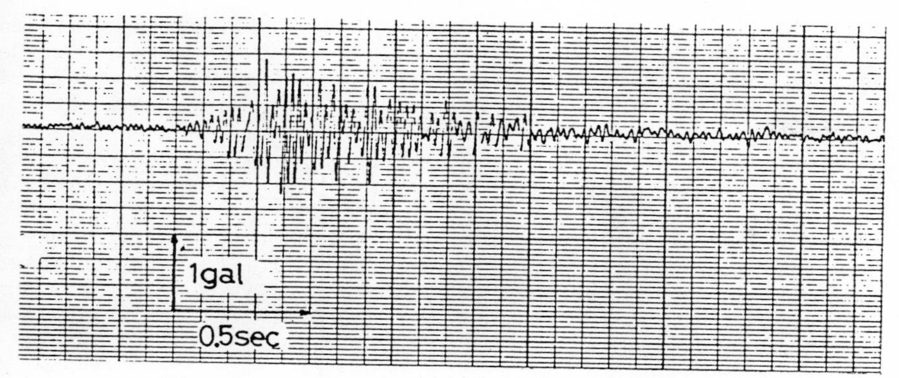

Fig. 9: Typical waveform of detected AE event

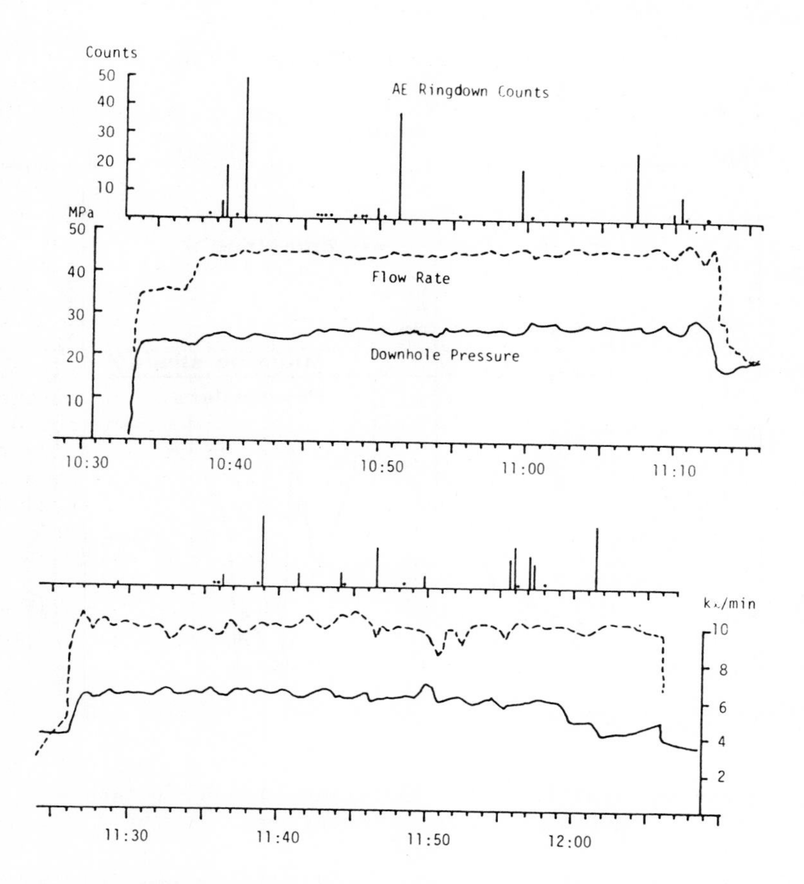

Fig. 10: Downhole pressure and flow rate versus time curve along with AE ringdown counts (point marks · indicate AE events detected by tone color monitoring)

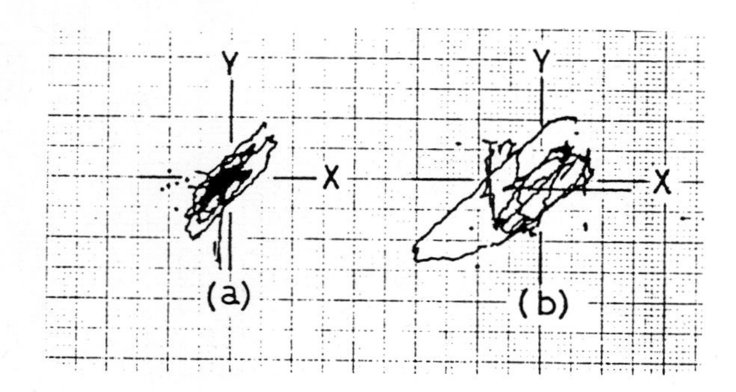

Fig. 11: Typical lissajou patern of X and Y components
(a) first 9 - 10 cycles (b) next 10 cycles

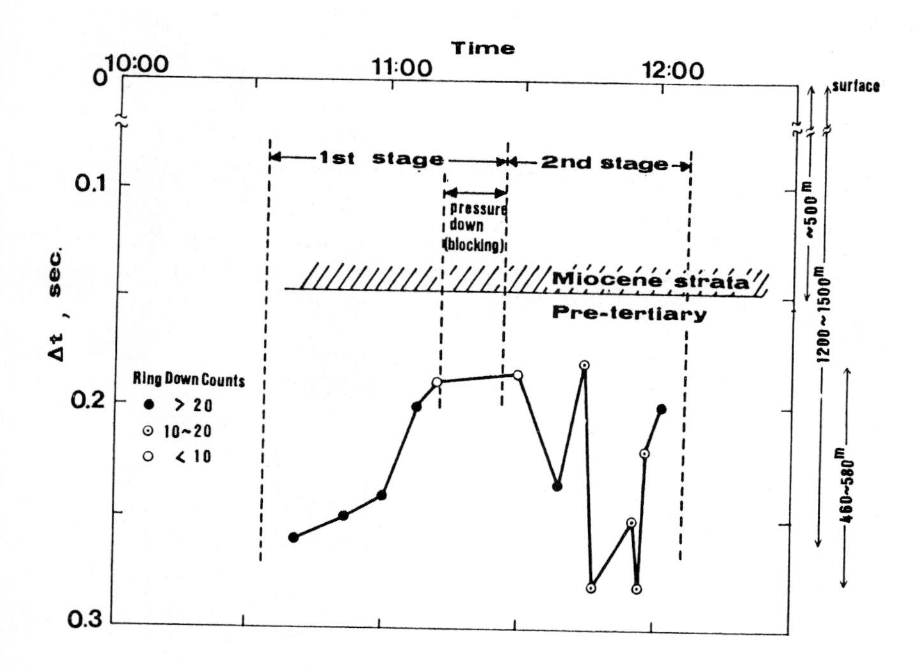

Fig. 12: Change of arrival time difference between P- and S-waves in AE event during hydraulic fracturing

DETECTION OF AE DURING A BUILD-UP TEST OF PRODUCTION WELL IN KAKKONDA GEOTHERMAL POWER PLANT

Acoustic emissions during a build-up test of production wells in the Kakkonda geothermal power plant, which is one of the largest geothermal plants in Japan and situated in the Iwate prefecture, were measured on July 15 - August 8, 1982. Figure 13 shows a profile of the subsurface structure of this area[4]. Eleven wells are being used for the production of 50MW electric power generation. The geothermal fluid is considered to be produced from fractures or faults in deeper reservoir rocks for the area[4]. Since the geothermal fluid is liquid-dominated, fourteen reinjection wells are used for the disposal of hot water.

The build-up test of production wells, which had been used for about one year, was carried out. The well-head valves of the four wells (E-1, C-5, B-5 and B-3) were closed successively within five hours. The crack extensions were expected during the test, as mentioned above.

The improved-type 4-arm AE sonde was fixed in a well of 15m depth, which had been drilled near the wells of CR-4 and C-3 in Fig. 13. The distance from the expected fracuture zone is 1000-1500m.

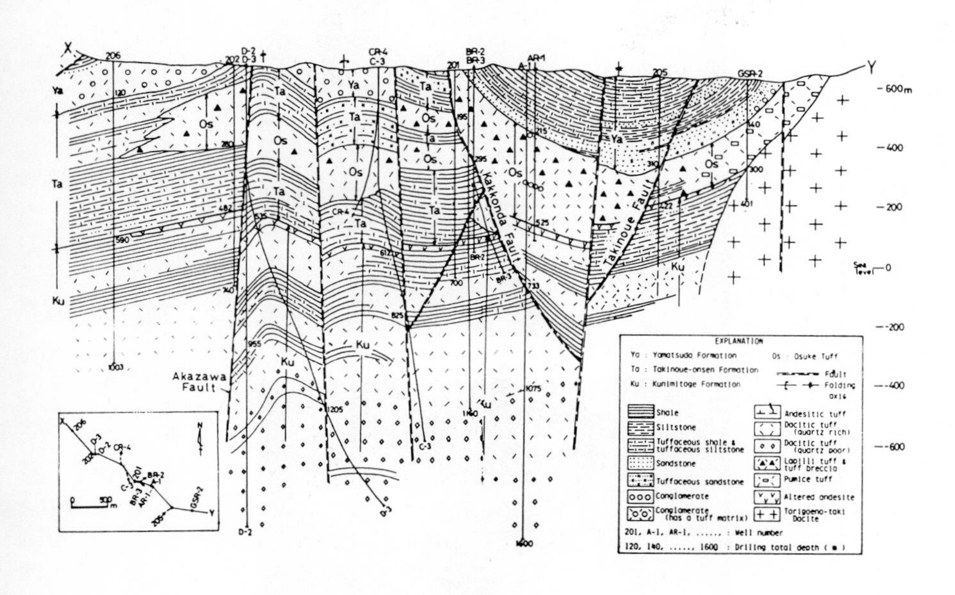

Fig. 13: Profile of the subsurface structure of Kakkonda geothermal field [4]

487 clear AE events were detected in six hours during the test. This fact leads to a conclusion that the subsurface crack extensions actually occur during the well-head valve operations. The maximum amplitude of the events was greater than 0.9gal (saturated). Figure 14 shows the typical waveform of the events. The signal is a damped wave with fundamental frequency of 45-95Hz. The difference between the P- and S-wave arrival times ranges from 65 to 212ms. The estimated distance from the AE sources to the sonde are 480-1590m, when the P- and S-wave velocities of 3000 and 5000m/s are used. This estimated result agrees fairly well with the expected distance. Figure 15 shows the event energy plotted with the time, where the open circles represent the individual AE events. A change of the well-head pressure and flow-rate in their wells are also shown. As shown in this figure, the generation of acoustic emission is closely related to the valve operation.

Thus the active motion of subsurface fractures during the valve operation has been confirmed, although detailed data analysis and mapping of the fracture zone are still now in progress. The AE measurement technique would provide a very useful tool for the evaluation of the integrity of geological structure in geothermal energy systems.

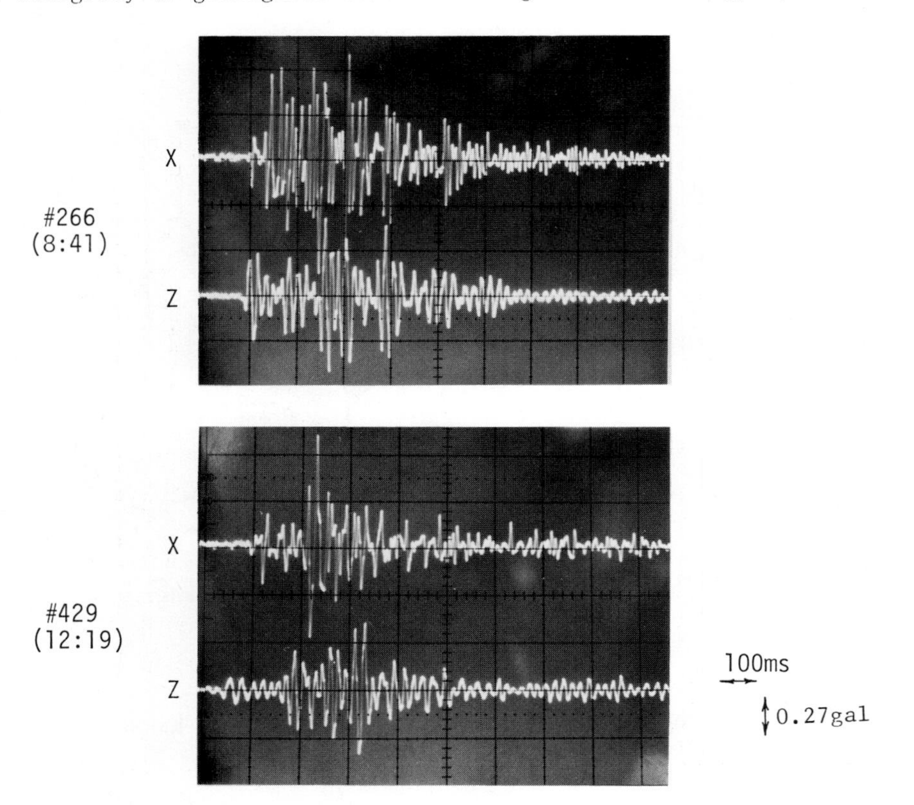

Fig. 14: Typical waveform of detected AE events during the build-up test of production well

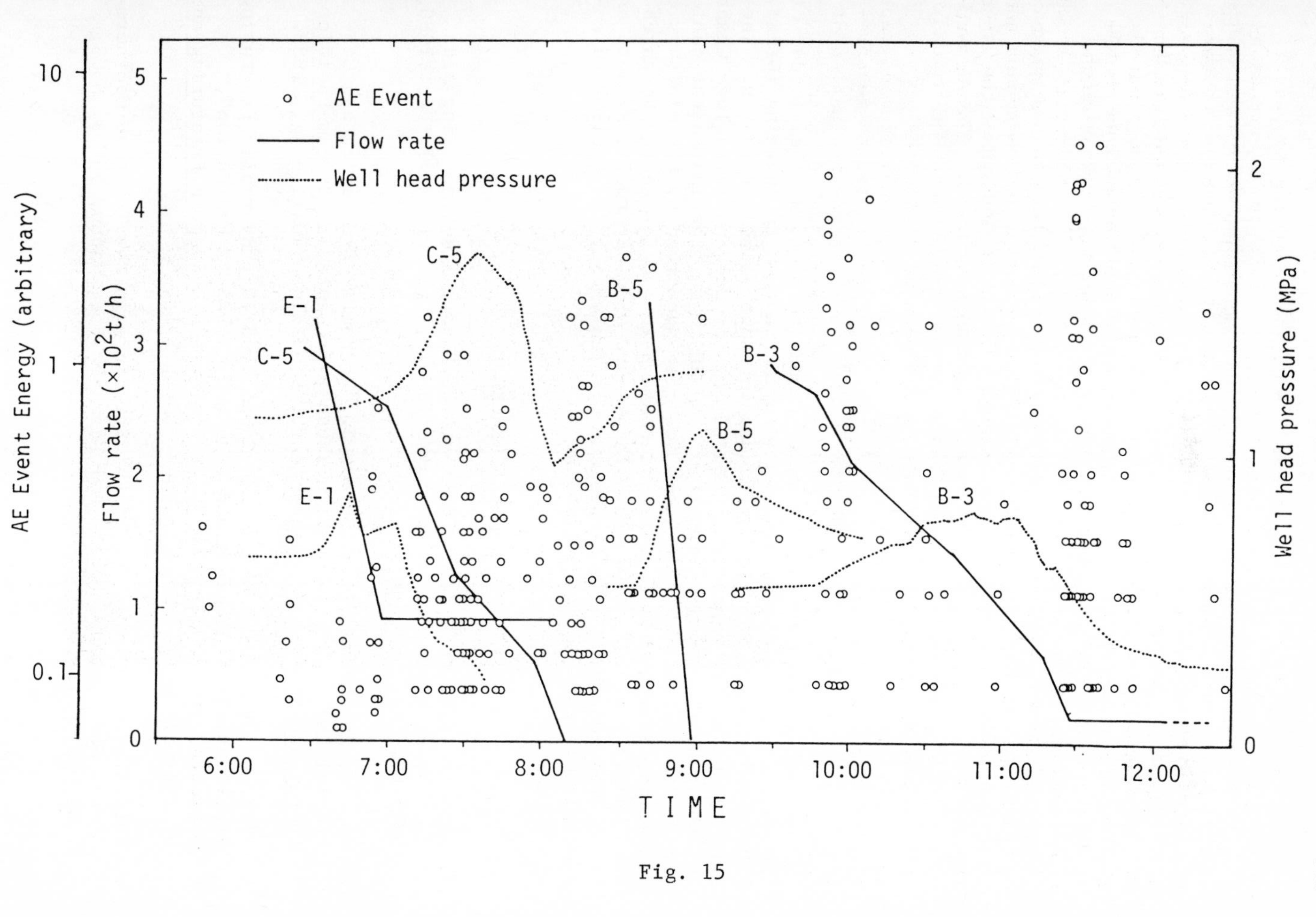

AE Event
Flow rate
Well head pressure
AE Event Energy (arbitrary)
10
1
0.1
Flow rate (×10²t/h)
0
1
2
3
4
5
Well head pressure (MPa)
0
1
2
6:00
7:00
8:00
9:00
10:00
11:00
12:00
TIME
E-1
C-5
C-5
E-1
B-5
B-5
B-3
B-3

Fig. 15

CONCLUSION

With special reference to an experimental verification of the fracture mechanics crack model for geothermal reservoir, the tri-axial acceleration-sense AE measurement system for monitoring of subsurface crack extension and its application to both man-made and natural hydraulic fracture monitoring has been described.

The feasibility of the AE measurement system for subsurface crack extension monitoring has been demonstrated by the experiment in the Nigorikawa geothermal field.

Many AE events were detected during the build-up test of the production well in the Kakkonda geothermal field. Acoustic emission rate was closely related to the valve operations. This fact indicates the active motion of subsurface structure in geothermal fields. The AE monitoring technique and fracture mechanics would provide a useful tool for controlling fractures in geothermal systems.

ACKNOWLEDGMENT

The authors wish to express their thanks to prof. H. Abé, leader of the research group of Geothermal Energy Extraction Engineering (GEEE), Tohoku University, for the continuous interest in this research. The work was partially supported by the Toray Science and Technology Grants for the research and development of the AE sonde, the instrument cable and the instrument car. The work was also supported by the Ministry of Education, Science and Culture under Grant-in-Aid for Energy Research Nos.505514, 56045015, 57045015. Thanks are also due to Dr. K. Sato and Mr. M. Takanohashi, Japan Metals and Chemicals Co., for their kind cooperation in the field experiments.

REFERENCES

1. Katagiri,K., Ott,W.K. and Nutley,B.G.,"Hydraulic Fracturing Aids Geothermal Field Development", World Oil, Vol.191, No.7, Dec. 1980, pp75-88.

2. Abé,J., Takahashi,H. and Suyama,J.,"Hydraulic Fracturing and Geothermal Energy Development in Japan", presented at Circum-Pacific Energy and Mineral Resources Conf., Honolulu, Hawaii, Aug., 1982.

3. Nakatsuka,K., Takahashi,H. and Takanohashi,M.,"Hydraulic Fracturing Experiment at Nigorikawa and Fracture Mechanics Evaluation", presented at 1st Japan US Joint Seminar on Hydraulic Fracturing and Geothermal Energy, Tokyo, Japan, Nov., 1982.

4. Sato,K.,"Analysis of Geological Structure in the Takinoue Geothermal Area", Jour. of the Geothermal Research Soc. Japan, Vol.3, No.3, 1982, pp135-148.

5. Hashida,T., Yuda,S., Tamakawa,K. and Takahashi,H.,"Determination of Fracture Toughness of Granitic Rock by Means of AE Technique", Proc. of 6th Intern. AE Symp., Tokyo, Japan, Nov. 1982 (to be published).

6. Takahashi,H.,"Fracture Toughness Evaluation in the Presence of Pressurized Water at Elevated Temperatures", presented at 1st Japan US Joint Seminar on Hydraulic Fracturing and Geothermal Energy, Tokyo, Japan, Nov. 1982.

7. Hayashi,K. and Abé,H.,"Opening of a Fault and Resulting Slip Due to Injection of Fluid for the Extraction of Geothermal Heat", Jour. of Geophysical Research, Vol.87, No.B2, Feb. 1982, pp1049-1054.

8. Dennis,B.R., Hill,J.H., Stephani,E.1., and Todd,B.E.,"Development of High-Temperature Acoustic Instrumentation for Characterization of Hydraulic Fractures in Dry Hot Rock", presented at the 22nd Intern. Instrum. Symp., San Diego, California, May, 1976, pp97-107.

9. Takahashi,H., Niitsuma,H., Tamakawa,K., Abé,H., Sato,R. and Suzuki, M.,"Detection of Acoustic Emission during Hydraulic Fracturing for Geothermal Energy Extraction", Proc. 5th Intern. AE Symp., Tokyo, Japan, 1980, pp443-453.

10. Nakatsuka,K., Niitsuma,H., Tamakawa,K., Takahashi,H., Abé,H. and Takanohashi,M.,"In-situ Measurement of the Extension of Hydraulically-formed Fracture in Geothermal Well by Means of Acoustic Emission" Jour. of the Mining and Metallurgical Institute of Japan, Vol.98, No.1129, 1982, pp209-214. (in Japanese).

11. Nakamura,H. and Sumi,K.,"Exploration and Development at Takinoue, Japan", Geothermal Systems: Principles and Case Histories, John Wiley & Sons Ltd, 1981, pp247-272.

EXPERIMENTAL STUDIES ON HEAT EXTRACTION FROM FRACTURED GEOTHERMAL RESERVOIRS

Paul Kruger

Stanford Geothermal Program
Stanford University
Stanford, CA 94305

ABSTRACT

Most of the available energy in hydrothermal systems and all of the available energy in hot igneous rock systems is present in the formation rock. Long-term commercial development of geothermal resources for electric power production will depend on optimum heat extraction from such systems. Studies underway in the Stanford Geothermal Program since 1972 to improve the efficiency of energy extraction from geothermal resources involve a combination of physical and mathematical models. Results of these studies include the development of a simple, lumped parameter heat extraction model to evaluate the potential for recharge-sweep production. Incorporated into the model are results from a series of tests on several rock loadings in a large physical model of a rechargeable hydrothermal reservoir, experiments on shape factor correlations for single, irregular-shaped rocks and assemblies of reservoir-shaped rocks, and experiments on the effect of thermal stressing on rock heat transfer properties. Current efforts involve the improvement of the model to include distributed-parameter analysis of heat extraction using a rock loading of known geometric shape and spacing.

INTRODUCTION

The United States electric utility industry estimates a geothermal electricity-generating capacity by the year 2000 to range from an announced 4480 MW to a possible 10,800 MW (Roberts and Kruger, 1982). Approximately 2000 to 4000 MW will be installed at The Geysers steam field in northern California. The rest must come from liquid-dominated hydrothermal and petrothermal resources which are not yet sufficiently developed as commercial resources in the United States.

Stimulation of geothermal energy production can occur by improved technologies in three areas (Kruger, 1976): (1) modes of utilization; (2) energy conversion; and (3) enhanced resource extraction. Improvements in the first area may be achieved by combining direct use of geothermal heat with electric power production. The possibility of also combining valuable mineral recovery could help alleviate the troublesome problems of scale deposition and corrosion. Improvements in the second area may be achieved with improved flash technology, with the binary fluid turbine-generator, or with wellhead total-flow converters, such as the rotary separator turbine or the helical screw expander.

The third area involves more efficient extraction of geothermal energy from hydrothermal and petrothermal resources. The potential for improving total energy recovery from hydrothermal reservoirs was estimated by Ramey, Kruger, and Raghavan (1973). The data in Table 1 compare the available energy content in hypothetical steam and hot water reservoirs at an initial temperature of 260°C, porosity of 25 percent over a field area 1230 m^3 in extent, with steam enthalpy of 2.33 MJ/kg for a useful life based on pressure decline from 4.7 MPa (at 260°C) to an abandonment pressure of 0.7 MPa (at 164°C). The results show that only 6 percent of the available energy in the steam reservoir is in the geofluid, while 94 percent is in the formation rock. They also noted that non-isothermal extraction of the heat by either flashing in the reservoir or by recycling spent fluid back into the reservoir could significantly enhance energy recovery.

The extraction efficiency of obtaining energy from impermeable hot igneous rock formations is small, but the potential is great (Smith et al., 1975). Kruger (1977) noted that the volumetric energy extractable from hot dry rock with assumed properties of 2600 kg/m^3 density, 1 percent porosity, and 1.30 kJ/kg·K heat capacity can be of the order of 4×10^{15} J/km^3 for a temperature reduction of 200°C with a heat recovery efficiency of 10 percent. The volumetric power extraction would be of the order of 1.4 MW/km^3 for 1 century. The technical challenge of the petrothermal resource is the ability to fracture such volumes of hot rock and to achieve an extraction efficiency of the order of 10 percent with an artificial circulation system.

Since 1972, the Stanford Geothermal Program has been investigating means of enhanced energy recovery from hydrothermal reservoirs. One of the key objectives has been to develop a relatively simple model to estimate the amount of extractable energy from hydrothermal resources under various production strategies. For this objective a physical model of a rechargeable hydrothermal reservoir was constructed. The model has been used with several rock loadings to test alternate modes of production and to develop numerical models of varying complexity for early estimation of energy extraction based on geologic information and thermodynamic properties of the formation.

The research has been accomplished in several phases. The first

Table 1

RELATIVE RECOVERY FROM HYDROTHERMAL RESERVOIRS*

	Steam Reservoir		Hot Water Reservoir	
	Rock	Fluid	Rock	Fluid
Reservoir Mass (kg)	$2.45x10^6$	7,330	$2.45x10^6$	242,100
Abandonment Content (kg)	--	885	--	28,260
Production (kg)	--	6,445	--	213,840
as Steam	--	6,445	--	168,740
as Water	--	0	--	45,100
Available Energy (GJ)	246	16	246	106
Recovery (%)				
of fluid mass	--	87.9	--	88.3
of available energy	--	6.1	--	99.1

* for a hypothetical reservoir of 260°C temperature, 25% porosity, $1230m^3$ volume, 2.33 MJ/kg steam enthalpy, and abandonment pressure of 0.69 MPa (at 164°C). Adapted from Ramey, Kruger, Raghavan (1973).

phase involved a lumped-parameter analysis of energy recovery using three non-isothermal production methods (Hunsbedt, Kruger, and London, 1978): (1) pressure reduction with in-place boiling; (2) reservoir sweep with injection of cold water; and (3) steam drive with pressurized fluid production. The second phase involved the development of a heat transfer model for a collection of irregular-shaped rocks with an arbitrary size distribution. The initial effort by Kuo, Kruger, and Brigham (1978) resulted in the development of shape factor correlations for single irregular-shaped rocks. Later efforts by Hunsbedt et al. (1979) extended the correlations to assemblies of fractured rock. The result was a one-dimensional model to analyze a hydrothermal rock system under cold water reinjection sweep using a single spherical rock of an "effective radius" with a "number of heat transfer units" parameter to predict heat recovery. The current phase of research seeks to improve the heat extraction model in distributed parameter form using a rock loading of known, regular geometric shape to evaluate the range of the "number of heat transfer units" parameter. During this phase, investigation is also underway of the effect of extended thermal stressing on heat transfer properties.

HEAT EXTRACTION BY NON-ISOTHERMAL PRODUCTION

The development of a simple model to evaluate the potential for heat extraction from fractured-rock hydrothermal resources is based on heat transfer properties of individual rock segments surrounded by circulating fluid. Heat transfer characteristics are expected to differ from reservoir to reservoir. A porous sedimentary rock may be in rapid, local thermal equilibrium with the geofluid, whereas in large, fractured igneous rocks, heat transfer is slow resulting in a large fraction of thermal energy left in the rock and rapid cooling of the surrounding fluid.

The size and shape of individual rock segments vary widely. Hunsbedt, Kruger, and London (1978) showed that heat extraction from large, irregular-shaped rocks can be modeled as if they were equivalent spherical-shaped rocks. The difference in temperature between a rock segment at a lumped mean temperature, $\overline{T}_r$, and the surrounding fluid at a temperature, T_f, for a linearly decreasing reservoir fluid temperature is

$$\overline{T}_r - T_f = \mu\tau[1-e^{-t/\tau}] \tag{1}$$

where μ = cooldown rate (°C/h)

$\tau = \frac{R_e^2}{3\alpha_r}[0.2+1/N_{Bi}]$ = time constant for the rock (h)

where R_e = equivalent rock radius (m)

α_r = thermal diffusivity of the rock (m^2/h)

N_{Bi} = Biot number of the rock

The thermal diffusivity is given by

$$\alpha = \frac{k}{\rho C} \tag{2}$$

where k = thermal conductivity (J/h·m·°C)

ρ = density (kg/m^3)

C = specific heat (J/kg·°C)

The Biot number of defined as

$$N_{Bi} = \frac{hR_e}{k} \tag{3}$$

where h = heat transfer coefficient (J/h·m^2·°C)

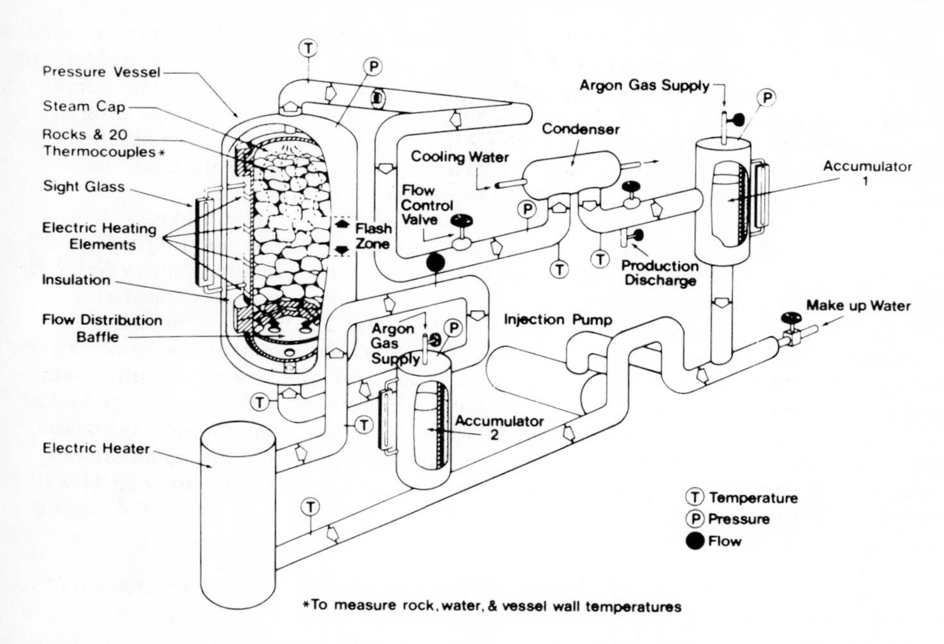

Figure 1. The physical model of a geothermal reservoir in the fluid production mode.

To evaluate the quantitative aspects of energy recovery from fractured rock in hydrothermal resources and to provide data for analytic model verification, a large physical model of a geothermal reservoir was constructed. The model was designed to contain sufficient rock mass and fluid at typical reservoir pressures and temperatures for sufficient production periods in the laboratory. The reservoir is a steel vessel approximately 1.5m high, 0.6m inner diameter, and a contained volume of 0.5 m^3. A schematic diagram of the system in the fluid production mode is given in Figure 1. Details of model design and operations were reported by Kruger and Ramey (1974).

The system can be brought to the desired initial reservoir temperature and pressure by circulating water through the 25 kW electric heater or by electric tape heaters located on the vessel outside wall. Maximum pressure is 5.5 MPa at 260°C. On attainment of thermal equilibrium of the rock-water-steel system, fluid production is initiated with the flow control valve. Fluid recharge rates and enthalpy can be controlled by the injection pump and the electric heater. Heat losses from the steel vessel can be compensated by control of the electric tape heaters. Locations of the 36 temperature, pressure, and flow sensors are noted in Figure 1. Thermocouples are placed in various locations to measure axial and radial temperature distribution, rock center temperatures, surrounding water temperatures, and steel vessel temperatures.

The early experiments to measure energy extraction efficiency were made for three production methods using two rock loadings of different rock type, fragment size, and bulk porosity. A summary of the rock system parameters is given in Table 2. The three orthogonal length dimensions and the mass of individual rocks were measured statistically. Porosity was determined by drainage. Permeability was considered "infinite" for these loosely-packed loadings.

The results of 35 production runs to test a lumped parameter model of energy recovery in the three modes of production were reported by Hunsbedt, Kruger, and London (1977). The ranges of experimental conditions for these experiments are given in Table 3. The first three were run without rock to measure the steel vessel cooldown rate. The next 22 runs were with the gabbro rock and the last 10 runs were with the larger granite rock. The major observed parameters included the temperature distribution in the model as functions of height and time, the extent of energy extraction from the rock, the rock-steam temperature difference, the fraction of total energy produced, the effect of liquid level height, and the effects of hot and cold water recharge on production.

Table 4 lists the three types of energy extraction experiments. The in-place boiling experiments consisted of steam production from a well penetrating a top producing zone. Experiments were run with and without fluid recharge at the bottom. The sweep experiments consisted of hot-water production from a top producing zone with cold-water injection at the bottom. The reservoir contained subcooled liquid during production. The steam drive experiments consisted of hot-water production from the bottom with no recharge. Steam and non-condensable gas pressure above the liquid/steam interface provided the steam drive.

Table 2

SUMMARY OF ROCK SYSTEM PARAMETERS*

Parameter	Rock Type	
	Gabbro	Granite
Mean Equivalent Diameter (cm)	2.52	6.73
Mean Thickness (cm)	1.88	4.11
Mean Thickness/Length Ratio	0.50	0.46
Mean Solid Density (g/cm^3)	2.79	2.61
Drainage Porosity (%)	44	35
Permeability (darcy)	∞	∞

*Adapted from Hunsbedt, Kruger, and London (1977).

Table 3

RANGE OF EXPERIMENT PARAMETERS*

Parameter	Range
Initial Pressure (MPa)	1.76-5.48
Final Pressure (MPa)	0.11-3.58
Initial Temperature (°C)	197 - 260
Final Temperature (°C)	47 - 216
Mean Fluid Production Rate (kg/h)	1.1 - 90
Mean Fluid Recharge Rate (kg/h)	0 - 89
Mean Cooldown Rate (°C/h)	2.4 - 35
Mean Produced Fluid Enthalpy (MJ/kg)	0.80-2.87
Mean Recharge Fluid Enthalpy (MJ/kg)	0.084-1.20

*Adapted from Hunsbedt, Kruger, and London, (1977).

Table 4

RESULTS OF EARLY HEAT EXTRACTION EXPERIMENTS

Production Method	Specific Energy Extraction (kJ/kg)	Energy Extraction Fraction (%)
In-Place Boiling	83 - 116	75 - 100
Sweep	145 - 175	80 - 86*
Steam Drive	21	22 - 27

*Based on steady-state water injection temperature. Others based on saturation temperature at final pressure. Adapted from Hunsbedt, Kruger, and London, (1977).

Table 4 shows the results for the three production methods. The specific energy extraction is defined as

$$\sigma = \frac{Q_r}{M_r} \tag{4}$$

where Q_r = energy extracted from rock (kJ)

M_r = mass of rock in reservoir (kg)

The energy extraction fraction is defined as

$$\eta = \frac{Q_r}{Q_r(\max)} \cong \frac{T_i - \overline{T}_f}{T_i - T_e} \qquad (5)$$

where $Q_r(\max)$ = energy stored in rock between initial rock temperature and lower temperature limit (kJ)

T_i = initial rock temperature (°C)

$\overline{T}_f$ = average final rock temperature (°C)

T_e = reference end fluid temperature (°C)

The reference end fluid temperature was adopted as the saturation temperature at abandonment pressure for the in-place boiling runs and as the inlet injection temperature for the sweep process.

Figure 2 shows a history of temperature profiles for an in-place boiling run without recharge (Hunsbedt, Kruger, and London, 1977). The successive time profiles show a continuous drop in the liquid level and a superheated steam zone above the liquid-steam interface. The uniform temperature in the liquid zone suggests that boiling is occurring throughout the liquid zone. The thermal energy extracted from the rock is represented by the area indicated by a-b-c-d-e-f-a. The maximum extractable energy corresponds to the area a-b-c-d-e-g-a, where line d-e-g represents the saturation temperature at the final pressure. The rock temperature symbols grouped at 1 hour show the thermal gradients achieved in the thermocoupled rocks of various sizes.

Figure 3 shows a history of temperature profiles for an in-place boiling experiment with recharge of cold water (Hunsbedt, Kruger, and London, 1977). One noted effect of the recharge was to maintain the liquid level close to the top thereby reducing markedly the height of the superheated zone. The figure also shows the development of a subcooled zone at the bottom of the reservoir. A second observation is the small rock-water temperature difference at 4 hours, showing the rapid energy extraction from the smaller-sized gabbro rock. The rock energy extraction factor for this run was 100%.

Figure 4 shows a history of temperature profiles for a sweep process run (Hunsbedt, Kruger, and London, 1978). The data show that a "cool" zone developed at the bottom while the "hot" zone at the top remained close to the initial temperature for 4 of the 5.5 hours of production. The data also show that a significant amount of energy remained in the top zone on abandonment. In this figure, the line d-f represents the recharge liquid temperature. The specific energy extraction for this run was 175 kJ/kg, more than a factor of two greater than the extraction of 83 kJ/kg obtained for the in-place boiling experiment in Figure 2. The energy extraction fraction in both cases was about 80%.

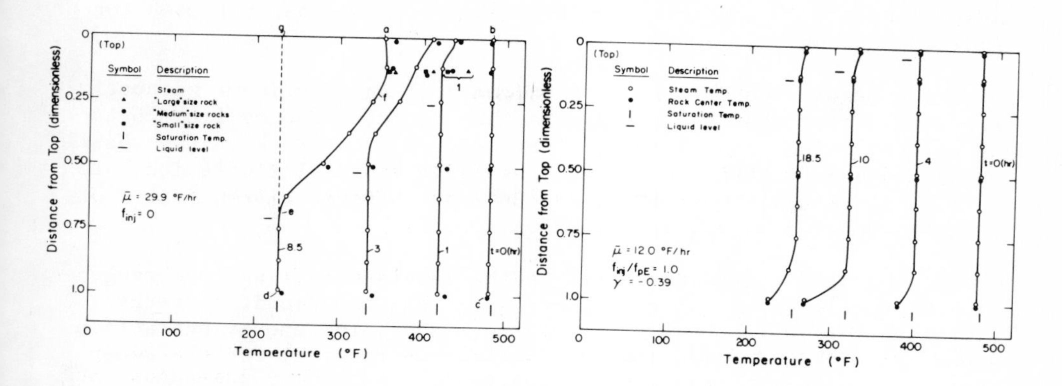

Figure 2. Temperature profiles for in-place boiling experiment without recharge.

Figure 3. Temperature profiles for in-place boiling experiment with recharge.

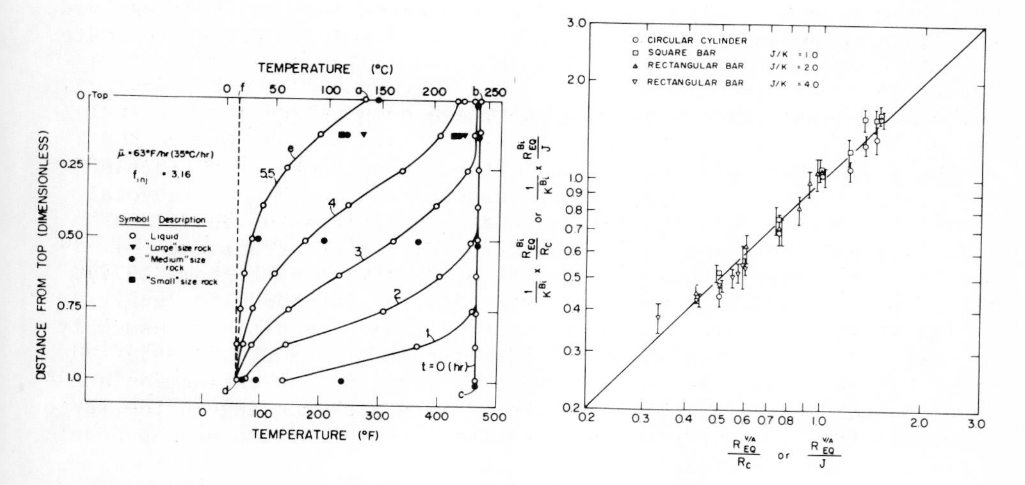

Figure 4. Temperature profiles for sweep drive experiment.

Figure 5. Correlation between equivalent Biot radius and surface-to volume radius.

The two steam-drive experiments showed considerably poorer results. The produced fluid was initially subcooled liquid with an average enthalpy of 963 kJ/kg. When liquid flow ceased, superheated steam was produced with an average enthalpy of 2606 kJ/kg. Although 99 percent of the in-place fluid was produced compared to about 60 percent in the in-place boiling runs, the energy extraction fraction was low, averaging only 21 percent. The steam-drive method has not been considered further.

The overall results for the in-place boiling experiment without recharge showed that the energy fraction produced ranged from 26 to 61 percent. The three boiling experiments without rock in the vessel showed an energy fraction produced of 14 percent. The data indicate that in-place boiling offers a production enhancement of a factor of 2 to 4½ compared to surface flashing.

The specific energy extracted in the in-place boiling runs ranged from 83 to 116 kJ/kg. For the sweep process, the specific energy extracted ranged from 145 to 175 kJ/kg. Thus, the energy extraction experiments showed a significant increased efficiency for the sweep process compared to the in-place boiling process. The degree of energy extraction in an actual large-scale hydrothermal reservoir will depend primarily on three factors: (1) the temperature distribution in the liquid on abandonment, (2) the temperature difference between rock and surrounding fluid, and (3) the temperature of the recharged fluid. Since the temperature difference between rock and liquid is proportional to the cooldown rate and the square of the rock dimension (see equation (1)), increased energy extraction will be achieved for fields with small mean rock size and with reduced production rate.

HEAT EXTRACTION FROM FRACTURED-ROCK RESERVOIRS

The development of a simple model for estimating heat extraction from a fractured-rock hydrothermal resource was achieved in several phases. The first was the analysis of shape factors of individual rock fragments as a means of predicting temperature transient behavior of irregular-shaped rocks (Kuo, Kruger, and Brigham, 1977). The second phase was a multiple rock heat transfer model for analysis of the parameters affecting heat transfer from a statistical assembly of rocks of distributed shape and size (Hunsbedt, et al., 1979). This model was examined for the sweep process of heat extraction and experimental data were collected for a third rock loading in the large physical model of a hydrothermal reservoir. Application of the model to large-scale reservoirs was examined.

Rock fragments in fractured geothermal resources are generally irregular in shape. Heat transfer behavior of large rock fragments of irregular shape in contact with recycled colder fluids is difficult to predict. Kuo, Kruger, and Brigham (1977) developed a set of shape factor correlations for heat transfer calculations by comparing the analytical solution for temperature transient behavior of known regular shaped rocks with that of spheres. From results of experiments using circular cylinders and rectangular bars with length to width aspect ratios ranging from 1 to 4, they obtained successful correlations for two shape parameters.

The first was an equivalent radius based on the volume to surface area ratio of a sphere, given by

$$R_{eq}^{V/A} = R_s^{V/A} = 3(\frac{V}{A}) \quad (6)$$

Analysis of the Biot numbers for the cylinders and bars as equivalent-radius spheres showed that the Biot number ratio, defined as the ratio of the Biot number of an equivalent sphere, R_{eq}^{Bi}, to the characteristic length of the rock, R_c, was approximately equal to the ratio of the volume-to-surface area equivalent radius, $R_{eq}^{R/A}$, to the same characteristic length. Thus from

$$\frac{R_{eq}^{Bi}}{R_c} \cong \frac{R_{eq}^{V/A}}{R_c} \quad (7)$$

it follows that, essentially independent of shape and aspect ratio,

$$R_{eq}^{Bi} \cong 3(\frac{V}{A}) \quad (8)$$

The correlation was further improved by introducing a sphericity parameter, ψ, defined by Leva (1947) and Pettyjohn and Christiansen (1948) as the area of a sphere divided by the surface area of a rock fragment having the same volume as the sphere; i.e.:

$$\psi = \frac{(A/V)_s}{(A/V)_{actual}} = \frac{4\pi R_s^2}{A_{actual}} \quad (9)$$

The effective radius of the rock fragment is

$$R_e = \psi[\frac{3V}{4\pi}]^{1/3} \quad (10)$$

where V = volume of the irregular-shaped rock fragment. Figure 5 shows the correlation obtained between Biot number and geometric equivalent radius for cylinders and bars with aspect ratios from 1 to 4.

Iregui et al. (1978) examined the correlations for a collection of unequal size rocks in a fractured geothermal reservoir. On the basis that the distribution of rock sizes can be described by a probability density function, a general heat transfer model can be derived from the heat transfer model for individual irregular-shaped rocks. The surface area to volume ratio for a collection of rocks can be estimated for each rock (i) by Equation (9) and summed for all rocks N as

$$\left(\frac{A}{V}\right)_c = 3\,\frac{\sum_{i=1}^{N} \frac{R_{s,i}^2}{\psi i}}{\sum_{i=1}^{N} R_{s,i}^3} \qquad (11)$$

For a probability density function $p(R_s) = \frac{n_i}{N}$ where n_i = number of equal sized rocks, an equivalent size sphere radius for the entire collection can be defined by

$$R_{e,c} = \overline{\psi}\,\frac{\sum_{j=1}^{N_L} p(R_{s,j})R_{s,j}^3}{\sum_{j=1}^{N_L} p(R_{s,j})R_{s,j}^2} \qquad (12)$$

where ψ = average sphericity

N_L = number of size groups in distribution

Iregui et al. (1978) reviewed the range of "effective radius" as defined in Equation (12) for a number of known distributions such as normal and exponential. The resulting "effective radius" allows the rock size collection to be treated as an equivalent single spherical rock for heat extraction estimation.

The multiple-rock "effective radius" concept was developed into a one-dimensional model of heat transfer to predict water temperature profiles in the reservoir during energy production with the cold-water sweep process. The problem of heat transfer from rock to circulating fluid had been studied previously by Schumann (1929) and Löf and Hawley (1948) for low Biot number conditions as occur for air flow through rock-filled bins. These solutions are not applicable to geothermal reservoir conditions of very large Biot numbers resulting from high heat transfer coefficients and large rock sizes.

Water temperature profiles in dimensionless space and time are expressed as $T_f^*(x^*,t^*)$ defined as

$$T_f^* = \frac{T_f(x,t)-T_e}{T_i-T_e} \qquad (13)$$

where $T_f(x,t)$ = water temperature at height (x) at time (t), referenced to initial temperature (T_i) and water injection temperature (T_e)

x^* = x/L, relative height from injection point to production level at height (L)

t^* = t/t_r, relative time, referenced to the fluid residence time (t_r)

The differential equation which describes the heat transfer from a collection of rock fragments with a single equivalent-sphere radius to surrounding fluid in linear sweep circulation is

$$\frac{\partial T_f^*}{\partial x^*} + \frac{\partial T_f^*}{\partial t^*} + \frac{1}{\gamma} N_{tu}(T_f^* - T_r^*) = q^* \tag{14}$$

with initial condition

$$T_f^*(x^*,0) = T_r^*(x^*,0) = 1 \quad , \; 0 \leq x^* \leq 1 \tag{15}$$

and boundary condition

$$T_f^*(0,t^*) = 0 \quad , \; t^*>0 \tag{16}$$

Thus, the water temperature profiles can be expressed as a function of three parameters:

$$T_f^*(x^*,t^*) = f(N_{tu},\gamma,q^*) \tag{17}$$

where N_{tu} = t_r/τ, the "number of heat transfer units" parameter

γ = modified storage ratio, the energy stored in the water relative to that in the rock, modified to include the effect of the steel vessel

q^* = normalized external heat transfer parameter to account for the wall effect of the vessel at mass flow rate

The "number of heat transfer units" parameter is a most significant one in the sweep model. The model indicates that when this parameter is small, e.g., $N_{tu} \lesssim 10$, a reservoir is heat transfer limited. In this case, the heat transfer rate from the rock is not sufficient to heat the circulating water, resulting in premature decline in water temperature and a relatively small fractional energy extraction.

The solution to Equation (15) was approximated by a Laplace transform equation of the form

$$T_f^*(x^*,t^*) = \mathcal{L}^{-1} \hat{T}_f(x^*,s) \tag{18}$$

$$\text{where } \hat{T}_f(x^*,s) = [\frac{1}{s} + \frac{q^*}{Ks^2}][1-e^{-Kx^*s}] \tag{19}$$

$$\text{and} \quad K = 1 + \frac{1}{\gamma}[\frac{1}{s/N_{tu} + 1}] \tag{20}$$

The inversion was accomplished numerically by the algorithm developed by Stehfest (1970) using a 10-coefficient calculation scheme given by

$$T_f^*(s^*,t^*) \doteq \frac{\ln 2}{t^*} \sum_{i=1}^{10} a_i \, \hat{T} \, [x^*, \frac{\ln 2}{t^*} i] \tag{21}$$

The 1-D sweep model as tested by Iregui was described by Hunsbedt et al. (1979), Nelson and Hunsbedt (1979), and Nelson, Kruger, and Hunsbedt (1980). The Piledriver granite rock loading was examined for two different values of the number of heat transfer units parameter, obtained from the original granite and by adding fine sand (80 to 100 mesh) to the original rock loading. Three of the rocks were instrumented with thermocouples, their model parameters are listed in Table 5. The input parameters to the two sweep experiments are summarized in Table 6. The rock only loading (with porosity of 42% and essentially infinite permeability) had an effective rock radius of 3.20 cm. The rock-sand loading (with porosity of 21% and permeability of $\sim$ 30 darcy) had the same effective rock radius. The respective number of heat transfer units parameter was 139 for the rock-only and 45 for the rock-sand loading.

The results of the measured and predicted rock center temperature for the rock-only sweep experiment are shown in Figure 6. The data show good agreement with prediction for the two upper rocks, but less satisfactory agreement for the rock closer to the injection site. This may be partly due to the dispersion of injected water temperature in the baffle inlet to the reservoir.

The results of the measured and calculated water temperatures at six locations in the rock-only loading are shown as a function of time in Figure 7. The observed fluid temperature is in good agreement with the calculated temperature at the bottom ($x^* = 0.08$) and at about

Table 5

MODEL PARAMETERS OF THE INSTRUMENTED ROCKS

Rock Number	Location x^*	Effective Radius R_e, (cm)	Sphericity ψ
1	0.08	7.39	0.92
2	0.48	7.80	0.80
3	0.79	5.79	0.95

Table 6

INPUT PARAMETER FOR THE 1-D SWEEP MODEL

Parameter (Units)	Value for Loading	
	Rock Alone (1)	Rock & Sand (2)
Initial Reservoir Temp, T_i (°C)	238	238
Injected Water Temp, T_e (°C)	21	16
Flow Rate, $\dot{m}$ (kg/h)	60	90
Reservoir Cross-Section, A (m^2)	0.30	0.30
Reservoir Height, L (m)	1.55	1.55
Rock Effective Radius, R_e (cm)	3.20	3.20
Water Residence Time, t_r (hr)	3.19	1.03
Heat Transfer Coefficient (W/m^2C)	53	53
Biot Number	22.5	22.5
Effective Time Constant, τ (hr)	0.023	0.023
Drainage Porosity (%)	42	21
Storage Ratio, γ	0.52	0.23
External Heat Transfer Parameter, q^*	0	0.026
Number of Heat Transfer Units, N_{tu}	139	44.8

(1) from Hunsbedt et al. (1979)

(2) from Nelson and Hunsbedt (1979)

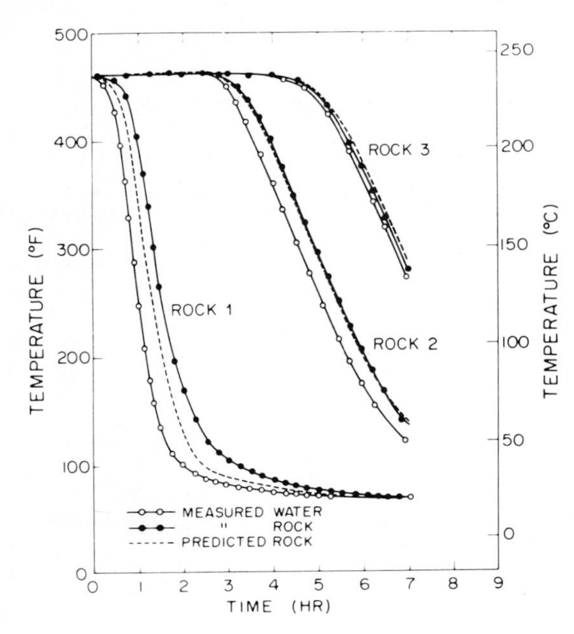

Figure 6. Measured and predicted rock center temperatures for sweep experiment.

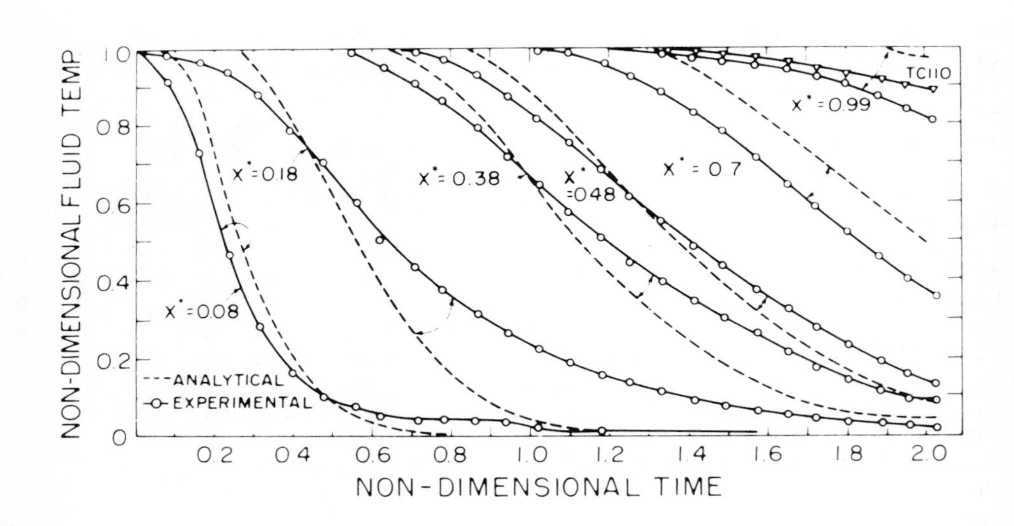

Figure 7. Water temperature profiles as a function of time in the sweep experiment.

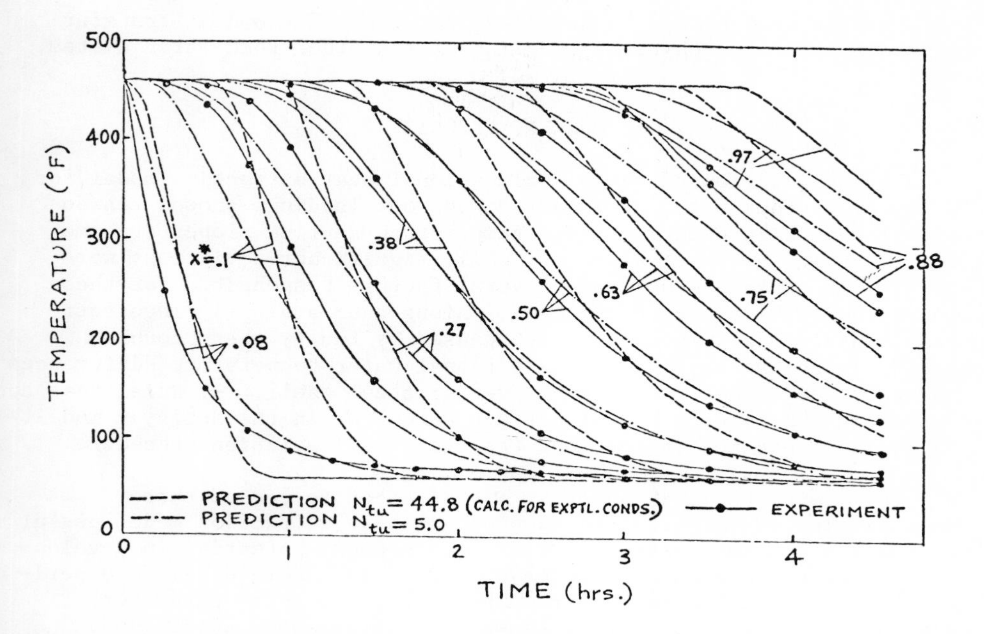

Figure 8. Comparison of observed and predicted water temperature profiles.

half-way up (x* = 0.48), but differs considerably in between (x* = 0.18). Hunsbedt et al. (1979) noted that this elevation corresponds to the location of the massive steel flange of the physical model. The results also show that the measured water temperature was always lower than the predicted one at x* = 0.7 and 0.99.

The results for the rock-sand loading sweep experiment are shown in Figure 8. Several observations may be made. Initially, the predicted water temperatures decline more slowly than the measured temperatures. At later times, the predicted water temperatures drop below the measured ones and remain lower, except at the top elevations (x* = 0.88 and 0.97) where the predicted temperatures are higher than the measured ones. One attempt to improve the agreement was to seek the best fit "number of heat transfer units" parameter. Figure 8 shows improvement achieved with a number of heat transfer units parameter of 5.0 instead of the value of 44.8 calculated from the experimental input parameters. The agreement is markedly improved at the middle elevation (x* = 0.5).

The conclusion of these studies indicated that a simple one-dimensional sweep model based on a limited number of reservoir parameters was in fair agreement with experimental results, but improvements were needed in numerical solution of the heat transfer transient or in the physical modeling of the steel-rock-water system.

CURRENT EFFORTS

Early application of the one-dimensional heat extraction model for the sweep process with irregular-shaped rock loadings showed reasonable agreement in water and rock temperature decline slopes at some elevations in the physical model, while significant deviations were observed at other elevations. Several factors responsible for these discrepancies have been considered. Among them are: (1) incorrect experimental rock temperature data caused by faulty thermocouple sealing in the rocks, (2) cross-sectional water temperature differences caused by uneven heating from the vessel steel shell, (3) axial heat conduction in the physical model not accounted for in the analytic model, and (4) inaccuracies in the heat transfer model solution procedure.

Current efforts in the program are directed to remove these uncertainties to make the one-dimensional heat extraction model useful for application in assessment of energy recovery potential in developing hydrothermal resources. A major effort is a more detailed experimental evaluation of the number of heat transfer units parameter, using a rock loading of large, known size and shape distribution. Figure 9, from Hunsbedt, Kruger, Nelson, Swenson, and Donaldson (1981), shows the rock loading configuration, consisting of 6 planes of 30 granite rock blocks 19 x 19 cm rectangular cross section and 24 triangular blocks. The blocks are 26.4 cm high and are spaced 0.64 cm vertically and 0.43 cm horizontally. The average fracture porosity of the matrix is 17.5 percent and the permeability is very large.

Three experiments have been run in the physical model to test the heat extraction as a function of the number of heat transfer units parameter; two of them at $N_{tu} \sim 2$ and $N_{tu} \sim 7$, considered to be in the "heat transfer limited" region and the third at a value of $N_{tu} \sim 15$. Experimental parameters for the first run, at $N_{tu} \sim 7$, are listed in Table 7. The results of this test are shown in Figure 10. Early analysis of these data indicate that water temperature at the inlet (I) plane is initially slightly hotter near the surface wall due to heating by the steel. The injected water approached a uniform, constant temperature of 15°C after about one hour. The data also show that the cross-sectional water temperatures were essentially uniform in each of the planes, with a maxiumum deviation of ± 2°C, well within the estimated uncertainty of thermocouple temperature difference of ± 3°C.

Figure 10 also shows several rock center temperature transients. The data show that the maximum rock center to surrounding water temperature difference of about 55°C developed during production, decreasing to small values toward the end of the experiment. These data indicate that the rock energy extraction was relatively complete

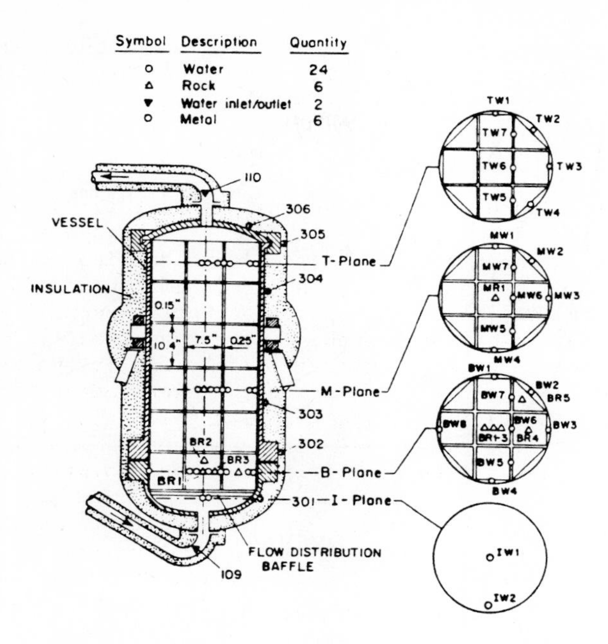

Figure 9. Experimental rock matrix configuration and thermocouple locations.

Table 7

EXPERIMENTAL PARAMETERS FOR N_{tu} = 7 TEST*

Parameter	Value
Initial Reservoir Temperature	239°C
Average Reservoir Pressure	0.38 MPa
Injection Water Temperature	15°C
Final Top Temperature	156°C
Final Bottom Temperature	19°C
Initial Water Mass	67 kg
Injected Water Mass	340 kg
Water Injection Rate	68 kg/hr
Production Time	5 hr

*from Hunsbedt, et al. (1979)

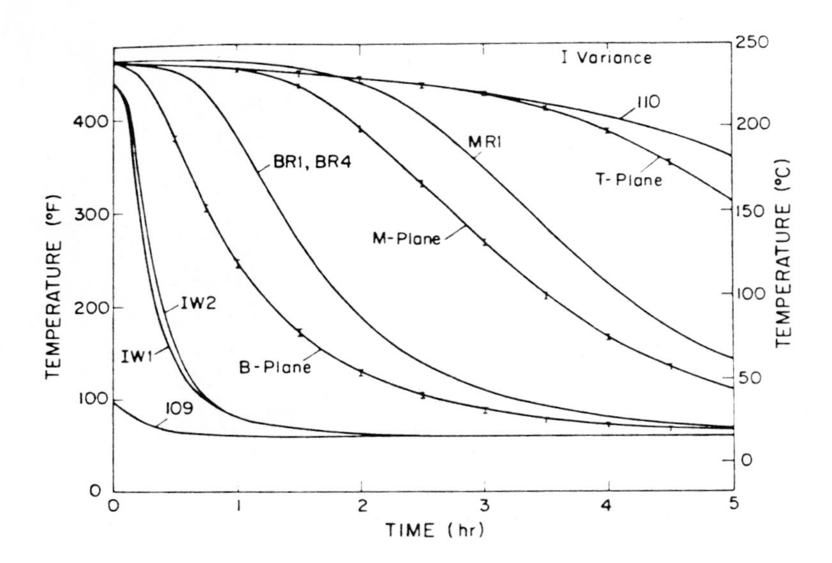

Figure 10. Rock and water temperature profiles for the known-geometry sweep experiment.

for this number of transfer units parameter, and the energy extracted from the rock resulted in a high, constant exit water temperature. Analysis of the latter two experiments is currently underway.

To assist in the analysis of these experiments, the data are being evaluated by several distributed parameter models in addition to the 1-D sweep model. One attempt was the development of a finite-element model (Swenson and Hunsbedt, 1981), in which the individual granite blocks can be represented as single elements. The model was expected to permit less restraint on element shapes compared to finite difference models and be more applicable to full size reservoirs. The model has not been completed and currently analysis of the data are being made with adaptation of the MINC numerical code developed by Karsten Pruess (1981) of the Lawrence Berkeley Laboratory. In Pruess's MINC model each granite block is divided into a number of concentric shells for greater sensitivity in examining the temperature gradient within the rock and at the rock-fluid boundary. Several computer runs with this model have been obtained for the first regular-geometry rock loading experiment. The results showed quite good agreement and several modifications to the input data to improve modeling accuracy have been identified. Further computer modeling is underway on the three N_{tu} experiments.

One additional factor that may influence the heat extraction history from a hydrothermal reservoir is the potential effect of thermal stressing on rock heat transfer properties. This possibility

Table 8

PROPERTIES OF THE GRANITE BLOCKS*

Parameter	Value
Density, ρ	2627 kg/m^3
Poisson's Ratio, ν	0.22
Modulus of Elasticity, E	4.92 mPa
Uniaxial Tensile Strength, σ	0.77 MPa
Specific Heat, C	918 J/kg·°C
Coefficient of Thermal Expansion, α	7.42×10^{-6}/°C
Thermal Conductivity, k	9770 J/h·m·°C

is suggested from the work of Nemat-Nasser and Ohtsubo (1978) and Murphy (1978) which predict that thermal stresses produced by fluid circulating in a hot dry rock geothermal reservoir are likely to initiate and propagate cracks in the rock. Thermally induced cracks in a fractured hydrothermal reservoir might augment the energy extracted if they cause a significant increase in effective heat transfer and flow channels. Nelson and Hunsbedt (1979) reported on preliminary experiments to observe the effects of severe thermal stressing on mechanical properties of granite blocks. Properties of this granite are listed in Table 8. Experimental blocks (11.3 x 11.3 x 25.4 cm in size) were heated slowly at 60°C/hr in a fabricated well-insulated oven with one removable wall to expose one face of the granite block. The exposed face was thermally stressed with water spray. To investigate the effect of stressing on the blocks, they were sliced into rectangular specimens (3.8 x 7.6 x 0.76 cm) and loaded to fracture in three point bending. Figure 11 shows the estimated thermal stress distribution in the granite block and the bending strength data for the 8 specimens taken from the blocks after 5 cycles of spray quenching. The data show a significant degradation in strength in the specimens near the quenched face where tensile thermal strength existed, and no loss of strength in specimens in depths of compressive stress. The loss of strength in the tensile stress zone is attributed to microfracturing in the granite block.

Models of heat transfer generally assume a constant set of thermal properties for the rock matrix. In hydrothermal systems undergoing cold water sweep energy extraction, changes in the thermal diffusivity and thermal conductivity by tensile thermal stressing from the cold fluid may be important in changing long-term heat extraction potential.

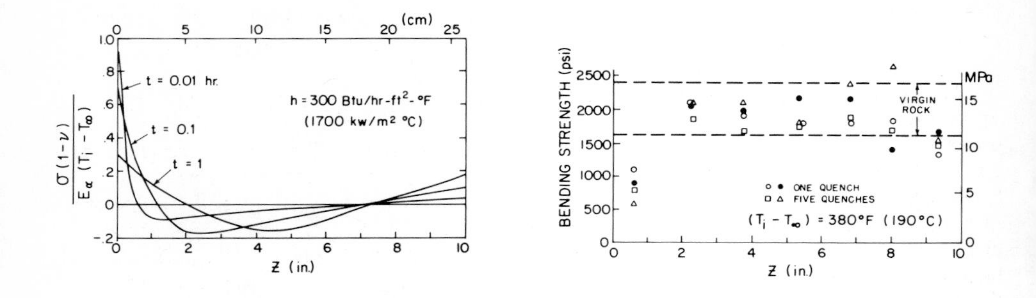

Figure 11. Thermal stress distributions with time in the granite block and bending strength measurements of specimens from the block.

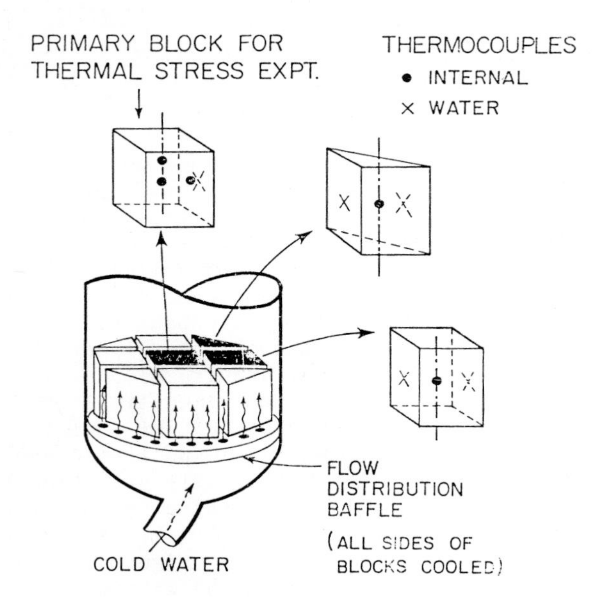

Figure 12. Instrumented block for the measurement of thermal stress effects after repeated sweep experiments.

As part of the current series of heat extraction experiments, the central block in the bottom layer in the physical model (see Figure 12) has been instrumented with 3 thermocouples to obtain an accurate temperature history of that block during all of the runs with this loading. At the conclusion of the experiments with this loading, this block will be examined for thermally induced changes in thermal properties compared to a similar granite block that was not part of the loading. It is anticipated that observed changes due to thermal stressing will be incorporated into the heat extraction model.

The overall objective of these studies is to develop an easily applied model for predicting the energy extraction potential for undeveloped hydrothermal resources. Future plans for this program are to test the model on newly developing hydrothermal resources. Possible cases include the forthcoming 50 MW binary conversion demonstration plant at Heber, California and the 20 MW rotary separator turbine-flash turbine plant at Roosevelt Hot Springs, Utah. The Heber field is a relatively hot field ($T \sim 180°C$) in a sedimentary formation and the Roosevelt Hot Springs is also relatively hot ($T > 235°C$) in an igneous rock formation. Efforts over the first few years of production can include pre-operation estimation of the geologic structure of the reservoir to provide an input data base for "effective rock radius" and "number of heat transfer units" parameter estimations. An energy extraction history can be predicted and matched against observed energy extraction history. Feedback can include post-operation improvements of the geologic structure affecting heat transfer and improved model development to lengthen the period of heat extraction history forecasts.

ACKNOWLEDGMENTS

This paper summarizes results from several dissertations by a number of students under the Stanford Geothermal Program, in particular Dr. Anstein Hunsbedt, Dr. Ming-Ching T. Kuo, Engr. Roberto Iregui, and current Ph.D. candidate Stephen Lam.

The Stanford Geothermal Program has been supported continuously by grants and contracts, originally from the National Science Foundation, later the United States Energy Research and Development Administration, and currently the United States Department of Energy's Division of Geothermal Energy. The author is grateful for this support.

REFERENCES

Hunsbedt, A., Kruger, P., and London, A. L., "Laboratory Studies of Non-Isothermal Fluid Production from Fractured Geothermal Reservoirs", Paper No. 77-HT-53, Proceedings, AIChE-ASME Heat Transfer Conference, August, 1977.

Hunsbedt, A., Kruger, P., and London, A. L., "Energy Extraction from a Laboratory Model Fractured Geothermal Reservoir", J. Petro. Tech. Vol. 30, No. 5, 712-718, May, 1978.

Hunsbedt, A., Iregui, R., Kruger, P., and London, A. L., "Energy Recovery from Fracture-Simulated Geothermal Reservoirs", Paper

79-HT-92, Proceedings, ASME/AIChE 18th National Heat Transfer Conference, San Diego, California, August, 1979.

Hunsbedt, A., Kruger, P., Nelson, D., Swenson, L., and Donaldson, I., "Heat Extraction from Fractured Hydrothermal Reservoirs", Proceedings, Third New Zealand Geothermal Workshop, November, 1981.

Kruger, P. and Ramey, H. J., "Stimulation and Reservoir Engineering of Geothermal Resources", Report No. SGP-TR-1, Stanford Geothermal Program, June, 1974.

Kruger, P., "Geothermal Energy", in Annual Review of Energy, Vol. 1, (Annual Reviews, 1976).

Kruger, P., "Stimulation of Geothermal Energy Resources", Proceedings, Il Fenomeno Geotermico e Sue Applicazioni, Accademia Nazionale dei Lince, Rome, 1977.

Kuo, M. C. T., Kruger, P., and Brigham, W. E., "Shape-Factor Correlations for Transient Heat Conduction from Irregular-Shaped Rock Fragments to Surrounding Fluid", Paper 77-HT-54, Proceedings, AIChE-ASME Heat Transfer Conference, Salt Lake City, Utah, August, 1977.

Leva, M., "Pressure Drop through Packed Tubes", Chem. Engr. Progr. 43, No. 10, 549 (1947).

Löf, G. O. G. and Hawley, R. W., "Unsteady State Heat Transfer between Air and Loose Fluids", Ind. & Engr. Chem. 40, No. 6, 1948.

Nelson, D. V. and Hunsbedt, A., "Progress in Studies of Energy Extraction from Geothermal Reservoirs", Proceedings, Fifth Workshop on Geothermal Reservoir Engineering, SGP-TR-40, Stanford University, December, 1979.

Nelson, D., Kruger, P., and Hunsbedt, A., "Geothermal Energy Extraction Modeling", Proceedings, Sixth Workshop on Geothermal Reservoir Engineering, SGP-TR-50, Stanford University, December, 1980.

Pettyjohn, E. S. and Christiansen, E. B., "Effect of Particle Shape on Free-Settling Rates of Isometric Particles", Chem. Engr. Progr. 44, No. 2, 157 (1948).

Pruess, K., "Heat Transfer In Fractured Geothermal Reservoirs with Boiling", Proceedings, Seventh Workshop on Geothermal Reservoir Engineering, SGP-TR-55, Stanford University, December, 1981.

Ramey, H. J., Kruger, P., and Raghavan, R., "Explosive Stimulation of Hydrothermal Reservoirs", Chapter 13 in P. Kruger and C. Otte, eds, Geothermal Energy (Stanford University Press, 1973).

Roberts, V., and Kruger, P., "Utility Industry Estimates of Geothermal Electricity", Geo. Res. Council Bull. 11, No. 5, 7-10, 1982.

Schumann, T. E. W., "Heat Transfer: A Liquid Flowing through a Porous Prism", J. Fluid Inv., September, 1929.

Smith, M. C., Aamodt, R. L., Potter, R. M., and Brown, D. W., "Man-Made Geothermal Reservoirs", in Proceedings, Second U.N. Symposium on the Development and Use of Geothermal Resources, San Francisco, 1975.

Stehfest, H., "Numerical Inversion of LaPlace Transforms. Algorithm No. 368", Comm. of the ACM, 13, No. 1, January, 1970.

Swenson, L. W. and Hunsbedt, A., "Experimental and Finite Element Analysis of the Stanford Hydrothermal Reservoir Model", Proceedings, Seventh Workshop on Geothermal Reservoir Egnineering, SGP-TR-55, Stanford University, December, 1981.

reservoir system with equations in the electric system. Rumble[5] et al. presented a reservoir analyzer study of the reservoir performance. An acceptable match of the production-pressure relationship in the East Texas Field was established on the analyzer and the analyzer was used to make predictions of the pressure in the field. McDowell[6] studied the performance of water drive reservoirs, including pressure maintenance by the use of an electrical reservoir analyzer.

In this paper, pressure drawdown and pressure buildup tests for synthesized fractured geothermal reservoirs are simulated with the numerical model[7] and the electrical model[8] developed by the authors. The results of these models are compared with those of other models.[2),3] The analysis of the transient pressure curves plotted on semi-log papers is also discussed.

THEORY

Numerical Model

The idealized model of a fractured reservoir is shown in Figure 1. The model consists of a set of horizontal matrix layers with fractures as the spacers and is based on some assumptions.[7] The basic equation governing fluid flow in this system is written as,

$$\frac{1}{r}\frac{\partial}{\partial r}\left(r\frac{K_r^*}{\mu}\frac{\partial P}{\partial r}\right)+\frac{\partial}{\partial z}\left(\frac{K_z^*}{\mu}\frac{\partial P}{\partial z}\right)-Q\delta\,(X-X_o)=\phi^* c^*\frac{\partial P}{\partial t} \tag{1}$$

Equation (1) is the modified version of the two-dimensional flow equation for slightly compressible fluids in radial coordinates. The effective coefficients of K_r^*, K_z^*, ϕ^* and c^* are defined as follows.

For matrix section;

$$K_r^* = K_{ma,r} \quad , \quad K_z^* = K_{ma,z} \quad , \quad \phi^* = \phi_{ma} \quad , \quad c^* = c_{ma} \tag{2}$$

For fracture-matrix section;

$$K_r^* = \frac{K_{f,r}b + K_{ma,r}h_{ma}}{b + h_{ma}} \quad , \quad K_z^* = \frac{b + h_{ma}}{\dfrac{b}{K_{f,z}} + \dfrac{h_{ma}}{K_{ma}}} \tag{3}_a$$

$$\phi^* = \frac{\phi_f b + \phi_{ma} h_{ma}}{b + h_{ma}} ,$$

$$c^* = \frac{c_f b + c_{ma} h_{ma}}{b + h_{ma}} \qquad (3)_b$$

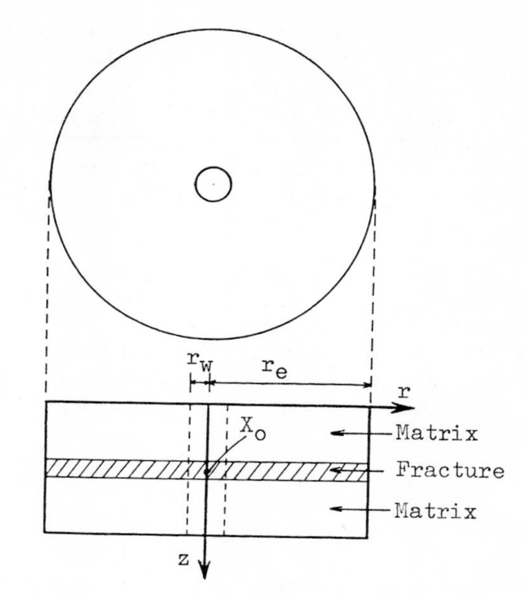

Fig. 1 Idealized Model of a Fractured Reservoir with Centrally Located Well

The initial and boundary conditions are expressed as,

$$P (r, z, 0) = P^*$$

$$\frac{\partial P}{\partial r} = 0 \; ; \quad 0 < z < h \; , \quad r = r_w \;\; \text{and} \;\; r = r_e \qquad (4)$$

$$\frac{\partial P}{\partial z} = 0 \; ; \quad r_w < r < r_e \; , \quad z = 0 \;\; \text{and} \;\; z = h$$

Electrical Model

An electrical reservoir analyzer is constructed based on the theory that electric current may be equivalent to fluid flow and electric potential may be equivalent to pressure, if certain functional relationships are established. To construct an electrical reservoir analyzer, a pie-shaped slice of a finite circular reservoir with centrally located well of Figure 2 is considered. The fluid capacitance and the resistance to fluid flow of the ik-th block can be calculated from the following equations.[8]

$$C_{Fik} = c \times \frac{\theta}{2\pi} \times h_k \times (\pi r_i^2 - \pi r_{i-1}^2) \times \phi_{ik} \qquad (5)_a$$

$$= \frac{1}{2} h_k \, \theta \, c \, \phi_{ik} \, (r_i^2 - r_{i-1}^2)$$

$$(R_{Fik})_r = \frac{\mu}{(K_{ik})_r} \int_{r_{i-1}}^{r_i} \frac{dr}{r\theta h_k}$$

$$= \frac{\mu}{(K_{ik})_r h_k \theta} \ln \frac{r_i}{r_{i-1}} \tag{5$_b$}$$

$$(R_{Fik})_z = \frac{\mu \, h_k}{(K_{ik})_z \times \frac{\theta}{2\pi} \times (\pi r_i^2 - \pi r_{i-1}^2)}$$

$$= \frac{2 \mu \, h_k}{(K_{ik})_z \, \theta \, (r_i^2 - r_{i-1}^2)}$$

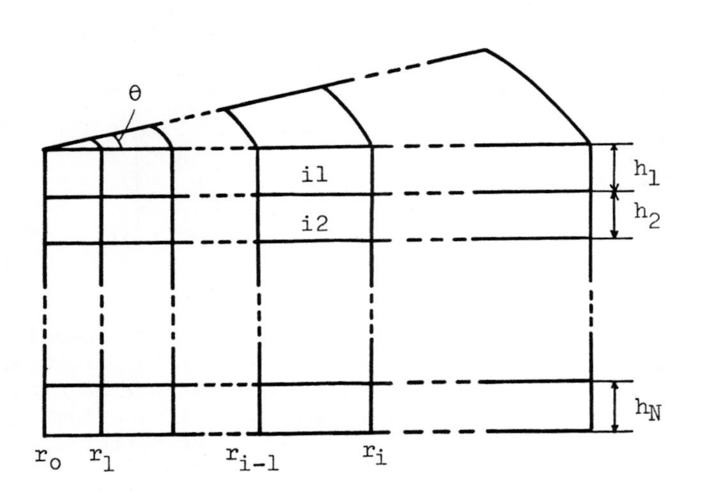

Fig. 2 Division of a Reservoir for Electrical Reservoir Analyzer

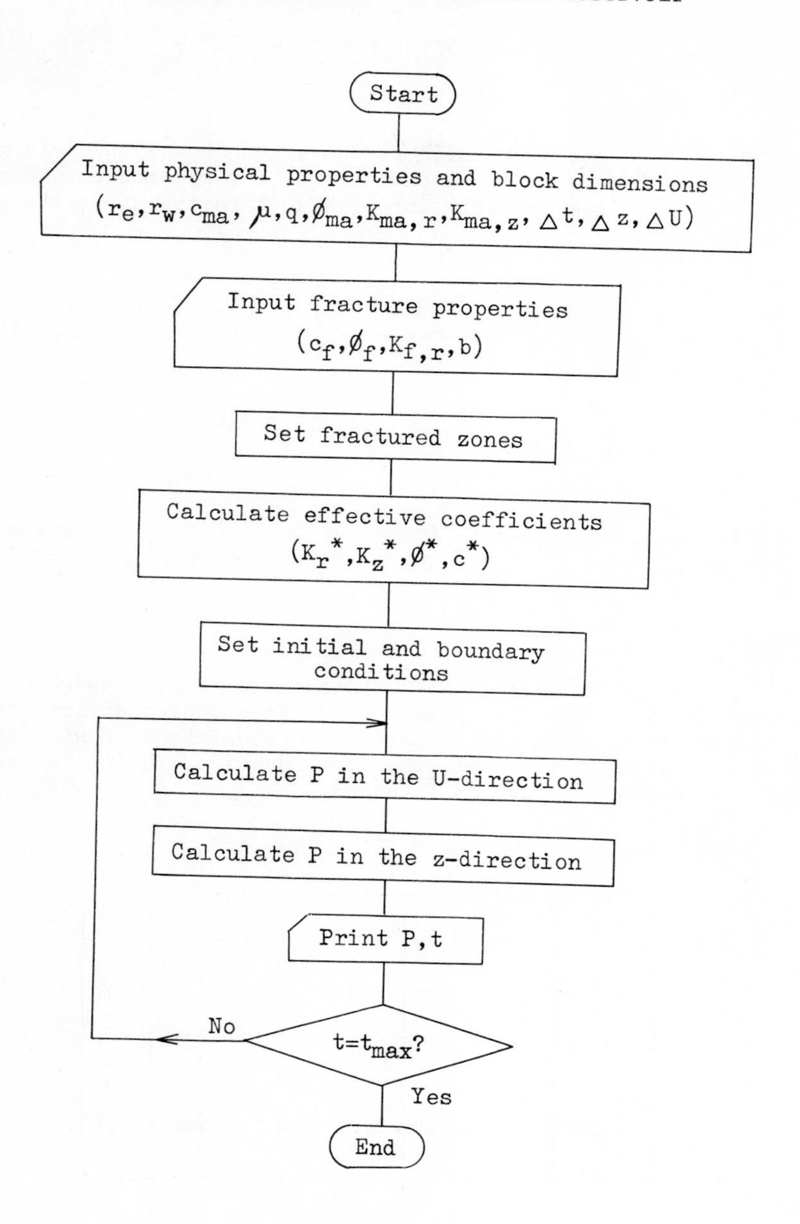

Fig. 3 Simplified Flow Chart

OUTLINE OF DEVELOPED SIMULATORS

Numerical Model

The logarithmic transformation to the space dimension, $U=\ln(r/r_W)$ was introduced, because the pressure tends to vary exponentially with the distance from the well bore. With this transformation, Equation (1) becomes

$$\frac{\partial}{\partial U}\left(\frac{K_r^*}{\mu}\frac{\partial P}{\partial U}\right) + e^{2U} r_W^2 \frac{\partial}{\partial z}\left(\frac{K_z^*}{\mu}\frac{\partial P}{\partial z}\right) - e^{2U} r_W^2 Q\delta(X-X_o)$$

$$= e^{2U} r_W^2 \phi^* c^* \frac{\partial P}{\partial t} \tag{6}$$

Equation (6) was expanded into finite-difference form and was solved by single-sweep ADI procedure, using mirror images at the boundaries. Figure 3 shows the calculation procedure.

Electrical Model

The fluid capacitance and the resistance to fluid flow were calculated by Equation (5). These values were converted into the electric capacitance and the resistance to electric flow with the analyzer proportionality constants. The layout of the resistances and the condensers is shown in Figure 4.

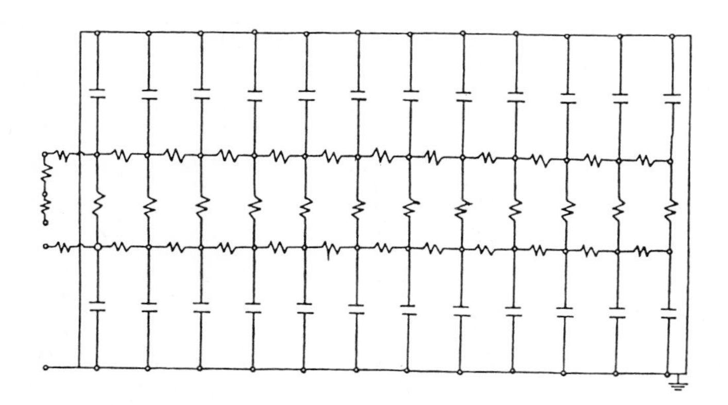

Fig.4 Graphic Layout of the Resistances and Condensers of Electrical Reservoir Analyzer

EXAMPLE AND DISCUSSION

The pressure drawdown and pressure buildup tests were simulated with the numerical model and the electrical model. Table 1 lists the synthesized cases of fractured geothermal reservoirs. They are very simple cases and are assumed that the reservoir contains only hot water and that the reservoir temperature is constant during the well tests. Figures 5 and 6 show the pressure drawdown curves and Figure 7 shows the pressure buildup curve for the synthesized cases. To compare these models with other models, the results of Kazemi's numerical model and those of de Swaan's analytical model are also plotted. The results of these models agree with those of other models. Two straight line segments are observed from these curves plotted on semi-log papers. This is the characteristic property of well pressure plot in a fractured reservoir.

The slope of the straight line segment of the pressure drawdown and pressure buildup curves is one of the important factors in analyzing the curves by the transient pressure techniques. In the

Table 1 The Synthesized Systems

Parameters		Case 1	Case 2
ϕ_{ma}		5.00×10^{-2}	8.00×10^{-2}
ϕ_f		4.50×10^{-1}	4.50×10^{-1}
h	(m)	2.76	2.98
b	(m)	7.60×10^{-3}	7.60×10^{-3}
K_{ma}	(m^2)	2.17×10^{-18}	9.87×10^{-16}
K_f	(m^2)	1.58×10^{-12}	5.45×10^{-12}
c_{ma}	(P_a^{-1})	1.45×10^{-9}	1.45×10^{-9}
c_f	(P_a^{-1})	1.45×10^{-9}	1.45×10^{-9}
P^*	(P_a)	2.76×10^{7}	3.44×10^{7}
q	(m^3/s)	1.67×10^{-4}	3.68×10^{-4}
μ	($P_a \cdot s$)	2.20×10^{-4}	2.35×10^{-4}
r_e	(m)	1.61×10^{3}	1.61×10^{3}
r_w	(m)	1.14×10^{-1}	1.14×10^{-1}
T	(k)	4.03×10^{2}	3.96×10^{2}

Remarks: These values are taken from Reference 2, although part of them are modified for geothermal reservoirs. They are converted into SI units by the authors.

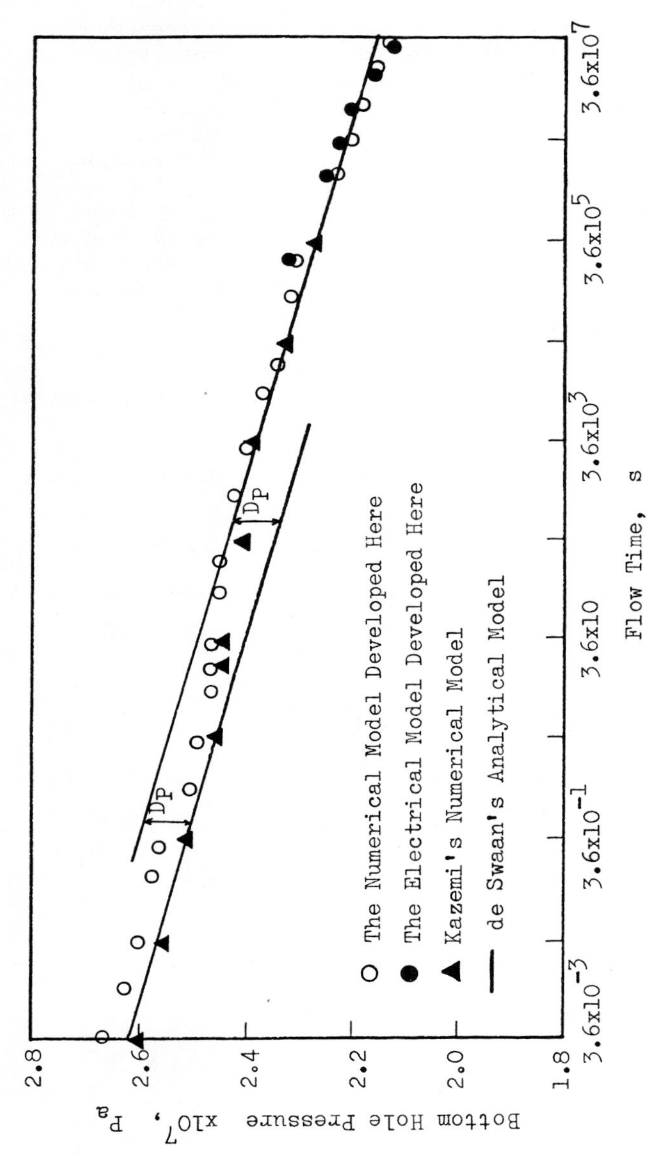

Fig. 5 Pressure Drawdown for Case 1

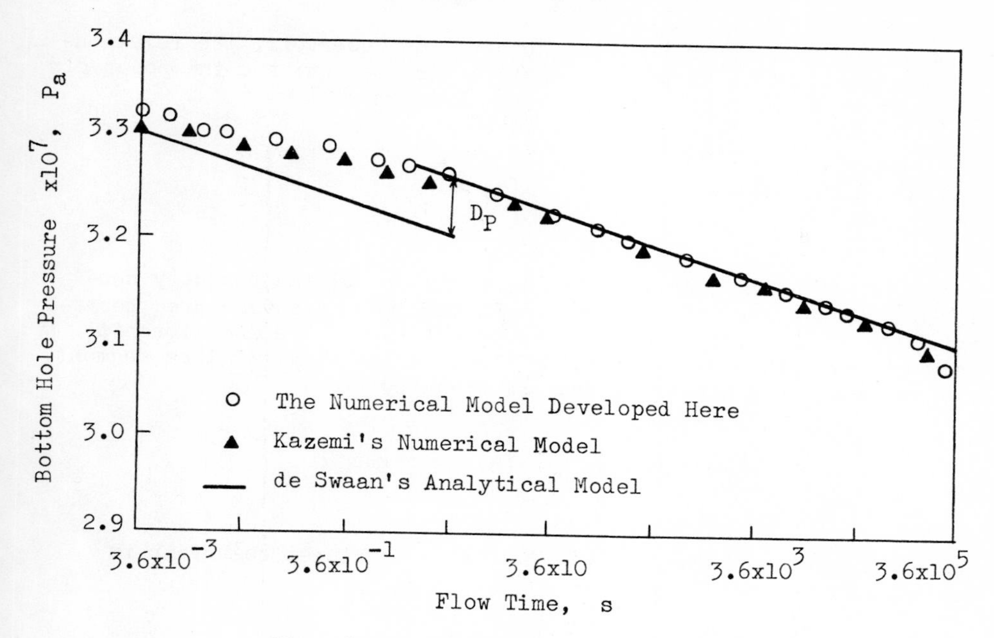

Fig. 6 Pressure Drawdown for Case 2

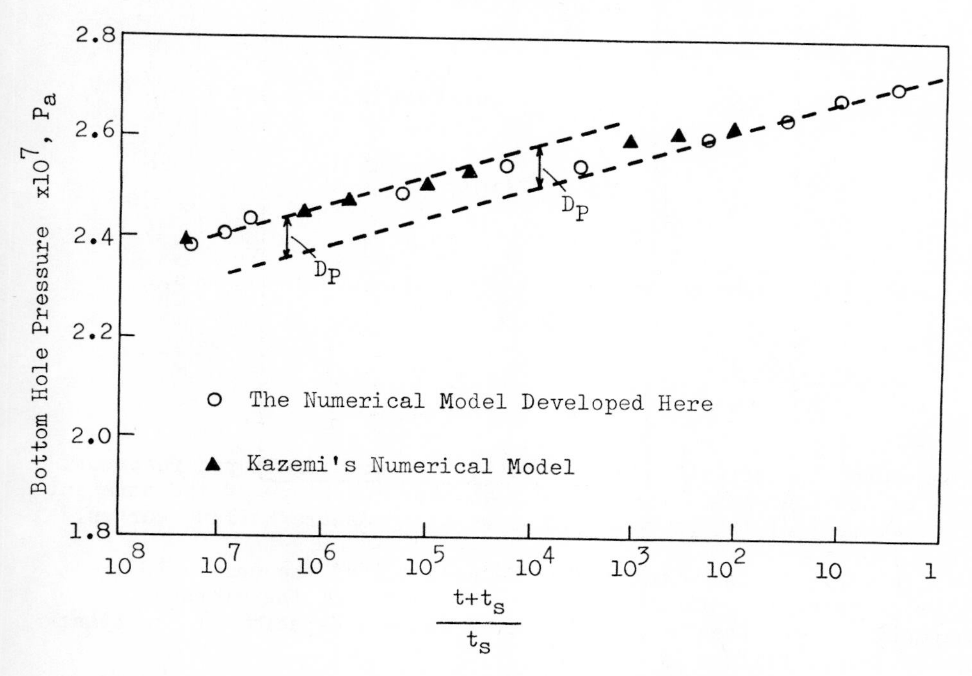

Fig. 7 Pressure Buildup for Case 1

case of one well in an infinite homogeneous reservoir, the relationship between the slope of the straight line segment and the reservoir properties is given by the following equation.[9)]

$$m = \frac{2.303\, q\, \mu}{4\pi\, Kh} \qquad (7)$$

Equation (7) holds even in a finite reservoir if the boundary conditions do not affect the curves. In the case of a fractured reservoir, a similar relationship can be obtained. The relationship between the absolute value of the slope of the straight line segment, m_f, and reservoir properties may be given by

$$m_f = \frac{2.303\, q\, \mu}{4\pi\, K_f b} \qquad (8)$$

The effective transmissibility, T_e, for a fractured reservoir is tentatively defined[7)] by

$$T_e = \frac{K_f \cdot b}{\mu} + \frac{K_{ma} \cdot h_{ma}}{\mu} \qquad (9)$$

If $K_f b$ is much larger than $K_{ma} h_{ma}$, Equation (9) reduces to

$$T_e = \frac{K_f b}{\mu} \qquad (10)$$

Equation (8) can be rewritten with Equation (10) as follows.

$$m_f = \frac{2.303\, q}{4\pi \cdot T_e} \qquad (11)$$

Therefore, the effective transmissibility for a fractured reservoir can be obtained approximately by analyzing the slope of the straight line segment of the pressure drawdown and pressure buildup curves.

The other factor obtained by the analysis of the curves is the storage capacity ratio,[1)] ω. It is the ratio of the storage capacity of the fracture to the total storage capacity of the reservoir, and is written by

$$\omega = \frac{\phi_2 c_2}{\phi_1 c_1 + \phi_2 c_2} \qquad (12)$$

For the reservoir model shown in Figure 1, Equation (12) is rewritten[2] by

$$\omega = \frac{b\phi_f c_f}{(h-b)\phi_{ma} c_{ma} + b\phi_f c_f} \qquad (13)$$

After the reliability study, Kazemi[2] suggested that the storage capacity ratio is related to the vertical separation of the two straight line segments, D_p.

$$\omega = e^{-2.303(D_p/m_f)} \qquad (14)$$

Table 2 lists the storage capacity ratio for the synthesized cases. The calculated results by Equation (14) are compared with the true values obtained from Equation (13). Table 2 shows that the approximate value of storage capacity ratio can be obtained from Equation (14) using the vertical separation and the slope, instead of Equation (13).

From Equations (11) and (14), it can be derived that small values of m_f and D_p give large values of T_e and ω, which may indicate a well fractured reservoir.

Table 2 Effective Transmissibility and Storage Capacity Ratio for the Synthesized Cases

	Case 1	Case 2
T_e $(m^2 \cdot m/P_a \cdot s)$	5.4×10^{-11}	1.9×10^{-10}
ω, from Eq.(13)	2.4×10^{-2}	1.4×10^{-2}
ω, from Eq.(14)	2.7×10^{-2}	1.9×10^{-2}

CONCLUSIONS

Pressure drawdown and pressure buildup tests for synthesized fractured geothermal reservoirs are simulated by the numerical model and the electrical model developed by the authors. The results of these models are in good agreement with those of other models. After some research on pressure drawdown and pressure buildup curves of hypothetical cases, it is found that the absolute value of the slope of the straight line segment of the curves is used to calculate the effective transmissibility for fractured reservoirs, and that the sufficient vertical separations of the two straight line segments give the storage capacity ratio. The combination of large values of effective transmissibility and storage capacity ratio may indicate a well fractured reservoir.

NOMENCLATURE

b = fracture aperture, m

c = compressibility, P_a^{-1}

c^* = effective compressibility, P_a^{-1}

C_F = fluid capacitance, m^3/P_a

D_p = vertical separation of parallel linear segment of pressure drawdown or pressure buildup curve, P_a

h = thickness, m

K = permeability, m^2

K^* = effective permeability, m^2

m = absolute value of slope of linear segment of pressure drawdown or pressure buildup curve, P_a/cycle

P = pressure, P_a

P^* = initial pressure, P_a

Q = volumetric rate of production per unit volume of reservoir, $m^3/s/m^3$, production is positive and injection is negative

q = production rate of well, m^3/s

r = radial coordinate, m

r_e = reservoir radius, m

r_i = radius of the i-th block, m

r_w = well radius, m

R_F = resistance to fluid flow, $P_a \cdot s/m^2 \cdot m$

T = temperature, K

T_e = effective transmissibility for a fractured reservoir, $m^2 \cdot m/P_a \cdot s$

t = flow time, s

t_s = shut-in time, s

U = transformed radial coordinate

$X(r,z)$ = coordinates of a point

X_o = coordinates of the production or injection well

z = vertical coordinate, m

$\delta(X-X_o)$ = Dirac-delta function
= 1 for $X=X_o$, 0 otherwise

θ = angle, rad

μ = viscosity, $P_a \cdot s$

ϕ = porosity, fraction

ϕ^* = effective porosity, fraction

ω = ratio of storage capacity of the fracture to total storage capacity

Subscripts

1 = matrix; attached to bulk matrix properties
2 = fracture; attached to bulk fracture properties
f = fracture; attached to point fracture properties
i = i-th block
ik = ik-th block
k = k-th block
ma = matrix; attached to point matrix properties
r = r-direction
z = z-direction

ACKNOWLEDGMENT

This work has been carried out by a grant-in-aid for developmental scientific research of the Ministry of Education, for which the authors are very grateful.

REFERENCES

1. Warren, J. E., and Root, P. J., "The Behavior of Naturally Fractured Reservoirs," Society of Petroleum Engineers Journal, Vol.3, Sept.1963, pp.245-255.

2. Kazemi, H., "Pressure Transient Analysis of Naturally Fractured Reservoirs with Uniform Fracture Distribution," Society of Petroleum Engineers Journal, Vol.9, Dec.1969, pp.451-462.

3. de Swaan, A. O., "Analytic Solutions for Determining Naturally Fractured Reservoir Properties by Well Testing," Society of Petroleum Engineers Journal, Vol.16, June 1976, pp.117-122.

4. Bruce, W. A., "An Electrical Device for Analyzing Oil Reservoir Behavior," Transactions, American Institute of Mining, Metallurgical, and Petroleum Engineers, Vol.151, 1943, pp.112-124.

5. Rumble, R. C., Spain, H. H., and Stamm, H. E. III, "A Reservoir Analyzer Study of the Woodbine Basin," Transactions, American Institute of Mining, Metallurgical, and Petroleum Engineers, Vol.192, 1951, pp.331.

6. McDowell, J. M., "Performance of Water Drive Reservoirs, Including Pressure Maintenance, as Determined by the Reservoir Analyzer," Transactions, American Institute of Mining, Metallurgical, and Petroleum Engineers, Vol.204, 1955, pp.73-78.

7. Hirakawa, S., and Yamaguchi, S., "Effective Transmissibility for Fractured Oil Reservoirs," Journal of the Japan Petroleum Institute, under consideration.

8. Yamaguchi, S., and Hirakawa, S., "Electrical Reservoir Analyzers of a Fractured Oil Reservoir," Journal of the Japan Petroleum Institute, under consideration.

9. Mattews, C. S., and Russel, D. G., "Pressure Buildup Analysis," Pressure Buildup and Flow Tests in Wells, Monograph Vol.1, AIME, New York, 1967, pp.18-34.

INFLUENCE OF PREEXISTING DISCONTINUITIES ON THE HYDRAULIC FRACTURING PROPAGATION PROCESS

J.-C. Roegiers[1], J. D. McLennan[2] and D. L. Murphy[3]

ABSTRACT

Appreciation of the mechanics of hydraulic fracture propagation in the vicinity of a preexisting discontinuity is of dramatic importance for economically maximizing resource recovery. The importance of the various controlling mechanisms, including the angle of incidence between the fracture and the discontinuity as well as incomplete penetration of fluid in the fracture, is synthesized and evaluated. Evaluations were performed qualitatively, analytically and numerically (using the Displacement Discontinuity Method).

INTRODUCTION

The success of a hydraulic fracturing treatment is usually measured in terms of increased productivity. In order to maximize the economic returns, an integrated design approach should incorporate reservoir characteristics and treatment fluid properties. Numerical design calculations are rapidly improving, embracing more realistic representations of the in-situ regime and the treatment process.

---

[1]Senior Associate Scientist, Dowell Division of Dow Chemical U.S.A., Tulsa, Oklahoma

[2]Senior Research Engineer, Dowell Division of Dow Chemical U.S.A., Tulsa, Oklahoma

[3]Far East Regional Marketing Manager, Dowell Schlumberger International, Inc., Killiney Road, P. O. Box 383, Singapore 9123

However, many of the numerical calculations do not adequately reflect the discontinuous nature of the medium. This contributes to unexpected performance, wherein post-treatment analyses strongly suggest a different fracture geometry than originally anticipated.

This paper highlights the influence of in-situ discontinuities on hydraulic fracturing, building on extensive published information. Additional considerations are outlined and some of the implications for treatment designs and back-analyses are discussed.

A fundamental issue is the behavior of a hydraulically induced fracture when approaching, or intersecting, a preexisting disjunction. Knowledge of the required conditions for fracture impediment exerts an overwhelming influence on fracture design. It should be indicated, however, that although fracture arrest may be desirable under certain circumstances, there are conditions when it is detrimental. For instance, fracture propagation across a boundary and into unproductive zones can drastically reduce the stimulation efficiency (i.e., ratio of increased production to treatment cost). On the other hand, when a hydraulic fracture propagates within the pay zone, it may encounter discontinuities. Deviations in the fracture path should be minimized as they may impede the fluid flow and cause obstructions for the proppants.

In addition, boundary conditions and material properties on each side of a discontinuity may not be identical. Joint characteristics (aperture, stiffness, infilling) are also important control parameters.

This paper is directed toward providing a qualitative review and sensitivity analysis of some of the pertinent parameters. Consequently, the mathematical considerations and numerical calculations discussed are two-dimensional, a situation which is strictly justifiable only after sufficient fracture extension has occurred.

PROPAGATION CRITERIA

As early as 1921, Griffith suggested that, based on thermodynamic equilibrium conditions, the amount of strain energy released during fracture propagation must equal the energy required to form new surface areas. This requires that the strain energy release rate be equal to a critical material property value. In 1957, Irwin stated that Griffith's approach could be expressed in terms of stress intensity factors for a linear elastic, homogeneous medium. This required the existence of an inverse root radius singularity in stress at the tip of the crack and assumed in-plane propagation. The situation in hydraulic fracturing is more complicated since coupled fluid flow must be considered. Moreover, only a portion of the propagating fracture is fluid-filled (Khristianovic and Zheltov, 1955) making the problem more complex.

Considering the case in Figure 1, Barenblatt (1962) derived the following expression for the stress intensity factor for a two-dimensional, plane strain situation.

$$K_I = 2P_o\left(\frac{\ell}{\pi}\right)^{1/2} \sin^{-1}\left(\frac{\ell_o}{\ell}\right) - \sigma_3 (\pi\ell)^{1/2} \tag{1}$$

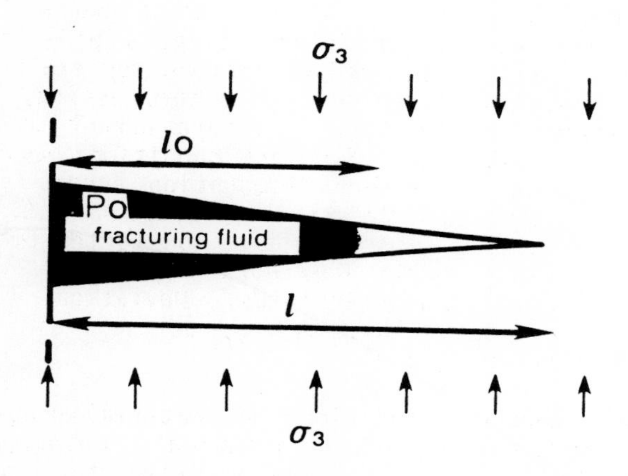

FIGURE 1

Partially Pressurized Hydraulic Fracture.

The second term on the right-hand side of Equation (1) becomes dominant as the fracture grows. Furthermore, under normal field conditions, fracture extension is extremely sensitive to the fluid pressure in the fracture. An increase in the fluid pressure by a small percentage of the field stress can increase the stress intensity factor significantly. Khristianovic and Zheltov's prediction of a "fluid lag" behind the fracture tip has certain implications. An alternate expression for Equation (1) at incipient extension is as follows.

$$\frac{\ell_o}{\ell} = \sin\left[\frac{\ell}{2P_o}\left(K_{IC}\sqrt{\frac{\pi}{\ell}} + \pi\sigma_3\right)\right] \tag{2}$$

For the restricted case of $K_{IC} = 0$, Figure 2 shows the characteristics of this equation. This case closely represents a long hydraulic fracture, the first term of Equation (2) becoming negligible as the length of the crack increases.

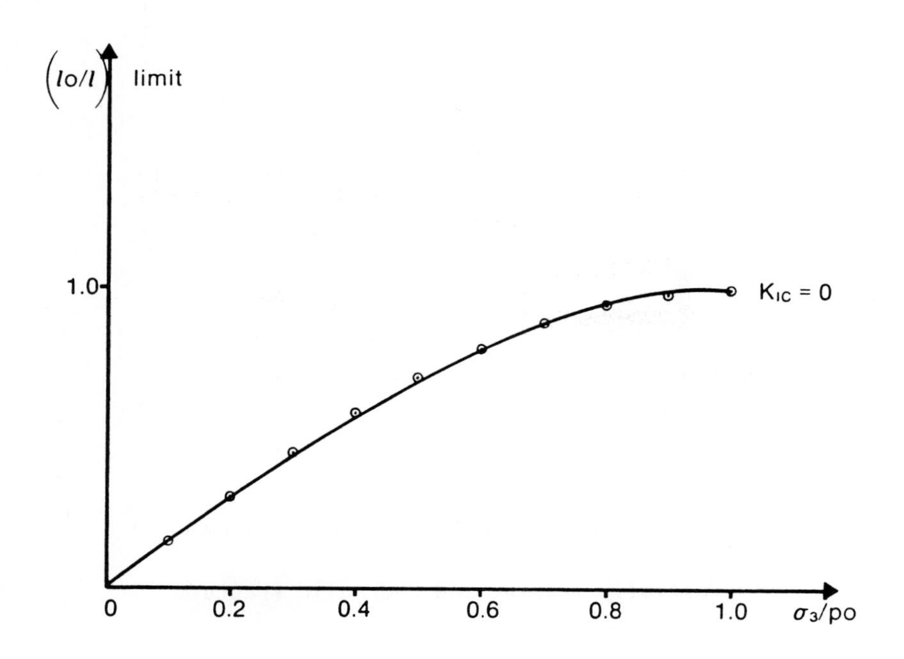

FIGURE 2

Fluid Lag as a Function of the Overpressure (Net Fracture Pressure).

One implication is that as the pressure in the fracture is kept closer to the far-field minimum principal stress, the degree of fluid penetration increases. In other words, the higher the pumping rate or the fluid viscosity, the farther behind the fluid front will remain. This may encourage more stable fracture propagation and explain why under laboratory conditions (where fracture lengths are small) different values of (ℓ_0/ℓ) are obtained depending on how well the fracture propagation process is regulated. In an uncontrolled situation, i.e., almost instantaneous propagation to the specimen boundaries, a large lag will be observed (Daneshy, 1978). The propagation characteristics are further modulated if the fracture approaches some form of interface.

FRACTURE APPROACHING AN INTERFACE

Consider the two-dimensional situation schematically shown in Figure 3.

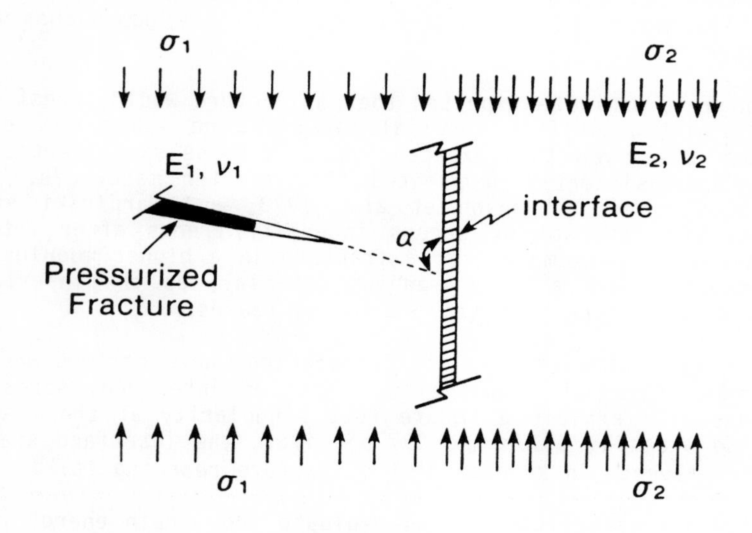

FIGURE 3

Schematic Representation of a Hydraulic Fracture Approaching a Preexisting Discontinuity.

The pertinent parameters influencing additional propagation are the following:

(1) the elastic properties;

(2) the far-field stresses;

(3) the interface characteristics;

(4) the angle of incidence, α; and

(5) the fluid lag (ℓ_0/ℓ).

These aspects are each described in subsequent sections.

Influence of Material Properties

Several authors (Cook and Erdogan, 1972; Erdogan and Biricikoglu, 1973; Ashbaugh, 1978; Hanson and Shaffer, 1980...) have considered the case of constant far-field stress across a bonded interface. The assumptions made in most of the published analyses were that the hydraulic fracture was perpendicular to the interface and that the entire fracture was uniformly pressurized. Typical results are shown in Figure 4a. It is indicated that the stress intensity factor decreases to zero when the fracture approaches a stiffer material and increases to infinity when the fracture approaches a less stiff material.

Although it would be expected that a fracture would arrest at the interface with a stiffer material, in-situ and laboratory experiments have disproved this expectation. In these experiments, fractures have consistently propagated into stiffer materials. Among the reasons cited by Hanson et al., 1978, and Warpinski et al., 1978, was the reversal of stress intensity trends after interface penetration. For example, for a fracture in a higher modulus material growing toward a lower modulus material, as the interface is penetrated the stress intensity factor decreases.

Furthermore, stress intensity formulations were derived under the assumptions of medium homogeneity. Near an interface, stresses do not necessarily exhibit a square root singularity at the crack tip (Goree and Venezia, 1977a and 1977b); i.e., the interface starts to exert an influence in advance of the fracture reaching it.

An alternative approach is to evaluate the strain energy release rate. Roegiers and Wiles, 1980, asserted that propagation trends in the vicinity of a bimaterial interface can be evaluated using --

$$G \propto \frac{\Gamma(\lambda + 1)\,\Gamma(\lambda)}{\Gamma(2\lambda + 1)}, \qquad (3)$$

where

Γ = Gamma function,

λ = constant dependent on the geometry and material properties (see Figure 5),

$\lambda = .5$ homogeneous material,

$\lambda = 0$ free surface, and

$\lambda = 1$ fixed surface.

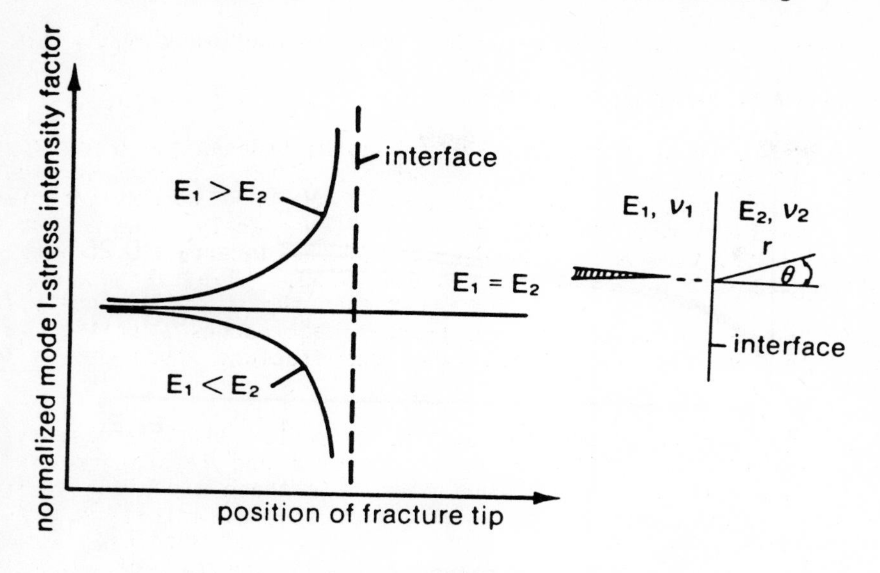

(a) after Hanson and Shaffer, 1980

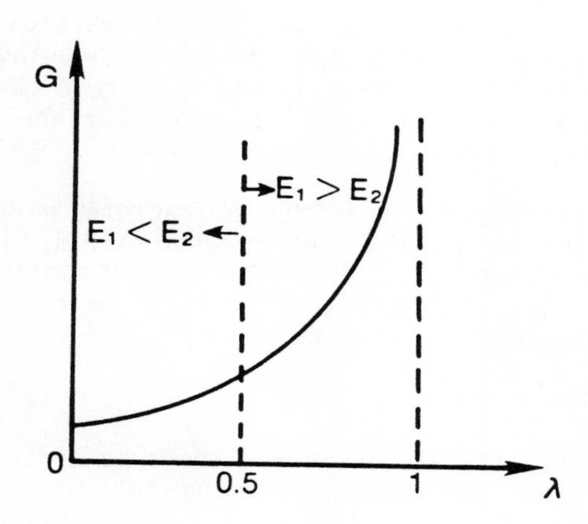

(b) after Roegiers and Wiles, 1980

FIGURE 4

Stress Intensity Factor and Strain Energy Release Rate Variations for a Fracture Approaching a Preexisting Discontinuity.

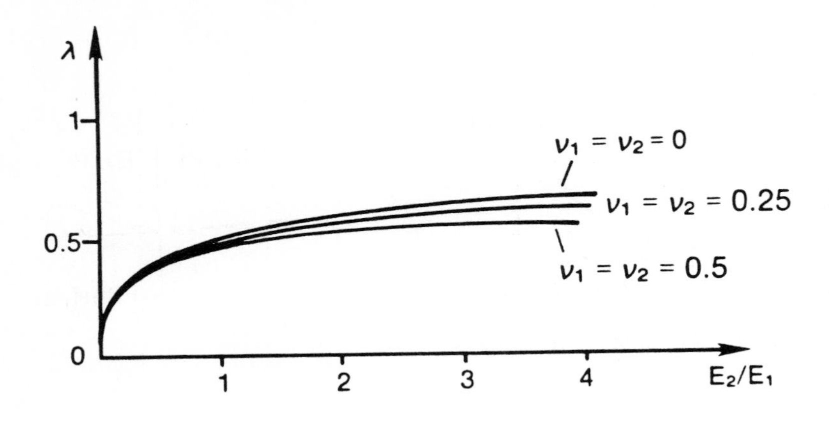

FIGURE 5

Variation of λ as a Function of the Material Properties.

This tendency is shown in Figure 4b for a fracture in Medium 1 approaching Medium 2. It is important to note that the strain energy release rate has a finite value at the interface, dependent on the stiffness ratio.

The indicated tendency is that for a fracture in a lower stiffness medium approaching a higher stiffness medium, arrest will be encouraged. If penetration occurs, this may not be the case. However, in stiffer zones, fracture aperture will be reduced (all other parameters being equal), impeding fluid flow.

Influence of Far-Field Stresses

Abou-Sayed et al., (1978) considered the case of an abrupt change in the in-situ stress magnitude, assuming intimate bonding conditions at the interface. They derived the following expression for the stress intensity factor.

$$K_I = 2(\sigma_b - \sigma_a)\left(\frac{\ell}{\pi}\right)^{1/2} \sin^{-1}\left(\frac{\ell_o}{\ell}\right) + (P_o - \sigma_b)(\pi\ell)^{1/2} , \qquad (4)$$

where

σ_a = far-field stress acting in the pay zone,

σ_b = far-field stress acting in the barrier.

ℓ = fracture half-length,

$2\ell_o$ = pay-zone thickness, and

P_o = fracture pressure.

This equation indicates the importance of stress contrast in arresting growth into more highly stressed layers. Additional propagation would require fluid pressure changes in the order of the in-situ stress contrast. An additional parameter of importance is an alteration in stress levels because of total stress changes owing to fluid penetration (Cleary, 1979). The fracture aperture is dependent on the stress level. Since the flow rate is very sensitive to the aperture, the in-situ stress conditions exert a strong limiting influence even if some penetration occurs.

The importance of stress contrasts has been demonstrated in field and laboratory situations and also by numerical modeling (Palmer and Carroll, 1982; Advani, 1980...).

Influence of Interface Characteristics

In reality, interfaces do not constitute perfect bonds between the two constitutive media. The strength and stiffness of the discontinuities play an important role in crack arrest phenomena. Several authors (Hanson et al., 1978; Teufel, 1979; Anderson, 1979; Hanson et al., 1979; Shaffer et al., 1979) have conducted experiments highlighting the influence of preexisting discontinuities. Teufel (1979) associated arrest with a complete lack of transmission of shear stress across the interface. Teufel's analyses were limited, however, to fractures propagating normal to the interface.

Roegiers and Wiles (1981) used conformal mapping to predict stresses around a partially pressurized fracture approaching an interface, along which slippage was prevented. Results are shown schematically in Figure 6, indicating the importance of joint properties. For example, for a gouge-filled feature exhibiting low shear strength (τ/σ), arrest will be encouraged except at large values of (σ_1/σ_3).

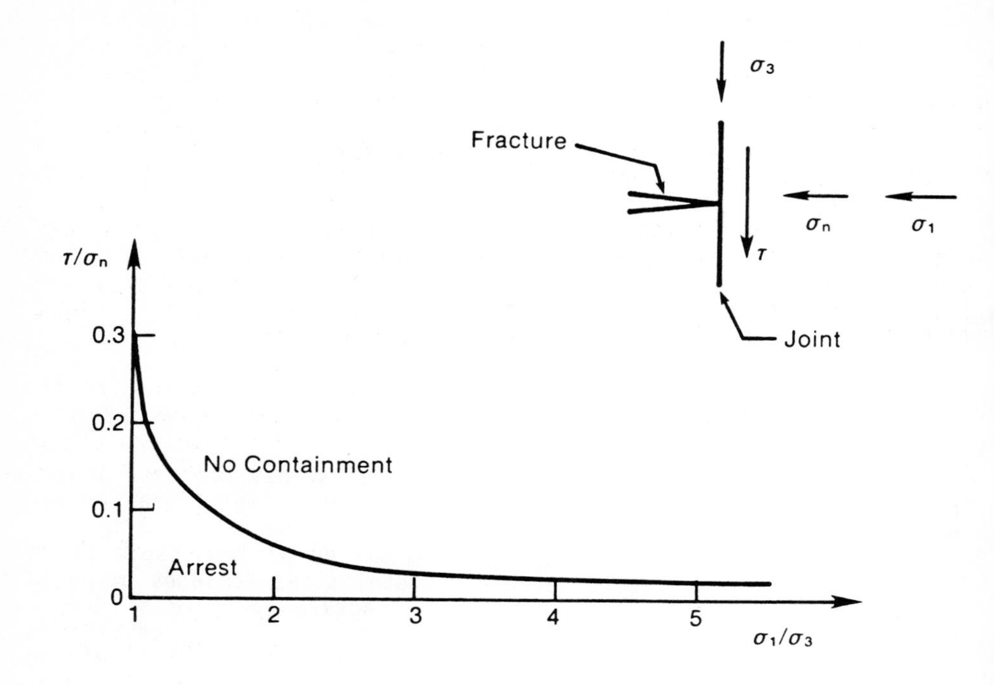

FIGURE 6

Frictional Strength Required for Fracture Containment as a Function of the Far-Field Principal Stress Ratio.

However, even if local slippage occurs, sufficient force may still be transmitted across the interface to reinitiate fracture growth if massive slip does not occur.

Influence of Angle of Incidence

In-situ discontinuities may not trend parallel to the maximum in-situ stress, depending on their mode of origin and the in-situ regime during their creation. The angle of incidence of a hydraulic fracture (opening perpendicular to the far-field minimum principal stress) is another parameter governing fracture arrest.

Laboratory experimentation (Blanton, 1982) strongly suggested the role of the angle of incidence:

> "The systematic experiments in hydrostone in particular show that hydraulic fractures tend to cross preexisting fractures

only under high differential stresses and high angles of approach. At intermediate and low-differential stresses and angles of approach, preexisting fractures tend either to open and divert fracturing fluid or arrest propagation of the hydraulic fracture... . It would be more likely to have fractures with wings diverted at different angles or with truncated wings of different lengths."

Blanton's observations are supported by numerical work by the authors. A suite of two-dimensional displacement discontinuity simulations was run for a uniformly pressurized fracture approaching a long discontinuity at various angles of incidence. Figure 7 highlights some important (although not unexpected) features.†

(1) For all angles of incidence, there is a local reduction in normal stress across the joint, in some cases below the level of the minimum in-situ stress. For inclined features, normal stress along the joint is reduced <u>ahead</u> of the fracture, encouraging joint inflation.

(2) These trends occur even before the joint intersects the feature.

(3) Figure 8 shows the variation of the ratio of shear-to-normal stress along the discontinuity for various angles of incidence indicating lower values of (τ/σ) for perpendicular features, suggesting less chance of arrest.

---

†Although these analyses were performed for a specific set of joint properties as well as specific stress state and material properties (invariant on either side of the discontinuity), qualitative observations should be significant.

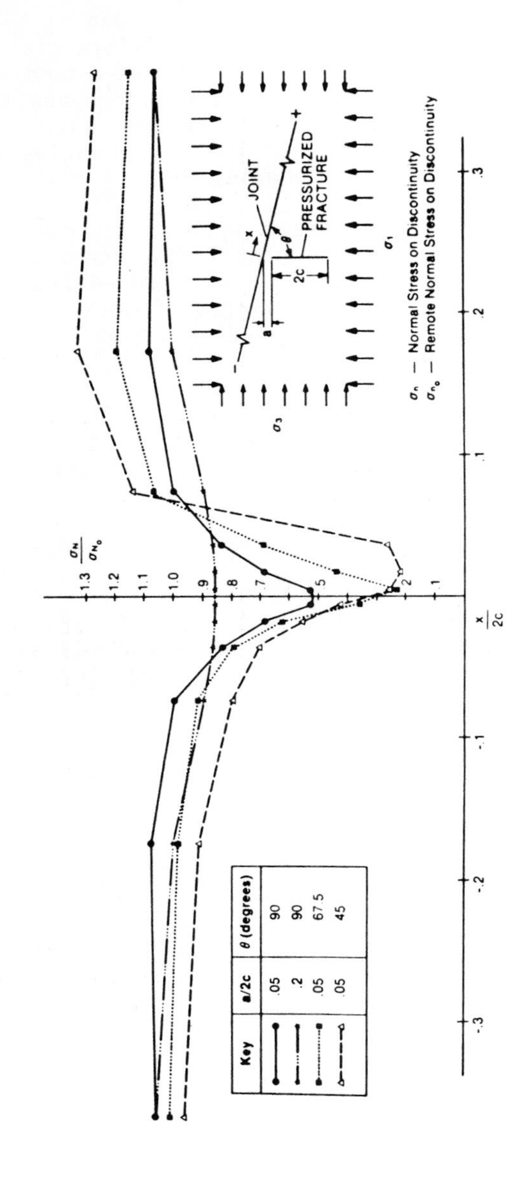

FIGURE 7

Normal Stress Profile along the Discontinuity as a Function of the Angle of Incidence of an Approaching Hydraulic Fracture.

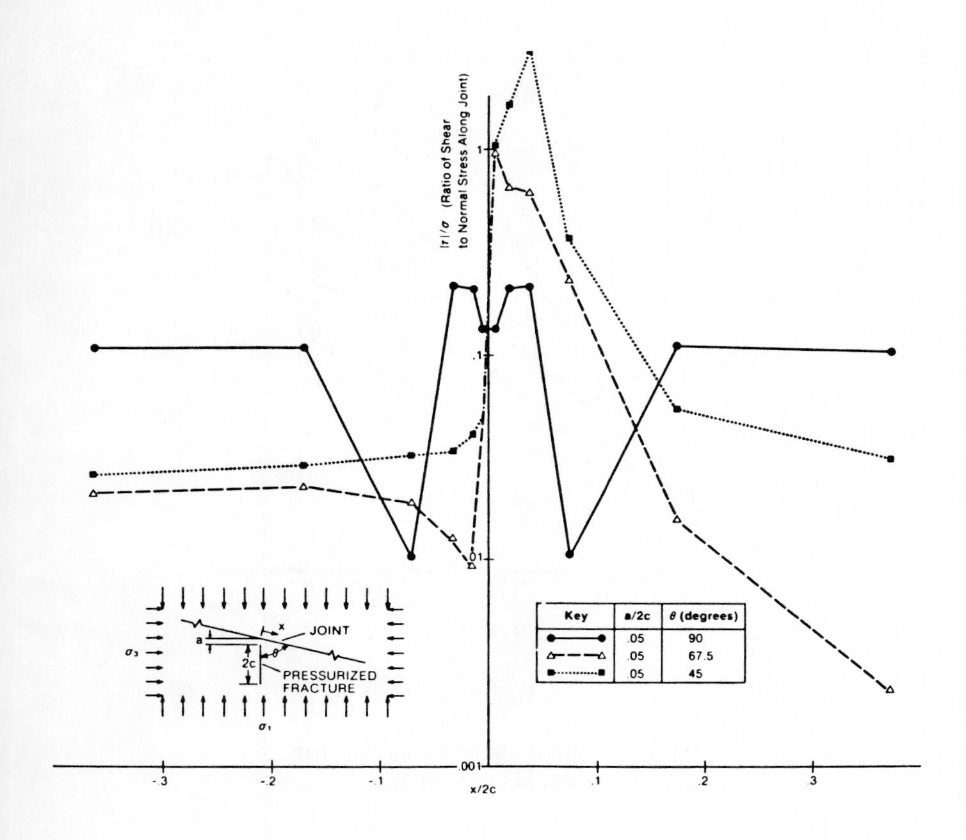

FIGURE 8

Discontinuity Shear Stress Profile as a Function of the Angle of Incidence of the Approaching Hydraulic Fracture.

Influence of Fluid Lag

The influence of fluid lag has been examined to a limited degree, largely because of the required numerical coupling between stress, fluid flow and fracture propagation. Figure 9 is a representation of the maximum shear stress (normalized with respect to the normal stress) as a function of fluid pressure for a no-slip interface. In general, (p_0/σ_1) will be relatively small so that fluid lag will not be a major containment consideration.

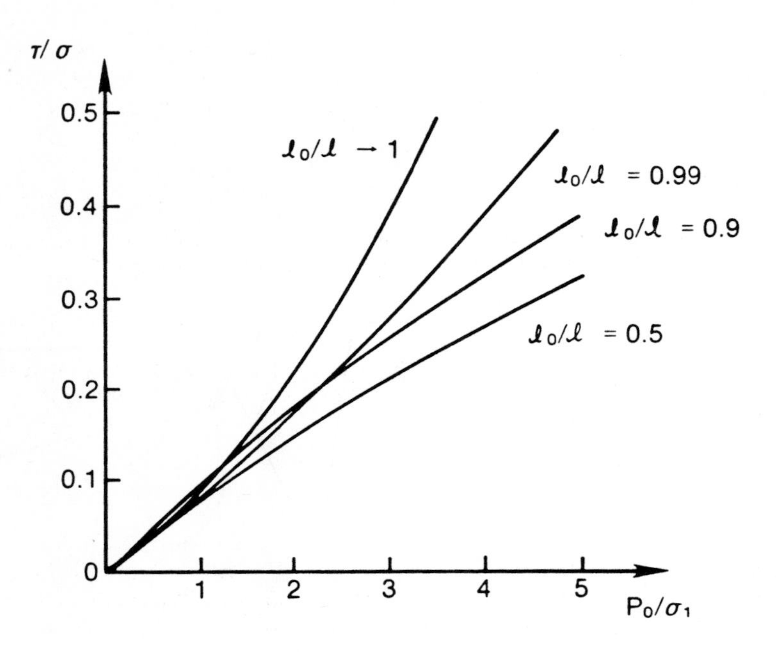

FIGURE 9

Maximum Shear Stress Ratio for Various Stress States.

OBSERVATIONS AND CONCLUSIONS

Several parameters influence the feasibility of fracture arrest because of a preexisting discontinuity.

Material Property Variations

Generally, material property contrasts are second-order controlling parameters, particularly when the realistic degrees of variability are appreciated. Material property variations influence propagation behavior but generally do not halt the process. Contrasts may encourage effective containment (aperture control).

In-Situ Stress Variations

Stress contrasts exert a dominant influence on containment. Pore pressure effects must be incorporated into containment considerations.

Material Interface

Fracture arrest is usually associated with massive shear slip on the interface. Consequently, interface strength properties are important. Such considerations are equally applicable to faults and joints as to lithologic boundaries.

Angle of Incidence

Hydraulic fractures skewed to in-situ discontinuities will probably have a higher opportunity for arrest or reorientation.

Fluid Lag

Under most conditions, fluid lag is probably a second-order consideration. Further investigation is required to establish this influence more precisely.

REFERENCES

1. Abé H., Mura, T. and Keer, L. M., "Growth Rate of a Penny-Shaped Crack in Hydraulic Fracturing of Rocks," Journal of Geophysical Research, V. 81, No. 29, 1976.

2. Abou-Sayed, A. S., Brechtel, C. E. and Clifton, R. J., "In-Situ Stress Determination by Hydrofracturing -- A Fracture Mechanics Report," Journal of Geophysical Research, V. 83, No. 36, pp. 2851-2862, 1978.

3. Advani, S. H., "Finite Element Model Simulations Associated with Hydraulic Fracturing," Soc. Pet. Eng. 8941, 1980.

4. Anderson, G. D., "Effect of Mechanical and Frictional Rock Properties on Hydraulic Fracturing Growth Near Unbonded Interfaces," U.S. Government Report UCRL-82456, July 1979.

5. Anderson, G. D. and Larson, D. B., "Laboratory Experiments on Hydraulic Fracture Growth Near an Interface," U.S. Government Report UCRL-80355, 1979.

6. Ashbaugh, N. E., "Stresses in Laminated Composites Containing a Broken Layer," Journal of Applied Mechanics, V. 40, p. 533, 1978.

7. Barenblatt, G. I., "The Mathematical Theory of Equilibrium Cracks in Brittle Fracture," Advances in Applied Mechanics, V. 7, p. 55, 1962.

8. Blanton, T. L., "An Experimental Study of Interaction between Hydraulically Induced and Preexisting Fractures,"

Unconventional Gas Recovery Symposium, Pittsburgh, SPE/DOE 10847, May 1982.

9. Chang, H. Y., "Development of Improved Design Criteria for Hydraulic Fracturing Operations," Ph.D. Thesis, West Virginia University, 1978.

10. Cleary, M. P., "Analysis of Mechanisms and Procedures for Producing Favourable Shapes of Hydraulic Fractures," Soc. Pet. Eng. 9260, 1980.

11. Cleary, M. P., "Primary Factors Governing Hydraulic Fractures in Heterogeneous Stratified Porous Formations," Energy Technology Conference and Exhibition, Houston, Texas, November 1978.

12. Cleary, M. .P., "Rate and Structure Sensitivity in Hydraulic Fracturing of Fluid-Saturated Porous Formations," 20th U.S. Symposium on Rock Mechanics, Austin, Texas, 1979.

13. Cook, T. S. and Erdogan, F., "Stress in Bonded Materials with a Crack Perpendicular to the Interface," International Journal Eng. Sci., V. 10, p. 677, 1972.

14. Daneshy, A. A., "A Study of Inclined Hydraulic Fractures," Soc. Pet. Eng. Journal, p. 61, April 1973.

15. Daneshy, A. A., "Hydraulic Fracture Propagation in Layered Formations," Soc. Pet. Eng. Journal, February 1978.

16. Erdogan, F. and Biricikoglu, V., "Two Bonded Half Planes with a Crack Going Through the Interface," International Journal Eng. Sci., V. 11, p. 745, 1973.

17. Goree, J. G. and Venezia, W. A., "Bonded Elastic Half-Planes with an Interface Crack and a Perpendicular Intersecting Crack that Extends into the Adjacent Material - I," International Journal Eng. Sci., V. 15, p. 1, 1977.

18. Goree, J. G. and Venezia, W. A., "Bonded Elastic Half-Planes with an Interface Crack and a Perpendicular Intersecting Crack that Extends into the Adjacent Material - II," International Journal Eng. Sci., V. 15, p. 19, 1977.

19. Griffith, A. A., "The Phenomena of Rupture and Flow in Solids," Phil. Trans. Royal Society London, V. A211, p. 163, 1921.

20. Griffith, A. A., "The Theory of Rupture," Proc. Int. Cong. Appl. Mech., 1924.

21. Hanson, M. E., Anderson, G. D. and Shaffer, R. J., "Effects of Various Parameters on Hydraulic Fracture Geometry," U.S. Government Report UCRL-83784, May 1980.

22. Hanson, M. E. and Shaffer, R. J., "Some Results from Continuum Mechanics Analyses of the Hydraulic Fracturing Process," Soc. Pet. Eng. Journal, April 1980.

23. Hanson, M. E., Anderson, G. D. and Shaffer, R. J., "Theoretical and Experimental Analyses of Hydraulic Fracturing and Some Reservoir Response to the Stimulation," U.S. Government Report UCRL-82578, 1979.

24. Hanson, M. E., Anderson, G. D., and Shaffer, R. J., "Theoretical and Experimental Research on Hydraulic Fracturing," U.S. Government Report UCRL-80558, 1978.

25. Irwin, G. R., "Analysis of Stresses and Strains near the Edge of a Crack Traversing a Plate," Trans. Am. Soc. Mech. Eng., Journal Applied Mech.

26. Irwin, G. R., "Relation of Stresses near a Crack to the Crack Extension Force," Proc. 9th Int. Cong. Appl. Mech., Brussels, p. 245, 1957.

27. Khristianovic, S. A. and Zheltov, Y. P., "Formation of Vertical Fractures by Means of Highly Viscous Fluid," Proc. 4th World Pet. Cong., V. 2, p. 579, 1955.

28. Mastrojannis, E. N., Keer, L. M. and Mura, T., "Growth of Planar Cracks Produced by Hydraulic Fracturing," Int. Journal Num. Methods, V. 15, p. 41, 1980.

29. Northrop, D. A., Warpinski, N. R. and Schmidt, R. A., "EGR Stimulation Research Project - Direct Observation of Hydraulic and Dynamic Fracturing," U.S. Government Report De-AC04-70P00789.

30. Palmer, J. D. and Carroll, H. B. Jr., "3-D Hydraulic Fracture Propagation in the Presence of Stress Variations," Unconventional Gas Recovery Symposium, Pittsburgh, SPE/DOE 10849, May 1982.

31. Roegiers, J.-C. and Wiles, T. D., "Hydraulic Fracturing of Rock Masses -- Influence of Preexisting Discontinuities," University of Toronto, Department of Civil Eng., Publication 81-04, ISBN 0-7727-2, April 1981.

32. Shaffer, R. J., Hanson, M. E. and Anderson, G. O., "Hydraulic Fracturing Near Interfaces," U.S. Government Report UCRL-83419, September 1979.

33. Simonson, E. R., Abou-Sayed, A. S. and Clifton, R. J., "Containment of Massive Hydraulic Fractures," Soc. Pet. Eng. Journal, Feb. 1978.

34. Teufel, L. W., "An Experimental Study of Hydraulic Fracture Propagation in Layered Rock," Ph.D. Thesis, Texas A&M University, 1979.

35. Van Elken, H., "Hydraulic Fracture Geometry: Fracture Containment in Layered Formation," Soc. Pet. Eng. 9261, 1980.

36. Warpinski, N. R., Schmidt, R. A., Northrop, D. C. and Tyler, L. D., "Hydraulic Fracture Behavior at a Geologic Formation Interface; Pre-Mineback Report," U.S. Government Report SAND78-1578, October 1978.

37. Warpinski, N. R., Schmidt, R. A. and Northrop, D. A., "In-Situ Stresses: The Predominant Influence on Hydraulic Fracture Containment," Soc. Pet. Eng. 8932, 1980.

38. Wiles, T. D. and Roegiers, J.-C., "Modeling Hydraulic Fractures Under Varied In-Situ Conditions Using a Displacement Discontinuity Approach," presented at 33rd Annual Tech. Meeting - CIM Paper 82-33-70, Calgary, June 1982.

GEOTHERMAL RESERVOIR SIMULATION FOR WATER DOMINANT FRACTURED RESERVOIR

by Seiichi Hirakawa, Saburo Wada and Tadao Tanakadate

Prof. Dr., The Faculty of Engineering
Univ. of Tokyo, Tokyo

Lecturer, The Faculty of Engineering, Univ. of Tokyo
Managing Director, Japan Oil Engineering Co., Ltd.
Ginza, Tokyo

Consulting Engineer, Japan Oil Engineering Co., Ltd.
Ginza, Tokyo

ABSTRACT

Geothermal reservoirs in Japan, except Matsukawa, are mostly the water dominated type, and their reservoir scales are comparatively small. Furthermore, since most of the geothermal areas are located near the national parks, environmental regulations prescribed by the government are very strict in Japan. On the other hand, reinjection of disposal water is an inevitable operation for the development of water dominated geothermal reservoirs.

It is very important to appraise the effect of reinjection of disposal water on production wells, and the simulation of geothermal reservoirs is applied here to this appraisal. It is found by now that the main factor governing the flow path of injected water to the producers is the presence of fracture in the reservoir rock. Analysis of tracer tests turned out to be an effective method to distinguish the fractured reservoir from those made of only porous media.

This paper mentions the application of tracer test analysis to classify fractured reservoirs and the method of geothermal reservoir simulation applied to fractured reservoirs in Japan.

INTRODUCTION

At present, five geothermal reservoirs under production (Otake, Onuma, Onikobe, Hachobaru, and Kakkonda), which supply the major share of geothermal power generation in Japan, are all the water dominated type. Most of them are located near or inside the national parks. The Japanese regulation to preserve the natural environment in these areas is so strict that most of the geothermal fields are limited by the border of national parks. As a result, their reservoir scales are comparatively small and the reinjection of disposal water back to underground is compulsory. The purpose of reinjection is mainly to dispose of the waste water after flashing back to underground, avoiding the surface discharge of harmful materials included in geothermal water.

Besides, the controversial effects of reinjection of the geothermal recovery have been discussed. Reinjection of disposal water prevents the loss of material from the reservoir and maintains the reservoir pressure, consequently improving resource recovery and preventing surface subsidence or formation collapse. On the other hand, since the temperature of waste water is lower than the reservoir temperature, reinjection causes the reduction of reservoir energy resulting in the loss of well productivity and in worse recovery. This controversial problem would be answered by a quantitative appraisal of the effects of reinjection upon geothermal production. The history of geothermal production from water dominant reservoirs is relatively short compared with that from vapor dominated reservoirs. At present, several reports are coming out on this subject from the field site(1)(2)(3).

This paper also tries to introduce one approach which is being examined in the Japanese fields. A new type of numerical reservoir simulation is applied here to monitor the observed performance of geothermal reservoir behavior realized by producing and reinjecting hot water.

The past observation of production injection cycles in geothermal reservoirs suggested that the main factor governing the flow path of injected water to the producers was the properties of reservoir rock, fractured or non-fractured. Methods to find the rock type need to be devised; a successful tracer test and its analysis method are introduced. Reservoir models are also required to simulate flows through both fractured and non-fractured porous media. The authors' approach to these two types of reservoir simulation brings out a combination of two reservoir models; a macroscopic wide area model and microscopic small sector model. The Wide Area Model covers the whole development area including the surrounding aquifer. Flow behavior of water is simulated macroscopically with the large size grid blocks. Unsteady inflow of aquifer water into the production injection area through non-fractured porous rock is computed with the pressure and temperature distribution at any moment. The role of the Wide Area Model is to reproduce a long-range performance of the whole geothermal reservoir by matching the performance of observation wells drilled in the surrounding aquifer area, and to give the boundary conditions to Sector Model at any required time stage. The Sector Model, with comparatively smaller grid blocks, covers the main production injection area. Flow through the fracture path from injector to producer is simulated by this model for the short period when the tracer test is conducted in the field. The simulation ends up with the computation of velocity and route of injected water flow into producer. The combination of these results with the tracer test analysis finally gives the permeability and porosity of the fracture path and the rate of contamination of producing water by reinjected water. The procedure described here provides a new tool for the quantitative assessment of the effects of reinjection of disposal water in geothermal fields.

ANALYSIS OF TRACER TEST

In Japan, many results of tracer tests are reported in the water dominated reservoirs. Most of them are conducted successfully with a sharp peak response at the producer within a short period of flow time. Some of them, however, failed without any detection of tracer ion at the producer. These tracer test results indicate that the flow path of reinjected water carrying the tracer ion from injector to producer is different in the two tests. A sharp peak response of tracer ion implies the travel of higher velocity through the fracture, while low velocity flow through non-fractured porous media results in no indication of peak response. The response of the tracer ion at a producer is thus interpreted qualitatively as an indication of the presence of a fracture path.

The authors present the quantitative examination of this interpretation by introducing a new analysis method of tracer test result.

Diffusion of tracer ion concentration in geothermal water is expressed by the general form of Fick's diffusion equation.

$$\nabla \cdot D_f \nabla C + \nabla \cdot Dp \nabla C - V \cdot \nabla C + Cad = \frac{\partial C}{\partial t} \qquad (1)$$

Here, the concentration change due to molecular diffusion and adsorption - desorption is comparatively minor, so that the first and fourth terms in the left-hand side of Eq. (1) are assumed negligible. Tracer ion concentration spreads out mainly because of dispersion and convection effects. The flow path now in consideration is a fracture connecting injector and producer. For simplicity, one dimensional flow is assumed and Eq. (1) is solved analytically with the initial and boundary conditions below.

$$\frac{\partial}{\partial x}\left(Dp\frac{\partial C}{\partial x}\right) - V\frac{\partial C}{\partial x} = \frac{\partial C}{\partial t} \qquad (2)$$

$$\begin{aligned} C &= Cl \quad \ldots\ldots \quad x = 0,\ 0 \leqq t < \Delta t \\ C &= Co \quad \ldots\ldots \quad x = 0,\ t \geqq \Delta t \\ C &= Co \quad \ldots\ldots \quad x = 0,\ t = 0 \end{aligned} \tag{3}$$

where the inital tracer ion concentration at the injector Cl is practically defined by

$$Cl = Co + \frac{I}{Gv\Delta t} \tag{4}$$

Solved for the concentration at location x and time t,

$$C(x,t) = Co + \frac{C_1 - C_0}{2}[\{\mathrm{erfc}(\frac{x-V.t}{2\sqrt{Dp.t}}) + e^{\frac{V\ x}{Dp}}\mathrm{erfc}(\frac{x+V.t}{2\sqrt{Dp.t}})\} - \{\mathrm{erfc}(\frac{x-V(t-\Delta t)}{2\sqrt{Dp(t-\Delta t)}}) + e^{\frac{V\ x}{Dp}}\mathrm{erfc}(\frac{x+V(t-\Delta t)}{2\sqrt{Dp(t-\Delta t)}})\}] \tag{5}$$

In Eq. (5), the concentration of tracer ion is expressed as a function of distance of location from injector (x), travel time (t), velocity of convective stream (V), and dispersion coefficient (D_p). Among these parameters, the dispersion coefficient is known to be proportional to convective velocity and the linear correlation between these two parameters is readily available by a laboratory test.

In the tracer test, the tracer ion is detected continuously at a producer which has a fixed distance (xo) from the injector. Therefore, the tracer ion response curve is expressed as the concentration change of tracer ion with time measured at a constant distance from the injector, i.e. the fixed length of flow path. The remaining parameter which mainly defines the shape of the tracer ion response curve is the velocity of travel (V). The determination of velocity depends on the hydraulic properties of travel route. Here Darcy's law is applied to calculate it.

$$V = \frac{k}{\phi\mu}\frac{\Delta P}{\Delta x} \tag{6}$$

By this expression, the tracer ion response curve is closely related to rock properties, i.e. permeability (k) and porosity (ϕ). The peak value of the concentration and its arrival time in the tracer ion response curve is utilized to calculate the rock properties of the flow path. It is presumed that the first and strongest arrival of the tracer ion at a producer travels through the shortest path (xo) or fracture path in the shortest arrival time (to) with a constant velocity (V). In the actual geothermal fields, the layout of injectors and producers is not simple. The Sector Model described below is operated to find out the shortest fracture path in the injector producer complex and the velocity of tracer travel. Since the Sector Model requires such fracture properties as permeability and porosity as an input, a trial and error procedure is repeated until the correct fracture properties are estimated with the best matching of tracer ion response curve.

WIDE AREA MODEL

In a water dominant geothermal reservoir, the mass flow rate expected for producers and injectors is fairly high. Therefore the wells are drilled in the more permeable area or highly fractured area to attain enough productivity, and a large aquifer area is also required to supply mass influx into this production injection area. The reservoir model has to take into account the mass and energy influx from the surrounding aquifer. It also has to reproduce the reservoir performance taking place in this surrounding area. The historical aquifer behavior which is monitored at the observation wells is simulated by this model.

The influx flow from the surrounding area into the injection production area has relatively low velocity, and the conventional model based on the flow through porous media is ready to be applied. The Wide Area Model, with large size grid blocks, is thus a conventional porous media model, which gives the boundary conditions as pressure and temperature distributions to the Sector Model at any time. The manner of areal division of the Wide Area Model and the Sector Model is demonstrated in Fig. 1.

Conservation of mass and energy, coupled with Darcy's law as a momentum balance, gives the basic mass and energy flow equations;

$$\nabla \frac{\rho k}{\mu}(\nabla P-\rho g \nabla D) - G = \phi \frac{\partial \rho}{\partial t} \tag{7}$$

$$\nabla\{\frac{\rho k h}{\mu}(\nabla P-\rho g \nabla D)\} + \nabla(K \nabla T) - G_L - G \cdot h = \frac{\partial}{\partial t}\{\phi \rho h + (1-\phi)\rho r C r T\} \tag{8}$$

These partial differential equations are solved numerically by a finite difference procedure. Solving Eqs. (7) and (8) simultaneously, pressure and temperature distribution in the reservoir are known at any moment,and the amount and direction of mass and energy influx provided from the surrounding area are also computed.

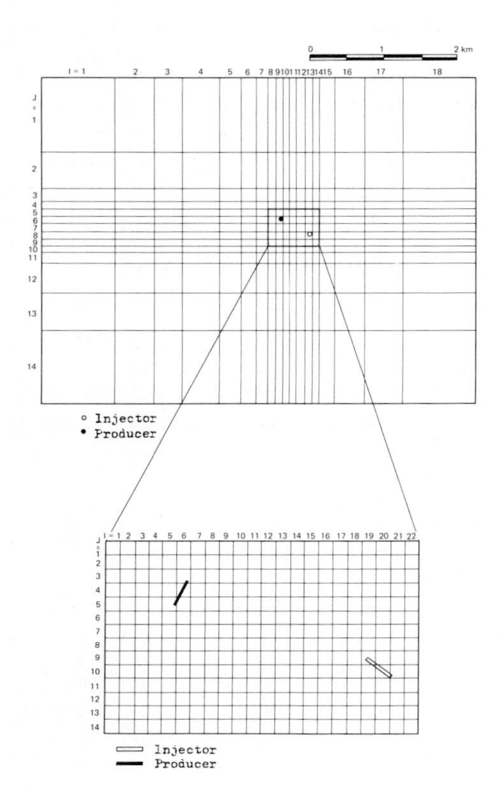

Fig. 1 Division of Grid Blocks for Wide Area Model(Above) and for Sector Model(Below)

SECTOR MODEL

While the hydrodynamic behavior taking place in the whole development area, which is simulated by the Wide Area Model, is slow in process and small in magnitude, the convective flow of injected water to producers occurs instantaneously with much higher velocity. By reinjecting the flashed water, pressure maintenance in the reservoir is quickly achieved and thermal drawdown at the producers is sooner or later observed. The highly permeable flow path, mostly a fractured zone connecting injectors and producers, has to be properly allocated inside the injection production area. The benefit and detriment of reinjection is quantitatively appraised well by well. The Sector

Model coupled with a tracer test analysis serves as a method to evaluate the effect of reinjection.

The theory of the Sector Model is based on the concept proposed by Warren and Root(4). For mass transportation this method has two basic assumptions.

First assumption is that there are two classes of porosity. Primary porosity is the matrix porosity and has high storage but low flow capacity. Secondary porosity consists of fracture and has a low storage but high flow capacity. The second assumption is that flow into the wellbore is only through the fracture network and this flow is unsteady. But flow between matrix and fracture is quasisteady.

Furthermore, the model to be employed in this investigation is based on the following general assumptions (See Fig. 2):

a) The material containing the primary porosity is homogeneous and isotropic, and is contained within a systematic array of identical, rectangular parallelepipeds.
b) All of the secondary porosity is contained within an orthogonal system of continuous, uniform fractures which are oriented so that each fracture is parallel to one of the principal axes of permeability; the fractures normal to each of the principal axes are uniformly spaced and are of constant width; a different fracture spacing of different width may exist along each of the axes to simulate the proper degree of anisotropy.
c) The complex of primary and secondary porosities is homogeneous though anisotropic; flow can occur between the primary and secondary porosities, but flow through the primary porosity elements cannot occur.

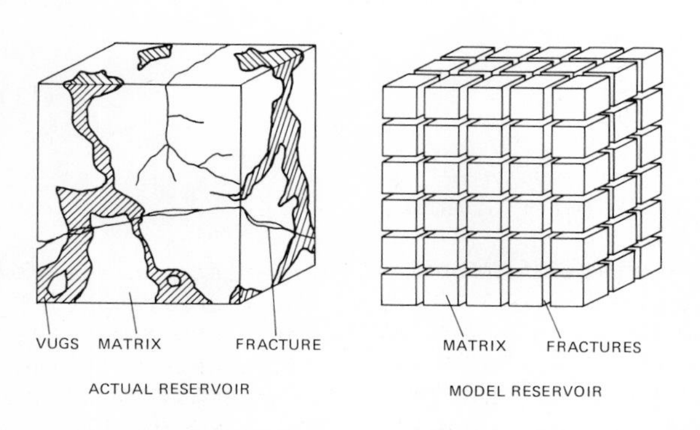

Fig. 2 Idealization of a Naturally Fractured Reservoir (after Warren and Root)

Based on these assumptions, the following mass balance equations are deduced:

$$\nabla\{\frac{\rho kf}{\mu}(\nabla Pf-\rho g\nabla D)\} - \frac{\rho km\sigma}{\mu}(Pf-Pm) - G = \frac{\partial(\phi f\rho)}{\partial t} \quad (9)$$

$$\frac{\rho km\sigma}{\mu}(Pf-Pm) = \frac{\partial(\phi m\rho)}{\partial t} \quad (10)$$

where, σ is a shape factor which reflects the geometry of the matrix elements.

For energy transportation three assumptions are made, corresponding to the assumptions for mass transportation. These are:

a) Matrix is composed of primary porosity and porous rock and has huge heat storage capacity, while fracture is secondary porosity and has high heat flow capacity.
b) Heat transmission in fracture occurs by conduction through the medium of water and by convection of moving water; but in the matrix heat transmission is only because of conduction through the medium of water in primary porosity and the medium of rock. These heat transmissions take place in an unsteady state.
c) Heat transmission between fracture and matrix consists of transfer by moving water and by forced convection on the interface. These transmissions occur in quasisteady state.

Energy balance equations are derived as follows:

$$\nabla\{\frac{\rho kfh}{\mu}(\nabla Pf-\rho g\nabla D)\} + \nabla(Kf\nabla Tf) - Hfm\beta(Tf - Tm) - Gfmhfm$$
$$- G_L - G\cdot h = \frac{\partial}{\partial t}(\phi f\rho h) \quad (11)$$

$$\nabla(Km\nabla Tm) + Hfm\beta(Tf-Tm) + Gfmhfm$$
$$= \frac{\partial}{\partial t}\{\phi m\rho h + (1-\phi m-\phi f)\rho rCrTm\} \quad (12)$$

where, β is a shape factor too.

Solving Eqs. (9)$\sim$(12) for pressure Pf and Pm and temperature Tf and Tm simultaneously, pressure and temperature distributions in the fracture and in the matrix are known. Fig. 3 shows the computation procedure as a flow chart, using a finite difference method, in order to solve these partial differential equations.

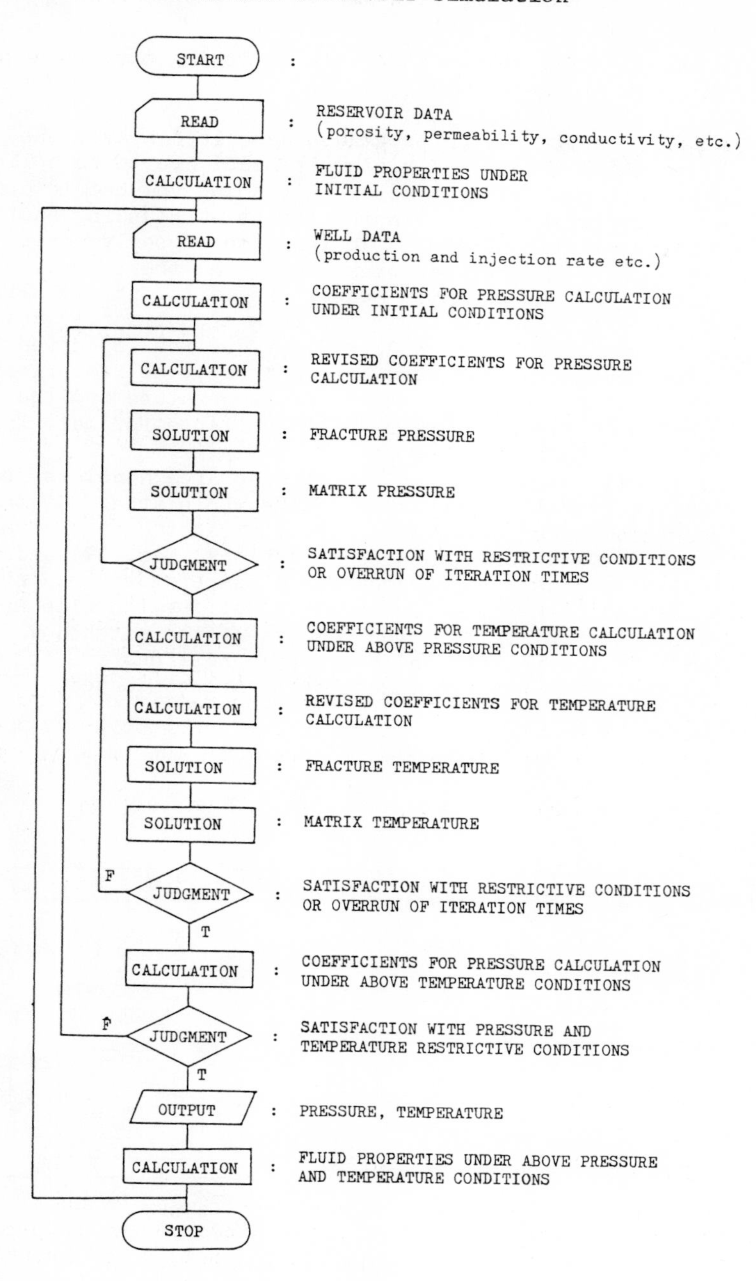
START :
READ : RESERVOIR DATA (porosity, permeability, conductivity, etc.)
CALCULATION : FLUID PROPERTIES UNDER INITIAL CONDITIONS
READ : WELL DATA (production and injection rate etc.)
CALCULATION : COEFFICIENTS FOR PRESSURE CALCULATION UNDER INITIAL CONDITIONS
CALCULATION : REVISED COEFFICIENTS FOR PRESSURE CALCULATION
SOLUTION : FRACTURE PRESSURE
SOLUTION : MATRIX PRESSURE
JUDGMENT : SATISFACTION WITH RESTRICTIVE CONDITIONS OR OVERRUN OF ITERATION TIMES
CALCULATION : COEFFICIENTS FOR TEMPERATURE CALCULATION UNDER ABOVE PRESSURE CONDITIONS
CALCULATION : REVISED COEFFICIENTS FOR TEMPERATURE CALCULATION
SOLUTION : FRACTURE TEMPERATURE
SOLUTION : MATRIX TEMPERATURE
F
JUDGMENT : SATISFACTION WITH RESTRICTIVE CONDITIONS OR OVERRUN OF ITERATION TIMES
T
CALCULATION : COEFFICIENTS FOR PRESSURE CALCULATION UNDER ABOVE TEMPERATURE CONDITIONS
F
JUDGMENT : SATISFACTION WITH PRESSURE AND TEMPERATURE RESTRICTIVE CONDITIONS
T
OUTPUT : PRESSURE, TEMPERATURE
CALCULATION : FLUID PROPERTIES UNDER ABOVE PRESSURE AND TEMPERATURE CONDITIONS
STOP

Fig. 3 Program Flow Chart for Sector Model

APPLICATION

Prior to applying the procedure described above, the parameters which define the reservoir model are to be collected in the field. Geological analysis gives the volumetric dimensions of the reservoir and its extension to the surrounding aquifer. The division of grid blocks is designed to properly cover the development area by the Wide Area Model, and the injection production area by the Sector Model. Porosity, permeability, and thermal properties of the reservoir rock are primarily determined by core analysis, well log analysis, production injection tests, and laboratory measurements. Besides these, the historical observation of reservoir pressure and temperature and the tracer test results at the wells have to be available for matching purposes.

At first, the simulation by the Wide Area Model is operated to match the overall performance of the whole area. Giving the required parameters including the past production injection data in their digitized forms to each grid block, Eqs. (7) and (8) are solved simultaneously for pressure and temperature. An example of pressure matching at an observation well by the Wide Area Model is demonstrated in Fig. 4. The results of the matched Wide Area Model provide the initial and boundary conditions for the Sector Model at the time when the tracer tests are conducted.

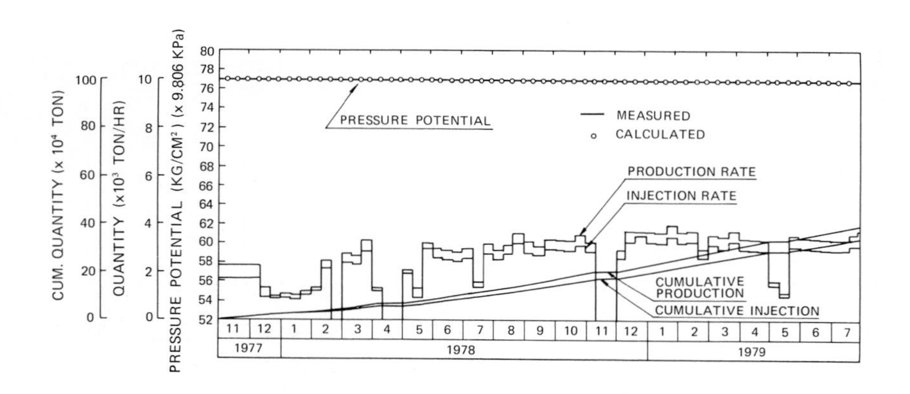

Fig. 4 Matching of Reservoir Pressure by Wide Area Model

The short duration of time after the tracer ion is dropped in an injector is the computation period of the Sector Model. The flow behavior of water inside the injection production area is investigated in detail. Giving the first approximation of the physical properties of fractured rock, the movement of injected water carrying the tracer ion to the production wells is simulated. With the smaller size grid blocks and shorter time interval, Eqs. (9)~(12) are solved simultaneously to give the pressure and temperature distribution in detail. The path of travel and its velocity are computed for each interference between injector and producer. The computed length of the shortest path and the velocity are input to Eq. (5) to reproduce the tracer ion response curve. The matching of the tracer ion response curve at a producer is demonstrated in Fig. 5. Until the best matching of this curve is obtained, the properties of the fracture path in the Sector Model are modified and the procedure is repeated.

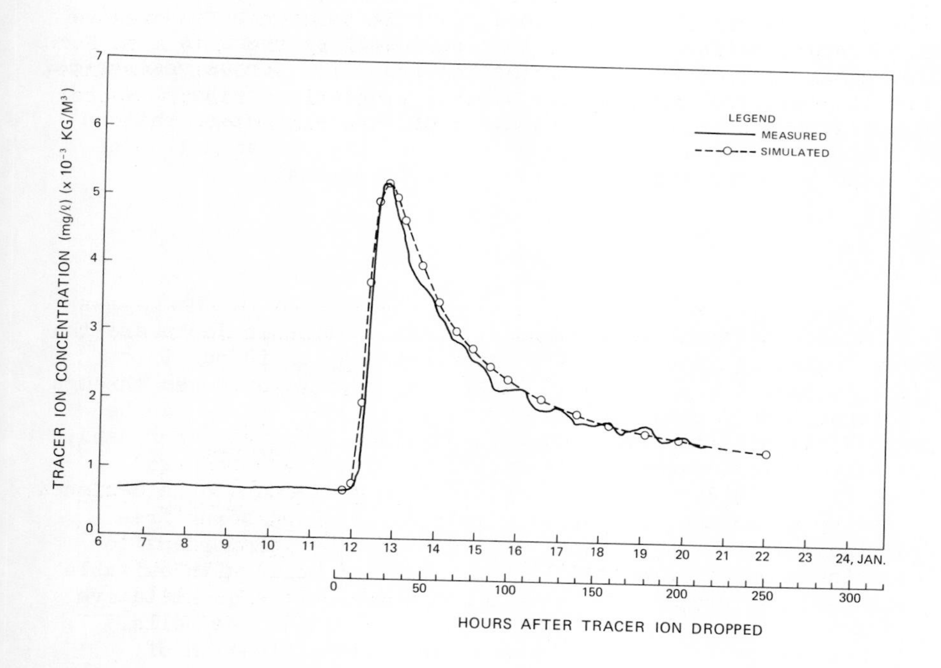

Fig. 5 Tracer Ion Response Curve Detected at a Producer

An example of tracer ion response curve detected at a producer is presented in Fig. 5, in which a sharp peak of tracer ion concentration is observed in the earlier arrival time. The producer is located at a horizontal distance of approximately 300 m from the injector, in which the tracer ion of 918 kg is dropped instantaneously in the reinjection stream at the rate of 170 t/h (47.2 kg/s). For this time stage, the Sector Model gives the pressure distribution of this area as in Fig. 6, and the stream lines of the reinjected water flow are drawn as in Fig. 7. By tracing the largest gradient of pressure difference block by block, the length of the shortest path is estimated at 312 m and the average velocity of travel is 8.4 m/h (0.0023 m/s).

The matched curve of tracer ion response by these values is also demonstrated in Fig. 5. The hydraulic properties of the fracture estimated by this procedure are permeability of 5 ~ 10 darcies (4.93 ~ 9.86 μm^2) and porosity of 2 ~ 2.5 %. The contamination of produced water by reinjected water from this injector is calculated as 5 %. Total contamination at this producer from all the injectors sums to 30 % at the stage of this tracer test. These values are not likely to be computed for the injector producer system composed of non-fractured porous media. By the order of these values, the fractured reservoir is distinctively classified from the non-fractured porous reservoirs.

In the actual application of this procedure to the reinjection system, the reallocation of the reinjection pattern is implemented based on these values, and the contamination of produced water by the reinjected water is minimal.

SUMMARY AND CONCLUSION

Most geothermal reservoirs under production in Japan are the water dominated type and waste water reinjection is put into practice inevitably. In most cases, injected water moves through fractures toward producer and the thermal interference can be very detrimental to the production of steam. On the other hand, hydraulic interference is beneficial in providing pressure support. Optimization of the reinjection cycle has to be designed by taking advantage of hydraulic interference and being free from thermal detriment. Accordingly, it is very important to determine an optimum allocation of injection wells with suitable distances from production wells or to appraise the quantitative effect of reinjection of disposal water on production wells.

For the purpose of this assessment, classification of reservoir rock, fractured or non-fractured, is important, and the tracer test analysis combined with the simulation study is an effective method of examination for this purpose.

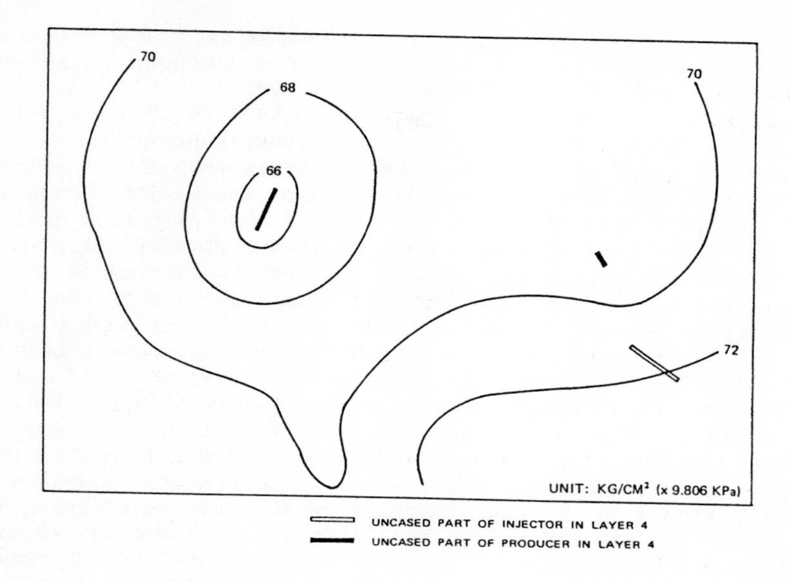

Fig. 6 Iso-Baric Contours in Injection Production Area

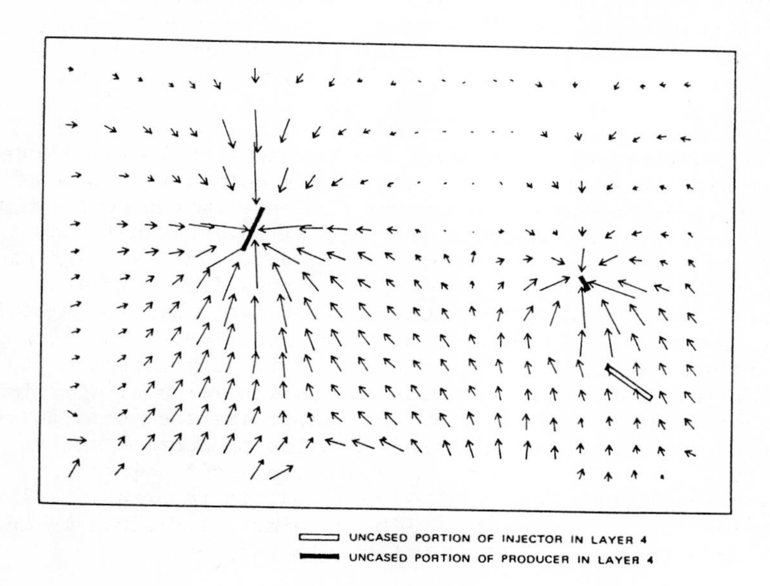

Fig. 7 Stream Line Vectors in Injection Production Area

This paper mentions the procedure to be followed in simulating the performance of a fractured reservoir and its application to optimize the reinjection system.

The following is a major summary and conclusion:

1. In order to simulate a fractured water-dominant reservoir, an effective combination of two simulation models offers a better insight to the flow mechanism of an injection production cycle and results in less expenses for assessment. The Wide Area Model simulates macroscopically the whole reservoir performance in a long time range, and the Sector Model does the microscopic interference between injectors and producers at any time stage.

2. The conventional concept of flow through porous media is applied to the Wide Area Model, while the Warren and Root concept of flow through a fracture matrix system is employed for the Sector Model. The simulation results by Wide Area Model are matched with the flow behavior of observation wells drilled in the surrounding aquifer area. Reservoir performance computed by the Sector Model matches that observed at the injectors and producers. Especially by the Sector Model, in order to evaluate the flow routes of injected water to producers and the physical properties of the fracture path, trial and error matching of tracer ion response curve by simulation with the actual test result is repeated until a satisfactory result is obtained.

3. The tracer test is a simple and economic reservoir engineering procedure for the appraisal of water-dominant geothermal reservoirs. Tracer test analysis provides not only the flow velocity of injected water to producers but also the characteristics of the fracture path between these wells. When the tracer ion response curve detected at a producer shows a sharp and high peak as demonstrated in Fig. 5, relatively large fractures are considered dominant between the injector and producer. Tracer test analysis, coupled with the Sector Model simulation, also provides a method to calculate the degree of contamination of produced water by injected water and to estimate the permeability and porosity of fractures.

NOMENCLATURE

$C(x,t)$: tracer ion concentration at distance x from injector and at time t after tracer ion dropped (kg/m^3)

C_1 : tracer ion concentration at injector (kg/m^3)

Co : background tracer ion concentration (kg/m^3)

Cad : concentration change due to adsorption - desorption ($kg/m^3/s$)
Cr : specific heat of rock (J/kg.K)
D : depth (m)
Df : coefficient of molecular diffusion (m^2/s)
Dp : coefficient of dispersion (m^2/s)
G : water production/injection rate (kg/s)
G_L : rate of heat loss ($J/m^3/s$)
H : heat transfer coefficient (W/m^2 K)
I : quantity of tracer ion (kg)
K : thermal conductivity (W/m K)
P : pressure (Pa)
T : temperature (°C)
V : flow velocity (m/s)
g : gravitational constant (m/s^2)
h : specific enthalpy (J/kg)
k : permeability (m^2)
t : time (s)
v : specific volume (m^3/kg)
x : distance (m)
β : shape factor ($1/m^2$)
Δt : time interval for dropping tracer ion (s)
μ : viscosity (Pa.s)
ρ : fluid density (kg/m^3)
ρ_r : rock density (kg/m^3)
σ : shape factor ($1/m^2$)
ϕ : porosity (fraction)

Subscript

f : fracture
m : matrix
fm : relationship between fracture and matrix

ACKNOWLEDGEMENT

The research and development on the subject of "Investigation of hot water reinjection mechanism" has been carried out by Japan Metals and Chemicals Co., Ltd. (JMC) for the past seven years under the contract comissioned by Agency of Industrial Science and Technology (AIST), Ministry of International Trade and Industry[5]. The first author, Dr. S. Hirakawa, has also been involved in the R & D as a head of the technical advisory comittee and the other authors, staff of Japan Oil Engineering Co., Ltd. (JOE), have been working with JMC by rendering their services of geothermal reservoir simulation.

We wish to thank JMC and Sunshine Project Promotion Headquarters of AIST for permission to publish this paper. Also we thank Mr. Omuta, reservoir engineer of JOE, for his assistance in writting this paper.

REFERENCE

1. Chin Fu Tsang, Gudmundur Bodvarsson, Marcelo J. Lippmann and Jesus Rivera R., "A Study of Alternative Reinjection Schemes for the Cerro Prieto Geothermal Field, Baja California, Mexico", Geothermal Resource Council, Trans. Vol. 2, July 1978.

2. Ito, J. et al, "On the Geothermal Water Flow of the Onuma Geothermal Reservoir", Jour. Japan Geoth. Energy Assoc. 14, 3, 139-151, 1977.

3. Hayashi, M., Mimura, T. and Yamasaki, T., "Geological Setting of Reinjection Wells in the Otake and the Hatchobaru Geothermal Field Japan", Geothermal Resource Council, Trans. Vol. 2, July 1978.

4. Warren, J.E., and Root, P.J., "The Behavior of Naturally Fractured Reservoirs", SPE Journal, Sept. 1963.

5. "Investigation of Hot Water Reinjection Mechanism", (The reports to Sunshine Project Promotion Headquarters), Japan Metals and Chemicals Co., Ltd., Mar., 1978 ~ 1980.

CRACK-LIKE RESERVOIR IN HOMOGENEOUS AND INHOMOGENEOUS HDR

by Hiroyuki Abé and Hideki Sekine

Professor, Department of Mechanical Engineering
Tohoku University, Sendai, Japan

Associate Professor, Department of Engineering Science
Tohoku University, Sendai, Japan

ABSTRACT

By circulating fluid through crack-like reservoirs created by hydraulic fracturing, an abundant amount of geothermal energy could be extracted from hot dry rocks. In the development of geothermal energy of this type, there is an obvious need to design and control the crack-like reservoirs in the earth's crust. This paper is concerned with two subjects, i.e., the characteristic of a crack-like reservoir and the behavior of a crack-like reservoir due to extraction of geothermal energy. The former subject includes the limit for the size of two-dimensional crack-like reservoirs, the intersection of a two-dimensional reservoir with a joint, and the stability of a penny-shaped reservoir.

INTRODUCTION

The extraction of geothermal energy from hot dry rocks in the earth's crust has received wide attention. An abundant amount of geothermal energy could be recovered from hot dry rocks by circulating fluid through crack-like reservoirs which are created by a hydraulic fracturing technique. In the development of geothermal energy extraction systems, the fracture mechanics study of problems for a fluid-filled reservoir should be made to design and control the crack-like reservoirs. On this subject many theoretical works have been done thus far[1-10].

In many regions of the earth's crust, the magnitude of all three principal stresses increases with depth. Weertman[4] suggested that if the reservoir length exceeds a certain value, a part of the vertical crack-like reservoir begins to close by the stress gradients. On the other hand, if the stress intensity factor at the border of the crack-like reservoir equals the fracture toughness of the rock, extension will begin at the border opposite to the one which tends to close. Thus the gradients in tectonic stress and internal hydrostatic pressure may play a dominant role in controlling the form and stability of large crack-like reservoirs. The theoretical analysis of extension and closure of vertical reservoirs subject to realistic stress gradients has been made by Pollard and coworkers [5-7] on the basis of the two-dimensional theory of elasticity, and they give an upper limit for the size of a stable reservoir. In those works, however, the influence of the earth's surface is disregarded.

It is reasonable to expect any large rock to contain joints. Thus the possibility exists that an extending crack-like reservoir might be stopped by a joint and that larger stable reservoirs can be created in the earth's crust. The intersection of a vertical crack-like reservoir with a horizontal joint is discussed by Weertman [8] and Keer and Chen [9].

In recent years, Hsu and Santosa [10] have discussed the effect of linear gradients in tectonic stress and internal hydrostatic pressure on the stability of a vertical penny-shaped reservoir. However, they disregard the partial closure of the reservoir.

Heat recovered through the reservoirs has been estimated by Gringarten et al. [11], Lowell [12], Wunder and Murphy [13], and Abé et al. [14]. During the extraction of geothermal energy, the surface of the crack-like reservoir is cooled by fluid, and thermal contraction of rock occurs. The thermal contraction of rock gives rise to an increase in the aperture of the crack-like reservoir. When the stress intensity factor at the border of the crack-like reservoir attains the fracture toughness of rock, the reservoir begins to extend at the border. A few such studies of crack-like reservoirs including thermoelastic effects have been made thus far; for example, Sekine and Mura [15].

In this paper, we are concerned with two subjects, i.e., the characteristic of a crack-like reservoir and the behavior of a crack-like reservoir due to extraction of geothermal energy. The former subject includes the limit for the size of two-dimensional crack-like reservoirs, the intersection of a two-dimensional reservoir with a joint, and the stability of a penny-shaped reservoir.

CHARACTERISTIC OF A CRACK-LIKE RESERVOIR

Limit for the Size of Two-Dimensional Crack-Like Reservoirs [16]

We are concerned with the limit for the size of an arbitrarily oriented crack-like reservoir near the earth's surface on the basis of the two-dimensional theory of elasticity. The surrounding rock is assumed to be a continuum that is homogeneous and isotropic with respect to elastic moduli and fracture toughness.

Consider a fluid-filled two-dimensional crack-like reservoir of length 2ℓ near the earth's surface ($y = 0$), as shown in Fig. 1. The reservoir makes an angle ϕ with the horizontal direction, and the distances from the upper reservoir tip and the reservoir center to the earth's surface are respectively denoted by d and h.

By neglecting the effects resulting from the flow of fluid (1,3), the hydrostatic pressure p in the crack-like reservoir is written as

$$p = p_u + \rho_w g(h - x_1 \sin\phi) \qquad (1)$$

where p_u is the pump pressure on the earth's surface, ρ_w the density of fluid, and g the gravitational acceleration.

The tectonic compressive principal stresses S_x and S_y in the x and y directions, due primarily to the overburden weight of the rock, are

$$S_x = \alpha\rho_r gy \ , \quad S_y = \rho_r gy \qquad (2)$$

where ρ_r is the density of the rock and α is the coefficient of the active rock pressure.

The well-known continuous dislocation method is adopted to solve this problem. Simulating the crack-like reservoir by continuous distributions of edge dislocations, we can readily set up a system of singular integral equations for dislocation densities with the Cauchy kernel.

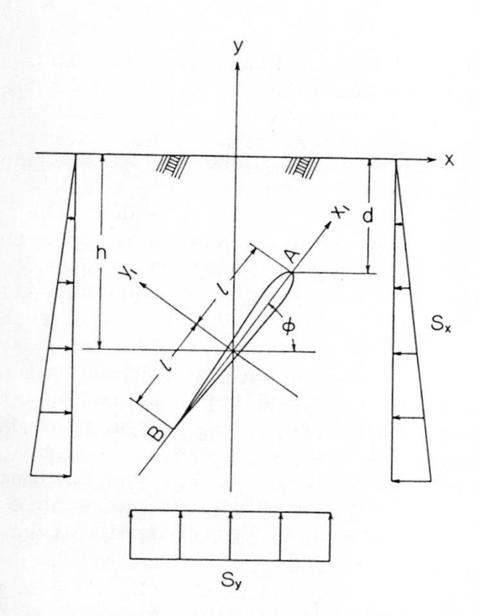

Fig. 1. An arbitrarily oriented crack-like reservoir and coordinate systems.

Some numerical calculations were performed. The values of the physical constants are taken as $\rho_w = 1.0 \times 10^3$ kg/m^3, $\rho_r = 2.5 \times 10^3$ kg/m^3, and $\alpha = 0.75$. In this case, the reservoir tends to close at the lower tip as the pump pressure p_u decreases. The maximum stress criterion of fracture proposed by Erdogan and Sih (17) is employed; the crack-like reservoir begins to extend at the tip when K^* attains the fracture toughness K_c of the rock:

$$K^* \equiv \cos\frac{\omega}{2}\left(K_I \cos^2\frac{\omega}{2} - \frac{3}{2}K_{II}\sin\omega\right) = K_c \qquad (3)$$

where K_I and K_{II} are the stress intensity factors for the opening mode and the in-plane shear mode, respectively, and ω is the initial angle of extension with respect to the plane of the reservoir which is determined from

$$K_I \sin\omega + K_{II}(3\cos\omega - 1) = 0 \tag{4}$$

The variation of K^* at the upper tip against the depth d/ℓ with keeping $K_{IB} = 0$ at the lower tip is shown in Fig. 2. In this figure, K^* is nondimensionalized by $\sigma_\infty\sqrt{\pi\ell}$, where

$$\sigma_\infty = (\alpha\rho_r - \rho_w)g\ell \tag{5}$$

It is known(5) that $\sigma_\infty\sqrt{\pi\ell}$ is the stress intensity factor K_I at the upper tip of a vertical reservoir, located extremely deep from the earth's surface, with $K_{IB} = 0$ at the lower tip. Fig. 2 reveals that for the vertical reservoir ($\phi = 90°$) K^* decreases monotonically and tends to $\sigma_\infty\sqrt{\pi\ell}$ as the reservoir is deeper, while for $\phi \neq 90°$ K^* takes the minimum value at a finite depth of the reservoir. Let K_{min} be the minimum value of $K^*/\sigma_\infty\sqrt{\pi\ell}$. Then, by equating K^* to K_c, the maximum half length ℓm for the stable crack-like reservoir is given by

$$\ell m = [K_c/\sqrt{\pi}\ (\alpha\rho_r - \rho_w)gK_{min}]^{\frac{2}{3}} \tag{6}$$

When the half length ℓ is less than ℓm, we can find, from this figure, the bounds of the depth for the crack-like reservoir to be stable.

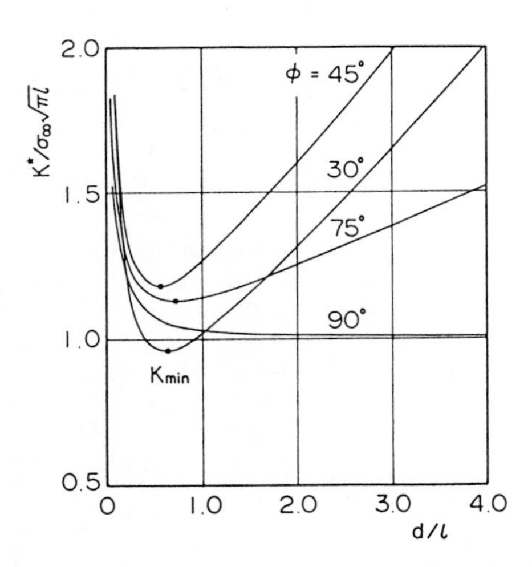

Fig. 2. Stress intensity factor K^* at the upper tip against the depth d/ℓ with keeping $K_{IB} = 0$ at the lower tip.

Intersection of a Crack-Like Reservoir With a Joint(18)

The problem of the intersection of a vertical crack-like reservoir with a joint is considered on the basis of the two-dimensional theory of elasticity. When a reservoir intersects a joint, an opening of the reservoir occurs at the juncture through separation and slip on the surface of the joint. The Coulomb law of friction is assumed to hold in the slip zone. The two-dimensional crack-like reservoir of length 2ℓ

makes an angle θ with the joint of length $\ell_1 + \ell_2$, as shown in Fig. 3. The earth's crust is supposed to be subject to a linear tectonic stress, and the distance of the reservoir from the earth's surface is assumed to be sufficiently large in comparison with the reservoir length.

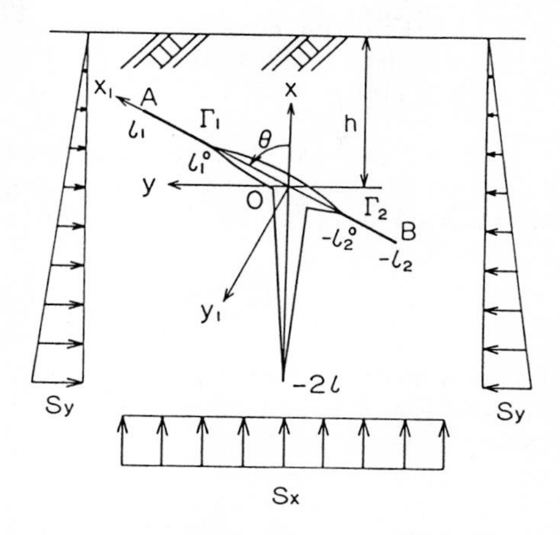

Fig. 3. A crack-like reservoir intersecting a joint.

The continuous dislocation method is also adopted to solve the problem. Numerical calculations were performed by taking $\rho_w = 1.0 \times 10^3$ kg/m³, $\rho_r = 2.5 \times 10^3$ kg/m³, $\alpha = 0.5$, $\mu_f = 0.5$, $h = 2000$ m and $K_c = 2.7\text{MPa}\cdot\text{m}^{1/2}$, where μ_f is the coefficient of friction.

Fig. 4 shows the variation of the stress intensity factors K_I, K_{II} and K^* at the joint tip A against the position of the tip ℓ_1/L_0 for $\theta = 45°$ and $\Delta p = 5.0$, where

$$L_0 = [K_c/\sqrt{\pi}(\alpha\rho_r - \rho_r)g]^{\frac{2}{3}} \quad (7)$$

and Δp is the parameter of the pump pressure defined as

$$\Delta p = [p_u - (\alpha\rho_r - \rho_w)gh] / \frac{3}{2}[K_c{}^2(\alpha\rho_r - \rho_w)g/\pi]^{\frac{1}{3}} \quad (8)$$

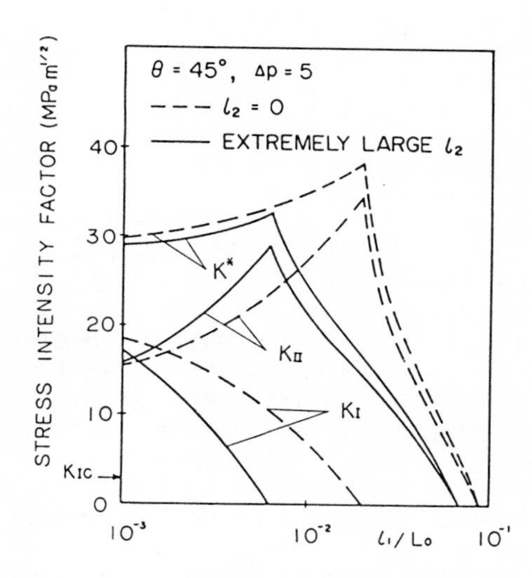

Fig. 4. Stress intensity factors K_I, K_{II} and K^* at the joint tip A against the position of the tip.

The solid and broken lines indicate, respectively, the results for the extremely large ℓ_2 and $\ell_2 = 0$. This figure reveals that if the closure does not occur at the joint tip A, i.e., $K_I \neq 0$, the stress intensity factor K_I decreases and the stress intensity factors K_{II} and K^* increase with increasing ℓ_1/L_0. On the other hand, when the closure area exists in the vicinity of the joint tip A, K_{II} and K^* decrease monotonically with increasing ℓ_1/L_0 because of the existence of resistance to slip by friction. In this case, we can obtain the critical value ℓ_1^* of the position of the joint tip A by setting K^* equal to K_c. It is easily understood that crack initiation does not occur at the joint tip when $\ell_1 > \ell_1^*$. Here it should be noted that the critical value ℓ_1^* for $\ell_2 = 0$ gives the maximum value of ℓ_1^*.

In the case of $\ell_2 = 0$, the maximum volume V of the stable reservoir is shown in Fig. 5. In this figure, the maximum volume V is nondimensionalized by

$$V_0 = (1 - \nu^2)K_c^{\,2}/E(\alpha\rho_r - \rho_w)g \tag{9}$$

where E is Young's modulus and ν is Poisson's ratio of the rock. The volume V_0 is the maximum volume of a vertical stable reservoir not intersecting with a joint. It is seen from this figure that larger stable reservoirs can be created in the earth's crust by intersecting with a joint.

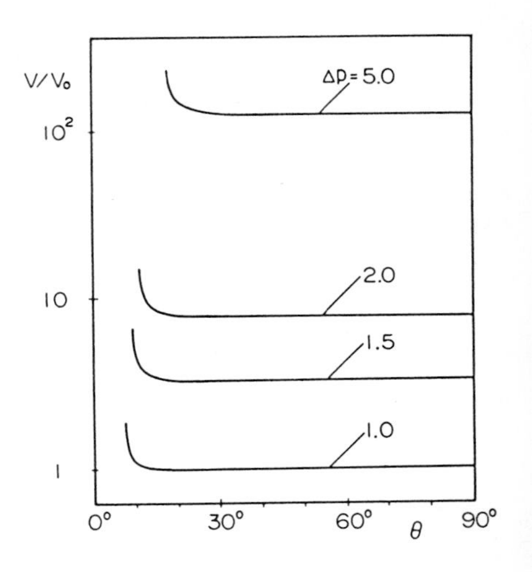

Fig. 5. Maximum volume of the stable reservoir intersecting with a joint.

Recently, the fault as a reservoir for geothermal energy has received attention; for example, Hayashi and Abé[19].

Consider a fluid-filled region opened by a hydrostatic pressure p on the interface of the fault as shown in Fig. 6. Before the fluid-filled region is created, the earth's crust is in a static equilibrium with stresses σ^0_{xx}, σ^0_{xy} and σ^0_{yy} (or the tectonic stresses). Once fluid is injected, a relative movement will take place along the interface of the fault due to the vanishing of the frictional forces in the opened region. Therefore, a newly created slipped region will appear. It is reasonable to consider that the tectonic stresses would vary spatially. Here, an example which gives a clear view over the problem is considered as shown in Fig. 6. The relationship between the half length a of the opened region and the given hydrostatic pressure p is presented in Fig. 7. The volume V of the opened region, i.e., the reservoir is also shown in the figure. The fluid-filled region could be used as a reservoir for the extraction of heat.

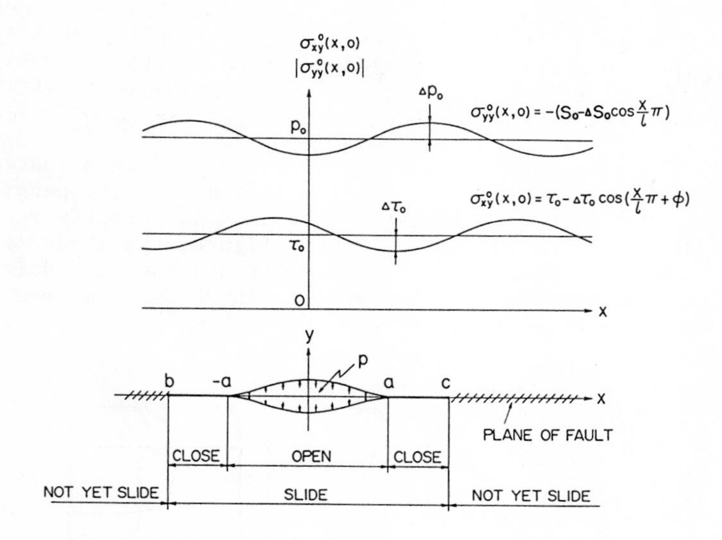

Fig. 6. Opening of a fault due to injection of fluid.

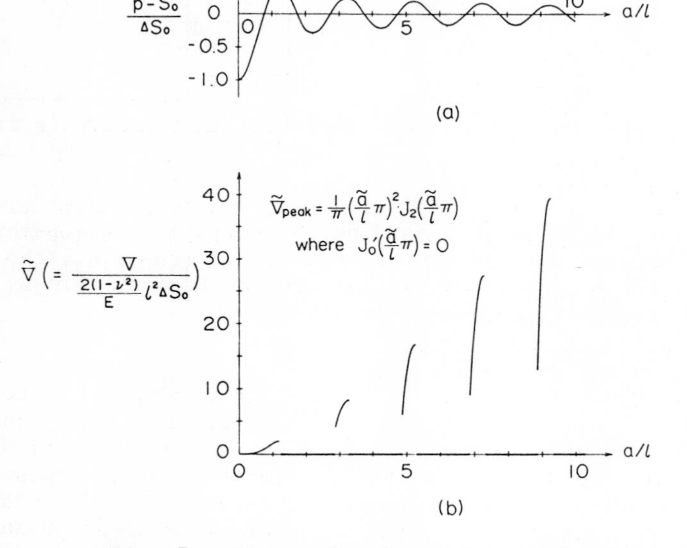

Fig. 7. Size and volume of opened region.

Stability of a Penny-Shaped Reservoir [20]

We are concerned with the theoretical analysis of stability and partial closure of a vertical penny-shaped reservoir in the earth's crust subject to linear tectonic stress gradients on the basis of the three-dimensional theory of elasticity.

A vertical penny-shaped reservoir with radius $\bar{a}$ is situated in the earth's crust, as shown in Fig. 8. The distance of the penny-shaped reservoir from the earth's surface is assumed to be sufficiently large in comparison with the radius $\bar{a}$. In the figure, the shaded part corresponds to the closure area of the reservoir and the reservoir profile given by Hsu and Santosa [10] disregarding the closure of the reservoir, is indicated by the broken line.

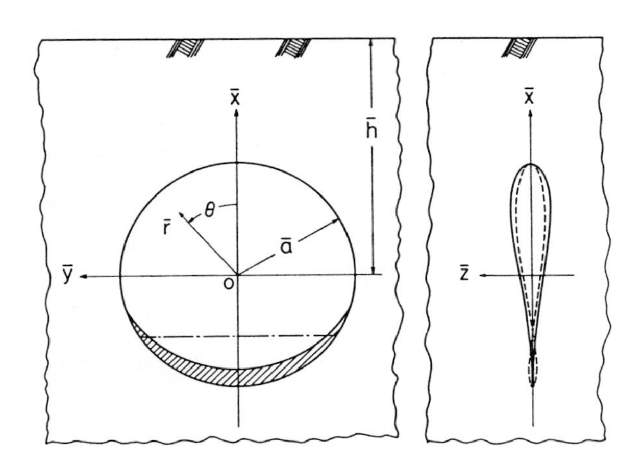

Fig. 8. A vertical penny-shaped reservoir in the earth's crust.

When a hydrostatic pressure p acts on the surface of the reservoir in the earth's crust under a compressive tectonic stress S_z, the boundary conditions of the problem of the disturbance stress field by the presence of the fluid-filled reservoir are written as

$$\sigma_z = p_s(\bar{r}, \theta) = -p(\bar{r}, \theta) + S_z(\bar{r}, \theta) \quad \text{(opening area)} \tag{10}$$

$$u_z = 0 \quad \text{(closure area)} \tag{11}$$

$$\tau_{zr} = \tau_{z\theta} = 0 \quad (0 \leqq \bar{r} \leqq \bar{a},\ \bar{z} = 0) \tag{12}$$

where σ_z, τ_{zr} and $\tau_{z\theta}$ are the stress components and u_z is the displacement component with respect to a cylindrical coordinate system ($\bar{r}$, θ, $\bar{z}$) whose origin is located at the reservoir center.

By using a Cartesian coordinate system ($\bar{x}$, $\bar{y}$, $\bar{z}$) whose origin is located at the reservoir center and the $\bar{x}$ axis perpendicular to the earth's surface, the hydrostatic pressure p and the tectonic stress S_z are written as

$$p = P_0 - A_0\bar{x} \tag{13}$$

$$S_z = S_{z0} - B_0\bar{x} \tag{14}$$

where P_0 and S_{z0} are the hydrostatic pressure and the tectonic stress at the reservoir center, respectively, and A_0 and B_0 denote the linear gradients in the hydrostatic pressure and the tectonic stress, respectively.

In order to find the solution satisfying the boundary conditions (10) to (12), the normal stress $p_s(\bar{r}, \theta)$ on the circular region ($0 \leqq \bar{r} \leqq \bar{a}$, $\bar{z} = 0$) is expanded in a double series

$$p_s(\bar{r}, \theta) = \sum_{m=0}^{\infty} \sum_{n=0}^{\infty} a_{mn} r^m \cos n\theta \tag{15}$$

where a_{mn} are unknown constants and $r = \bar{r}/\bar{a}$. Then, the opening displacement $w(\bar{r}, \theta)$ and the stress intensity factor for the opening mode are given by

$$w(\bar{r}, \theta) = -\frac{4(1-\nu^2)\bar{a}^{m+1}}{E} \sum_{m=0}^{\infty} \sum_{n=0}^{\infty} a_{mn} \frac{\Gamma([m+n+2]/2)}{\Gamma([m+n+3]/2)} \times \int_r^1 \frac{t^{m-n+1}dt}{(t^2-r^2)^{\frac{1}{2}}} \times r^n \cos n\theta \tag{16}$$

$$K_I = -\bar{a}^{\frac{1}{2}} \sum_{m=0}^{\infty} \sum_{n=0}^{\infty} a_{mn} \frac{\Gamma([m+n+2]/2)}{\Gamma([m+n+3]/2)} \cos n\theta \tag{17}$$

where $\Gamma(\ )$ is the gamma function.

When there is no partial closure area of the penny-shaped reservoir, the normal stress $p_s(\bar{r}, \theta)$ can be expressed by

$$p_s(\bar{r}, \theta) = -(P_0 - S_{z0}) - (B_0 - A_0)\bar{r} \cos\theta \tag{18}$$

Then, the stress intensity factor K_I are[10]

$$K_I = 2\left(\frac{\bar{a}}{\pi}\right)^{\frac{1}{2}} \left[(P_0 - S_{z0}) + \frac{2}{3}(B_0 - A_0)\bar{a}\cos\theta\right] \tag{19}$$

By setting equal the maximum value of the stress intensity factor K_I along the border of the reservoir to the fracture toughness of the rock K_c, the upper critical pressure is obtained in the form

$$P_0 - S_{z0} = \frac{K_c}{2}\left(\frac{\pi}{\bar{a}}\right)^{\frac{1}{2}} - \frac{2}{3}|B_0 - A_0|\bar{a} \tag{20}$$

In view of (19) the minimum pressure before the closure of the reservoir is given by

$$P_0 - S_{z0} = \frac{2}{3}|B_0 - A_0|\bar{a} \tag{21}$$

The analysis will be applied to clarify the upper critical pressure and partial closure of a vertical penny-shaped reservoir in the earth's crust under a linear tectonic stress.

The solution is found by using the boundary collocation technique. The values of the linear gradients in the hydrostatic pressure and the tectonic stress, and the fracture toughness of rock are taken as[10] $A_0 = \rho_w g = 8709\text{Pa/m}$, $B_0 = c_1 A_0 + 8664\text{Pa/m}$, $K_c = 2.76\text{MPa}\cdot\text{m}^{1/2}$ where c_1 $(0.1 \leq c_1 \leq 0.9)$ is the coefficient due to the effective pore pressure. In this case, the extension starts at the upper border because the stress intensity factor attains its maximum at $\theta = 0$, and the reservoir tends to close at the lower border.

Fig. 9 shows the variation of the upper critical pressure of $P_0 - S_{z0}$ against the radius of the reservoir $\bar{a}$ for $c_1 = 0.1$, 0.5 and 0.9.

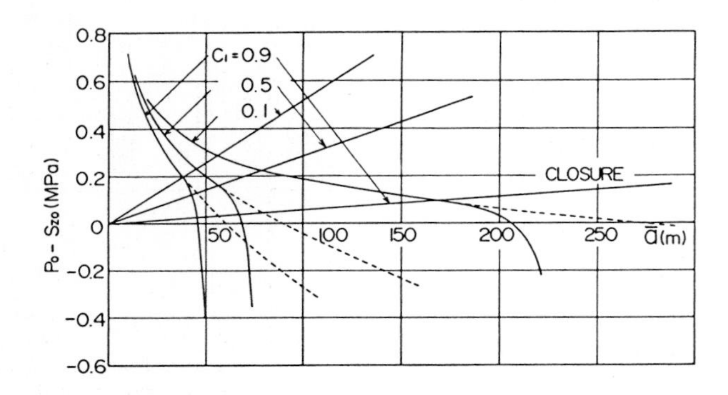

Fig. 9. Upper critical pressure of $P_0 - S_{z0}$ against radius $\bar{a}$.

The closure curves, given by (21), are shown in the straight lines. If $\bar{a} > \bar{a}_i$, where $\bar{a}_i$ is the radius at the intersection point of the upper critical pressure curve and the closure curve, the partial closure area appears in the reservoir. The broken lines indicate the results disregarding the closure. The shape of the closure area is shown for c_1 = 0.5 in Fig. 10.

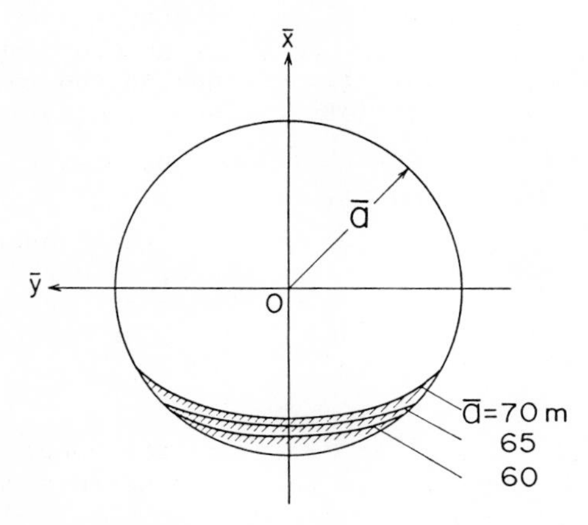

Fig. 10. Shape of closure area (shaded part).

MECHANICAL BEHAVIOR OF A CRACK-LIKE RESERVOIR DUE TO EXTRACTION OF GEOTHERMAL ENERGY

Consider a vertical two-dimensional crack-like reservoir with an inlet and an outlet in the earth's crust under a linear tectonic stress, as shown in Fig. 11. The distance from the inlet to the outlet is 2ℓ, and the upper and lower tips of the crack-like reservoir are $x_1 = L_1$ and $x_1 = -L_2$. The distance of the reservoir from the earth's surface is supposed to be very large in comparison with the reservoir length.

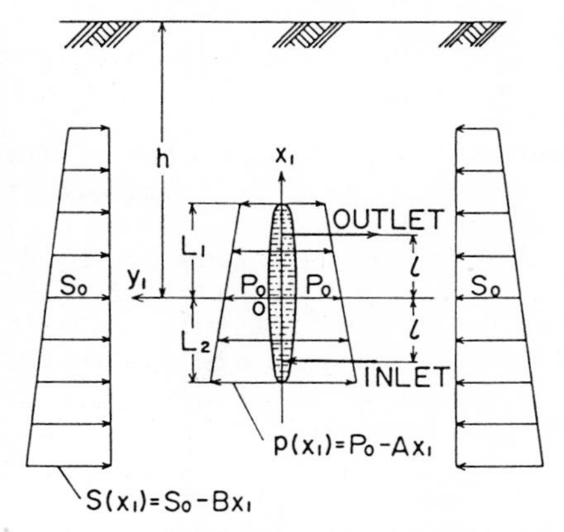

Fig. 11. A vertical two-dimensional crack-like reservoir and coordinate system.

Let $T_w(x_1, t)$ and $T_r(x_1, y_1, t)$ be, respectively, the fluid and rock temperatures, and t is time. On the reservoir surface, they satisfy the following conditions:

$$T_w(x_1, t) = T_r(x_1, 0, t) \quad (|x_1| < \ell, \ t \geqq 0) \tag{22}$$

$$T_w(-\ell, t) = T_r(-\ell, 0, t) = T_i \quad (t \geqq 0) \tag{23}$$

where T_i is the constant fluid temperature at the inlet. The fluid temperature $T_w(x_1, t)$ must satisfy the energy equation of the fluid:

$$c_w \rho_w Q T_{w,1} = -q(x_1, t) \quad (|x_1| < \ell, \ t \geqq 0) \tag{24}$$

where c_w is the specific heat of the fluid, Q is the mass flow rate and $q(x_1, t)$ is the heat flux through the reservoir surface. The rock temperature $T_w(x_1, y_1, t)$ is governed by the heat conduction equation for the rock:

$$\kappa_r \nabla^2 T_r = \frac{\partial T_r}{\partial t} \tag{25}$$

where κ_r is the thermal diffusivity of the rock.

In the analysis the singular point method is used on the basis of the quasi-static theory of thermoelasticity. The stress intensity factors at the tips of the crack-like reservoir are obtained from the equations of thermoelasticity[21]. When the stress intensity factors at the tips of the reservoir attain the fracture toughness of rock, the reservoir starts to extend at the tips. Therefore, for the extending crack-like reservoir, the following conditions must be satisfied:

$$K_I(L_1^*) = K_c \ , \qquad K_I(-L_2^*) = K_c \tag{26}$$

where L_1^* and $-L_2^*$ are the positions of the upper and lower tips of the extending crack-like reservoir, respectively.

The following values are used in the numerical calculations: $E = 3.5 \times 10^{10}$Pa, $\nu = 0.12$, $\gamma = 1.2 \times 10^{-5}$°K^{-1}, $K_c = 2.84$MPa·m$^{1/2}$, $B_0 - A_0 = 6.05 \times 10^3$Pa·m^{-1}, $\ell = 50$ m, $T_\infty - T_i = 180$°K and $\alpha = c_w \rho_w Q/\lambda_r = 200$, where γ, T_∞ and λ_r are the thermal expansion coefficient, the constant temperature of the rock at $t = 0$, and the thermal conductivity of the rock, respectively.

The positions of the upper and lower tips of the extending crack-like reservoir against time are shown in Fig. 12. As heat begins to recover through the reservoir, the length of the reservoir increases with time. The tendency is marked by increasing the hydrostatic pressure at the origin P_0 in the case of S_0 being constant. When $p_0 - S_0 =$

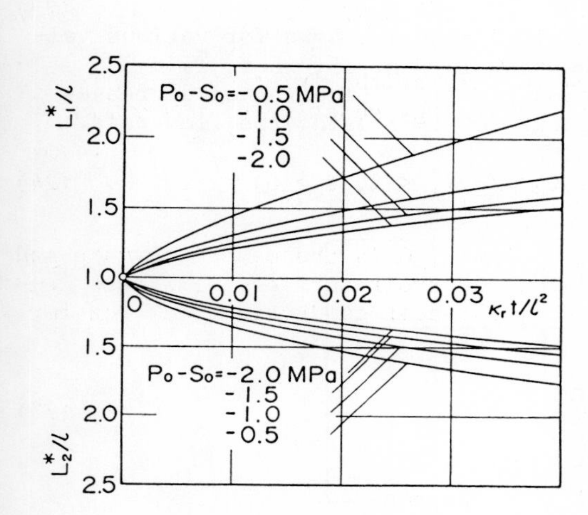

Fig. 12. Positions of the tips of the extending crack-like reservoir against time.

Fig. 13. Cross-sectional shape of the crack-like reservoir.

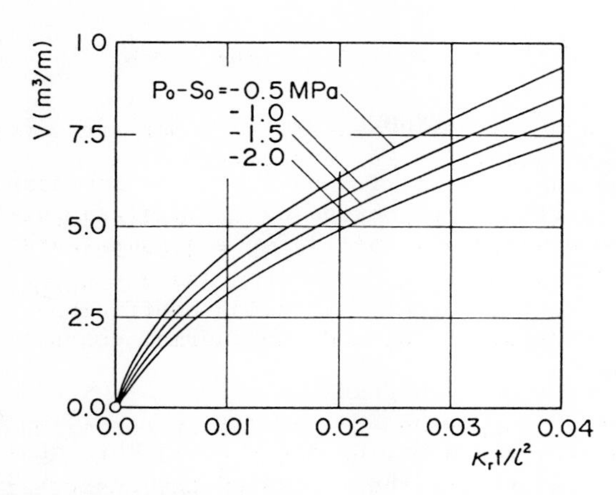

Fig. 14. Volume of the reservoir against time.

-1.0 MPa, the cross-sectional shape of the reservoir is shown in Fig. 13. The opening displacement of the reservoir becomes large with time and is almost constant from the inlet to the outlet. Fig. 14 shows the variation of the volume of the reservoir against time. The volume also becomes large with time.

The outlet fluid temperature against time is shown for various values of the parameter α of the flow rate in Fig. 15, where $\bar{T}_0 = [T_w(-\ell, t) - T_i]/(T_\infty - T_i)$. The outlet fluid temperature decreases with time, and the variation is marked by increasing the flow rate.

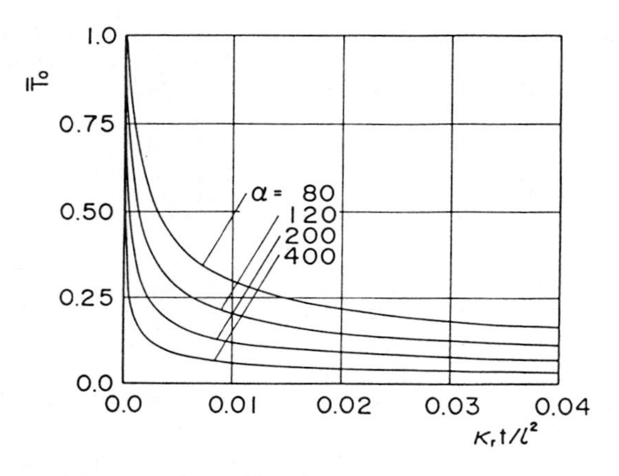

Fig. 15. Outlet fluid temperature against time.

ACKNOWLEDGMENTS

The authors wish to thank Mr. T. Shōji for his assistance. The work presented here was partially supported by the Ministry of Education, Science and Culture under Grant-in-Aid for Energy Research Nos. 505514, 56045015 and 57045014.

REFERENCES

1. Abé, H., Mura, T., and Keer, L. M., "Growth Rate of a Penny-Shaped Crack in Hydraulic Fracturing of Rocks," Journal of Geophysical Research, Vol. 81, No. 29, Oct. 1976, pp. 5335-5340.

2. Abé, H., Keer, L. M., and Mura, T., "Growth Rate of a Penny-Shaped Crack in Hydraulic Fracturing of Rocks, 2," Journal of Geophysical Research, Vol. 81, No. 35, Dec. 1976, pp. 6292-6298.

3. Weertman, J., and Chang, S. P., "Fluid Flow Through a Large Vertical Crack in the Earth's Crust," Journal of Geophysical Research, Vol. 82, No. 5, Feb. 1977, pp. 929-932.

4. Weertman, J., "Theory of Water-Filled Crevasses in Glaciers Applied to Vertical Magma Transport Beneath Oceanic Ridges," Journal of Geophysical Research, Vol. 76, No. 5, Feb. 1971, pp. 1171-1183.

5. Secor, D. T., Jr., and Pollard, D. D., "On the Stability of Open Hydraulic Fractures in the Earth's Crust," Geophysical Research Letters, Vol. 2, No. 11, Nov. 1975, pp. 510-513.

6. Pollard, D. D., and Muller, O. H., "The Effect of Gradients in Regional Stress and Magma Pressure on the Form of Sheet Intrusions in Cross Section," Journal of Geophysical Research, Vol. 81, No. 5, Feb. 1976, pp. 975-984.

7. Pollard, D. D., "On the Form and Stability of Open Hydraulic Fractures in the Earth's Crust," Geophysical Research Letters, Vol. 3, No. 9, Sep. 1976, pp. 513-516.

8. Weertman, J., "The Stopping of a Rising, Liquid-Filled Crack in the Earth's Crust by a Freely Slipping Horizontal Joint," Journal of Geophysical Research, Vol. 85, No. B2, Feb. 1980, pp. 967-976.

9. Keer, L. M., and Chen, S. H., "The Intersection of a Pressurized Crack With a Joint," Journal of Geophysical Research, Vol. 86, No. B2, Feb. 1981, pp. 1032-1038.

10. Hsu, Y. C., and Santosa, F., "The Stability of a Large Open Hydraulic Penny-Shaped Fracture (Crack) Near the Earth's Surface," Report ME-89(78)LASL-494-1, Los Alamos Scientific Laboratory, 1978.

11. Gringarten, A. C., Witherspoon, P. A., and Ohnishi, Y., "Theory of Heat Extraction From Fractured Hot Dry Rock," Journal of Geophysical Research, Vol. 80, No. 8, Mar. 1975, pp. 1120-1124.

12. Lowell, R. P., "Comments on 'Theory of Heat Extraction From Fractured Hot Dry Rock' by A. C. Gringarten, P. A. Witherspoon, and Y. Ohnishi," Journal of Geophysical Research, Vol. 81, No. 2, Jan. 1976, p. 359.

13. Wunder, R., and Murphy, H., "Thermal Drawdown and Recovery of Singly and Multiply Fractured Hot Dry Rock Reservoirs," Report LA-7219-MS, Los Alamos Scientific Laboratory, 1978.

14. Abé, H., Keer, L. M., and Mura, T., "Theoretical Study of Hydraulically Fractured Penny-Shaped Cracks in Hot, Dry Rocks," International Journal for Numerical and Analytical Methods in Geomechanics, Vol. 3, No. 1, Jan.-Mar. 1979, pp. 79-96.

15. Sekine, H., and Mura, T., "Characterization of a Penny-Shaped Reservoir in a Hot Dry Rock," Journal of Geophysical Research, Vol. 85, No. B7, Jul. 1980, pp. 3811-3816.

16. Abé, H., Sekine, H., Ishino, T., and Kamata, Y., "On the Limit for the Size of Hydraulic Fractures Near the Earth's Surface," International Journal of Fracture, Vol. 18, No. 2, Feb. 1982, pp. R17-R21.

17. Erdogan, F., and Sih, G. C., "On the Crack Extension in Plates Under Plane Loading and Transverse Shear," Transactions of the ASME, Journal of Basic Engineering, Vol. 85, No. 4, Dec. 1963, pp. 519-527.

18. Arima, S., "Fracture Mechanics Study on the Stability of a Branched Crack in the Earth's Crust," M. Eng. Thesis, Tohoku University, 1982.

19. Hayashi, K., and Abé, H., "Opening of a Fault and Resulting Slip Due to Injection of Fluid for the Extraction of Geothermal Heat," Journal of Geophysical Research, Vol. 87, No. B2, Feb. 1982, pp. 1049-1054.

20. Abé, H., Sekine, H., and Kitada, S., "Stability of a Penny-Shaped Geothermal Reservoir in the Earth's Crust," Transactions of the ASME, Journal of Energy Resources Technology, Vol. 104, No. 2, Jun. 1982, pp. 93-95.

21. Abé, H., Sekine, H., and Shibuya, Y., "Thermoelastic Evaluation of a Two-Dimensional Crack for Extraction of Geothermal Energy," Transactions of the Japan Society of Mechanical Engineers, Vol. 48, No. 431, Jul. 1982, pp. 899-903 (in Japanese).

Applications of Rock Fracture Mechanics

by Michael P. Cleary

Associate Professor of Mechanical Engineering
Massachusetts Institute of Technology
Cambridge, Massachusetts 02139

ABSTRACT

A survey is given of the many and varied areas of engineering and science which involve the existence, creation or prevention of fracture in rock-like materials. These encompass central activities like quarrying and construction, safety and excavation for tunnels and mines, drilling and fracturing for extraction of oil, gas or thermal energy; other applications include specialty (e.g., ornamental) use of rock or the phenomenology associated with structural geology and earthquake regions, illustrating the vast scale variations in rock studies. Scale and stress/material state are emphasised heavily in contrasting specimens used for laboratory measurements of fracture toughness with the state of representative samples in rock masses; it is even shown that material decohesion plays a negligible role in the overall energy balance of many processes, especially when most work is done, directly or frictionally, to overcome large confining stresses. The relative (energy) efficiency of various rock-breaking techniques is evaluated and some innovations are described. The status of analysis and design procedures for rock fracturing applications is discussed: the potential of finite element and surface integral techniques is put in proper perspective and ongoing/needed developments are outlined, in the context of primary applications.

INTRODUCTION AND BACKGROUND ON ROCK FRACTURING ACTIVITIES

A dramatic increase in activity has ocurred, in the past decade especially, in the area of rock mechanics and particularly in rock fracture. This has arisen from a variety of application areas but the main driving force has been the prospect of shortages in energy and material resources (1,2): this has led to extensive technology development in deep underground fracturing (3), in excavation and mining techniques/machinery (4,5) and in quarrying methods (6). Other developing areas of endeavor are the large programs on earthquake hazards reduction (8), chemical and nuclear waste disposal (8,9) and underground defense chambers (8). As well, some more traditional areas have seen a burgeoning of new interest--as in structural analysis and design for (discontinuous) rock masses around structures, tunnels and surface cuttings for transportation or open-pit mining (11,12).

The major components of the scientific and engineering development associated with these areas have been the recognition of fracturing details and coupled processes, such as fluid/heat flow and mechanical deformation, as central to the phenomenology involved. Resulting space and time scale effects have since been working themselves into the design considerations and major new ideas have evolved: more relevant laboratory testing is being done and analysis codes are incorporating (even predicting) much more realistic versions of the physical observations made. As well, this proper rationalisation of the systems involved has led to a spate of innovative ideas, tools and methodologies which are rapidly bearing fruit, especially in the author's specialties of rock-breaking and oil/gas/mineral reservoir exploitation.

Considerable experience has been gained in various aspects of rock fracturing over the years. Beginning with man's first success in forming sharp tools by breaking off suitable pieces of hard rock, increasingly sophisticated techniques have been developed to fragment rock structures: impact with hammer or projectile became by far the most common method--one might say the only practically efficient method--until the discovery of explosives relieved the tedium of quarrying and mining by making possible the fragmentation of large rock segments at once. Almost simultaneously, less dramatic wear processes (--which, it is often forgotten, also involve fracture at some level--) were being developed, starting from the filing of tools into shape by primitive man: industrial machine tools quickly led to rock-cutting machinery, which soon was applied to mining of soft rock, and modern developments (e.g., water-jets, drilling tools of all types) have advanced this process to a high level of sophistication, in machine design if not fundamental understanding.

The purpose of this paper is to review some of the more central aspects of developments in recent years. We begin with an overview of various techniques for breaking rock and then comment on the converse desire, of actually preventing rupture, in the context of creating stable openings on the surface or below ground. The principal ingredients of any design procedure are identified and suitable methods of analysis are described, together with the primary required parameters. Determination of these parameters is considered in the context of laboratory experimentation and field testing, with special emphasis on scales of applications--some of which are then discussed in more detail, with the foregoing as background; the actual lecture will present pictures of equipment.

MECHANISMS AND TECHNIQUES OF ROCK FRACTURE

There are many ways to create fractures, above and below ground; some broad categories may be identified as follows:

1. Mechanical Application of Stress to the Rock

Direct contact with solid machine elements. This approach is typical of conventional excavation and rock breakage equipment (e.g., drill-bits, indentation or drag cutters, even water-jets and the common pick-axe) and has seen limited improvement mainly in people's recognition of rock's low tensile (vs. compressive) strength, but not extensively in equipment developed.

Use of fluid pressure to exploit the tensile weakness. This approach has many advantages and much unexplored potential; it really should be divided into four main types of processes: 1) Hydraulic fracturing by external injection of fluid into a propagating crack, one already existing or one created for the purpose. This technique has seen extensive use by industry (10). 2) Pore-pressure-induced cracking (PPIC) due to creation of excess internal pressure p in the rock pores, exploiting the effective stress law so that $p-\sigma_c$ exceeds the tensile strength of the material, causing a crack to propagate under the pressure of inflowing fluid from the pores. This may be classified (3) as a hydrafrac process but we distinguish it by the source of the pore-fluid driving the crack propagation. 3) Fluid impact and penetration into surface cracks, e.g., in water-jets which have been used both for direct cutting and consist of mechanical indenters. 4) Explosive generation of high fluid pressures in a confined cavity (e.g., borehole), which induce transient tensile stresses high enough to open cracks in the rock as stress-waves pass by. The main drawbacks of this methodology, apart from complexity of analysis, are the (delaying) need for great safety precautions during the operation and the re-closure of the cracks (under the ever-present confining earth stresses) when the dynamic event has passed--indeed, there are often deleterious "compression cages" which can wipe out permeability in critical regions (e.g., around a borehole) (See refs. 3,12,22).

2. Induction of Thermal Stresses by Temperature Alteration

Since reservoirs are hot, there is always the possibility of exploiting the tensile stresses induced by cooling around cavities (e.g., wellbores); most simply, this requires the injection of cold fluid--either into the body of pore-space or along a hydraulic fracture. Cooling of the primary fracture walls may produce enough secondary cracking (e.g., ref. 13) and overcome "scum" effects but bulk cooling produces the best large-scale crack pattern. Injection of hot fluid or heat is also common, e.g., in steam drives and fire flooding for heavy oil extraction. This produces compression in the heated region but, for high enough temperatures, it may drive ahead of it an array of cracks which never fully close after the front sweeps by them: this mechanism may well explain the apparent success of recent field trials with a (downhole) rocket motor propelling superheated steam into some closely spaced reservoir wellbores.

Another major aspect of thermal processing is the phase-transformation and potentially very high pore-pressure, which can be induced by heat supply, thereby leading to the PPIC mentioned previously. An example of this is the conversion of solid kerogen to various gaseous and liquid hydrocarbons, which we are studying in the laboratory, using oil shale samples from Colorado (3,7).

3. Chemical Alteration of Microstructure and Induced Stresses

The oil and gas industry is quite expert at finding new chemical treatments (e.g. acidisation) to clean up wellbores and enhance permeability: well-known chemical reactions of fluids with rock minerals govern this process. Less attention has been paid to two other aspects: the feasibility for the chemical treatment to drastically change the stress state in the reservoir and the enormous potential for induction of high pore-pressures, e.g., great enough to maintain reaction products in the condensed state at reservoir temperatures, and to produce PPIC.

4. Other Chemophysical Processes and Technology

Many other techniques have been suggested for cracking rock, varying from laser beams (which might sometime be considered for pre-notching the walls of a borehole) to flow of electricity (which has little effect except due to resistive heating) to dielectric heating by HF electromagnetic waves; these may merit more attention.

Among these various processes, many have played a role both in nature and in human technology: over-pressured reservoirs have become naturally fractured, thermal and shrinkage cracks are every-where, and magmatic intrusions (dykes and sills) are fascinating examples of the earliest (inverse) hydrafracs--fortunately (usually) stabilised by the cooling of molten rock as it approaches earth's surface. Indeed, many of these same mechanisms play a role in deciding the converse question of structural stability against severe fracture onset: as a primary example, the difficulty of breaking rock under compressive loading (discussed previously) becomes an advantage, in requiring much higher stored energy (hence compressive strengths) for the dominant frictional work than the small amount needed for the fracture process itself (of which tensile energies/strengths are indicative). This frictional energy dissipation is also much more predictable than the very state- and scale-dependent tensile response, so that limit-loads can sometimes be well bounded; however, if the stress-drop caused by fracture is too great (relative to purely frictional resistance against slippage), the actual rupture mechanism has to be traced correctly in the design model, e.g., as a shear-band growing through the structure (14).

PROCEDURES AND PARAMETERS FOR TRACING FRACTURE EVOLUTION

The correct incorporation of the various foregoing mechanisms and observations requires quite careful modelling with effective numerical procedures (14). The models must contain at least the following ingredients:

a) Sufficiently accurate representations of the constitutive behavior in homogeneous deformation fields, including non-linearities and time-dependence. Examples of critical parameters here are yield strengths σ_y, hardening moduli h and relaxation times t_R. This task is most effectively accomplished by widely-available finite-element schemes, provided the deformation involved is smoothly varying on the scale of element size (11,21).

b) Incorporation of (growing) fracture discontinuities, which we have found to be most effectively achieved by surface integral methods, without remeshing around the fracture (14). Dominant parameters here are elastic modulus $\bar{E}$, required fracture energy G_c, failure strength τ_F and frictional resistance coefficient μ on any closed sliding surfaces. Structure features, such as joints, are important.

c) Account of fluid flow and induced stresses, around and in the fracture, which can be modelled by surface integrals (and finite elements in combination, for nonlinearities and heterogeneities): primary parameters are permeability k to and diffusivity c of the surrounding reservoir fluid, the poroelastic coefficients A and B (of mutually-induced stresses and pore-pressure) and (thermo-) rheology of fluid in the fracture, e.g., characterised by effective viscosity $\bar{\eta}$. Multiphase flow also often dominates.

d) Calculation of heat transfer around the fracture, by methods analogous to (c), involving parameters of conductivity κ, diffusivity D and surface convection H, plus thermal expansion coefficients α.

e) Modelling of other processes, such as chemical reactions and complex fluid rheology (including dispersed solids, such as suspended fines and proppant for crack wedging).

Although a completely comprehensive computer program, which accounts for all of these processes, does not yet exist, the development of the many components and their gradual combination as needed, has indeed been undertaken by us (3)--motivated especially by the major application to hydraulic fracturing technology. A great amount of physical data is needed as input to these programs, and this has led to parallel efforts in the area of laboratory and field investigations as follows, but we note that the computer models can even go beyond such data--allowing the identification and elimination of redundant or second-order variables for any particular application.

LABORATORY AND FIELD TESTING

A broad spectrum of activity has been generated by the need for model parameters and field applications. This ranges from deformation and fracture toughness testing in the laboratory (e.g., 6-9,15), to small-scale field testing (e.g., 8,12,16,17) and even to very large-scale assessment of results obtained by various designs (e.g., 18,19).

Laboratory testing has focused on measurement of properties on a small-scale, especially stress-strain relations and fracture behavior. Great emphasis has been placed on a detailed characterisation of linear, nonlinear, creep and rupture response of samples--with appropriate emphasis on the role of confining pressure, requiring the use of triaxial (pressure vessel) test systems (e.g., 23). As well, careful studies have been made (e.g., 7,15) of resistance to fracture growth (so-called fracture toughness), in a manner entirely analogous to procedures now carefully specified for testing of metals and other fabricated materials. While such research clearly has great merit in fundamental scientific terms, it has tended to defer three major aspects from the viewpoint of applications:

i) Fractures will tend to follow weak pathways in real heterogeneous rock structures (e.g., 12);

ii) The actual energy expended on fracturing is usually much larger than that measured, because of multiple aborted branched pathways, frictional absorption of energy and the dominant role of confining stresses (see refs. 10,21 for example);

iii) The small-scale static/dynamic properties need not be related to the large-scale dynamic/static properties relevant to the application in question. Because the discrepancy in fracture processes can be also regarded in the same light, we regard this question of size-effect and the need for large-scale testing as central to activities in the geomechanics area and therefore we focus on it for discussion here (see also refs. 16,20).

Large-Scale Testing in Geomechanics

One of the central difficulties in the experimental process is the presence of important heterogeneities or (material resonse) mechanisms in the system on a scale larger than the sample size. A very familiar example is the presence or onset of a fracture or deformation-localisation mode of rupture which may not be represented (e.g., because of its size-dependency) in the smaller test. Such features (e.g., fissures, joints, striations, interfaces) can dominate the response of a rock mass on the large scale of many (natural and artificial) underground processes, but they may be misrepresented (or not even be considered) in the modelling. Simply put, for example, a mine-roof may fall in pieces even when a precise stress analysis predicts a safe response to excavation e.g., with all stresses below a measured tensile strength, based on moduli and yield/frictional parameters measured in the laboratory.

The engineering community has reacted in various ways to these problems, which are also perceived in various ways. A surprisingly persistent point of view has been that we really should be making measurements in the laboratory on a much larger scale and using these observations in our models, perhaps both as input and verification of output. This approach has again recently led to a variety of proposals and some actual development work (e.g., on use of large-scale test frames, true triaxial confinement and centrifuges). Although some of this work can be fruitful (e.g., application of centrifuges to testing scaled models of gravity structures composed of relatively scale-invariant materials such as sands and clays), it seems that much of it may be misdirected in terms of both feasibility and applicability. As well, there seem to be much more fruitful approaches, as we now discuss.

The most promising, potentially efficient and general approach is the (by now) classical method of measuring properties on the scale of the most elementary components of the system and then integrating these to describe the whole system. Thus, one adopts the properties of the largest element size which contains a representative distribution of microstructure and uses them to describe (in a continuum or discrete sense) each of the separate blocks of material which make up the system--according to the remaining invariant laws of geometric compatibility and momentum conservation. This whole approach has become more realistic with the aid of ever-increasing computational power and it may now be that any problems not amenable to computer analysis are hardly amenable to characterisation in large-scale testing (unless conducted with the whole real system being studied or designed). Indeed, the latter can probably only fulfill the role of verification for the analysis (which incorporates the component parameters of the test and then, having been vindicated, can switch to those of the real system if necessary).

In this respect, a small-scale test may be adequate for verification, if it has features similar to (but not necessarily scale-models of) the real system. As well, we must note that increases of size from a few inches to a few feet (i.e., from comfortable to uncomfortable in laboratory terms) will not usually yield a dramatic change from partial to complete representativeness of microstructure. Also, behavior of special features (like cracks and faults) can usually be studied just as well (or badly) on either scale. The (computer) model should actually be able to incorporate and vary both the microscopic and macroscopic features (e.g., growing fractures, varying contact geometries) which are dependent on the location and character of the test, allowing testing on the smaller scale to serve for confident extension to the massive scale of many operations (without necessarily always requiring verification tests on that scale).

However, a large-scale test can have some vital merits: especially, it can test the response of a conglomerate of elements acting together in a way that requires complex computation for its description (e.g., irregular contact boundaries, multiple but not simply statistically distributed arrays of constituent features). Although a small-scale model can be made to represent such complexities, it will not exactly simulate the conglomerate response and, as stated, a computational model (to pick up the test and transfer function) is assumed prohibitive. Hence we might properly rationalize the need for large-scale tests, but *not* Large-Scale Laboratory Tests (LSLT). These still have an artificiality and unwieldiness about them which is hardly justified by the information gained. The rock or soil sample is tested after removal from its ambient environment (hence it is disturbed from the history it might see in the real application)--and it requires a very major endeavor to extract and transport and test a large enough sample. As well, such a large sample may often simply not be accessible, especially in the context of extraction through wellbores only from deep underground--which is surely the primary objective of any rational plan for the (perhaps far) future of resource mining, and even today pervades the massive hydrocarbon industry. Thus, while lab testing of large samples might have some short-term phenomenological interest, it would contribute little to long-term practical applications. Nevertheless, it might be pursued if informative large-scale in-situ testing were impossible (or much more difficult and costly) to undertake, as LSLT advocates seem to maintain.

In fact, for determination of most relevant parameters and phenomenology, Large-Scale In-Situ Testing and Verification (LSISTV) seems to us to be quite feasible--if undertaken in the proper way with imaginative design of test configurations, simple loading schemes and response instrumentation. The simplest of these, which has been examined and tested sufficiently to convince us of complete feasibility (e.g., 17) involves borehole access and suitable hydraulic (or other mechanical) fracturing on any scale, with careful monitoring of fluid pressure/flow and nearby deformation, temperature, electric potential, pore-pressure, etc. Wave propagation and electromagnetic sensing provide complementary techniques. The overall methodology is well within the realm of technology (some already existing) which the hydrocarbon industry can develop with suitable guidance from analysis of the processes involved. It can be applied at all depths and all scales, covering the complete realm of resource extraction, tunneling, etc. Instead, geotechnical engineers seem to have concentrated on rather massive loading machines (e.g., plate/wedge indentation, floor-to-roof bracing, flatjacks, etc. (16)). which, of course, cannot even be put in place for many applications (e.g., wellbore extraction from deep reservoirs).

SUMMARY AND CONCLUSIONS

As in any engineering endeavor, a number of approaches are possible in pursuit of understanding and designs for phenomenology and operations occuring (usually on the large-scale) in the earth beneath us. These may be broken up into three broad divisions:

1. Measurements of physical quantities, parameters and material properties.
   - A. On a small scale in the laboratory.
   - B. On a (small or) large scale in the field.
   - C. On a large scale in the laboratory.
2. Development of theoretical models.
   - A. Qualitative characterisation of the phenomenology.
   - B. Approximate (analytical) formulae for the relations between variables/parameters involved.
   - C. More precise (numerical) analysis of the relations between variables/parameters involved.
3. Development of experimental models.
   - A. Analogue models based on the governing mathematical equations, using some more convenient physical system.

B. Small-scale models of the physical processes involved, usually in the laboratory.

C. Part or full-scale models in the field; these may be sample real (prototype) situations and include the monitoring of many real operating systems.

To avoid unreasonable expense, most engineering rests heavily on a combination of 1A with 2C (or 2B, if knowledge of parameters doesn't justify 2C), and use of 3C for verification. There obviously can be a large gap between measurements in 1A and 1B and it is quite difficult to develop 3B toward the complete realism which would allow confident self-contained prediction of 3C, hence providing verification (or even bypass) of the theoretical modelling in 2C (e.g., with parameters extrapolated from 1A). As well, since 3A merely represents the mathematical equivalency or equation structure, it will not pick out new features in basic material response--although it may sometimes avoid intractable analysis and show (unexpected) features in overall system response.

It seems the community would do well now to pursue techniques with more long-term potential and direct application to the processes which are the (major) motivation for the studies proposed. These could both provide raw data for design modelling on the very large scale and also test out the predictions of models on the intermediate scale (e.g., based on parameters determined from small-scale cores). Thus, they could fill in the remaining gaps in the logical progression from small-scale testing (1A) to modelling (2B/C) to large-scale testing (1B) to design of full-scale operations (3C)--which could also be instrumented similarly with 1B to give a final (perhaps initially expensive) verification of the procedures.

Acknowledgements

Research related to the work discussed here has been supported by grants from the National Science Foundation and a C. Richard Soderberg Career Development Chair of Engineering at the Massachusetts Institute of Technology, Cambridge, Massachusetts. The work is part of an overall RESOURCE EXTRACTION LABORATORY which has been developed by the author at MIT. A major ongoing project is MIT UFRAC (Ref. 3) for which a flow chart is provided in figure P1: note that the numerals and letters there denote sections of Ref. 3 not the subdivisions in the SUMMARY AND CONCLUSIONS.

REFERENCES

1. Abelson, P.H. and A.L. Hammond (Eds.), "Materials: Renewable and non-Renewable Resources," pub. by Amer. Association Advancement Science, 1976. (See also National Geographic Special Report on Energy, February 1981).

2. "Rock Mechanics Research Requirements," pub. by U.S. National Committee on Rock Mechanics, Nat'l Academy of Sciences and Engineering, 1981.

3. Cleary, M.P., et al., "Theoretical and Laboratory Simulation of Underground Fracturing Operations," First Annual Report of the M.I.T. UFRAC Project, August 1981.

4. Jones, A.H.(Ed.), "Mining Technology for Energy Resources," pub. by Amer. Soc. Mech. Eng., 1978.

5. Sikarsie, D.(Ed.), "Rock Mechanics Symposium," pub. by Amer. Soc. Mech. Eng., 1973.

6. Rossmanith, H.P.(Ed.), Rock Fracture Mechanics, Springer-Verlag, to appear 1983. (Proc. Summer School at Udine, Italy, July, 1982; contributing authors: M.P. Cleary, W.L. Fourney, A. Ingraffea, F. Ouchterlony and R. Schmidt).

7. Cleary, M.P., "Some Deformation and Fracture Characteristics of Oil Shale," Proc. 19th U.S. Symp. Rock Mech., 1978. (See also P. Switchenko, "Thermomechanical Response of Oil Shale," M.S. Thesis, M.I.T., 1979.)

8. "Proceedings of 23rd U.S. Rock Mechanics Symposium," pub. by Univ. California, Berkeley, 1982. (See also Proc. 22nd U.S. Rock Mechanics Symposium," pub. by M.I.T., Cambridge, Mass., 1981).

9. Angino, E.E., "High-Level and Long-Lived Radioactive Waste Disposal," Science, vol. 198, p. 885, Dec. 1977.

10. Cleary, M.P., "Mechanisms and Procedures for Producing Favourable Shapes of Hydraulic Fractures," Paper SPE 9260, 1980. (See also SPE 9259, 1980).

11. Goodman, R., "Methods of Geological Engineering," West Pub. Co., 1976. (See also Roberts, W.J. and H.H. Einstein, "Comprehensive Model for Rock Discontinuities", ASCE Jour. Geotech. Div., pp. 553-569, May 1978.)

References (cont'd)

12. Hendron, A.J., "Engineering of Rock Blasting on Civil Projects," in Structural and Geotechnical Mechanics, pub. by Prentice-Hall, 1977. (See also W.L. Fourney, D.B. Barker, and D.C. Holloway, "Explosive Fragmentation of Jointed Brittle Media," Int. Jour. Rock Mech. and Mining Sciences, to appear 1982.)

13. Barr, D.T., "Thermal Cracking in Nonporous Geothermal Reservoirs," M.S. Thesis, M.I.T., 1980.

14. Cleary, M.P. and J.L. Dong, "Analyses of Deformation and Failure in Geological Materials," Proc. Int'l Conf. on Constitutive Relations, Tucson, AZ, January 1983.

15. Ouchterlony, F., "Review of Fracture Toughness Testing of Rock," SM Archives, 7, 131-211, pub. by Martinus Nijhoff, The Hague, 1982.

16. Stagg, K.G. and O.C. Zienkiewicz, "Rock Mechanics in Engineering Practice," pub. by Wiley, N.Y., 1968.

17. Smith, M.B., "Stimulation Design for Short Precise Hydraulic Fractures," Paper No. SPE 10313, 1981.

18. Smith, M.D., G.B. Holman, C.R. Fast and R.J. Covlin, "The Azimuth of Deep Penetrating Fractures in the Wattenberg Field," J. Pet. Tech., February 1978.

19. Warpinski, N.R., R.A. Schmidt and D.A. Northrop, "In-Situ Stresses: The Predominant Influence on Hydraulic Fracture Containment," Paper SPE/DOE8932, 1980. (Jour. Pet. Tech., 653-664, March 1982).

20. Jaegar, J.C. and N.G.W. Cook, "Fundamentals of Rock Mechanics," Chapman & Hall, 1979.

21. Cleary, M.P., "Fracture Discontinuities and Structural Analysis in Resource Recovery Endeavors," Paper No. 77-Pet-32, Amer. Soc. Mech. Eng., 1977.

22. Young, C., "Evaluation of Simulation Technologies in the Eastern Gas Shales Project," pps. G5/1-16, Proc. DOE Symposium on Enhanced Oil Recovery, vol. 2, 1978.

23. Abou-Sayed, A.S., "Fracture Toughness K_{IC} of Triaxially-Loaded Limestones," pp. 2A3, 1-8 in Proc. 18th Symposium on Low Permeability Gas Reservoirs, Denver, May 1981.

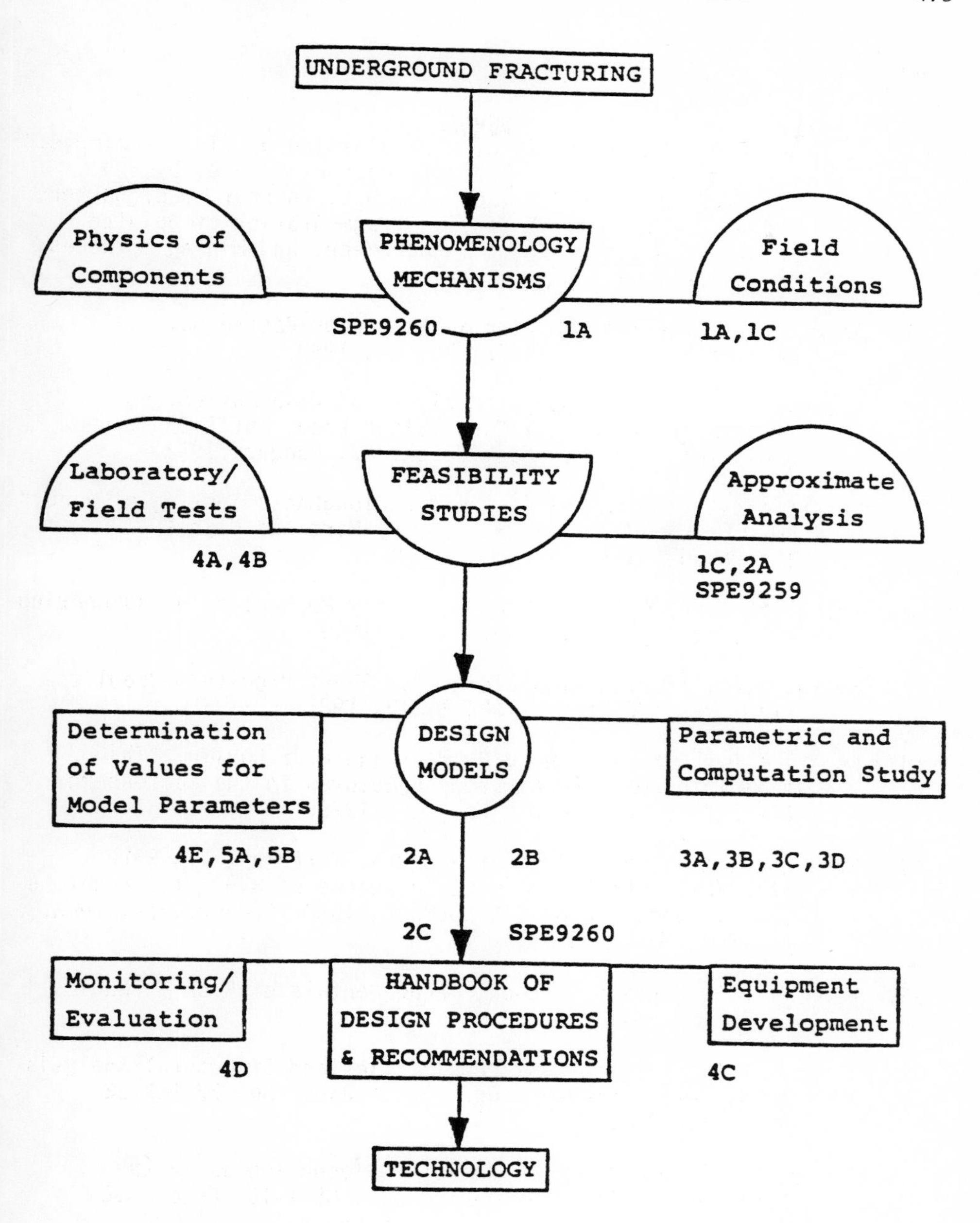

Figure P1. Flowchart of context for the various activities being undertaken in the MIT UFRAC Project.

A RESERVOIR CREATED BY INJECTION OF FLUID ON A FAULT FOR THE EXTRACTION OF GEOTHERMAL HEAT

by Kazuo Hayashi and Hiroyuki Abé

Research Associate, Tohoku University
Sendai, Japan

Professor, Tohoku University
Sendai, Japan

ABSTRACT

On the basis of the two-dimensional theory of elasticity, the opening of a fault and the resulting slip due to injection of fluid are studied for the extraction of geothermal heat from hot dry rock masses. A fault is modeled by a plane of frictional contact between two elastic bodies. Once fluid is injected between the two surfaces of a fault plane, it will be opened up over a region where the pressure acting on the fault plane before the injection is locally smaller than that of neighboring regions. The size of the opened region, i.e., the fluid-filled region, is compared with that of a vertical, fluid-filled crack within the earth's crust to show that the fluid-filled region could be used as a reservoir for the extraction of geothermal heat. As the opening takes place, slip spreads simultaneously along the fault plane due to the reduction of frictional forces in the fluid-filled region. This slip is analyzed by the use of two models for the fault plane, i.e., an interface with frictional resistance only, and an interface with both frictional resistance and fracture toughness for the sliding mode. It is shown that the two models do not exhibit any significant difference in the size of the slipped region when the fluid-filled region is of dimensions of practical interest.

INTRODUCTION

A fluid-filled crack within the earth's crust is considered to be a potential candidate for the reservoir in the scheme of the extraction of geothermal heat from hot, dry rock masses, and it has been employed in many theoretical studies to determine the feasibility of the scheme; brief surveys are presented in the papers by Sector and Pollard (1), Abé, et al. (2)-(5), Hsu and Santosa (6) and Sekine and Mura (7).

A crack which is filled with fluid that is less dense than the surrounding rock can rise to the earth's surface by growing at the upper crack tip while simultaneously squeezing shut at its lower tip (Weertman (8) and Keer and Chen (9)). From this mechanism, it is easily inferred that the fluid-filled crack would encounter a prefractured plane, such as a fault or a joint, and, consequently, fluid would leak into the prefractured plane. Albright and Pearson (10), on the other hand, have observed microearthquakes in successive massive injections of water at the Department of Energy Hot Dry Rock Geothermal Energy demonstration site at Fenton Hill, New Mexico, and have suggested that the microearthquakes result from shear failure, probably on preexisting planes of weakness that intersect or make up the main hydraulic system. These suggestions indicate the possibility that a fault which is a prefractured plane or a plane of weakness in the earth's crust, may serve as a reservoir or a part of a main reservoir for the extraction of geothermal heat.

The main objective of this paper is to investigate the behavior of a fault with a fluid-filled region which is opened by hydraulic pressure. The two rock masses facing each other across a fault are in temporal equilibrium under frictional forces acting on their surfaces. A fault itself would not have any resistance against a tensile load applied perpendicularly to the surfaces. In other words, it can support a compressive load normal to and a shearing load parallel to the surfaces. The ability for carrying a shearing load would depend primarily on the geometries of asperities (or the roughness of the surfaces) in addition to their material properties. Once fluid is injected into a fault, a part would be opened up and the resulting slip would spread until the two rock masses reach a new state of equilibrium. Under these considerations, a fault is modeled by an interface across which two elastic bodies are in frictional contact.

The problem on a partially opened interface is easily reduced to solving a set of singular integral equations, where the unknown functions are the derivatives of the displacement-discontinuities across the interface with respect to the direction along the interface. The size of the opened region is determined by the condition that the stresses are finite at the ends of the opened region. As to the slip, two cases are considered: (i) the surfaces of the interface are fairly smooth and the stresses caused by the slip are finite everywhere, and (ii) the interface is rough enough, so that the slip spreads with shear failure along the interface. In the second case, the concept of linear fracture mechanics should be applicable. Namely, the interface itself has its own fracture toughness for the sliding mode, or alternatively, the stresses are allowed to have singularities at the ends of the slipped region.

Results show that a fault could be used as a reservoir for the extraction of geothermal heat provided that the fault has some regions where the pressure due to the tectonic stresses on the fault plane is smaller than that on the neighboring regions. The size of the slipped region depends strongly on the magnitude of the coefficient of friction for the two cases just mentioned. However, the differences between the sizes for the two cases are small when the fluid-filled region is of dimensions of practical interest.

STATEMENT OF PROBLEM AND BASIC EQUATIONS

A fault is modeled by an interface across which two elastic bodies are in frictional contact. Consider an oblique fault which is intersecting the earth's surface with an angle θ. The two rock masses facing each other across the fault plane are in contact under the condition of plane strain. A Cartesian coordinate system (ξ, η) is located on the fault plane as shown in Fig. 1. On the fault plane, there is a fluid-filled region, between the points (a, 0) and (b, 0), opened by a hydraulic pressure p_w. Before the fluid-filled region was created by the injection of fluid, the two rock masses were in static equilibrium with stresses $\sigma^0_{\xi\xi}$, $\sigma^0_{\eta\eta}$ and $\sigma^0_{\xi\eta}$ (or the tectonic stresses). Once fluid is injected, a relative movement will take place along the fault plane due to the reduction of frictional forces in the opened region. The newly created, slipped region is assumed to be lying between the points (c, 0) and (d, 0). In the following, it is assumed for simplicity that effects of the earth's surface are negligibly small and the elastic properties of the two rock masses are equal to each other.

A subsidiary Cartesian coordinate system (x, y) is located on the earth's surface as shown in Fig. 1. Then, the stresses σ^0_{xx}, σ^0_{yy} and σ^0_{xy}, referred to the subsidiary coordinate system, before the injection of fluid would be expressed as

$$\left.\begin{aligned} \sigma^0_{xx} &= -S_x + \tilde{\sigma}_x \\ \sigma^0_{yy} &= -S_y + \tilde{\sigma}_y \\ \sigma^0_{xy} &= S_{xy} + \tilde{\sigma}_{xy} \end{aligned}\right\} \qquad (1)$$

Here, the first terms in (1) are the fractions varying linearly with respect to the depth and are assumed to be given by

$$\left.\begin{aligned} S_y &= -\rho_r gy \\ S_x &= \alpha S_y \quad (0 < \alpha < 1) \\ S_{xy} &= 0. \end{aligned}\right\} \qquad (2)$$

where ρ_r is the density of the rock masses and g is the acceleration due

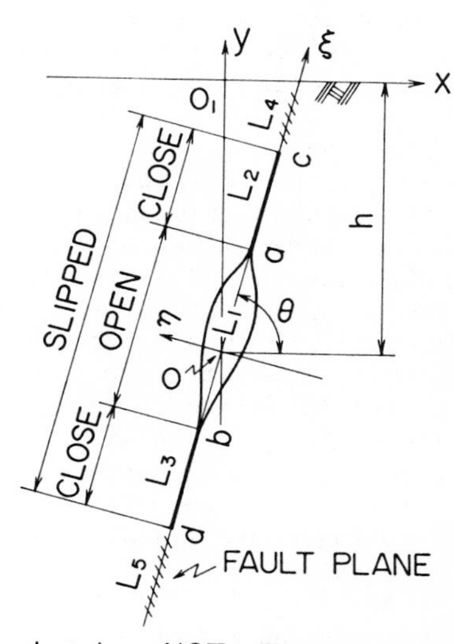

Fig. 1. Geometry and coordinate systems.

to gravity. The second terms in (1) are introduced to express the spatial variation of the tectonic stresses other than the linear variation with respect to the depth. Therefore, before the injection of fluid, the stresses referred to (ξ, η) are

$$\left.\begin{aligned} \sigma^0_{\xi\xi} &= -S_y(\sin^2\theta + \alpha\cos^2\theta) + \tilde{\sigma}_{\xi\xi} \\ \sigma^0_{\eta\eta} &= -S_y(\alpha\sin^2\theta + \cos^2\theta) + \tilde{\sigma}_{\eta\eta} \\ \sigma^0_{\xi\eta} &= S_y(\alpha - 1)\sin\theta\cos\theta + \tilde{\sigma}_{\xi\eta} \end{aligned}\right\} \qquad (3)$$

where the second terms stand for the components derived from the second terms in (1) by the coordinate transformation. The pressure of the fluid is given by

$$p_w = p_{w0} - \rho_w g\xi\sin\theta \qquad (4)$$

where p_{w0} denotes the pressure at the depth $y = -h$ and ρ_w the density of the fluid.

Employing the Coulomb law of friction and denoting the stresses after the injection of fluid by $\sigma_{\xi\xi}$, $\sigma_{\eta\eta}$ and $\sigma_{\xi\eta}$, and also the displacements by u_ξ and u_η, the boundary conditions are given as follows:

(a) On the opened region L_1 $(b < \xi < a)$,

$$\sigma_{\eta\eta} = -p_w \; (<0), \quad \sigma_{\xi\eta} = 0 \qquad (5)$$

(b) On the regions L_2 $(a < \xi < c)$ and L_3 $(d < \xi < b)$,

$$\Delta u \equiv u_\eta\big|_{\eta\to 0^+} - u_\eta\big|_{\eta\to 0^-} = 0 \qquad (6)$$

$$\sigma_{\xi\eta} = f\sigma_{\eta\eta} \qquad (7)$$

$$\sigma_{\eta\eta} \leqq 0 \qquad (8)$$

(c) Outside the newly slipped regions, i.e. on L_4 $(c < \xi)$ and L_5 $(\xi < d)$,

$$\Delta v \equiv u_\xi\big|_{\eta\to 0^+} - u_\xi\big|_{\eta\to 0^-} = 0, \quad \Delta u = 0 \qquad (9)$$

(d) Far from the opened region,

$$\sigma_{\xi\xi} = \sigma^0_{\xi\xi}, \quad \sigma_{\eta\eta} = \sigma^0_{\xi\eta}, \quad \sigma_{\xi\eta} = \sigma^0_{\xi\eta}, \quad \text{as } \sqrt{\xi^2 + \eta^2} \to \infty \qquad (10)$$

Here, f represents the coefficient of friction.

The gap Δu on L_1 and the shift Δv on $L_1 + L_2 + L_3$ induce additional stresses, which will be denoted by $t_{\xi\xi}$, $t_{\eta\eta}$ and $t_{\xi\eta}$. Modeling both the opened and the slipped regions as continuous distributions of edge dislocations, we readily arrive at

$$t_{\eta\eta}(\xi,\ 0) = \frac{1}{\pi\lambda}\int_b^a \frac{B_\eta(s)}{\xi-s}\,ds\ ,\quad t_{\xi\eta} = \frac{1}{\pi\lambda}\int_d^c \frac{B_\xi(s)}{\xi-s}\,ds \tag{11}$$

where

$$\lambda = \frac{2(1-\nu)}{\mu} \tag{12}$$

μ is the shear modulus and ν is Poisson's ratio. The function $B_\xi(\xi)$ and $B_\eta(\xi)$ are related to Δv and Δu as follows:

$$B_\xi(\xi) = -\frac{d}{d\xi}\Delta v\ ,\qquad B_\eta(\xi) = -\frac{d}{d\xi}\Delta u \tag{13}$$

The stresses after the injection of fluid are expressed as the sum of the two sets of stresses $(\sigma^0_{\xi\xi},\ \sigma^0_{\xi\eta},\ \sigma^0_{\eta\eta})$ and $(t_{\xi\xi},\ t_{\xi\eta},\ t_{\eta\eta})$:

$$\sigma_{\xi\xi} = \sigma^0_{\xi\xi} + t_{\xi\xi}\ ,\quad \sigma_{\xi\eta} = \sigma^0_{\xi\eta} + t_{\xi\eta}\ ,\quad \sigma_{\eta\eta} = \sigma^0_{\eta\eta} + t_{\xi\eta} \tag{14}$$

On the basis of (11) and (14), the boundary conditions (5) and (7) lead immediately to the following integral equations for $B_\xi(\xi)$ and $B_\eta(\xi)$:

$$\frac{1}{\pi\lambda}\int_b^a \frac{B_\eta(s)}{\xi-s}\,ds = -p_{w0} + p_0(\alpha\sin^2\theta + \cos^2\theta)$$
$$+ p_0\frac{\xi}{h}\left\{\frac{\rho_w}{\rho_r} - (\alpha\sin^2\theta + \cos^2\theta)\right\}\sin\theta - \tilde{\sigma}_{\eta\eta}(\xi,\ 0)$$
$$(b < \xi < a) \tag{15}$$

$$\frac{1}{\pi\lambda}\int_d^c \frac{B_\xi(s)}{\xi-s}\,ds - fH(\xi)\frac{1}{\pi\lambda}\int_b^a \frac{B_\eta(s)}{\xi-s}\,ds$$
$$= -p_0\left\{(\alpha-1)\sin\theta\cos\theta + fH(\xi)(\alpha\sin^2\theta + \cos^2\theta)\right\}$$
$$+ p_0\frac{\xi}{h}\left\{(\alpha-1)\sin\theta\cos\theta + fH(\xi)(\alpha\sin^2\theta + \cos^2\theta)\right\}\sin\theta$$
$$+ fH(\xi)\tilde{\sigma}_{\eta\eta}(\xi,\ 0) - \tilde{\sigma}_{\xi\eta}(\xi,\ 0)\qquad (d < \xi < c) \tag{16}$$

where

$$p_0 = \rho_r gh \tag{17}$$

and

$$H(\xi) = \begin{cases} 1 & \text{for } \xi \in L_2 + L_3 \\ 0 & \text{for } \xi \in L_1 + L_4 + L_5 \end{cases} \tag{18}$$

Furthermore, the single-valuedness of displacements requires

$$\int_b^a B_\eta(s)ds = 0 \ , \qquad \int_d^c B_\xi(s)ds = 0 \tag{19}$$

The solutions of (15) and (16) under the condition (19) are uniquely fixed, if asymptotic properties of $B_\eta(\xi)$ and $B_\xi(\xi)$ at the ends of the respective regions are specified. In view of (13), it is clear that these properties can be determined from the conditions imposed on the stresses at the ends.

The spatial variations of the tectonic stresses other than the linear dependence on the depth are not known yet. Here, for an example which is fairly simple and gives a clear view over the problem, we put

$$\tilde{\sigma}_{\eta\eta}(\xi,\ 0) = \Delta p_0 \cos\frac{\xi\pi}{\ell}\ , \quad \tilde{\sigma}_{\xi\eta}(\xi,\ 0) = -\ \Delta\tau_0 \cos(\frac{\xi\pi}{\ell} + \phi) \tag{20}$$

Such a field might have little practical interest. However, the results obtained from (20) would give an intrinsic insight into the phenomenon under consideration, since any spatially periodic field can be expressed by the Fourier series and (20) is the typical one of the series. Although, because of the non-linear properties of the problem, the length parameter a, b, c and d for spatially periodical fields other than (20) cannot be obtained by superposition of those for the individual terms in the Fourier series, the method employed in the following can readily be applicable to such fields.

OPENING OF THE FAULT

The stresses caused by the gap Δu must be bounded, since a fault itself does not have any resistance against tensile loads acting perpendicularly to the fault plane. Namely, in view of the second equation of (13), $B_\eta(\xi)$ is bounded at ξ = a and b, and the solution of (15), which can be readily obtained by the usual inversion technique (Erdogan (11)), is given by

$$B_\eta(\xi) = \lambda\ [p_0 \frac{\ell}{h} \{\frac{\rho_w}{\rho_r} - (\alpha\sin^2\theta + \cos^2\theta)\} \sin\theta \sqrt{(a-\xi)(\xi-b)/\ell^2}$$

$$- \Delta p_0 \sum_{n=1}^{\infty} C_n(\varepsilon) J_n(\frac{\delta\pi}{\ell}) U_n(\frac{\xi-\varepsilon}{\delta})] \tag{21}$$

where

$$\varepsilon = \frac{a + b}{2} , \qquad \delta = \frac{a - b}{2} \tag{22}$$

and

$$\left.\begin{aligned} C_{2k-1}(\varepsilon) &= (-1)^k \, 2\sin \frac{\varepsilon\pi}{\ell} \\ C_{2k}(\varepsilon) &= (-1)^k \, 2\cos \frac{\varepsilon\pi}{\ell} \end{aligned}\right\} \quad (k = 1, 2, 3, \cdots) \tag{23}$$

The function $J_n(t)$ is the Bessel function and $U_n(t)$ is the Chebyshev polynomial of the second kind defined by

$$U_n(t) = \sin(n \cos^{-1} t) \qquad (|t| \leqq 1) \tag{24}$$

The solution (21) must comply with the so-called consistency condition (Erdogan (11)), which yields

$$p_{w0} - p_0(\alpha\sin^2\theta + \cos^2\theta) - p_0 \frac{\ell}{h} \frac{\varepsilon}{\ell} \{\frac{\rho_w}{\rho_r} - (\alpha\sin^2\theta + \cos^2\theta)\}\sin\theta$$
$$+ \Delta p_0 J_0(\frac{\delta\pi}{\ell}) \cos \frac{\varepsilon\pi}{\ell} = 0 \tag{25}$$

Furthermore, the condition for the single-valuedness of the displacement u_η, i.e. the first equation of (19), leads to

$$p_0 \frac{\ell}{h} \frac{\delta}{\ell} \{\frac{\rho_w}{\rho_r} - (\alpha\sin^2\theta + \cos^2\theta)\}\sin\theta$$
$$+ 2\Delta p_0 J_1(\frac{\delta\pi}{\ell}) \sin \frac{\varepsilon\pi}{\ell} = 0 \tag{26}$$

The location of the ends, ξ = a and b, of the opened region is determined from (25) and (26).

With the aid of (21) and the second equation of (13), the expressions for the volume V of the opened region measured per unit thickness is obtained as follows:

$$V = \frac{\pi\lambda}{2} \Delta p_0 \delta^2 J_2(\frac{\delta\pi}{\ell}) \cos \frac{\varepsilon}{\ell} \pi \tag{27}$$

SLIP ON THE FAULT PLANE

The integral equation for $B_\xi(\xi)$ is easily obtained from (16) with the aid of (21). Now, let us consider the following two models: (i) the stresses caused by the slip are bounded everywhere (model I), and (ii) the fault has its own fracture toughness (K_{IIC}) for the sliding mode (model II). The former is a typical model used usually in contact problems of elastic bodies. The latter, on the other hand, is appropriate for an interface made up of asperities interlocking one another. In such a case, shear failure takes place along the interface as slip spreads.

In the case of model I, $B_\xi(\xi)$ must be bounded at $\xi = c$ and d, and therefore, the solution can be obtained by the same method as employed in the previous section. However, its explicit expression will not be presented here for brevity. The conditions to determine the location of the ends of the slipped region are

$$\text{If } \frac{\ell}{h}\frac{\delta}{h}\int_d^c \frac{D(\xi)H(\xi)}{z(\xi)}d\xi - f\frac{\Delta p_0}{p_0}\sum_{n=1}^{\infty} C_n(\varepsilon)J_n(\frac{\delta\pi}{\ell})$$

$$\times\int_d^c \frac{\{D(\xi)\}^n H(\xi)}{z(\xi)}d\xi + J\pi - L\pi\frac{\ell}{h}\frac{\beta}{\ell} - Kf\{\pi - Q(c, d)\}$$

$$+ Mf\frac{\ell}{h}\{\frac{\beta}{\ell}\pi + \frac{z(a)-z(b)}{\ell} - \frac{\beta}{\ell}Q(c, d)\}$$

$$+ f\frac{\Delta p_0}{p_0}\int_d^c \frac{H(\xi)\cos(\xi\pi/\ell)}{z(\xi)}d\xi + \frac{\Delta\tau_0}{p_0}\int_d^c \frac{\cos(\xi\pi/\ell+\phi)}{z(\xi)}d\xi = 0 \quad (28)$$

$$\text{If } \frac{\ell}{h}\frac{\delta}{\ell}\int_d^c D(\xi)H(\xi)\frac{\xi-\beta}{\ell}\frac{d\xi}{z(\xi)} - f\frac{\Delta p_0}{p_0}\sum_{n=1}^{\infty} C_n(\varepsilon)J_n(\frac{\delta\pi}{\ell})$$

$$\times\int_d^c \{D(\xi)\}^n H(\xi)\frac{\xi-\beta}{\ell}\frac{d\xi}{z(\xi)} + Kf\frac{z(a)-z(b)}{\ell} - L\frac{\pi}{2}\frac{\ell}{h}(\frac{\gamma}{\ell})^2$$

$$+ Mf\frac{\ell}{h}[\frac{1}{2}(\frac{\gamma}{\ell})^2\{\pi - Q(c, d)\} + \frac{a+\beta}{2\ell}\frac{z(a)}{\ell} - \frac{b+\beta}{2\ell}\frac{z(b)}{\ell}]$$

$$+ f\frac{\Delta p_0}{p_0}\int_d^c H(\xi)\frac{\xi-\beta}{\ell}\cos(\frac{\xi\pi}{\ell})\frac{d\xi}{z(\xi)}$$

$$+ \frac{\Delta\tau_0}{p_0}\int_d^c \frac{\xi-\beta}{\ell}\cos(\frac{\xi\pi}{\ell} + \phi)\frac{d\xi}{z(\xi)} = 0 \quad (29)$$

where

$$J = (1 - \alpha)\sin\theta\cos\theta, \quad K = \alpha\sin^2\theta + \cos^2\theta$$

$$I = (\rho_w/\rho_r - K)\sin\theta \ , \qquad L = J\sin\theta \ , \qquad M = K\sin\theta \tag{30}$$

$$D(\xi) = \frac{\xi-\varepsilon}{\delta} - \mathrm{sgn}\left(\frac{\xi-\varepsilon}{\delta}\right)\sqrt{\left(\frac{\xi-\varepsilon}{\delta}\right)^2 - 1} \tag{31}$$

$$\beta = \frac{c+d}{2} \ , \qquad \gamma = \frac{c-d}{2} \tag{32}$$

$$Q(c,\ d) = \sin^{-1}\frac{b-\beta}{\gamma} - \sin^{-1}\frac{a-\beta}{\gamma} \tag{33}$$

$$z(\xi) = \sqrt{(c-\xi)(\xi-d)} \tag{34}$$

Here, (28) is the consistency condition of the integral equation and (29) is the condition for the single-valuedness of the displacement u_ξ.

In the case of model II, $B_\xi(\xi)$ is allowed to have singularities at the ends, $\xi = c$ and d. Therefore, we choose $1/z(\xi)$ as the characteristic function of the integral equation for $B_\xi(\xi)$. Then, the solution, which satisfies the second equation of (19), is given by

$$B_\xi(\xi) = \lambda p_0 \ell \frac{\Phi(\xi)}{z(\xi)} + p_0\left(K - M\frac{\ell}{h}\frac{\xi}{\ell}\right)\lambda\frac{f}{\pi}$$
$$\times \log\left|\frac{b-\xi}{a-\xi}\cdot\frac{(\beta-\xi)(a-\xi) + z(\xi)\{z(\xi)-z(a)\}}{(\beta-\xi)(b-\xi) + z(\xi)\{z(\xi)-z(b)\}}\right| \tag{35}$$

where

$$\Phi(\xi) = I\frac{f}{\pi}\frac{\ell}{h}\frac{\delta}{\ell}\int_d^c \frac{D(s)H(s)}{\ell(s-\xi)}z(s)ds - \frac{f}{\pi}\frac{\Delta p_0}{p_0}\sum_{n=1}^{\infty} C_n(\varepsilon)J_n\left(\frac{\delta\pi}{\ell}\right)$$
$$\times\int_d^c \frac{\{D(s)\}^n H(s)}{\ell(s-\xi)}z(s)ds - J\frac{\xi-\beta}{\ell} + K\frac{f}{\pi}\left\{\pi\frac{\xi-\beta}{\ell} + \frac{z(a)-z(b)}{\ell}\right.$$
$$\left. - \frac{\xi-\beta}{\ell}Q(c,\ d)\right\} - \frac{L}{2}\frac{\ell}{h}\left(\frac{\gamma}{\ell}\right)^2 + L\frac{\ell}{h}\frac{\xi}{\ell}\frac{\xi-\beta}{\ell}$$
$$+ M\frac{f}{\pi}\frac{\ell}{h}\left[\frac{\pi}{2}\left(\frac{\gamma}{\ell}\right)^2 - \pi\frac{\xi}{\ell}\frac{\xi-\beta}{\ell} - \left\{\frac{1}{2}\left(\frac{\gamma}{\ell}\right)^2 - \frac{\xi}{\ell}\frac{\xi-\beta}{\ell}\right\}Q(c,\ d)\right.$$
$$\left. - \frac{z(a)}{\ell}\left(\frac{\xi}{\ell} + \frac{a-\beta}{2\ell}\right) + \frac{z(b)}{\ell}\left(\frac{\xi}{\ell} + \frac{b-\beta}{2\ell}\right)\right] + \frac{f}{\pi}\frac{\Delta p_0}{p_0}$$
$$\times\int_d^c \frac{H(s)\cos(s\pi/\ell)}{\ell(s-\xi)}z(s)ds + \frac{1}{\pi}\frac{\Delta\tau_0}{p_0}\int_d^c \frac{\cos(s\pi/\ell+\phi)}{\ell(s-\xi)}z(z)ds \tag{36}$$

In view of (11) and (14), $\sigma_{\xi\eta}$ on the ξ-axis near the points ξ = c and d are expressed asymptotically as

$$\sigma_{\xi\eta} \sim \frac{p_0}{\pi} \ell \int_d^c \frac{\phi(s)}{z(s)(\xi-s)} ds \sim \begin{cases} p_0 \dfrac{\Phi(c)}{\sqrt{2\gamma(\xi-c)/\ell^2}} & \text{as } \xi \to c^+ \\ -p_0 \dfrac{\Phi(d)}{\sqrt{2\gamma(d-\xi)/\ell^2}} & \text{as } \xi \to d^- \end{cases} \tag{37}$$

Hence, the stress intensity factors K_{II} are given by

$$K_{II}(c) = p_0\sqrt{\pi h}\sqrt{\frac{\ell/h}{\gamma/\ell}}\,\Phi(c) \;, \quad K_{II}(d) = -\, p_0\sqrt{\pi h}\sqrt{\frac{\ell/h}{\gamma/\ell}}\,\Phi(d) \tag{38}$$

Here, $K_{II}(c)$ and $K_{II}(d)$ are defined by

$$\left.\begin{aligned} K_{II}(c) &= \lim_{\xi\to c^+} \{\sqrt{2\pi(\xi-c)}\;\sigma_{\xi\eta}(\xi,\, 0)\} \\ K_{II}(d) &= \lim_{\xi\to d^-} \{\sqrt{2\pi(d-\xi)}\;\sigma_{\xi\eta}(\xi,\, 0)\} \end{aligned}\right\} \tag{39}$$

The location of the ends of the slipped region, ξ = c and d, is determined from the following condition:

$$|K_{II}(c)| = |K_{II}(d)| = K_{IIC} \tag{40}$$

RESULTS AND DISCUSSION

The following values have been used (Abé, et al. (12)):

$$\rho_r = 2.65 \text{ g/cm}^3 \;, \quad \rho_w = 1.0 \text{ g/cm}^3 \;, \quad \alpha = 0.49$$

The half length δ of the opened region, which is evaluated from (25) and (26), is presented in Fig. 2, for θ = 75°, ℓ/h = 0.1 and $\Delta p_0/p_0$ = 0.1, as a typical example. The parameter p_n in the ordinate is given by

$$p_n = p_{w0} - p_0(\alpha\sin^2\theta + \cos^2\theta) \tag{41}$$

where p_{w0} and p_0 are the pressure of fluid and the overburden pressure at the depth y = − h, respectively. Namely, the parameter p_n represents the surplus of the pressure of fluid over the pressure which was acting on the plane of the oblique fault before the injection of fluid. As

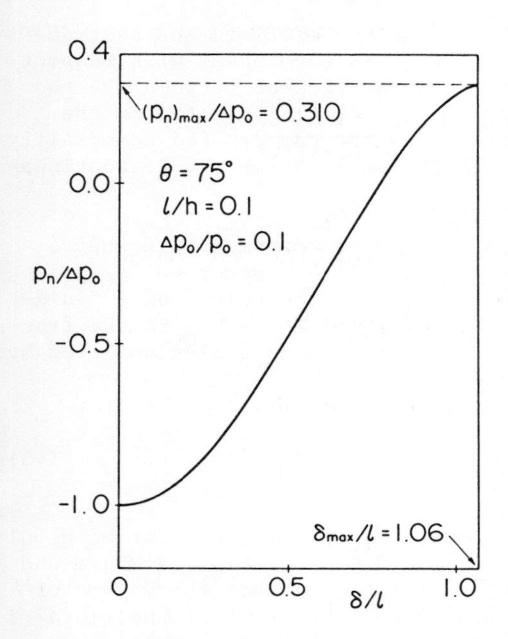

Fig. 2. Size of the opened region.

shown in Fig. 2, the opened region, i.e., the reservoir, grows up as the pressure of the fluid becomes higher. However, there exists an upper limit for the size of the reservoir, which is denoted by δ_{max}. When the pressure of the fluid is in excess of the corresponding upper limit, the fluid leaks out from the upper end of the reservoir towards the earth's surface. The value δ_{max} is presented in Fig. 3 for various $\Delta p_0/p_0$. This figure reveals that the reservoir of nearly the same size as ℓ can be created on the plane of an oblique fault, even if Δp_0 is fairly small compared with the overburden pressure p_0. For an illustrative example, suppose that $\ell/h = 0.1$, $\theta = 75°$ and $\Delta p_0/p_0 = 0.05$, then $\delta_{max}/\ell \cong 1.0$ as shown in Fig. 3 and, therefore, $\delta_{max} \cong 200$m for $h = 2000$m. Now, consider a vertical, water-filled crack within the earth's crust. The crack is a typical reservoir for the extraction of geothermal heat from a hot dry rock mass. As Weertman (8) pointed out, there exists an upper limit for the size of the crack. In two-dimensional cases, it is of the order of 40 - 200m in practical cases (Keer and Chen (9)). Therefore, it can be concluded that, if there is a region where the pressure due to the tectonic stresses is locally smaller than that in the neighboring regions, then a fluid-filled region can be created on a fault plane and it could serve as a reservoir for the extraction of geothermal heat. Another important conclusion can also be deduced from Fig. 3. The points of intersection between the curve and the abscissa indicate the lower limit for Δp_0, which is given by

$$\frac{\Delta p_0}{p_0} = \left| \frac{1}{\pi} \frac{\ell}{h} \left\{ \frac{\rho_w}{\rho_r} - (\alpha \sin^2\theta + \cos^2\theta) \right\} \sin\theta \right| \quad (42)$$

When $\Delta p_0/p_0$ is less than the

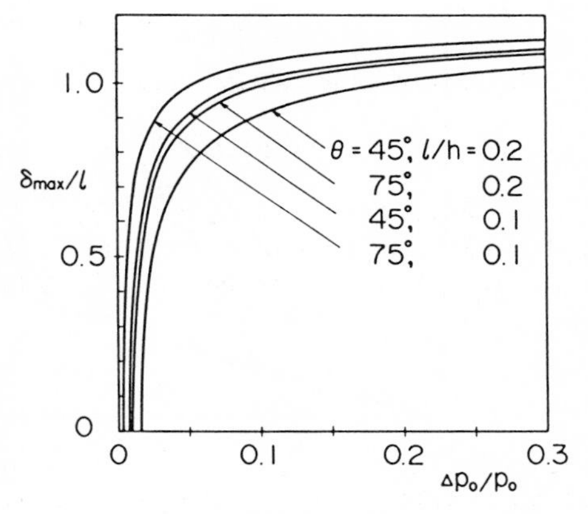

Fig. 3. Maximum size of the opened region.

limit, then the difference between the pressure of fluid and that due to the tectonic stresses on the fault plane is monotonous with respect to the depth, and, consequently, fluid cannot be kept between the two surfaces of the fault plane. In such a case, fluid injected on the fault plane will spread away infinitely and the gap created temporalily on the fault plane becomes infinitesimally small after the discontinuation of the injection.

The volume V of the opened region attains its maximum V_{max} when δ reaches δ_{max}. In Fig. 4, V_{max} are presented with respect to $\Delta p_0/p_0$, where V_c in the ordinate denotes the maximum of the volume of a fluid-filled, two-dimensional crack in the earth's crust and $\bar{K}_{IC}$ is the fracture toughness of the rock for the opening mode, non-dimensionalized by $p_0\sqrt{\pi h}$. The volume V_c is given by (Weertman (8))

$$V_c = \frac{\lambda K_{IC}^2}{4g(\alpha\rho_r - \rho_w)} \tag{43}$$

The non-dimensionalized fracture toughness $\bar{K}_{IC}$ is of the order of 0.001 for practical cases. For example, it is 0.0007 when $K_{IC} \sim 3$ MPa$\sqrt{m}$ and $h \sim 2000$m. Therefore, it is concluded from Fig. 4 that the volume of the reservoir on a fault plane is much larger than that of the fluid-filled crack except for the case that $\Delta p_0/p_0$ is extremely small.

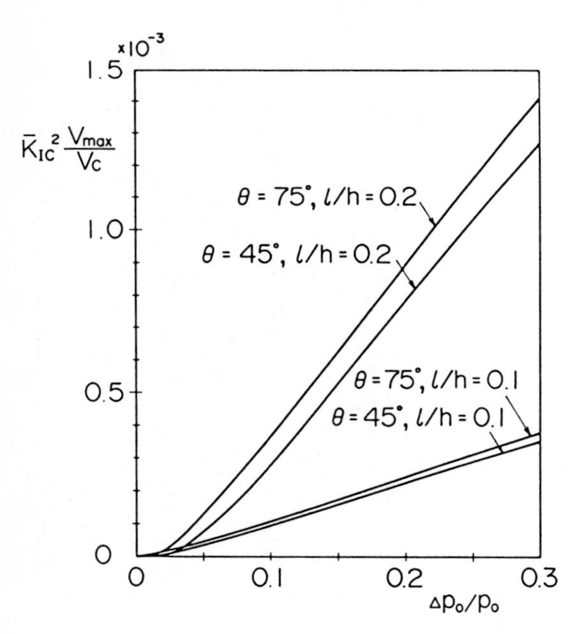

Fig. 4. Volume of the opened region.

It is readily seen from (25) and (26) that δ/ℓ is a multi-valued function of $p_n/\Delta p_0$ and, therefore, (25) and (26) have generally many roots for a given $p_n/\Delta p_0$. However, for the values of $\Delta p_0/p_0$, ℓ/h, ρ_w/ρ_r and α used here, it has been revealed numerically that, unless $\theta \cong 0°$, the solution corresponding to each root other than the minimum one is physically trivial, i.e., Δu is negative on some parts of the opened region. For $\theta \cong 0°$, on the other hand, there are many roots which give non-trivial solutions. Let us consider the case $\theta = 0°$, that was discussed in detail in the previous paper (Hayashi and Abé (13)). In this case, the solution of (26) is $\varepsilon = 0$ and the volume V is given by

$$V = \frac{\lambda}{2}\,\ell^2\Delta p_0\,\frac{1}{\pi}\,\left(\frac{\delta\pi}{\ell}\right)^2 J_2\left(\frac{\delta\pi}{\ell}\right) \tag{44}$$

In Fig. 5, V is plotted against δ/ℓ. The bold lines on the abscissa in the figure indicate the regions where Δu is negative on some parts of the reservoir. It is conluded from the figure that an arbitrarily large reservoir can be created on a horizontal fault if there exist spatially periodical tectonic stresses on the fault plane.

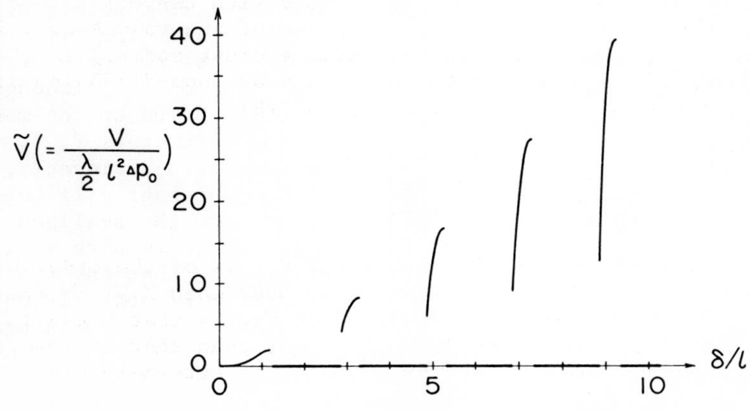

Fig. 5. Volume of the opened region for $\theta = 0°$.

In order to evaluate the size of the slipped regions the quantities $\Delta p_0/p_0$ and $\Delta\tau_0/p_0$ must be specified. However, the magnitude of these quantities are not known yet, so that we tentatively put $\Delta p_0/p_0 = \Delta\tau_0/p_0 = 0.1$ in the evaluation of the size of the slipped regions since the spatial variation of the tectonic stresses would be slight. In Fig. 6, the location of the ends of the slipped regions is presented for model I with the coefficient of friction being fixed to 0.5. The dashed lines in the figure indicate the location of the ends of the opened region. The slipped regions are about twice the size of the opened region at most. The size depends strongly on f as shown in Fig. 7. It decreases rapidly as f becomes larger. For model II, it is necessary to specify K_{IIC} in order to perform the numerical calculations.

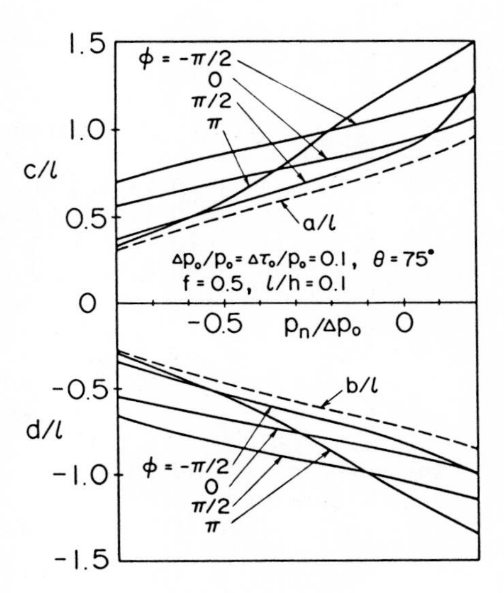

Fig. 6. Location of the ends of the slipped regions.

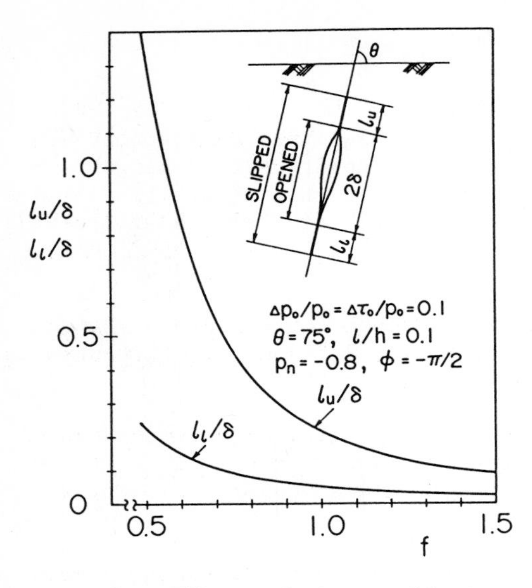

Fig. 7. Effect of the coefficient of friction on the size of the slipped regions.

For granite, K_{IIC} is 2 MPa$\sqrt{m}$ approximately (Takahashi, et al. (14)). Therefore, with h = 2000m, $\bar{K}_{IIC} = 0.0005$ is choosen as an example, where $\bar{K}_{IIC}$ is the fracture toughness non-dimensionalized by $p_0\sqrt{\pi h}$. In Table 1, the location of the ends of the slipped regions is presented for model II together with that for model I. The differences between the two sets of results are very small. Although the results depend on the magnitude of the parameters f, ϕ, θ, $\Delta p_0/p_0$ and $\Delta\tau_0/p_0$ evidently, the above statement will hold generally since the smallness of the differences results mainly from the reason that the fracture toughness of rocks is much smaller than $p_0\sqrt{\pi h}$ in practical cases.

TABLE 1. Size of the slipped region in model II, $f = 0.5$, $\phi = -\pi/2$, $\Delta p_0/p_0 = \Delta\tau_0/p_0 = 0.1$, $K_{IIC}/(p_0\sqrt{\pi h}) = 0.0005$, $\theta = 75°$.

$p_n/\Delta p_0$	a/ℓ	b/ℓ	c/ℓ		d/ℓ	
			model I	model II	model I	model II
-0.8	0.3094	-0.2773	0.7018	0.6920	-0.3433	-0.3400
-0.4	0.5624	-0.5200	0.9104	0.9043	-0.5845	-0.5815
0	0.8021	-0.7348	1.0935	1.0886	-0.8263	-0.8223
0.2	0.9650	-0.8582	1.2220	1.2171	-1.0077	-1.0010

ACKNOWLEDGMENTS

The authors wish to thank Mr. T. Shōji for his assistance. The work presented here was supported by the Ministry of Education, Science and Culture under Grant-in-Aid for Energy Research No. 505514, 56045015, 57045015.

REFERENCES

1. Sector, D.T. Jr., and Pollard, D.D., "On the Stability of Open Hydraulic Fractures in the Earth's Crust," Geophysical Research Letters, Vol. 2, No. 1, Jan. 1975, pp. 510 - 513.

2. Abé, H., Mura, T. and Keer, L.M., "Growth Rate of a Penny-Shaped Crack in Hydraulic Fracturing of Rocks," Journal of Geophysical Research, Vol. 81, No. 29, Oct. 1976, pp. 5335 - 5340.

3. Abé, H., Keer, L.M. and Mura, T., "Growth Rate of Penny Shaped Crack in Hydraulic Fracturing of Rocks, 2," Journal of Geophysical Research, Vol. 81, No. 35, Dec. 1976, pp. 6292 - 6298.

4. Abé. H., Sekine, H. and Kitada, S. "Stability of a Penny Shaped Geothermal Reservoir in the Earth's Crust," Transactions of the ASME, Journal of Energy Resources Technology, Vol. 104, No. 2, June 1982, pp. 93 - 95.

5. Abé. H., et al., "On the Limit for the Size of Hydraulic Fractures near the Earth's Surface," International Journal of Fracture, Vol. 18, No. 2, Feb. 1982, pp. R17 - R21.

6. Hsu, Y.C. and Santosa, F., "The Stability of a Large Open Hydraulic Penny-Shaped Fracture (Crack) near the Earth's Surface," Rep. ME-89(78)LASL-494-1, Los Alamos Scientific Laboratory, Los Alamos, N. Mex., 1978.

7. Sekine, H. and Mura, T., "Characterization of a Penny-Shaped Reservoir in a Hot Dry Rock," Journal of Geophysical Research, Vol. 85, No. B7, July 1980, pp. 3811 - 3816.

8. Weertman, J., "The Stopping of a Rising, Liquid-Filled Crack in the Earth's Crust by a Freely Slipping Horizontal Joint," Journal of Geophysical Research, Vol. 85, No. B2, Feb. 1980, pp. 967 - 976.

9. Keer, L.M. and Chen, S.H., "The Intersection of Pressurized Crack With a Joint," Journal of Geophysical Research, Vol. 86, No. B2, Feb. 1981, pp. 1032 - 1038.

10. Albright, J.N. and Pearson, C.F., "Location of Hydraulic Fractures Using Microseismic Techniques," SPE 9509, presented at the 55th Annual Fall Technical Conference and Exhibition of the Society of Petroleum Engineers of AIME, Dallas, Texas, 1980.

11. Erdogan, F., "Mixed Boundary Value Problems in Mechanics," Mechanics Today, Vol. 4, Nemat-Nasser, S., ed., Pergamon Press, 1978, pp. 1 - 86.

12. Abé, H., Keer, L.M. and Mura, T., "Theoretical Study of Hydraulically Fractured Penny-Shaped Cracks in Hot, Dry Rocks," International Journal for Numerical and Analytical Methods in Geomechanics, Vol. 3, No. 11, Nov. 1979, pp. 79 - 96.

13. Hayashi, K. and Abé, H., "Opening of a Fault and Resulting Slip Due to Injection of Fluid for the Extraction of Geothermal Heat," Journal of Geophysical Research, Vol. 87, No. B2, Feb. 1982, pp. 1049 - 1054.

14. Takahashi, H., Ohtani, S., Tamagawa, K., Shoji, T., Yuda, S. and Hashida, T., "Application of Fracture Mechanics to Hot Dry Rock Geothermal Energy Extraction: (B) Fracture Toughness and Microfracture Mechanisms of Rocks," Abstracts of Research Project, Grant-in-Aid for Energy Research 1980, Kimura, K., ed., the Ministry of Education, Science and Culture, Tokyo, 1981, pp. 129 - 131 (in Japanese).

APPLICATION OF LABORATORY FRACTURE MECHANICS DATA TO HYDRAULIC FRACTURING FIELD TESTS

by F. Rummel and R.B. Winter

Professor
Research Associate
Institute of Geophysics, Ruhr-University Bochum, FRG

ABSTRACT

A fracture mechanics approach to hydraulic fracturing is presented which takes into account the fluid pressure distribution within the fracture. The problem can be treated analytically in most cases. The theoretical results are in agreement with the classical theory of hydraulic fracturing which assumes an elastic unfractured rock. The fracture mechanics approach permits to use laboratory fracture parameters such as fracture toughness, tensile strength or crack length data obtained from rock specimens to interprete hydrofrac field data and to understand fracture growth.

INTRODUCTION

The classical theory predicts hydraulic fracture initiation at the borehole wall in an ideally homogeneous, isotropic, elastic and impermeable rock, when the net tangential stress on the borehole wall, induced by the far-field stresses and the fluid pressure within the borehole, is equal to the tensile strength of the wall rock (maximum tensile strength theory). For the case of a biaxial stress field the fracture initiation pressure p_c (breakdown pressure) is given by the relation

$$p_c = 3\,S_2 - S_1 + T, \tag{1}$$

which reduces to

$$p_c = 2\,p_e + T \tag{2}$$

if $S_1 = S_2 = p_e$ ($S_1 \geqq S_2$ far field principal stresses). The latter equation is applicable to fracture initiation in internally pressurized cylindrical rock samples subjected to a constant external confining pressure $p_m = p_e$. However, as shown by numerous experimental investigations, the observed fracture gradients are generally much smaller than the theoretical stress concentration factors used in the classical approach (2 in the symmetric case). Obviously this is because the classical theory neglects that rocks per definitionem contain pre-existing cracks or fractures.

Thus a more realistic approach is to investigate the stability of pre-existing fractures using the principles of linear elastic fracture mechanics. For the case of hydraulic fracturing we may confine the consideration on mode I frac growth, where instability occurs if the stress intensity factor K_I approaches a critical value, the fracture toughness K_{IC}. This parameter may be considered as an intrinsic rock property which can be measured experimentally.

In the following analysis some solutions assuming various fluid pressure distributions within the fracture are presented. The teoretical results are applied to the results of hydraulic fracturing field tests using numerical data obtained from laboratory tests on rock samples.

FRACTURE MECHANICS APPRAOCH

Recently, various fracture mechanics models related to hydraulic fracturing have been suggested (e.g. 1-6). The following model essentially is based on an analysis presented by Hardy (1), which assumes a symmetric double crack of half length a extending from a circular hole of radius R (Fig. 1). The tangential stress $\sigma_y(x,o)$ along the

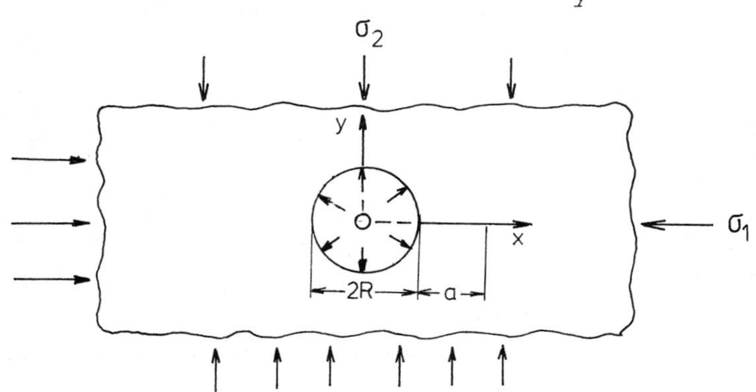

Fig. 1. Slit-type crack of half length a extending from a circular hole of radius R in an infinite plate subjected to the far-field stresses σ_1, σ_2.

x-axis is induced by the far-field stresses σ_1, σ_2 and by the fluid pressure p(x) acting within the hole and the crack. By superposition of the three stress fields the resulting stress intensity factor at the crack tip is given by

$$K_I(\sigma_1, \sigma_2, p) = K_I(\sigma_1) + K_I(\sigma_2) + K_I(p), \quad (3)$$

where

$$K_I(\sigma_1) = -\sigma_1 \cdot f^* \cdot R^{1/2} \quad (4)$$

$$K_I(\sigma_2) = -\sigma_2 \cdot g^* \cdot R^{1/2} \quad (5)$$

The dimensionless functions f^* and g^* are given as

$$f^* = -2 \left[(b^2-1)/\pi b^7 \right]^{1/2} \quad (6)$$

$$g^* = b^{1/2} (\pi^{1/2} - 2\pi^{-1/2} \text{ arc sin } b^{-1}) + 2(b^2+1) \left[(b^2-1)/\pi b^7 \right]^{1/2} \quad (7)$$

with $b = 1 + a/R$. f^* and g^* are presented graphically in Fig. 2 as

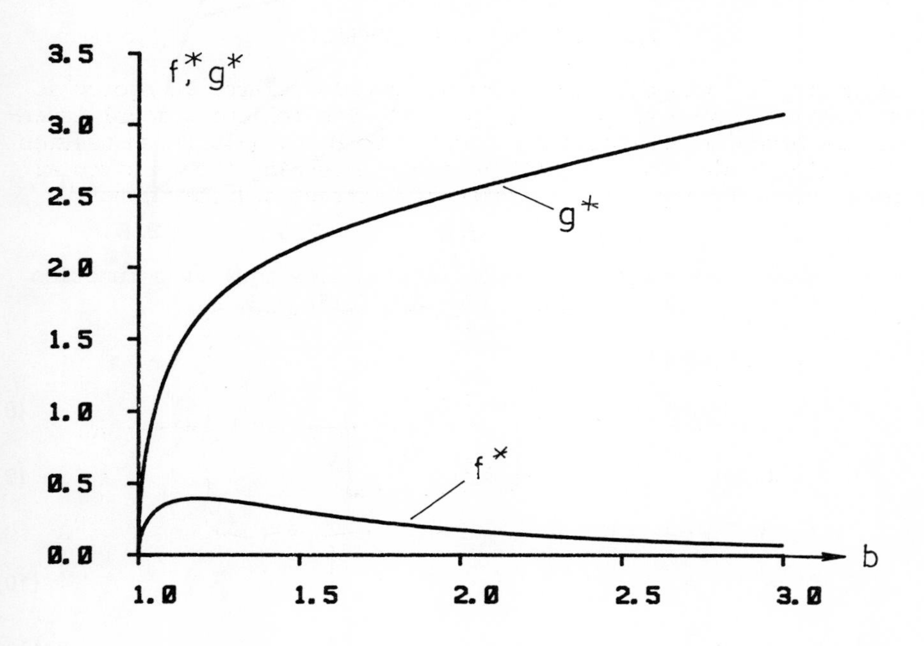

Fig. 2. Dimensionless functions f^* and g^* as a function of the normalized crack length $b = 1 + a/R$.

a function of the normalized crack length b. The detailed derivation of $K_I(\sigma_1)$ and $K_I(\sigma_2)$ is presented in (7). The derivation is based on Kirsch's formula on the stress distribution around a circular hole due to a biaxial stress field (8), and on the stress intensity factor given by Paris and Shi (9) for a slit-type crack in an infinite plate. In the derivation the presence of the hole is neglected for the integration.

In deriving the component $K_I(p)$ in eq. (3), 6 different fluid pressure distributions p(x) in the crack are considered (Fig. 3).

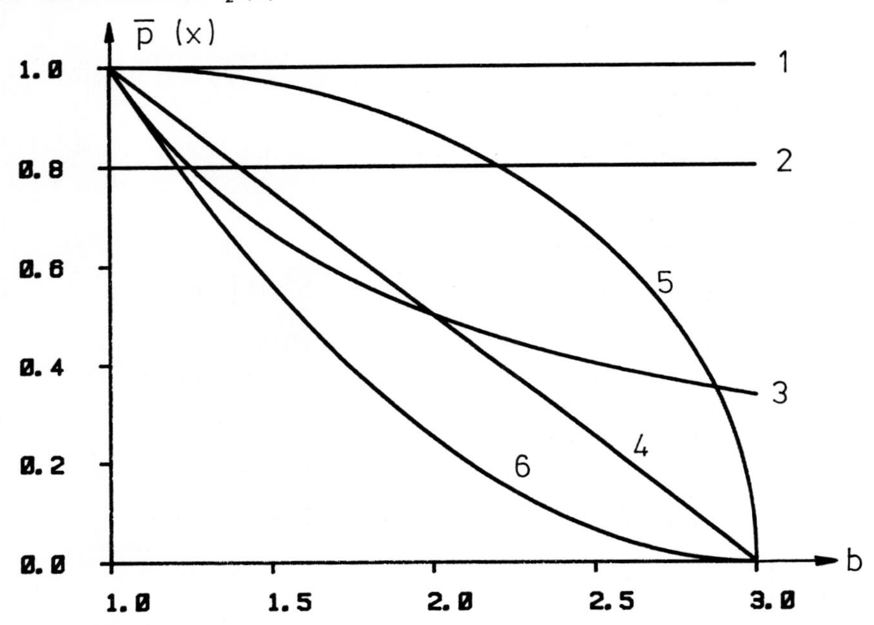

Fig. 3. Normalized fluid pressure distributions p(x) as a function of the normalized crack length b for b = 3.

(i) $p(x) = p = \text{const.}$

$$K_I(p) = p \cdot h_1^* \cdot R^{1/2} \quad (8)$$

$$h_1^* = (\pi b)^{1/2} (1-2\pi^{-1} \text{ arc sin } b^{-1}) \quad (9)$$

(ii) $p(x) = w \cdot p;\ 0 \leq W \leq 1$

$$K_I(p) = p \cdot h_2^* \cdot R^{1/2} \quad (10)$$

$$h_2^* = W \cdot h_1^* \quad (11)$$

(iii)

$$p(x) = \begin{cases} -p\,\frac{R}{x} \text{ for } -(R+a) \leqq x \leqq -R \\ p\,\frac{R}{x} \text{ for } R \leqq x \leqq R+a \end{cases}$$

$$K_I(p) = p \cdot h_3^* \cdot R^{1/2} \tag{12}$$

$$h_3^* = 2 \cdot (\pi b)^{-1/2} \ln\left[b + (b^2-1)^{1/2}\right] \tag{13}$$

(iv)

$$p(x) = \begin{cases} p(d+\frac{x}{a}) \text{ for } -(R+a) \leqq x \leqq -R \\ p(d-\frac{x}{a}) \text{ for } R \leqq x \leqq R+a \end{cases}$$

$$K_I(p) = p \cdot h_4^* \cdot R^{1/2} \tag{14}$$

$$h_4^* = d(\pi b)^{1/2}(1-2\pi^{-1} \text{arc sin } b^{-1}) - 2\left[b\pi^{-1} \cdot (b+1)/(b-1)\right]^{1/2} \tag{15}$$

$$d = 1 + R/a \tag{16}$$

(v)

$$p(x) = \begin{cases} p\left[1-(\frac{x+R}{a})^2\right]^{1/2} \text{ for } -(R+a) \leqq x \leqq -R \\ p\left[1-(\frac{x-R}{a})^2\right]^{1/2} \text{ for } R \leqq x \leqq R+a \end{cases}$$

$$K_I(p) = p \cdot h_5^* \cdot R^{1/2} \tag{17}$$

h_5^* is calculated numerically.

(vi)

$$p(x) = \begin{cases} p(\frac{a+R+x}{a})^2 \text{ for } -(R+a) \leqq x \leqq -R \\ p(\frac{a+R-x}{a})^2 \text{ for } R \leqq x \leqq R+a \end{cases}$$

$$K_I(p) = p \cdot h_6^* \cdot R^{1/2} \tag{18}$$

$$h_6^* = (\pi^{-1} b)^{1/2} (b-1)^{-2} \cdot \left[3b^2(\pi/2-\text{arc sin } b^{-1}) + (1-4b) \cdot (b^2-1)^{1/2}\right] \tag{19}$$

The 6 expressions for h^* are graphically presented in Fig. 4 as functions of the normalized crack length b. The plot demonstrates the

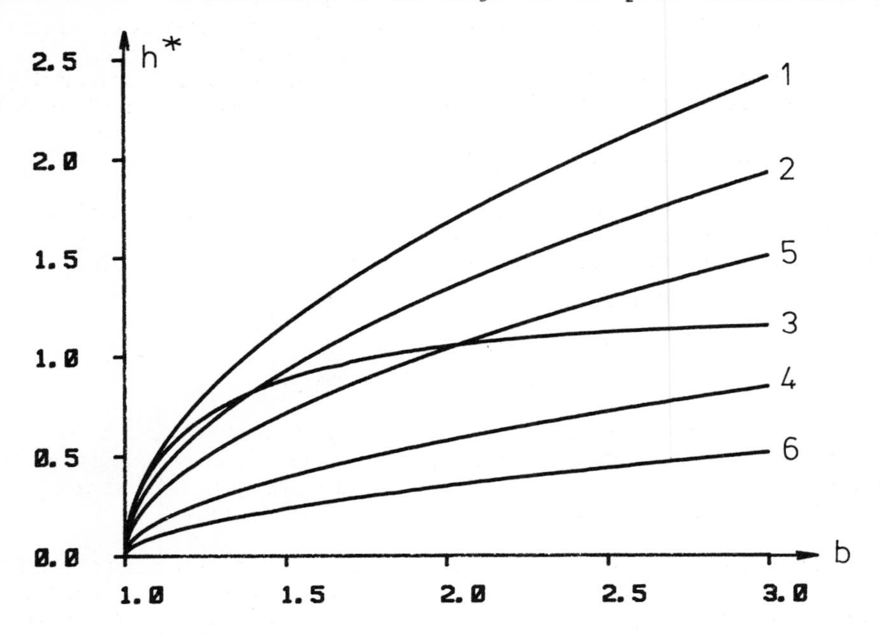

Fig. 4. Dimensionless functions h_n^* for fluid pressure distributions as indicated in Fig. 3.

effect of the fluid pressure distribution within the fracture on the stress intensity factor in the vicinity of the crack tip and therefore on the critical pressure for frac growth:

$$p_c = (h^*)^{-1} \, (K_{IC} \, R^{-1/2} + \sigma_1 f^* + \sigma_2 g^* \tag{20}$$

where

$$p_{co} = K_{IC}/(h^* R^{1/2}) \tag{21}$$

corresponds to the hydraulic fracturing tensile strength of the material at $\sigma_1 = \sigma_2 = 0$. For a lithostatic biaxial stress field ($\sigma_1 = \sigma_2 = p_e$) the frac gradient is given by

$$dp_c/dp_e = (f^* + g^*)/h^* \tag{22}$$

which yields the classical stress concentration factor 2 for zero crack length and a frac gradient of 1 for large cracks, respectively.

APPLICATION TO FIELD TESTS

The validity of this analysis is demonstrated for two case histories, where hydraulic fracturing stress measurements have been conducted in vertical boreholes. For both cases the fracture properties of the rock have been measured in extensive laboratory investigations.

Example I: Hydrofrac Tests in Sandstone

The field tests were conducted in a 100 m deep borehole drilled into a medium-grained sandstone. The borehole radius was R = 48 mm. The following in-situ frac data were obtained:

P_c = 12.9 MPa (at 60 m depth)

P_{si} = 3.3 MPa (shut-in pressure)

P_{co} = 8.6 MPa (in-situ hydrofrac tensile strength)

Laboratory hydraulic fracturing tests on 3 cm diameter mini-cores containing a 2.5 mm diameter axial hole and subjected to confining pressures up to 50 MPa yield a hydraulic fracturing tensile strength of $p_{co} \approx 26$ MPa and a frac gradient $dp_c/dp_m \approx 1.32$ (10).

The laboratory frac gradient indicates an initial intrinsic crack length of the rock of a = 1.8 mm. This value appears to be realistic if one considers the grain size to be a measure of the crack length. This crack length and the value of p_{co} permit to estimate the fracture toughness as $K_{IC} = 1.9$ MPa · $m^{1/2}$ for the case of a constant fluid pressure distribution (i). Alternatively, the in-situ tensile strength P_{oc} and the laboratory K_{IC}-value result in an intrinsic rock mass crack length of 18 mm within the "intact" test interval.

Using eq. (22) and the appropriate in-situ values for the function f^*, g^*, h^*, and for $\sigma_2 = P_{si}$ the maximum horizontal stress (σ_1 in eq. (22) is $S_H = 6.6$ MPa.

Example II: Hydrofrac Tests in Granite

Extensive hydraulic fracturing tests have been conducted in several 300 m deep boreholes within the Falkenberg Granite Massif (11). The following in-situ frac data were obtained:

P_c = 15 MPa

P_{si} = 4.6 MPa

P_{co} = 7.8 MPa

R = 48 mm

The laboratory experiments on granite specimens revealed the following data:

p_{co} = 13.8 MPa

$dp_c/dp_m = 1.1$

$r = 125$ mm

$K_{IC} = 0.8$ MPa $\cdot$ $m^{1/2}$ (three-point bending tests).

Using a similar procedure as applied to example I gives an intrinsic rock specimen crack length a $\approx$ 1 mm for the granite. In comparison, the crack length of the "intact" rock in the in-situ test intervals is about 4 mm which seems reasonable for the coarse-grained granite. Finally, the values $h^* = 0.47$, $f^* = 0.36$, $g^* = 1.3$ and $\sigma_2 = p_{si}$ result in a maximum horizontal compression of $S_H = 7.2$ MPa.

ACKNOWLEDGEMENT

We are grateful to H.J. Alheid, J. Baumgärtner, B. Kappernagel and Th. Wöhrl who participated in the laboratory and field studies. The work was supported by grants from the Federal Ministry of Research and Technology, Project No. ET 4150, 3023 and 3068. Project management was provided by PLE/KFA Jülich and DGMK Hamburg.

REFERENCES

1. Hardy, M.P., "Fracture Mechanics Applied to Rock," PhD Thesis, 1973, Univ. Minnesota, Minneapolis, USA.

2. Abou-Sayed, A.S., Brechtel, C.S., and Clifton, R.J., "In-Situ Stress Determination by Hydraulic Fracturing: A Fracture Mechanics Report," J.G.R., 83/36, 2851-2862, 1978.

3. Schmidt, A.R., "A Microcrack Model and its Significance to Hydraulic Fracturing and Fracture Toughness Testing," 21st U.S. Symp. Rock Mech., Rolta, Missouri, 1980.

4. Hanson, M.E., and Shaffer, R.J., "Effects of Various Parameters on Hydraulic Fracturing Geometry," SPE/DOE 8942, 163-172, 1980.

5. Advani, S.H., "Finite Element Model Simulations associated with Hydraulic Fracturing," SPE/DOE 8941, 151-161, 1980.

6. Abou-Sayed, A.S., Ahmed, U., and Jones, A., "Systematic Approach to Massive Hydraulic Fracturing Treatment Design," SPE/DOE 8977, Denver, 1981.

7. Rummel, F., and Winter, R.B., "Fracture Mechanics as Applied to Hydraulic Stress Measurements," J. Earthquake Prediction Res., 1982, (in press).

8. Kirsch, G., "Die Theorie der Elastizität und die Bedürfnisse der Festigkeitslehre," Zeitschr. Verein Deutscher Ing. (VDI), 42, 113, 1898.

9. Parish, P.C., and Sih, G.C., "Stress Analysis of Cracks," ASTM Spec. Techn. Publ. STP 381, 30-83, Philadelphia, 1965.

10. Rummel, F., "Hydraulic Fracturing Spannungsmessungen in der Bohrung KB 28/5 (NBS H/W Süd, Mühlbergtunnel)," RU Report, Bochum, 1980.

11. Rummel, F., Baumgärtner, J., and Alheid, H.J., "Hydraulic Fracturing Stress Measurements along the Eastern Boundary of the SW-German Block," U.S.G.S. Open-File Report on the Earthquake Research Program, Menlo Park, California, Sept. 1982, and U.S. National Committee for Rock Mechanics Report, Washington, 1983 (in press).

FINITE ELEMENT TREATMENT OF SINGULARITIES FOR THERMALLY INDUCED CRACK GROWTH IN HDR

by Yoichi Sumi

Department of Naval Architecture and Ocean Engineering
Yokohama National University
Yokohama, Japan

ABSTRACT

Theoretical and numerical investigations are made for the interactive behavior of thermally induced secondary cracks in hot dry rock masses. Fundamental features of the crack growth pattern, i.e. equally growing cracks and arrest of certain cracks, are indentified by using a minimum principle of the total potential energy. Then a numerical solution scheme, which combines analytical and finite-element methods, is introduced to analyze the possible crack growth pattern. Singularities associated with the stationary crack tips and also with the quasi-statically moving ones are expressed analytically, while the regular part of the solution is approximated by the finite element method. Numerical results are first compared with model experiments, and then a quantitative estimation is made for a possible crack growth pattern of thermally induced cracks in hot dry rock masses.

INTRODUCTION

An attempt has been made to extract energy from hot dry rock masses by Los Alamos National Laboratory, New Mexico, where an experimental plant has been constructed and operated successfully over a period of several years[1]. The basic problems regarding this system[2,3,4] are (i) to generate a large fractured zone by hydraulic fracturing, (ii) to circulate water under pressure through the fractured zone, and (iii) the cooling of the rock, and consequent thermally induced secondary cracks and the possible growth of the primary fracture. The primary fracture is almost vertical and penny-shaped, when the overburden pressure is large compared with the two horizontal principal tectonic stresses. As the circulation of water continues, cooling of the rock masses occurs in the neighbourhood of the primary fracture surfaces. Consequently, thermally induced secondary cracks are formed almost vertical and perpendicular to the primary fracture surfaces.

Thermally induced cracks have been modeled as parallel edge cracks in two-dimensional plane strain conditions, and first treated by Nemat-Nasser, Keer and their associates,[5,6] who showed the highly interactive growth behavior of these cracks. A rather complete theoretical treatment of the problem has been described by Sumi, Nemat-Nasser and Keer[7] by using a minimum principle of the total potential energy, by which equally growing cracks and arrest of certain growing cracks due to the interaction are identified.

In this paper a minimum principle of the total potential energy is applied to the system of two equally spaced edge cracks in a unit cell, and the combined analytical and finite-element solution scheme is presented for the analysis of singularities[8] associated with the colinear growth of thermally induced secondary cracks under the mixed mode condition. Numerical results are first compared with the model experiment performed by Geyer and Nemat-Nasser[9], and then a possible crack growth behavior is estimated for thermally induced secondary cracks in hot dry rock masses.

INTERACTIVE GROWTH BEHAVIOR OF THERMALLY INDUCED SECONDARY CRACKS

Basic Assumptions

When circulation of water starts in the reservoir, the cooled region begins to extend in the neighbourhood of the surface of the primary fracture. If we consider a sudden uniform cooling on the surface of the semi-infinite region, it causes a one-dimensional temperature profile (Fig.1) given by

$$T(x_1;\Lambda)= \frac{2}{\sqrt{\pi}}(To-Ts) \int_0^{\sqrt{3}x_1/\Lambda} \exp(-u^2)\,du + Ts \tag{1}$$

where T_0 and T_s are respectively, the initial and surface temperatures of the body. Λ is a specific penetration depth of the cooled region defined as

$$\Lambda=\sqrt{tk/\rho C_V} \tag{2}$$

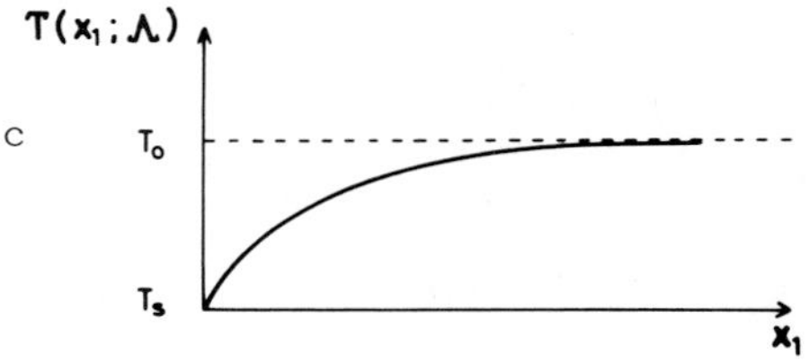

Fig.1. Temperature profile.

where t: time, k: heat transfer coefficient, ρ: mass density, and Cv: specific heat at constant volume.

Since the thermal contraction is constrained by the surrounding hot dry rock masses, a locally tensile layer is formed, and expressed as

$$\sigma_{11} = \sigma_{12} = 0 \tag{3}$$

$$\sigma_{22}(x_1;\Lambda) = -\frac{\alpha E}{1-\nu}[T(x_1;\Lambda) - T_o] \tag{4}$$

where α: linear thermal expansion coefficient, E: Young's modulus, and ν: Poisson's ratio. Consequently, thermally induced brittle crackings are formed (Fig.2). The crack initiation process may be dynamic, and the sizes and spacings of those cracks are highly dependent upon the distribution of initial defects or the surface roughness of the primary fracture. However, we pass through this problem, because our attention is focused on the quasi-static growth of the thermally induced cracks due to the development of the cooled region.

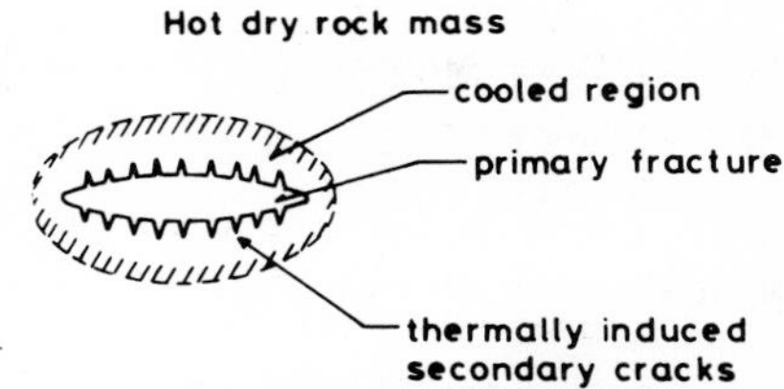

Fig.2. Thermally induced secondary cracks.

We consider equally spaced edge cracks being perpendicular to the primary fracture surface as shown in Fig.3. Although all cracks may interact with each other during their growth process, we introduce a unit cell consisting of two equally spaced edge cracks having lengths H_1 and H_2 (Fig.4), to avoid the complexities of analysis and also to obtain the fundamental feature of the interactive growth of cracks. This model may lead to unequal crack lengths of period 2b, where b is the crack spacing. As will be seen later, the numerical result obtained by using this model is in good agreement with a model experiment.

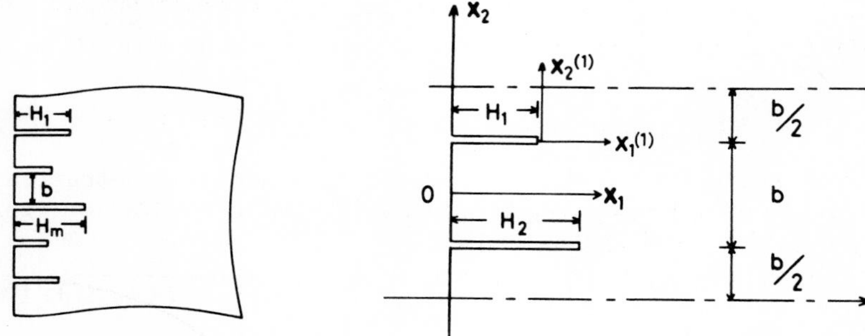

Fig.3. Equally spaced edge cracks. Fig.4. Unit cell consisting of two cracks.

Equal Crack Growth and Stable Bifurcation Point

We shall first introduce a minimum principle of the total potential energy for the system of two interacting cracks. The total potential energy is defined as[5,7]

$$\Pi[H_1,H_2;\Lambda] \equiv U[H_1,H_2;\Lambda] + S[H_1,H_2] \tag{5}$$

where crack lengths H_1 and H_2, strain energy U, and surface energy S are expressed as

$$H_m = H_m(\Lambda) \qquad m=1,2 \tag{6}$$

$$U = U[H_1, H_2; \Lambda] \tag{7}$$

$$S = S[H_1, H_2] \tag{8}$$

The surface energy is assumed to satisfy the following conditions

$$\left.\frac{\partial S}{\partial H_m}\right|_+ = 2\gamma_m > 0 \quad (9) \qquad \left.\frac{\partial^2 S}{\partial H_m \partial H_n}\right|_+ = \delta_{mn} \left.\frac{\partial^2 S}{\partial {H_m}^2}\right|_+ \tag{10}$$

where γ_m is the surface energy per unit length for the m-th crack.

Crack growth behavior is determined by the minimization of the total potential energy (5) subjected to the constraints

$$H_m' = \frac{dH_m}{d\Lambda} > 0 \qquad \text{for growing cracks} \tag{11}$$

$$\left.\begin{array}{l} H_m' = 0 \\ S_{,m} > \Pi_{,m} > 0 \end{array}\right\} \qquad \text{for stationary cracks} \tag{12}$$

where comma followed by indices denotes partial differentiation with respect to the crack length , i.e.

$$S_{,m} \equiv \frac{\partial S}{\partial H_m} \qquad \text{and} \qquad \Pi_{,m} \equiv \frac{\partial \Pi}{\partial H_m} \tag{13}$$

Assume an arbitrary initial state, at which the thermal penetration depth is Λ^0 and the corresponding equilibrium crack lengths are $H_m(\Lambda^0)$, m=1,2. We shall consider a small crack growth after this point.

$$\Lambda = \Lambda^0 + \lambda \tag{14}$$

$$H_m = H_m(\Lambda^0) + h_m(\lambda) \qquad m=1,2 \tag{15}$$

where λ and h_m's are the increase of the thermal penetration depth and the crack growth lengths, respectively. Using the Taylor expansion with respect to λ, the crack length (15) can be represented as

$$H_m = H_m(\Lambda^0) + h_m'(0)\lambda + \frac{1}{2!}h_m''(0)\lambda^2 + \frac{1}{3!}h_m'''(0)\lambda^3 + O(\lambda^4) \tag{16}$$

The total potential energy during this process is denoted by $V(h_1, h_2; \lambda)$, and given by

$$\begin{aligned} V[h_1, h_2; \lambda] = {} & \Pi^0 + [\Pi^0_{,m}\, h_m' + \Pi^{0\prime}]\lambda + [\tfrac{1}{2}\Pi^0_{,m}\, h_m'' + \tfrac{1}{2}\Pi^0_{,mn}\, h_m'\, h_n' \\ & + \Pi^0_{,m}{}'\, h_m' + \tfrac{1}{2}\Pi^{0\prime\prime}]\lambda^2 + [\tfrac{1}{6}\Pi^0_{,m}\, h_m''' + \tfrac{1}{2}\Pi^0_{,mn}\, h_m'\, h_n'' \\ & + \tfrac{1}{2}\Pi^0_{,m}{}'\, h_m'' + \tfrac{1}{6}\Pi^0_{,mnl}\, h_m'\, h_n'\, h_l' + \tfrac{1}{2}\Pi^0_{,mn}{}'\, h_m'\, h_n' \\ & + \tfrac{1}{2}\Pi^0_{,m}{}''\, h_m' + \tfrac{1}{6}\Pi^{0\prime\prime\prime}]\lambda^3 + O(\lambda^4) \end{aligned} \tag{17}$$

The first and second variation of V with respect to the crack growth lengths h_1 and h_2 are given by

$$\delta V = \Pi^0_{,m}\delta h_m'\lambda + [\frac{1}{2}\Pi^0_{,m}\delta h_m'' + \{\Pi^0_{,mn}h_n' + \Pi^0_{,m}'\}\delta h_m']\lambda^2$$
$$+ [\frac{1}{6}\Pi^0_{,m}\delta h_m''' + \frac{1}{2}\{\Pi^0_{,mn}h_n' + \Pi^0_{,m}'\}\delta h_m'' + \frac{1}{2}\{\Pi^0_{,mn}h_n''$$
$$+ \Pi^0_{,mnl}h_n' h_l' + 2\Pi^0_{,mn}{}' h_n' + \Pi^0_{,m}''\}\delta h_m']\lambda^3 + O(\lambda^4) \quad (18)$$

$$\delta^2 V = \Pi^0_{,mn}\delta h_m'\delta h_n'\lambda^2 + [\frac{1}{2}\Pi^0_{,mn}\delta h_m'\delta h_n''$$
$$+ \frac{1}{2}\{\Pi^0_{,mnl}h_l' + \Pi^0_{,mn}{}'\}\delta h_m'\delta h_n']\lambda^3 + O(\lambda^4) \quad (19)$$

If $\Pi^0_{,mn}$ is positive definite, that is $\delta^2V > 0$, the stationary condition $\delta V=0$ gives the crack growth behavior as natural conditions of the variational principle. In this case we have

$$\Pi^0_{,m} = 0 \quad (20)$$

$$\Pi^0_{,mn}h_n' + \Pi^0_{,m}{}' = 0 \quad (21)$$

$$\Pi^0_{,mn}h_n'' + \Pi^0_{,mnl}h_n'h_l' + 2\Pi^0_{,mn}{}'h_n' + \Pi^0_{,m}{}'' = 0 \quad (22)$$

where Eq.(20) gives the equilibrium condition of the initial state, while Eqs.(21) and (22) give the first and second derivatives of crack growth length with respect to the thermal penetration depth.

If we choose the initial state, Λ^0, as a very early stage of the crack growth, the crack spacing is relatively large compared with their lengths. Consequently one can assume the situation, where the interaction between the two cracks is very small, and where the two cracks grow almost independently, having equal crack length $H_1=H_2$. Then we have

$$\Pi^0_{,11} = \Pi^0_{,22} > \Pi^0_{,12} = \Pi^0_{,21} > 0 \quad (23)$$

$$\Pi^0_{,1}{}' = \Pi^0_{,2}{}' \quad (24)$$

Eq.(23) leads to the positive definiteness of $\Pi^0_{,mn}$ and the equal crack growth rate

$$h_1' = h_2' = -\Pi^0_{,1}{}'/(\Pi^0_{,11} + \Pi^0_{,12}) \quad (25)$$

is obtained by solving Eq.(21).

As the two cracks continue to grow at an equal rate, the interaction between these cracks may increase. Then the possibility exists that the crack growth behavior reaches a critical state, at which the first inequality of Eq.(23) does not hold. If this critical state is reached at $\Lambda=\Lambda^c$, then

$$\Pi^c_{,11} = \Pi^c_{,12} > 0 \ ; \ \text{i.e.} \quad \det|\Pi^c_{,mn}| = 0 \quad (26)$$

In this case the minimization of the total potential energy (17) of order λ^2 gives a pair of singular equations

$$\Pi^c_{,mn}h_n' + \Pi^c_{,m}{}' = 0 \qquad m=1,2 \quad (27)$$

by which one cannot determine the crack growth rate, and has the following linear dependent relation

$$h_1' + h_2' = -\Pi^c_{,1}'/\Pi^c_{,11} \tag{28}$$

Therefore, one must determine the crack growth rate by the minimization of the total potential energy of order λ^3, which is denoted by V^c_3 and given by

$$V^c_3 = \frac{1}{6}\Pi^c_{,mnl}\, h_m'h_n'h_l' + \frac{1}{2}\Pi^c_{,mn}'h_m'h_n' + \frac{1}{2}\Pi^c_{,m}''h_m' + \frac{1}{6}\Pi^{c}{}''' \tag{29}$$

Using Eq.(28) we can eliminate h_2' in Eq.(29), which then becomes quadratic with respect to h_1'

$$V^c_3 = \frac{1}{2}\{(\Pi^c_{,112}-\Pi^c_{,111})(\Pi^c_{,1}'/\Pi^c_{,12}) + 2(\Pi^c_{,11}'-\Pi^c_{,12}')\}[(h_1')^2 + (\Pi^c_{,1}'/\Pi^c_{,12})h_1' + \{-\Pi^c_{,111}(\Pi^c_{,1}'/\Pi^c_{,12})^3 + 3\Pi^c_{,11}'(\Pi^c_{,1}'/\Pi^c_{,12})^2 - 3\Pi^c_{,1}''(\Pi^c_{,1}'/\Pi^c_{,12}) + \Pi^{c}{}'''\} / 3\{(\Pi^c_{,112}-\Pi^c_{,111})(\Pi^c_{,1}'/\Pi^c_{,12}) + 2(\Pi^c_{,11}'-\Pi^c_{,12}')\}] \tag{30}$$

where the admissible range of h_1' is given by

$$0 \leq h_1' \leq -\Pi^c_{,1}'/\Pi^c_{,11} \tag{31}$$

As can easily be seen, the coefficient of $(h_1')^2$ in Eq.(30) is negative [7], and the variation of V^c_3 with respect to h_1' becomes convex as shown in Fig.5. The stationary point of V^c_3 becomes maximum, and the minimum value is attained at both ends of the admissible region, i.e.

$$\left.\begin{aligned} &h_1' = 0 \quad \text{and} \quad h_2' = -\Pi^c_{,1}'/\Pi^c_{,12} \\ \text{or} \\ &h_1' = -\Pi^c_{,1}'/\Pi^c_{,12} \quad \text{and} \quad h_2' = 0 \end{aligned}\right\} \tag{32}$$

After this state the equal crack growth given by Eq.(25) is terminated, and every other crack becomes stationary, while the crack growth rate is doubled for the remaining active cracks (Figure 6).

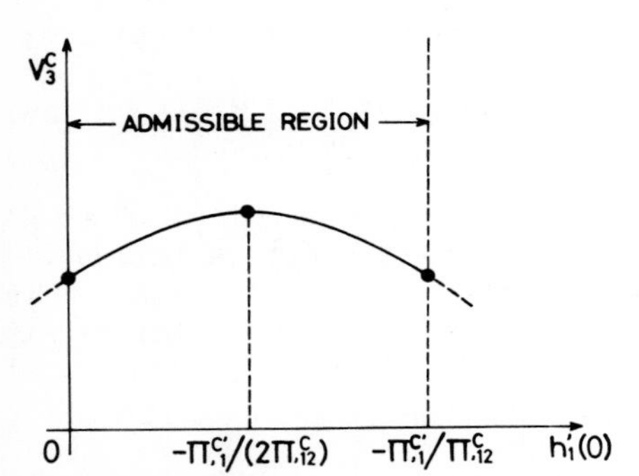

Fig.5. Variation of the third order potential energy V^c_3.

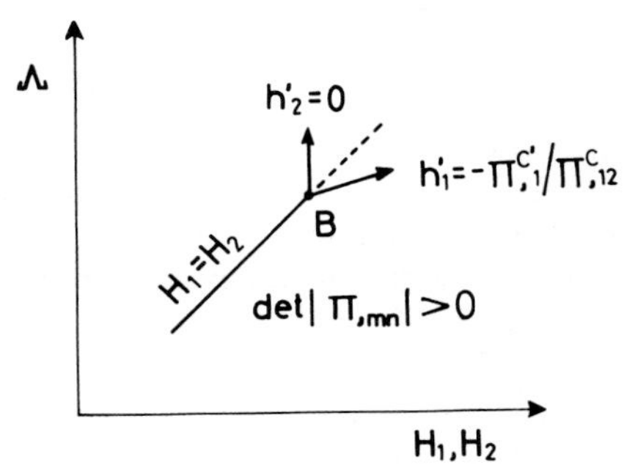

Fig.6. Stable bifurcation point.

In the case of an elastic brittle solid, derivatives of the total potential energy are expressed as

$$\Pi_{,m} = -\frac{1-\nu^2}{E}(K_{Im}^{\ 2} + K_{IIm}^{\ 2}) + 2\gamma_m \tag{33}$$

$$\Pi_{,mn} = -\frac{2(1-\nu^2)}{E}\left(K_{Im}\frac{\partial K_{Im}}{\partial H_n} + K_{IIm}\frac{\partial K_{IIm}}{\partial H_n}\right) + 2\frac{\partial \gamma_m}{\partial H_n} \tag{34}$$

where K_{Im} and K_{IIm} are the stress intensity factors of Mode I and Mode II at the m-th crack tip. From Eq.(34), it is clear that not only the analysis of stress intensity factors but also the derivatives of those quantities play an essential role in the numerical estimation of the interactive growth of the cracks.

FINITE ELEMENT TREATMENT OF SINGULARITIES ASSOCIATED WITH THE GROWTH OF THERMALLY INDUCED CRACKS

Basic Boundary Value Problem

Basic equations of the boundary value problem are expressed by

$$\sigma_{ij,j} = 0 \qquad \text{in } V \tag{35}$$

$$\sigma_{ij}n_j = t_i \qquad \text{on } S_t \tag{36}$$

$$u_i = v_i \qquad \text{on } S_u \tag{37}$$

where V denotes the domain considered, S_t and S_u are the parts of the boundary where surface traction t_i and surface displacement v_i are prescribed, and n_j denotes the component of the exterior unit normal of the boundary. In the present problem, surface tractions are determined in such a manner that they cancel out the thermally induced stresses in an intact body. Those are given by

$$t_1 = t_2 = 0 \qquad \text{on } x_1 = 0 \tag{38}$$

$$t_1 = 0, \quad t_2 = \begin{cases} \sigma_{22}(x_1;\Lambda) & \text{on upper surfaces} \\ -\sigma_{22}(x_1;\Lambda) & \text{on lower surfaces} \end{cases} \quad \text{on } 0 \leq x_1 \leq Hm \text{ and } x_2 = \pm b/2 \tag{39}$$

For the analysis of the stress field near the m-th crack tip, it is convenient to introduce local Cartesian coordinates ($x_1^{(m)}, x_2^{(m)}$), and also the local polar coordinates (r_m, θ_m), where

$$x_1^{(m)} = r_m \cos\theta_m \qquad x_2^{(m)} = r_m \sin\theta_m \tag{40}$$

The asymptotic behavior of stresses around the m-th crack tip is expressed as

$$\sigma_{ij}^{(m)} = \sum_{k=1}^{5} a_m^{(k)} \ \sigma_{mij}^{(k)}(r_m,\theta_m) + \sigma_{mij}^{(0)} + O(r_m) \tag{41}$$

where $\sigma_{mij}^{(0)}$ is determined from the boundary tractions at the root of the m-th crack tip, and

$$\sigma_{mij}^{(1)} = \frac{1}{\sqrt{2\pi r_m}} \cos(\theta_m/2) \begin{cases} 1-\sin(\theta_m/2)\sin(3\theta_m/2) & i=j=1 \\ 1+\sin(\theta_m/2)\sin(3\theta_m/2) & i=j=2 \\ \sin(\theta_m/2)\cos(3\theta_m/2) & i\neq j \end{cases} \quad (42)$$

$$\sigma_{mij}^{(2)} = \frac{1}{\sqrt{2\pi r_m}} \begin{cases} -\sin(\theta_m/2)\{2+\cos(\theta_m/2)\cos(3\theta_m/2)\} & i=j=1 \\ \sin(\theta_m/2)\cos(\theta_m/2)\cos(3\theta_m/2) & i=j=2 \\ \cos(\theta_m/2)\{1-\sin(\theta_m/2)\sin(3\theta_m/2)\} & i\neq j \end{cases} \quad (43)$$

$$\sigma_{mij}^{(3)} = \begin{cases} 1 & i=j=1 \\ 0 & i=j=2 \\ 0 & i\neq j \end{cases} \quad (44)$$

$$\sigma_{mij}^{(4)} = \sqrt{\frac{r_m}{2\pi}} \cos(\theta_m/2) \begin{cases} 1+\sin^2(\theta_m/2) & i=j=1 \\ 1-\sin^2(\theta_m/2) & i=j=2 \\ -\sin(\theta_m/2)\cos(\theta_m/2) & i\neq j \end{cases} \quad (45)$$

$$\sigma_{mij}^{(5)} = \sqrt{\frac{r_m}{2\pi}} \begin{cases} \sin(\theta_m/2)\{2+\cos^2(\theta_m/2)\} & i=j=1 \\ -\sin(\theta_m/2)\cos^2(\theta_m/2) & i=j=2 \\ \cos(\theta_m/2)\{1+\sin^2(\theta_m/2)\} & i\neq j \end{cases} \quad (46)$$

In Eqn.(41) $K_{Im} \equiv a_m^{(1)}$ and $K_{IIm} \equiv a_m^{(2)}$ represent the stress intensity factors at the m-th crack tip, and the coefficients $T_m \equiv a_m^{(3)}$, $b_{Im} \equiv a_m^{(4)}$, and $b_{IIm} \equiv a_m^{(5)}$ can also be determined from the boundary conditions (38) and (39).

At the initial stage, the relation between the thermal penetration depth Λ and the equal crack length $H_1 = H_2$ is determined in such a manner that Eq.(20), i.e. $\Pi_{,m} = 0$, holds at the crack tips. Then the derivatives of the stress intensity factors with respect to the crack length are calculated, and the stability condition (23) is checked. If the inequality (23) is violated, i.e. the condition (26) holds, the stable bifurcation point is obtained. After this state, one of the two cracks in a unit cell is stopped, and the crack growth behavior is traced for the remaining crack.

Analysis of Stress Intensity Factors

As is shown in Eq.(41), the stress distribution ahead of the m-th crack tip is characterized by the coefficients K_{Im}, K_{IIm}, T_m, b_{Im}, and b_{IIm}. Those quantities can be obtained by the method of superposition of analytical and finite-element solutions.[10,11] We consider the following problems and solve them by the finite element method,

$$\sigma_{ij,j} = 0 \qquad \text{in } V \tag{47}$$

$$\sigma_{ij}n_j = -\sigma_{ij}^{(k)}(r_m, \theta_m)n_j \qquad \text{on } S_t \tag{48}$$

$$u_i = -u_i^{(k)}(r_m, \theta_m) \qquad \text{on } S_u \tag{49}$$

where $k=1,\cdots,5$ and $u_i^{(k)}$ is the displacement field which gives rise to $\sigma_{ij}^{(k)}$. Let the finite-element solution of the original problem (35) -(37) and those of (47)-(49) be denoted, respectively, by u_i^e; σ_{ij}^e and $u_{mi}^{(k)e}$; $\sigma_{mij}^{(k)e}$, $k=1,\cdots 5$, $m=1,2$. Then we seek to obtain the solution of the original problem in the following form, which combines the analytical and finite-element solutions,

$$\sigma_{ij} = \sum_{m=1}^{2}\sum_{k=1}^{5} a_m^{(k)}\{\sigma_{mij}^{(k)} + \sigma_{mij}^{(k)e}\} + \sigma_{ij}^{e} \tag{50}$$

Since Eq.(41) holds at the m-th crack tip, condition

$$\sum_{m=1}^{2}\sum_{k=1}^{5} a_m^{(k)}\{\sigma_{mij}^{(k)e} - (1-\delta_{mn})\sigma_{mij}^{(k)}\} + \sigma_{ij}^{e} - \sigma_{nij}^{(0)} = O(r_n) \qquad n=1,2 \tag{51}$$

is derived for the n-th crack tip. Disregarding the right-hand side term, we shall determine the unknown constants $a_m^{(k)}$ $k=1,\cdots 5$; $m=1,2$ by solving Eq.(51) at five collocation points properly chosen near each crack tip. To this end let us consider the case where the collocation points are chosen at the centroids P, Q, R, and S of the four elements around the crack tip, and at the point T (Fig.7).

A pair of the average stresses of the four collocation points P,Q,R,and S are defined by

$$\left.\begin{aligned} \sigma_{ij}(P+S) &\equiv \frac{1}{2}\{\sigma_{ij}(P)+\sigma_{ij}(S)\} \\ \text{and} \\ \sigma_{ij}(Q+R) &\equiv \frac{1}{2}\{\sigma_{ij}(Q)+\sigma_{ij}(R)\} \end{aligned}\right\} \tag{52}$$

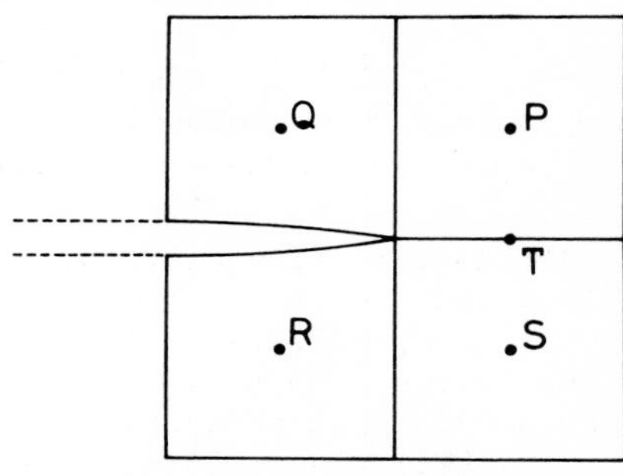

Fig.7. Collocation points.

Then we can use the following condition at each crack tip,

$$\sum_{m=1}^{2}\sum_{k=1}^{5} a_m^{(k)}\{\sigma_{m22}^{(k)e}(P+S) - (1-\delta_{mn})\sigma_{m22}^{(k)}(P+S)\} = -\sigma_{22}^{e}(P+S) + \sigma_{22}^{(0)}(P+S)$$

$$\sum_{m=1}^{2}\sum_{k=1}^{5} a_m^{(k)}\{\sigma_{m22}^{(k)e}(Q+R) - (1-\delta_{mn})\sigma_{m22}^{(k)}(Q+R)\} = -\sigma_{22}^{e}(Q+R) + \sigma_{22}^{(0)}(Q+R)$$

$$\sum_{m=1}^{2}\sum_{k=1}^{5} a_m^{(k)}\{\sigma_{m11}^{(k)e}(T)-(1-\delta_{mn})\sigma_{m11}^{(k)}(T)\}=-\sigma_{11}^{e}(T)$$

$$\sum_{m=1}^{2}\sum_{k=1}^{5} a_m^{(k)}\{\sigma_{m12}^{(k)e}(P+S)-(1-\delta_{mn})\sigma_{m12}^{(k)}(P+S)\}=-\sigma_{12}^{e}(P+S)+\sigma_{12}^{(0)}(P+S)$$

$$\sum_{m=1}^{2}\sum_{k=1}^{5} a_m^{(k)}\{\sigma_{m12}^{(k)e}(Q+R)-(1-\delta_{mn})\sigma_{m12}^{(k)}(Q+R)\}=-\sigma_{12}^{e}(Q+R)+\sigma_{12}^{(0)}(Q+R)$$

$$n=1,2 \qquad (53)$$

where the first two and last two equations are evaluated for σ_{22} and σ_{12}, respectively. Those are the dominant components of stresses being mainly symmetric and asymmetric with respect to the crack line, while the σ_{11} component of stresses is evaluated at the point T. Then the unknown ten coefficients $a_m^{(k)}$ $k=1,\cdots 5;\ m=1,2$ are obtained by solving Eq.(53).

Differentiated Field and Derivatives of Stress Intensity Factors with Respect to Crack Growth Length

If we differentiate the original problem (35)-(37), with respect to the crack growth length h_n, the following boundary value problem is obtained for the differentiated field,

$$\left(\frac{\partial\sigma_{ij}}{\partial h_n}\right)_{,j}=0 \qquad \text{in } V \qquad (54)$$

$$\left(\frac{\partial\sigma_{ij}}{\partial h_n}\right)n_j=\frac{\partial t_i}{\partial h_n} \qquad \text{on } S_t \qquad (55)$$

$$\left(\frac{\partial u_i}{\partial h_n}\right)=\frac{\partial v_i}{\partial h_n} \qquad \text{on } S_u \qquad (56)$$

where n=1,2. The near tip field solution is represented by differentiating Eq.(41) with respect to h_n,

$$\frac{\partial\sigma_{ij}^{(m)}}{\partial h_n}=\begin{cases} K_{Im}\sigma_{mij}^{(f1)}+K_{IIm}\sigma_{mij}^{(f2)}+\left(\dfrac{\partial K_{Im}}{\partial h_n}-\dfrac{1}{2}b_{Im}\right)\sigma_{mij}^{(1)} \\ \qquad +\left(\dfrac{\partial K_{IIm}}{\partial h_n}-\dfrac{1}{2}b_{IIm}\right)\sigma_{mij}^{(2)}+O(1) & m=n \\ \dfrac{\partial K_{Im}}{\partial h_n}\sigma_{mij}^{(1)}+\dfrac{\partial K_{IIm}}{\partial h_n}\sigma_{mij}^{(2)}+O(1) & m\neq n \end{cases} \qquad (57)$$

where

$$\sigma_{mij}^{(f1)}=\frac{1}{4r_m\sqrt{2\pi r_m}}\begin{cases} 2\cos(3\theta_m/2)-3\sin\theta_m\sin(5\theta_m/2) & i=j=1 \\ 2\cos(3\theta_m/2)+3\sin\theta_m\sin(5\theta_m/2) & i=j=2 \\ 3\sin\theta_m\cos(5\theta_m/2) & i\neq j \end{cases} \qquad (58)$$

and

$$\sigma_{mij}^{(f2)}=\frac{1}{4r_m\sqrt{2\pi r_m}}\begin{cases}-4\sin(3\theta_m/2)-3\sin\theta_m\cos(5\theta_m/2) & i=j=1\\ 3\sin\theta_m\cos(5\theta_m/2) & i=j=2\\ 2\cos(3\theta_m/2)-3\sin\theta_m\sin(5\theta_m/2) & i\neq j\end{cases} \quad (59)$$

As is seen from Eq.(57), the solution of the boundary value problem has singularities of order $r_m^{-3/2}$, whose strain energy cannot be integrable in the vicinity of the crack tip. However, these singular terms are easily removed, because their strengths K_{Im} and K_{IIm} have already been obtained. To this end we introduce new variables,

$$\left.\begin{aligned}\bar{\sigma}_{nij}&=\frac{\partial\sigma_{ij}}{\partial h_n}-K_{In}\sigma_{nij}^{(f1)}-K_{IIn}\sigma_{nij}^{(f2)}\\ \bar{u}_{ni}&=\frac{\partial u_i}{\partial h_n}-K_{In}u_{ni}^{(f1)}-K_{IIn}u_{ni}^{(f2)}\end{aligned}\right\} \quad (60)$$

where $u_{ni}^{(f1)}$ and $u_{ni}^{(f2)}$ represent the displacement fields corresponding to the stress fields $\sigma_{nij}^{(f1)}$ and $\sigma_{nij}^{(f2)}$. After substituting Eq.(60) into Eqs.(54)-(56), the following boundary value problem is obtained for $\bar{\sigma}_{nij}$ and $\bar{u}_{ni}$,

$$\bar{\sigma}_{nij,j}=0 \qquad \text{in } V \quad (61)$$

$$\bar{\sigma}_{nij}n_j=\frac{\partial t_i}{\partial h_n}-K_{In}\sigma_{nij}^{(f1)}n_j-K_{IIn}\sigma_{nij}^{(f2)}n_j \qquad \text{on } S_t \quad (62)$$

$$\bar{u}_{ni}=\frac{\partial v_i}{\partial h_n}-K_{In}u_{ni}^{(f1)}-K_{IIn}u_{ni}^{(f2)} \qquad \text{on } S_u \quad (63)$$

Since the higher order singularities are removed, stress intensity factors are obtained by the same method used in the previous subsection.

Suppose the stress intensity factors of the problem (61)-(63) are $\bar{k}_{Im}$ and $\bar{k}_{IIm}$ m=1,2 at each crack tip, then from Eq.(57) we have

$$\bar{k}_{Im}=\begin{cases}\frac{\partial K_{Im}}{\partial h_n}-\frac{1}{2}b_{Im} & m=n\\ \frac{\partial K_{Im}}{\partial h_n} & m\neq n\end{cases} \quad (64) \qquad \text{and} \qquad \bar{k}_{IIm}=\begin{cases}\frac{\partial K_{IIm}}{\partial h_n}-\frac{1}{2}b_{IIm} & m=n\\ \frac{\partial K_{IIm}}{\partial h_n} & m\neq n\end{cases} \quad (65)$$

Therefore, the derivatives of the stress intensity factors are obtained as

$$\frac{\partial K_{Im}}{\partial h_n}=\begin{cases}\bar{k}_{Im}+\frac{1}{2}b_{Im} & m=n\\ \bar{k}_{Im} & m\neq n\end{cases} \quad (66) \quad \text{and} \quad \frac{\partial K_{IIm}}{\partial h_n}=\begin{cases}\bar{k}_{IIm}+\frac{1}{2}b_{IIm} & m=n\\ \bar{k}_{IIm} & m\neq n\end{cases} \quad (67)$$

NUMERICAL RESULTS

Numerical Result Compared with Model Experiment

A numerical simulation of thermally induced crack growth is compared with a model experiment performed by Geyer and Nemat-Nasser[9] by using glass plates. We shall use the following material properties and thermal conditions,

$E=6.9\times10^4$MPa, $\nu=0.23$, $\alpha=8.5\times10^{-6}$ $1/^\circ c$,$K_{Ic}=0.5$MPa $m^{1/2}$,

$T_s=-78^\circ c$, $T_0=52^\circ c$,

where K_{Ic} is the critical stress intensity factor for Mode I loading. Initial crack spacing is assumed to be 25mm, and the crack growth process is illustrated in Fig.8 . All cracks grow at an equal rate along the path A_1 B_1, until the stable bifurcation condition holds at the point B_1, where $\Lambda=35.1$mm and $H_1= H_2=30.7$mm. After this state every other crack ceases to grow, and the crack growth behavior will bifurcate to the path B_1 B_1^*, along which the effective crack spacing becomes doubled, i.e. 50mm. This crack growth continues until it reaches another stable bifurcation point B_2, where $\Lambda=74.2$mm and $H_1=H_2=68.7$mm.

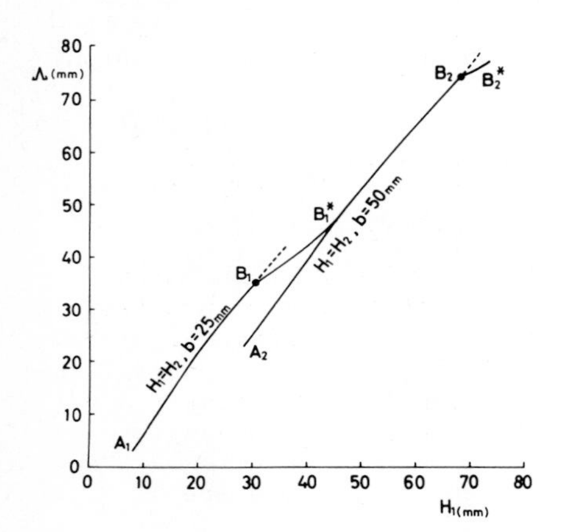

Fig.8. Crack growth behavior in glass.

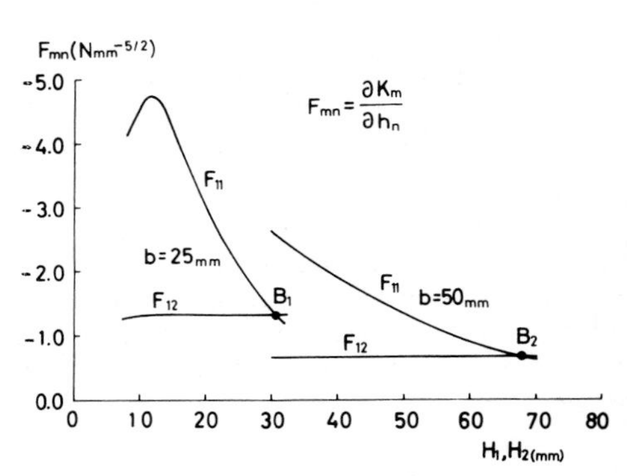

Fig.9. Derivatives of stress intensity factors.

Only every fourth crack spaced out 100mm will grow along the path $B_2B_2^*$ after this point. The change of the derivatives of the stress intensity factors and the stable bifurcation points are illustrated in Fig.9 .

Experimental observations showed that the spacing of the growing cracks was actually doubled during their growth process, and the levels of the crack lengths corresponding to the stable bifurcation points averaged 13 mm and 53 mm, while the numerically estimated values are 30.7mm and 68.7mm. Considering the assumptions made in the theory, the agreement is fairly good. It is also shown that the theoretical estimation of the crack lengths at the stable bifurcation points are longer than the experimental ones. For the more precise quantitative estimation it would be better to take into account the effects of imperfections. There are indeed possibilities that material inhomogeneities[12] and/or geometrical irregularities[13] may cause 20-30 percent reduction of the crack length, at which every other crack ceases to grow.

Secondary Crack Growth in Hot Dry Rock Masses

An interactive behavior of thermally induced secondary cracks is considered, where the initial crack spacing is assumed to be 0.4m, and the temperature drop $T_0-T_s=50^oc$. Here we assume the material properties of granite as $E=3.76x10^4$MPa, $\nu=0.3$, $\alpha=8.0x10^{-6}$ $1/^oc$ and $K_{Ic}=3.8$ MPam$^{1/2}$, which are often used for the analysis. The crack growth behavior is illustrated in Fig.10 .

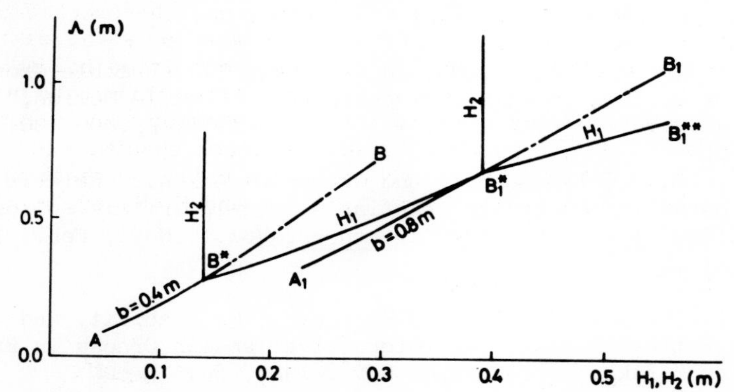

Fig.10. Secondary crack growth behavior in HDR .

The equal crack growth continues along the path A B^*, as the thermal penetration depth increases. When it reaches the stable bifurcation point at B^*, every other crack ceases to grow and the effective crack spacing becomes 0.8m along the path $B^*B_1^*$. This kind of process will be repeated during the course of their growth, and the effective crack spacing will continuously be increased with increasing thermal penetration depth.

ACKNOWLEDGEMENT

The author would like to express his sincere gratitude to Professors S. Nemat-Nasser and L.M. Keer for their helpful discussions and support during his visiting appointments at Northwestern University (1978-80 and 1982). Numerical calculations were performed by using HITAC M200 of Computor Center of the University of Tokyo through Yokohama National University Computing Center.

REFERENCES

1. Cremer,G.M.,"Hot Dry Rock Geothermal Energy Development Program Annual Report Fiscal Year 1980," LA-8855-HDR, July 1981, Los Alamos National Laboratory, Los Alamos, New Mexico.

2. Harlow,F.and Pracht,W.,"A Theoretical Study of Geothermal Energy Extraction," J. Geo. Res., Vol.77, 1972, pp.7038-7048.

3. Smith,M.,Potter,R.,Brown,D.,and Aamodt,R.L.,"Induction and Growth of Fractures in Hot Rock," P.Kruger and C.Otto, ed., Chap.14 in Geothermal Energy, Stanford University Press, Stanford, 1973.

4. Nemat-Nasser,S.,"Geothermal Energy: extracting heat from hot dry rock masses," Proceedings, 16th Midwestern Mechanics Conference 1979, pp.285-297.

5. Nemat-Nasser,S.,Keer,L.M.,and Parihar,K.S.,"Unstable Growth of Thermally Induced Interacting Cracks in Brittle Solids," Int. J. Solids Structures, Vol.14, No.6, June, 1978, pp.409-430.

6. Keer,L.M.,Nemat-Nasser,S.,and Oranratnachai,A.,"Unstable Growth of Thermally Induced Interacting Cracks in Brittle Solids:Further Results," Int. J. Solids Structures, Vol.15,No.2, Feb., 1979, pp.111-126.

7. Sumi,Y.,Nemat-Nasser,S.,and Keer,L.M.,"On Stability and Postcritical Behavior of Interactive Tension Cracks in Brittle Solids," ZAMP, Vol.31,No.6, Nov., 1980, pp.673-690.

8. Sumi,Y.,Nemat-Nasser,S.,and Keer,L.M.,"A New Combined Analytical and Finite-Element Solution Method for Stability Analysis of the Growth of Interacting Tension Cracks in Brittle Solids," Int. J. Engng. Sci., Vol.18,No.1, Jan., 1980, pp.211-224.

9. Geyer,J.F. and Nemat-Nasser,S.,"Experimental Investigation of Thermally Induced Interacting Cracks in Brittle Solids," Int. J. Solids Structures, Vol.18,No.4, Apr., 1982, pp.349-356.

10. Yamamoto,Y. and Tokuda,N.,"Determination of Stress Intensity Factors in Cracked Plates by the Finite Element Method," Int. J. Numer. Meth. in Engng., Vol.6, 1973, pp.427-439.

11. Yamamoto,Y.,Tokuda,N.,and Sumi,Y.,"Accuracy Considerations for Finite Element Calculations of Stress Intensity Factors by the Method of Superposition," S.N.Atluri, R.H.Gallagher, and O.C. Zienkiewicz,ed. Hybrid and Mixed Finite Element Methods, John Wiley, in press.

12. Nemat-Nasser,S.,Sumi,Y., and Keer,L.M.,"Unstable Growth of Tension Cracks in Brittle Solids:Stable and Unstable Bifurcations, Snap-Through, and Imperfection Sensitivity," Int. J. Solids Structures, Vol.16,No.11, Nov., 1980, pp.1017-1035.

13. Oranratnachai,A., Nemat-Nasser,S., and Keer,L.M.,"Effect of Geometric Imperfection on Stability of the Growth Regime of Thermally Induced Interacting Edge Cracks," in preparation.

GROWTH OF MICROCRACKS IN ROCKS, AND LOAD-INDUCED ANISOTROPY

by S. Nemat-Nasser and H. Horii

Northwestern University
Evanston, Illinois 60201 USA

ABSTRACT

The effect of microcracks on overall mechanical properties of rocks and other brittle solids is considered, emphasizing their influence on the overall moduli, on the load-induced anisotropy, and on the failure of the solid by the formation of tension cracks at tips of sliding microcracks under the overall farfield compression.

INTRODUCTION

Rocks generally contain microcracks, fissures, and other inhomogeneities, which profoundly affect their overall mechanical response. For example, growth of microcracks under farfield compression may lead to the phenomena of axial splitting, rockburst, surface spalling, and exfoliation; see Holzhausen [1], Holzhausen and Johnson [2], and Nemat-Nasser and Horii [3]. Another interesting feature of rocks with microcracks is their overall stress-induced anisotropy. Under applied loads, some microcracks may close, some may undergo frictional sliding and some may extend in a preferred direction, resulting in a highly anisotropic response as well as failure modes.

In this paper some recent studies by the authors on the overall response and on the local failure mechanism of rocks with microcracks are reviewed; see Nemat-Nasser and Horii [3] and Horii and Nemat-Nasser [4, 5].

OVERALL MODULI

Let a sample of rock of volume V and exterior surface S contain a set of cavities of total volume V_Ω and surface S_Ω. The matrix of volume $V - V_\Omega$ is linearly elastic with strain (ε_{ij})— stress (σ_{ij}) relations

$$\varepsilon_{ij} = D_{ijk\ell}\,\sigma_{k\ell}, \qquad (1)$$

where a fixed rectangular Cartesian coordinate system with coordinate axes x_i, $i = 1,2,3$, is used, and summation on repeated indices is implied; this convention is employed throughout this paper. In (1), $D_{ijk\ell}$ is the elastic compliance tensor of the matrix of the rock which may or may not be isotropic.

Assume the rock sample is subjected to a self-equilibrating set of overall tractions, T_i^0, in such a manner that $T_i^0 = \bar{\sigma}_{ij}n_j$ on S, where n_j are the components of the exterior unit normal on S, and where the average stress is defined by

$$\bar{\sigma}_{ij} = \frac{1}{V}\int_V \sigma_{ij}(\underset{\sim}{x})dV. \qquad (2)$$

It is easy to show that the average strain

$$\bar{\varepsilon}_{ij} = \frac{1}{V}\int_V \varepsilon_{ij}(\underset{\sim}{x})dV \qquad (3)$$

is then given by

$$\bar{\varepsilon}_{ij} = D_{ijk\ell}\bar{\sigma}_{k\ell} + \frac{1}{V}\int_{S_\Omega} \frac{1}{2}(u_in_j + u_jn_i)dS, \qquad (4)$$

where u_i are the displacement components, and the integration is performed over the surface of all cavities; Horii and Nemat-Nasser [5]. Since no tractions are applied on the surface of cavities, it follows that

$$\frac{1}{V}\int_{S_\Omega} \frac{1}{2}(u_in_j + u_jn_i)dS = H_{ijk\ell}\,\bar{\sigma}_{k\ell}, \qquad (5)$$

and then the overall compliance, $\bar{D}_{ijk\ell}$, becomes

$$\bar{D}_{ijk\ell} = D_{ijk\ell} + H_{ijk\ell}. \qquad (6)$$

Equations (5) and (6) are rather general and provide a powerful means for estimating the overall moduli of any elastic solid with cavities. They also apply to solids with cracks which may be closed and may undergo frictional sliding, provided that the shear stress transmitted across any crack is a linear function of the corresponding normal stress.

Self-Consistent Approach

The self-consistent approach considers a single cavity embedded in an unbounded homogeneous solid having the overall effective moduli. Using this model, the tensor $H_{ijk\ell}$ is calculated and then the overall moduli are obtained from (6). The method has been discussed and used by Hershey [6], Kröner [7,8], Budiansky [9], Hill [10, 11], and others in various contexts; see Mura [12] for references. It has been applied to rocks with randomly distributed open cracks by Budiansky and O'Connell [13].

When some cracks in an elastic solid (which contains randomly distributed cracks) are closed or undergo frictional sliding, the overall response of the solid may become anisotropic, depending on the loading path. To apply the self-consistent method to this problem, a single crack in an anisotropic homogeneous solid must be considered, as discussed by Horii and Nemat-Nasser [5]. The anisotropy in general couples all three fracture modes; see Hoenig [14]. For plane strain case, however, only Modes I and II occur. The solution for this case has been given by Sih, Paris, and Irwin [15], and has been used by Horii and Nemat-Nasser [5] to calculate the tensor $H_{ijk\ell}$ under the assumption that across a closed crack a constant shear stress proportional to the corresponding constant normal stress is transmitted. It is not difficult to improve on this result by considering piecewise constant shear and normal stresses across a closed crack, but if microscopic cracks are involved then this refinement may not be necessary.

The mathematical details for both open and closed cracks are presented by Horii and Nemat-Nasser [5]. Here we illustrate some of the final results.

Open Penny-Shaped Cracks

The assumption of random distribution of cracks leads to overall isotropic response when all cracks are open. Let $\bar{E}$ and $\bar{\nu}$ be the corresponding overall Young modulus and Poisson ratio, respectively, and set

$$f = a^3N, \tag{7}$$

where a is the average crack radius, and N is the average number of cracks per unit volume. Then either by the method of Budiansky and O'Connell [13] or from (6) it follows that

$$\frac{\bar{E}}{E} = 1 - f\,\frac{16(1-\bar{\nu}^2)(10-3\bar{\nu})}{45(2-\bar{\nu})}, \tag{8}$$

$$f = \frac{45(\nu-\bar{\nu})(2-\bar{\nu})}{16(1-\bar{\nu}^2)[10\nu-\bar{\nu}(1+3\nu)]},$$

where E and ν are the matrix Young modulus and Poisson ratio, respectively. In this simple case, the tensor $H_{ijk\ell}$ is given explicitly by

$$H_{1111} = H_{2222} = H_{3333} = \frac{16(1-\bar{\nu}^2)(10-3\bar{\nu})}{45(2-\bar{\nu})}\,\frac{1}{\bar{E}}\,a^3N,$$

$$H_{1122} = H_{2233} = H_{3311} = -\frac{16\bar{\nu}(1-\bar{\nu}^2)}{45(2-\bar{\nu})}\frac{1}{\bar{E}}a^3N, \tag{9}$$

$$H_{1212} = H_{2323} = H_{3131} = \frac{16(1-\bar{\nu}^2)(5-\bar{\nu})}{45(2-\bar{\nu})}\frac{1}{\bar{E}}a^3N,$$

with $H_{ijk\ell} = H_{jik\ell} = H_{ij\ell k} = H_{k\ell ij}$; otherwise $H_{ijk\ell} = 0$.

Open and Closed Cracks in Plane Problems

When some cracks are open while others are closed, the instantaneous moduli depend on the load-history. In a proportional loading, the same set of cracks remains either closed or open. The overall response will be linear if there is no frictional sliding. For the non-proportional loading, an incremental calculation is necessary, since the population of closed cracks changes during the course of deformation; see Horii and Nemat-Nasser [5] for an illustrative example.

For proportional loading, we define the overall shear modulus, $\bar{G}$, by

$$\gamma = \frac{1}{\bar{G}}\tau, \tag{10}$$

and the overall bulk modulus, $\bar{\kappa}$, by

$$\bar{e}_1 + \bar{e}_2 = \frac{1}{2\bar{\kappa}}(\bar{\sigma}_1 + \bar{\sigma}_2), \tag{11}$$

where $\bar{\sigma}_1 = \bar{\sigma}_2 = p$ is the mean stress, $\tau > 0$ is the maximum shear stress, and $\bar{e}_1$, $\bar{e}_2$, and γ are the corresponding overall strains.

When $p/\tau \leq -1$ all cracks are closed, and for $p/\tau \geq 1$ all cracks are open. If the coefficient of friction, μ, is zero, then all closed cracks undergo sliding, whereas if μ is very large, there is no sliding. Figure 1 gives the overall shear modulus in terms of f for indicated p/τ and for the proportional loading. The corresponding overall bulk modulus is given in Fig. 2. Figures 3 and 4 correspond to Figs. 1 and 2, but are calculated using the results for a crack embedded in an unbounded solid with the matrix moduli rather than the overall ones. Unlike the self-consistent results, the overall moduli do not vanish at f slightly greater than 0.3, when all cracks are open.

Stress-Induced Anisotropy

To illustrate load-induced anisotropy in a rock sample with randomly distributed microcracks, consider a sample subjected to overall uniform pressure p, and let the overall shear stress τ be gradually increased from zero (plane problem). If the coefficient of the fric-

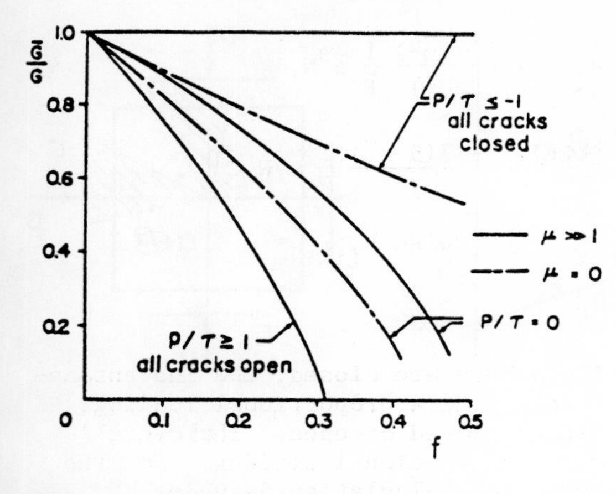

Fig. 1. Overall shear modulus estimated by the self-consistent method; from Horii and Nemat-Nasser [5].

1.0
0.8
0.6
0.4
0.2
0
0.1
0.2
0.3
0.4
f
T/P ≤ 1
T/P = 5
T/P = 10

Fig. 2. Overall bulk modulus estimated by the self-consistent method; from Horii and Nemat-Nasser [5].

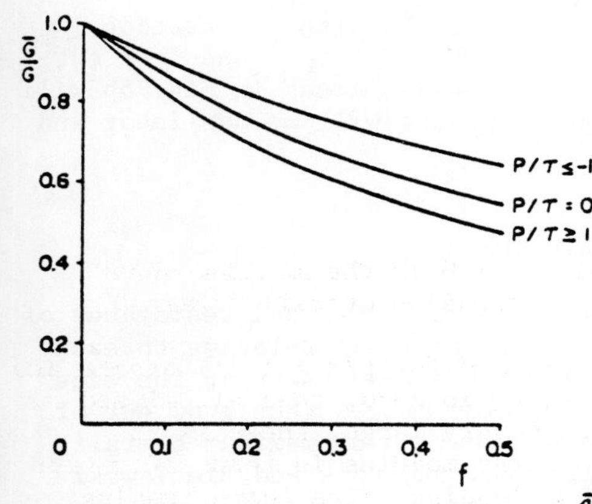

Fig. 3. Overall shear modulus estimated ignoring interaction effects; from Horii and Nemat-Nasser [5].

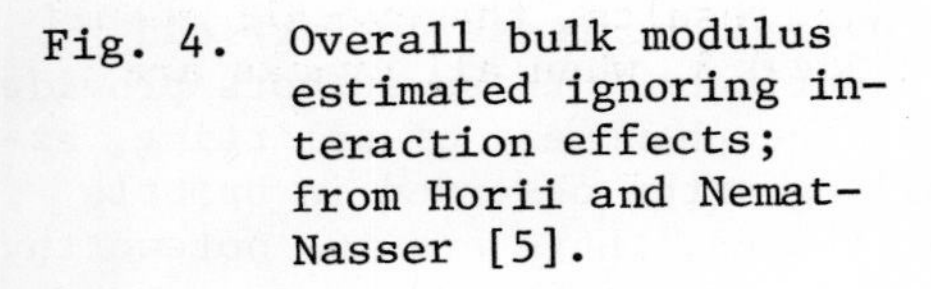

Fig. 4. Overall bulk modulus estimated ignoring interaction effects; from Horii and Nemat-Nasser [5].

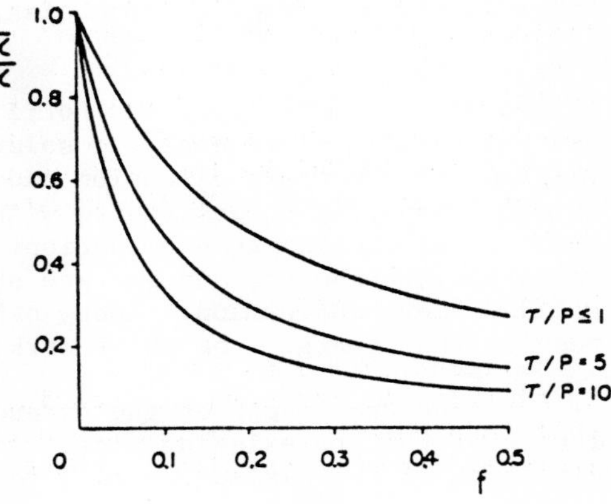

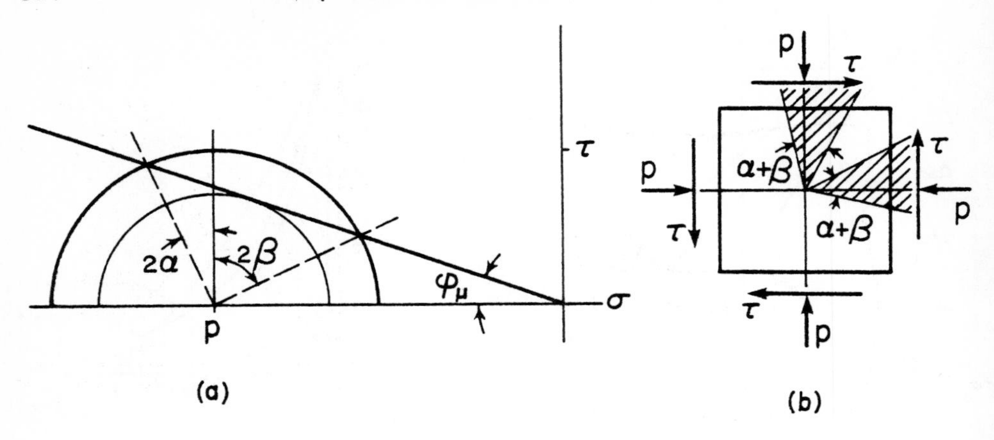

Fig. 5

tional sliding is μ, then no sliding takes place as long as

$$\mu \equiv \tan \phi_{\mu} > \tan \phi \equiv \tan\left\{\sin^{-1} \frac{\tau}{|p|}\right\}; \tag{12}$$

see Fig. 5a. Sliding first occurs across cracks with orientation $\frac{1}{2}\phi_{\mu}$ relative to the maximum shear plane, i.e. when $\phi = \phi_{\mu}$. When $\phi > \phi_{\mu}$, then all cracks with a unit normal in the shaded areas in Fig. 5b will undergo frictional sliding. Hence the response will be nonlinear and anisotropic for $\phi > \phi_{\mu}$.

LOCAL FAILURE

When the resolved shear stress exceeds the frictional resistance of a closed microcrack, the crack faces begin to slide relative to each other. This may lead to the formation of tension cracks at the tips of the pre-existing microcrack. The tension cracks then grow out of the sliding plane and curve toward the direction of maximum overall compression; see Brace and Bombolakis [16], and Hoek and Bieniawski [17].

Recently, Nemat-Nasser and Horii [3] and Horii and Nemat-Nasser [4] have given a complete analytic solution (in two dimensions) for an out-of-plane curved crack extension under overall compressive farfield loads. Comparing their results with experiments, those authors provide some rather convincing explanations for the phenomena of splitting, exfoliation, rockburst, and surface spalling often observed in brittle solids under compression. Among other things, there are two noteworthy results that emerge from their work, as discussed below.

Let ℓ be the length of the extended tension crack, and 2c be the length of the pre-existing microcrack which makes the angle γ with the direction of the maximum overall compression σ_1; see Fig. 6. Then for

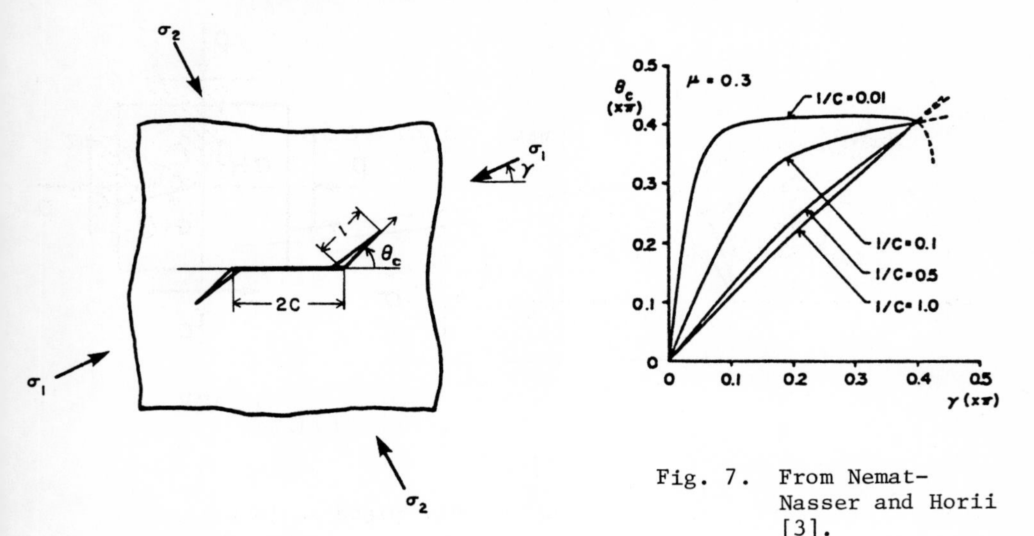

Fig. 7. From Nemat-Nasser and Horii [3].

Fig. 6

a straight out-of-plane extension, the critical "kink angle," θ_c, is essentially constant over a wide range of γ when ℓ/c is small, but equals γ for $\ell/c \simeq 1$ and larger. A typical example is shown in Fig. 7. Hence, with a suitable orientation, a pre-existing microcrack extends first by kinked out-of-plane extension at an angle close to 70° relative to the initial straight direction, and then curves into a position essentially parallel to the direction of maximum compression. This result has been further verified by Horii and Nemat-Nasser [4] by a more rigorous analysis.

The growth of the out-of-plane extension of the microcrack under overall compression is first stable in the sense that an incremental growth requires an increase in the maximum axial compression. However, after a critical ℓ/c is attained, then the crack growth becomes unstable in the sense that ℓ/c becomes unbounded if some slight lateral tension also exists. An example is shown in Fig.8; see Nemat-Nasser [18].

Although the quantitative results depend on the fracture criterion used, this dependency is rather insignificant, since various commonly used fracture criteria seem to yield essentially the same results; see Nemat-Nasser and Horii [3] for a discussion and further references.

It therefore appears that, in the absence of large lateral compressions, axial compression of rocks and other brittle solids leads to axial splitting by the kinked-curved extension of suitably oriented microcracks. This is illustrated in Fig. 9 which shows the failure mode of a plate of Columbia Resin CR39 under axial compression. The plate contained "randomly distributed" pre-existing cracks. The failure by axial splitting resulted from the growth of tension cracks at the tips of the pre-existing ones.

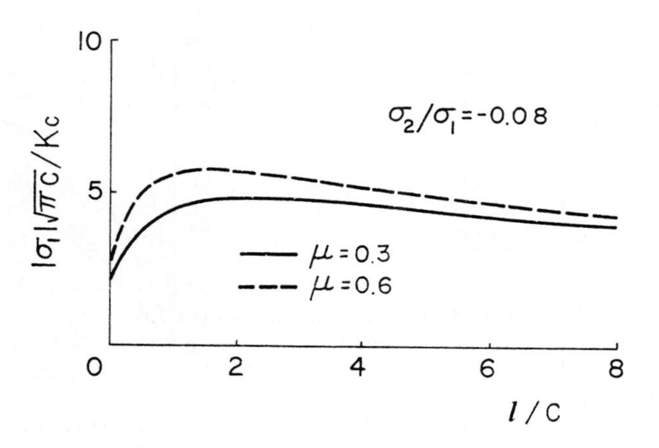

Fig. 8. From Nemat-Nasser [18].

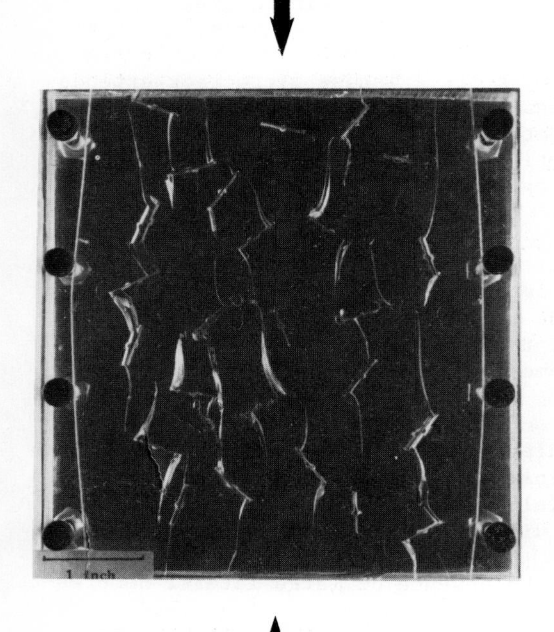

Fig. 9. From Nemat-Nasser and Horii [3].

Nemat-Nasser and Horii [3] report several other test results where the failure mode seems to be "tensile" rather than "shear," as long as small lateral compressions are involved. Under suitably large lateral compressions, say, up to 20% of the maximum compression, however, rocks fail in the shear mode. This, however, may not be due to shear crack growth, but rather due to the formation of localized bands of high density microtension cracks; see Hallbauer, Wagner, and Cook [19]. Here, again, one may expect that microtension cracks remain dominant for a wide range of loading regimes. The present authors are now investigating this problem.

ACKNOWLEDGMENT

This work has been supported by the United States Air Force Office of Scientific Research under Grant No. AFOSR-80-0017 to Northwestern University.

REFERENCES

1. Holzhausen, G. R., "Sheet Structure in Rock and Some Related Problems in Rock Mechanics," Ph.D. Thesis, Stanford University, Stanford, CA, 1978.

2. Holzhausen, G. R., and Johnson, A. M., "Analyses of Longitudinal Splitting of Uniaxially Compressed Rock Cylinders", Int. J. Rock Mech. Min. Sci. Geomech. Abstr. 16, pp. 163-177, 1979.

3. Nemat-Nasser, S., and Horii, H., "Compression Induced Crack Kinking and Curving with Application to Splitting, Exfoliation, and Rockburst," Earthquake Research and Engineering Laboratory Technical Report No. 81-11-44, Department of Civil Engineering, Northwestern University, Evanston, IL, November 1981; "Compression-Induced Nonplanar Crack Extension with Application to Splitting, Exfoliation, and Rockburst," J. of Geophysical Research 87 pp. 6805-6821, 1982.

4. Horii, H., and Nemat-Nasser, S., "Curved Crack Growth in Brittle Solids Under Far-Field Compression," 1982 Advances in Aerospace Structures and Materials, ASME Publication No. AD-03, pp. 75-82, ASME, New York, NY, 1982.

5. Horii, H., and Nemat-Nasser, S., "Overall Moduli of Solids with Microcracks: Load-Induced Anisotropy," Earthquake Research and Engineering Laboratory Technical Report No. 82-6-46, Department of Civil Engineering, Northwestern University, Evanston, IL, June 1982; J. of the Mechanics and Physics of Solids, 1982, to appear.

6. Hershey, A. V., "The Elasticity of an Isotropic Aggregate of Anisotropic Cubic Crystals," J. of Applied Mechanics 21, pp. 236-240, 1954.

7. Kröner, E., "Berechnung der elastischen Konstanten des Vielkristalls aus den Konstanten des Einkristalls," Z. Phys. 151, pp. 504-518, 1958.

8. Kröner, E., "Elastic Moduli of Perfectly Disordered Composite Materials," J. of the Mechanics and Physics of Solids 15, pp. 319-329, 1967.

9. Budiansky, B., "On the Elastic Moduli of Some Heterogeneous Materials," J. of the Mechanics and Physics of Solids 13, pp. 223-227, 1965.

10. Hill, R., "Continuum Micro-Mechanics of Elastoplastic Polycrystals," J. of the Mechanics and Physics of Solids 13, pp. 89-101, 1965.

11. Hill, R., "A Self-Consistent Mechanics of Composite Materials," J. of the Mechanics and Physics of Solids 13, pp. 213-222, 1965.

12. Mura, T., Micromechanics of Defects in Solids, Martinus Nijhoff Publ., The Hague-Boston, 1982.

13. Budiansky, B., and O'Connell, R. J., "Elastic Moduli of a Cracked Solid," Int. J. of Solids and Structures 12, pp. 81-97, 1976.

14. Hoenig, A., "Near-Tip Behavior of a Crack in a Plane Anisotropic Elastic Body," Engineering Fracture Mechanics 16, pp. 393-403, 1982.

15. Sih, G. C., Paris, P. C., and Irwin, G. R., "On Cracks in Rectilinearly Anisotropic Bodies," Int. J. of Fracture Mechanics 1, pp. 189-203, 1965.

16. Brace, W. F., and Bombolakis, E. G., "A Note on Brittle Crack Growth in Compression," J. of Geophysical Research 68, pp. 3709-3713, 1963.

17. Hoek, E., and Bieniawski, Z. T., "Brittle Fracture Propagation in Rock Under Compression," Int. J. of Fracture Mechanics 1, pp. 137-155, 1965.

18. Nemat-Nasser, S., "Non-Coplanar Crack Growth," Earthquake Research and Engineering Laboratory Technical Report No. 83-1-49, Department of Civil Engineering, Northwestern University, Evanston, IL, January 1983; Proc. ICF Int. Symp. on Fracture Mechanics, Beijing, China, Nov. 22-25, 1983, in press.

19. Hallbauer, D. K., Wagner, H., and Cook, N. G. W., "Some Observations Concerning the Microscopic and Mechanical Behaviour of Quartizite Specimens in Still, Triaxial Compression Tests," Int. J. Rock Mech. Min. Sci. Geomech. Abstr. 10, pp. 713-726, 1973.